UNDERSTANDING PHYSICAL CHEMISTRY

UNDERSTANDING PHYSICAL CHEMISTRY

Dor Ben-Amotz

Purdue University
West Lafayette, IN

WILEY

VP & Publisher: Kaye Pace
Associate Publisher and Editor: Petra Recter
Editorial Assistant: Ashley Gayle
Marketing Manager: Kristine Ruff
Marketing Assistant: Andrew Ginsberg
Designer: Kenji Ngieng
Associate Production Manager: Joyce Poh
Production Editor: Jolene Ling

This book was set by MPS Ltd, Chennai, India. Cover and text printed and bound by Courier Kendallville.

This book is printed on acid free paper. ♾

Founded in 1807, John Wiley & Sons, Inc. has been a valued source of knowledge and understanding for more than 200 years, helping people around the world meet their needs and fulfill their aspirations. Our company is built on a foundation of principles that include responsibility to the communities we serve and where we live and work. In 2008, we launched a Corporate Citizenship Initiative, a global effort to address the environmental, social, economic, and ethical challenges we face in our business. Among the issues we are addressing are carbon impact, paper specifications and procurement, ethical conduct within our business and among our vendors, and community and charitable support. For more information, please visit our website: www.wiley.com/go/citizenship.

Library of Congress Cataloging-in-Publication Data

Ben-Amotz, Dor, 1954–
Understanding physical chemistry / Dor Ben-Amotz, Purdue University, West Lafayette, IN.
pages cm
Includes index.
ISBN 978-1-118-29815-2 (pbk. : acid-free paper) 1. Chemistry, Physical and theoretical. I. Title.
QD453.3.B46 2014
541—dc23

2012039489

Printed in the United States of America

10 9 8 7 6 5 4 3 2 1

To Stephanie, Zeno, and Jonah
from whom I have learned what is really important

Brief Contents

Contents

Preface

Physical chemistry has traditionally been understood as that field of study that links chemistry and physics, with a particular emphasis on mathematically modeling chemical processes. Nowadays, physical chemistry is rapidly evolving to include a fantastic variety of subjects ranging from biotechnology and quantum computing to nonlinear optics and nanoscience. Although these are clearly new directions, they are not inconsistent with the famous pronouncement that G.N. Lewis long ago made when he defined physical chemistry as "anything that is interesting." I also find it useful to think of physical chemistry as a subject whose primary goal is to *physically* understand the fundamental principles underlying the fascinating variety of chemical phenomena taking place in the world around us (and inside our bodies). Inspiring such a physical understanding is the principal aim of this book.

While writing this book I kept two guiding principles firmly in mind – I only included material that I judged to be either too important or too interesting to exclude. This philosophy has produced a book that is both shorter and deeper than most other physical chemistry textbooks. I hope you will find that *Understanding Physical Chemistry* succeeds in stimulating an appreciation for the beauty and elegance of this fundamentally important and practically useful subject.

Although *Understanding Physical Chemistry* does not sacrifice mathematical rigor, it also does not assume that students have anything more than a general knowledge of algebra and basic calculus. Applications of partial differential equations in thermodynamics and matrix algebra in quantum mechanics are fully introduced and explained in this book, without assuming any prior proficiency in these mathematical subjects. Moreover, extensive footnotes provide mathematical and historical information that is not critical to understanding the core ideas described in each chapter, but may be useful to students who wish to dig deeper.

Understanding Physical Chemistry is primarily intended to accompany a full-year undergraduate physical chemistry sequence, but is readily adaptable to shorter courses of various sorts (as further described on page xii and illustrated in the diagram on page xiv). A key feature of this book is its integrated and cohesive structure, which strikes a new balance in the presentation of quantum and classical descriptions of chemical systems. The first chapter entitled *The Basic Ideas* introduces foundational concepts such as energy, force, quantization, and thermal distributions. This overview is followed by chapters containing more detailed investigations of chemical thermodynamics, quantum mechanics, spectroscopy, and kinetics.

Understanding Physical Chemistry stresses core ideas underlying physical chemistry, such as the entropic forces that drive all chemical processes and the quantum states that dictate the structures and colors of atoms and molecules. This book aims to demystify these core concepts by explaining where they come from, why they make sense, and how they may be applied to understanding topics ranging from molecular spectroscopy and chemical reactivity to biological self-assembly and liquid computer simulation strategies. The premise of this approach is that any student who emerges from a physical chemistry course with such an understanding will be ready to take on any conceptual challenges they are likely to encounter as an academic or industrial research scientist.

Two categories of homework problems are provided at the end of each chapter: *Problems That Illustrate Core Concepts* and *Problems That Test Your Understanding*. The first category consists of problems designed to help students actively learn the key ideas presented in that chapter. The second category contains problems intended to help students prepare for exams (with answers provided in Appendix A). These homework problems could be used in various other ways, either by assigning only problems from the first category and leaving the second category for independent study, or by assigning a mix of problems from both categories, perhaps supplemented by additional problems designed to complement the instructor's individual teaching and exam writing style.

Understanding Physical Chemistry may be used in its entirety for a full-year (two-semester or three-quarter) physical chemistry sequence. Alternatively, some or all of Chapters 1–5 may be used in a shorter course focused primarily on chemical thermodynamics (with an introduction to quantum phenomena and statistical mechanics), while Chapter 1 as well as some or all of Chapters 6–9 may be used in a course focused primarily on quantum chemistry. Chapter 10, which describes chemical (and photon-molecule) reaction equilibria and kinetics, may be included in either of the above one-semester courses. It is also possible to use this book for a lower-level survey of physical chemistry by, for example, covering Chapter 1 (Sections 1.1–1.3 only), all of Chapter 2, Chapter 3 (part of Section 3.4, only to introduce $H = U + PV$ and $G = H - TS$ as definitions), Chapter 4 (Sections 4.1 and 4.2 only), Chapter 6 (Sections 6.1–6.4 only), Chapter 7 (Sections 7.1 and 7.2 only), Chapter 8 (Section 8.1 only), Chapter 9 (Sections 9.1 and 9.2 only), and Chapter 10 (Sections 10.1 and/or 10.2 only).

Even in an in-depth course it is possible to skim over (or skip) some of the more advanced material appearing at the end of some chapters. For example, one could briefly summarize (or skip) much of Section 1.5, which presents a derivation of the equipartition theorem; Section 4.4 on the consequences of irreversibility; some or all of Chapter 5 about non-ideal gases, liquids, and theorems used in molecular dynamics simulations; Section 6.5 on the formal postulates of quantum mechanics; some or all of Sections 7.3, 7.4, and 8.2 on raising and lowering operators and matrix mechanics; as well as the introduction to *ab initio* methods in Section 9.5. However, even if these (or other) sections are not covered in detail, it may be useful to present some of the key results obtained in those sections (without the derivations) because of their fundamental importance and practical relevance. For example, although it is not necessary to present the mathematical derivations of the Maxwell-Boltzmann velocity distribution (Eq. 1.32), equipartition theorem (Eqs. 1.43 and 1.44), and the virial theorem (Eq. 1.46), it is important for students to be aware of these results, as they are made use of at various points throughout the book.

Understanding Physical Chemistry owes much to the authors of many other excellent textbooks – well-worn copies of which adorn my bookshelf. For example, both the title and the pedagogical perspective of this book are inspired in part by two wonderful books entitled *Understanding Chemistry* (an introductory chemistry textbook written in the early 1970s by George Pimentel and Richard Spratley) and *Understanding Molecular Simulations* (an introduction to current computer simulation methods by Daan Frenkel and Berend Smit). There are many other books that I appreciate and admire, such as Feynman's *Lectures on Physics*; Widom's *Statistical Mechanics, a Concise Introduction for Chemists*; Chandler's *Introduction to Modern Statistical Mechanics*; Golden's *Introduction to Theoretical Physical Chemistry*; Lindsay and Margenau's *Foundations of Physics*; Callen's *Thermodynamics*; Reiss's *Methods of Thermodynamics*; Hill's *An Introduction to Statistical Thermodynamics*; Ratner and Schatz's *Introduction to Quantum Mechanics in Chemistry*; and Laidler's *The World of Physical Chemistry*, to name just a few, as well as fine physical chemistry textbooks by Atkins and de Paula, McQuarie and Simon, Engel and Reid, Castellan, Moore, Levine, Berry, Rice and Ross, Moelwyn-Hughes, and others. Numerous people, including Ben Widom, Chris Jarzynski, Adam Wasserman, Eitan Geva, and Mark Sellke (as well as several anonymous reviewers), have provided comments and suggestions that have improved this book.

The other people who have inspired me to write this textbook are far too numerous to list. These include all the teachers I have had, both in and out of school, from whom I have learned both what ideals to strive for and what pitfalls to avoid, as a teacher and as a person. My most important teachers have been my students, because teaching is a two-way conversation.

It is often said that there is no better way to learn than to teach, which I have certainly found to be the case. I have also found that the converse is equally true, as the best classroom experiences are ones in which student comments, questions, and discussions inspire a synergy of collective understanding.

Understanding Physical Chemistry

Relationship between the chapters and alternative presentation pathways

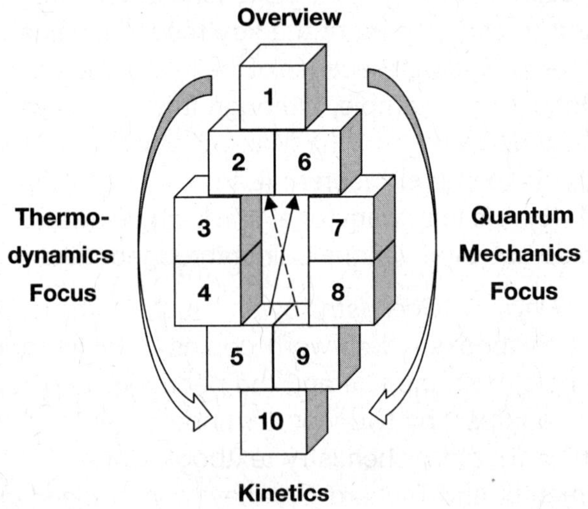

The Basic Ideas

CHAPTER 1

This chapter contains a summary of some of the most important ideas that underlie all of physical chemistry. In other words, it could be subtitled *Ingredients in a Physical Chemist's Cookbook* or *Tools in a Physical Chemist's Workshop*. These ideas are ones that physical chemists frequently refer to when they are having conversations with each other. So, you could think of this chapter as a quick-start guide to thinking, talking, and walking like a physical chemist. Having these basic ideas in mind can help make physical chemistry less confusing by providing a broad overview of how various pieces of nature's puzzle fit together to produce a big beautiful picture.

1.1 Things to Keep in Mind

Physical Chemistry Is a Conversation

Science is sometimes incorrectly envisioned as a static and impersonal body of knowledge – in fact, it is much more like an interesting conversation that evolves in endlessly surprising ways. This multifaceted conversation often takes place between good friends, over lunch or coffee (or some other beverage), while taking a break in the lab, or during a walk in the woods. It often includes people who live in very different places (and times). For example, it can take place over the Internet, on the phone, or at scientific meetings. Communication also includes publishing and reading journal articles, both in the latest issues and in archives extending back many years, spanning centuries, and drawing on memories that reach into the foggy depths of recorded history, and beyond.

A classroom is one of the main places in which scientific conversations happen. A classroom, of some kind or other, is where every single scientist throughout history has come to find out more about the most interesting discussions and realizations that other scientists have had. The best classroom experiences are themselves conversations in which students and teachers struggle to improve their individual and collective understandings by working hard to clearly communicate and think in new ways.

Like any good conversation, scientific progress requires an open-minded attitude. Obviously, having a conversation also requires speaking the same language and sharing a common body of knowledge and experience. However, the preconceptions that inevitably come along with any body of knowledge can also be among the greatest impediments to scientific progress, or, for that matter, any other kind of productive exchange of ideas. So, the feeling of confusion or disorientation that may at times overtake you while you are struggling to learn physical chemistry is not necessarily a bad thing – that is often how it feels when an interesting conversation is on the verge of a breakthrough.

Longing for Equilibrium

All changes in the world appear to be driven by an irresistible longing for equilibrium. Although this longing is not the same as a subjective feeling of longing, the effect can be much the same. Any change in the world clearly implies the existence of an underlying driving force. Moreover, our experience suggests that some changes can and do often occur spontaneously while others are highly improbable or even impossible. These ideas are best illustrated by some simple examples.

Consider a boulder situated very comfortably up on the side of a mountain. Although this boulder may remain in more or less the same spot for many years, if the ground holding it gives way, the boulder will spontaneously careen down into the valley below – dramatically converting its potential energy into a great burst of kinetic energy. However, our experience also tells us that under no circumstances would the boulder ever spontaneously roll uphill, unless significant work were expended to push it.

Understanding the universal principle that drives all systems towards equilibrium can be of great practical value. For example, one can build a waterwheel or a hydroelectric generator to perform useful work, such as mechanically grinding wheat into flour or generating horsepower in the form of electricity. Similarly, the tendency of electrons to flow downhill in potential energy from one chemical compound to another may be used to produce batteries and fuel cells, as well as to flex muscles and create brain storms.

The Sun is another good example of the importance of disequilibrium. Hydrogen atoms are just like boulders sitting high up on a hill, where they can remain in a very stable state for many years. However, given the right circumstances (such as the very high pressures and temperatures inside the Sun) hydrogen atoms can be dislodged to undergo fusion reactions such as $4H \rightarrow He + 2e^- + 2e^+$, releasing a great burst of energy in the form of sunlight.[1]

The longing for equilibrium is of keen interest not only to physical chemists but also to engineers and mathematicians – whose research expenses are often subsidized by investors anxious to capitalize on nature's tendencies. Although the above examples illustrate how some spontaneous processes may be converted to useful work, the general analysis of nature's proclivity for equilibration is a deep and subtle subject that motivated the development of thermodynamics.

Among the most remarkable results of thermodynamics is the discovery of a function, called entropy, which expresses the longing of all systems for equilibrium in rigorous mathematical terms. This function may be used to predict whether a given process can or cannot occur spontaneously. Even more importantly, entropy can be used to predict the maximum amount of work that can be obtained from any spontaneous process (or conversely, the work required to drive a non-spontaneous process).

Invariants, Constraints, and Symmetry

A recurring theme underlying all of physical chemistry (and other branches of science) is the search for universal principles, or fundamental quantities, which give rise to all observed phenomena. The search for such invariant properties of nature has ancient roots, tracing back at least to the Ionian school of Greek philosophy, which thrived in the sixth century BC, and whose adherents, including Thales and Anaximander, postulated that all things are composed of a single elemental substance. This school of thought also influenced a young Ionian named Democretus, who proposed that everything in the world is composed of atoms that are too small to be visible with the naked eye.

The idea that some quantities are conserved in the course of chemical processes seems pretty obvious. For example, although a chemical reaction may produce dramatic changes in color, texture, and other measurable properties, one might expect the products of a reaction to weigh the same amount as the reactants. Careful experimental measurements

---

[1] Note that the reaction of four H ($^1_1$H) atoms to form a He ($^4_2$He) atom makes use of the fact that a proton may decompose into a neutron plus a positron. There are also other lower-order reactions that can produce helium from heavy isotopes of hydrogen (deuterium and tritium), such as $^2_1H + ^3_1H \rightarrow ^4_2He + ^1_0n$, which is among the processes that may someday form the basis of environmentally safe nuclear fusion power plants on Earth.

demonstrate that mass is indeed conserved during chemical reactions, to within the accuracy of a typical analytical balance. However, it turns out that mass is not, in fact, perfectly conserved! This discrepancy is linked to the even more fundamental principle of energy conservation, as we will see.

The invariant properties of a system are also intimately linked to the constraints and symmetries that characterize the system. A good example of the connection between invariants, constraints, and symmetry emerges from considering the motion of an object in a central force field – such as the Earth moving in the central gravitational force field of the Sun, or an electron moving in the central coulombic (electrostatic) force field of a proton. In the seventeenth century Johannes Kepler demonstrated that a planet that is constrained to move under the influence of the Sun's gravitational force must sweep out a constant (invariant) area per unit time. This is a special case of the principle of conservation of angular momentum, which is a general consequence of the spherical symmetry of a central-force constraint, and so also applies to an electron in a quantum mechanical orbit around an atomic nucleus.

Similar connections between invariants, constraints, and symmetry underlie the conservation of linear momentum in a system with translational invariance, such as objects moving in free space, or billiard balls rolling on a pool table. Thinking about these connections led Einstein to develop the special theory of relativity, which is a consequence of the experimentally observed invariance of the speed of light, independent of the relative velocity of an observer (and the corresponding invariance of Maxwell's electromagnetic equations). The theory of relativity leads to all sorts of surprising predictions about the interrelations between light, space, time, energy, and mass. Among these is a prediction that mass cannot be perfectly conserved in any chemical reaction that either releases or absorbs energy (as further discussed in Section 1.2).

Constraints and symmetries also play an important role in thermodynamics. For example, as first clearly demonstrated by Joseph Black, an eighteenth-century Scottish professor of medicine and chemistry, two chemical systems that are constrained in such a way that they cannot physically mix but can exchange heat (e.g., because they are separated by a partition made from copper or some other good thermal conductor), invariably evolve to an equilibrium state of the same temperature. Similarly, two systems, separated by a constraint that can freely translate (such as a movable piston), will evolve to an equilibrium state of the same pressure. As another example, if we remove a constraint that separates two different kinds of gases (i.e., by opening a stopcock between the containers that hold the two gases), then they will evolve to a state in which they are uniformly mixed.

A common theme underlying the above examples is that systems tend to evolve toward a state of maximum symmetry, to the extent allowed by the constraints imposed on the system. Note that in the gas-mixing example we implicitly assumed that the two gases do not react with each other and that there is no difference in potential energy between one part of the system and another. However, even when molecules can react or when there is a potential energy variation within the system, Willard Gibbs brilliantly demonstrated that one can nevertheless identify a quantity called the chemical potential whose invariance (i.e., uniformity throughout the system) is assured at equilibrium (as we will see).

1.2 Why Is Energy So Important?

Conservation of Energy

The principle of energy conservation, which is closely related to the first law of thermodynamics, identifies energy as the one quantity that is invariably conserved during any process, chemical or otherwise. This principle also leads to the recognition of different forms of energy, including kinetic and potential energy, as well as work and heat, which represent means of exchanging energy between a system and its surroundings.

The connection between kinetic, K, and potential, V, energies may be illustrated by considering an apple falling off of a tree. If the apple of mass, m, is initially hanging at a height, h, then its potential energy (relative to the ground) is $V = mgh$ (where $g = 9.8$ m/s^2 is the acceleration due to gravity). Once the apple hits the ground all of its potential energy will be converted to kinetic energy, $K = \frac{1}{2}mv^2 = mgh = V$ (where v is the velocity of the apple). Similar expressions could be used to obtain the increase in kinetic energy of any object that results when it freely falls through a given potential energy drop, or to calculate how high up an object could go if it is launched with a particular initial value of kinetic energy. In the next section we will see why it is that these two kinds of energy have the functional forms that they have.

Kinetic and potential energies can also be related to work, W, which is defined as the product of the force, F, experienced by an object multiplied by the distance, x, over which that force is imposed. Thus, the work associated with an infinitesimal displacement is

$$dW = F\,dx \tag{1.1}$$

and so the total work exchanged during a given process is

$$W = \int dW = \int F\,dx \tag{1.2}$$

where both integrals are performed from the starting point to the end point of the path of interest. For example, if a constant force is used to accelerate an object over a distance Δx, then a total work of $W = F\Delta x$ will be performed. Moreover, the work that is done will produce an increase in kinetic and/or potential energy, which is exactly equal to W. We can also use Eq. 1.2 to calculate the work associated with many other types of processes, such as compressing a gas, or moving an electron through a voltage gradient, or breaking a chemical bond.

The way in which heat is related to other forms of energy is a subtle and interesting issue. From a macroscopic perspective, the heat exchanged between a system and its surroundings may be *defined* as any change in the energy of the system, other than that due to the performance of work on the system. From a microscopic perspective, one may identify heat exchange with energy that is deposited into molecular (thermally equilibrated) degrees of freedom.[2]

The irreversible loss of useful macroscopically organized energy into less useful random molecular energy is intimately linked to the concept of entropy. Given that this concept has required centuries to develop, we should not be surprised if it takes us some time and effort to fully comprehend its significance and implications. One of the primary aims of physical chemistry is to attain such an understanding by revisiting this and the other key ideas from various different perspectives. Just as the different perspectives provided by our two eyes are required to produce a three-dimensional image of the world, so too are different perspectives required in order to better visualize the world of physical chemistry.

The Hamiltonian

The significance of energy conservation may be further illuminated by considering the interactions of particles moving on a flat potential surface (such as billiard balls colliding on a pool table) or objects that move under the influence of external forces (such as those produced by magnetic, electric, or gravitational fields). In the late seventeenth century, Isaac Newton formulated his famous second law, which applies to all such processes

$$\boxed{F = ma} \tag{1.3}$$

---

[2] It is sometimes useful to think of thermodynamics work as resulting from forces imposed on macroscopic objects, while heat exchanges result from forces that change the kinetic and/or potential energy molecules. However, recent theoretical and experimental studies involving work exchanges performed on single molecules make it clear that thermodynamic work need not be macroscopic, and energy exchanges with molecular degrees of freedom are not always heat exchanges. We will further investigate these and other important issues associated with work, heat, and energy exchanges in the next four chapters, and particularly in Sections 3.1 and 5.5.

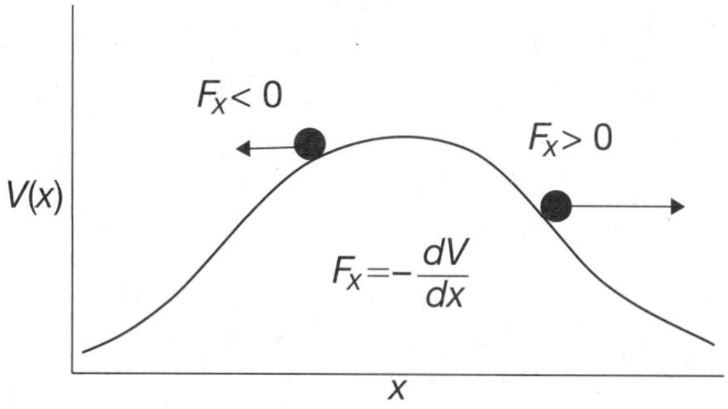

FIGURE 1.1 A ball on a hill feels a force that is opposite in sign to the slope of the hill. In other words, when the slope is positive, the ball is pushed backwards, while when the slope is negative, the ball is pushed forward.

where F is the force acting on an object of mass, m, and $a = dv/dt$ is the acceleration it experiences as a result. The force on an object may also be related to the slope of a potential energy function. For example, consider a car parked on a hill. If you release the brakes, then the car will tend to accelerate down the hill. The force that produces this motion is proportional to the slope of the hill. More specifically, if some object is moving along the x-direction under the influence of a potential energy function, $V(x)$, then it will experience the following force of F_x (along the x-direction).

$$F_x = -\left[\frac{dV(x)}{dx}\right] \tag{1.4}$$

The minus sign simply indicates that a potential function (hill), which goes up when you move forward, will exert a force that pushes you back (down the hill), as illustrated in Figure 1.1.

Exercise 1.1

Consider a ball on a hill that is inclined at an angle θ with respect to the horizontal axis and recall that its potential energy is $V(h) = mgh$, where m is the mass of the ball, g is the acceleration due to gravity, and h is the ball's height.

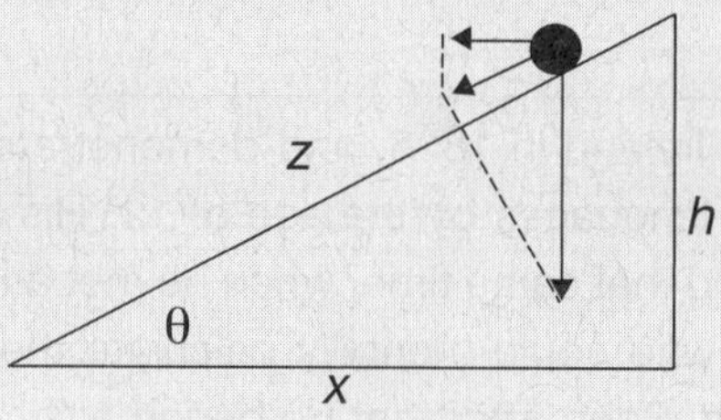

- Obtain an expression for the gravitational force experienced by the ball in the vertical direction, and explain why the sign of the force makes physical sense.

Solution. The gravitational force in the vertical h-direction is $-dV/dh = -mg$, as indicated by the vertical arrow in the above figure. The negative sign of this force is consistent with our physical experience, as it indicates that the force of gravity is directed downwards (towards more negative h values).

- Project the above gravitational force onto the z-direction to obtain F_z, the net force along the slope of the hill.
  Solution. Note that all the right triangles in the above figure are similar, as they each have exactly the same three angles, so the projection of the gravitational force along the z-direction is $F_z = -(dV/dh)\sin\theta = -mg\sin\theta$.
- Project F_z onto the x-axis to obtain the component of the net force along the horizontal direction.

  Solution. The horizontal component of the force is $F_x = F_z\cos\theta = -mg\sin\theta\cos\theta$.

The total energy of any system is defined as the sum of its kinetic, K, and potential, V, energies, and is also referred to as the Hamiltonian function, H.

$$\boxed{H = K + V = \text{ Total Energy}} \tag{1.5}$$

This function is named after William Rowan Hamilton, a leading nineteenth-century physicist.[3]

The usefulness of the Hamiltonian function can be illustrated by considering a particle moving in the x-direction under the influence of a potential energy function, $V(x)$. The kinetic energy of the particle is $K = \frac{1}{2}mv^2$, and so the Hamiltonian of such a system is

$$H = \frac{1}{2}mv^2 + V(x) \tag{1.6}$$

---

[3] Hamilton was born in Ireland in 1805, and demonstrated an early brilliance by learning more than ten languages by the age of 12. His interest in mathematics apparently began around that same time, when he met an American child prodigy named Zorah Colburn who could mentally calculate the solutions of equations involving large numbers. Soon after that he began avidly reading all the mathematical physics books he could get his hands on, included Newton's *Principia* and Laplace's *Celestial Mechanics*, in which he uncovered a key error. He published a highly influential paper on optics while he was still an undergraduate, and was appointed a professor of Astronomy at Trinity College at the age of 21.

If we take the time-derivative of both sides of Eq. 1.6, we discover a very interesting property of the Hamiltonian function.

$$\begin{aligned}
\frac{dH}{dt} &= \frac{d}{dt}\left[\frac{1}{2}mv^2 + V(x)\right] \\
&= \frac{1}{2}m\left[2v\left(\frac{dv}{dt}\right)\right] + \left[\frac{dV(x)}{dt}\right] \\
&= mv\left(\frac{dv}{dt}\right) + \left[\frac{dV(x)}{dx}\right]\left(\frac{dx}{dt}\right) \\
&= mva - Fv \\
&= (ma - F)v = (0)v \\
&= 0
\end{aligned}$$

This result clearly indicates that H is time-independent, and so the total energy of the system is conserved! In other words, we have shown that Newton's law implies the conservation of energy.[4] Notice that this derivation also demonstrates why we define $K \equiv \frac{1}{2}mv^2$, as this is the quantity that combines with potential energy to form a Hamiltonian that is time-independent.

Although the above derivation only considered a single particle moving in the x-direction, the result can be generalized to show that the Hamiltonian of any *isolated* system, no matter how complicated, must also be time-independent. Note that an isolated system is defined as one from which nothing can leave (or enter). Thus, the entire universe is one example of an isolated system, which implies that the energy of the universe must be conserved!

---

[4] While Newton formulated classical mechanics in terms of forces (which may have a complicated time-dependence), the Hamiltonian formulation of classical mechanics is founded on a time-independent (conserved) property – the total energy. Hamilton also demonstrated that all of classical mechanics could be obtained from what is now called Hamilton's principle, which states that $\delta \int (K - V)dt = 0$. In other words, he demonstrated that the path followed by any mechanical system is one that minimizes the time integral of the difference between its kinetic and potential energies. This principle is closely related to Fermat's principle of least time, which applies to the path followed by light in a medium of varying refractive index. Hamilton's principle also played a central role in both Schrödinger's and Feynman's twentieth-century contributions to the development of quantum mechanics.

Exercise 1.2

Consider an object of mass $m = 0.02$ kg that is confined by harmonic potential of the form $V(x) = \frac{1}{2}fx^2$, whose force constant has a value of $f = 2$ J/m^2. The object is initially held (stationary) at $x = 0.5$ m, and is then released so that it is free to move under the influence of the potential (with no friction).

- What is the initial force on the object (as soon as it is released)?

  Solution. We can obtain the force from the derivative of the potential energy $F = -(dV/dx) = -\frac{1}{2}f(2x) = -fx = -2$ J/m^2 $\times$ 0.5 m $= -1$ N. Note that a force of 1 N is equivalent to 1 J/m. The fact that the initial force is negative means that the object will initially move towards smaller x values (thus lowering its potential energy and increasing its kinetic energy).

- What is the value of the object's total energy?

  Solution. Since the object is initially at rest, and so has no kinetic energy, its Hamiltonian (total energy) is equal to its initial potential energy, $H = K + V = 0 + V(0.5) = \frac{1}{2}fx^2 = \frac{1}{2} \times 2$ J/m$^2 \times 0.5^2$ m$^2 = 0.25$ J. Since the Hamiltonian is time-independent (so the total energy of the system remains constant) $H = 0.25$ J at all times.

- What is the value of the object's maximum kinetic energy?

  Solution. Because the object's total energy is constant, its kinetic energy will be maximized when its potential energy has reached its lowest value, which occurs when $x = 0$ (and $V = 0$ J). Thus, the maximum value of the kinetic energy is the same as the maximum value of the potential energy, $K_{max} = V_{max} = 0.25$ J. Note that the potential energy reaches its maximum value when $x = \pm 0.5$ m, while the kinetic energy is maximized when $x = 0$ m.

- What is the maximum velocity of the object?

  Solution. The object's maximum velocity may be obtained from its maximum kinetic energy (and mass). $K_{max} = \frac{1}{2}mv_{max}^2$ and so $v_{max} = \sqrt{2K_{max}/m} = \sqrt{2 \times 0.25/0.02}$ (J/kg)$^{1/2} = 5$ m/s. Note that 1 J $=$ 1 kg m^2/s^2, so (J/kg)$^{1/2}$ is equivalent to m/s.

Our experience tells us that the energy of some subsystems within the universe may *not* be conserved. For example, a car dissipates energy when it drives, and so it is clearly not an isolated mechanical system. This is also why a car is valuable, because it can use chemical energy to drive up hills and speed along a highway for many miles at a steady clip, in spite of frictional drag and wind resistance.

One of the simplest examples of a non-isolated system is an object that experiences a frictional force that is proportional to its velocity, $F_{\text{friction}} = -fv$. Thus, the total force on such an object can be expressed as the sum of this frictional force plus the force arising from its potential energy.

$$F = ma = -\left(\frac{dV}{dx}\right) - fv \tag{1.7}$$

Notice that this equation can also be obtained by equating the time-derivative of H with $-fv^2$.

$$\begin{aligned}
\frac{dH}{dt} = \frac{d}{dt}\left[\frac{1}{2}mv^2 + V(x)\right] &= -fv^2 \\
\left[ma + \left(\frac{dV}{dx}\right)\right]v &= -(fv)v \\
ma + \left(\frac{dV}{dx}\right) &= -fv \\
ma &= -\left(\frac{dV}{dx}\right) - fv
\end{aligned}$$

This indicates that the Hamiltonian of such a system is *not* constant (since $dH/dt = -fv^2$). In other words, a frictional force has the effect of dissipating the total energy of the system at a rate of $-fv^2$. But where does this energy go? Our experience tells us that friction is often accompanied by an increase in temperature, such as that which you feel when you rub your hands together rapidly. So, heat is evidently closely related to dissipative energy loss.

In summary, the Hamiltonian (total energy) of any isolated system is time-independent (conserved), while that of a non-isolated system may not be. However, since the entire universe (which includes the system and all of its surroundings) is itself isolated, the universe must have a fixed amount of total energy. Thus, any energy that leaves a system is not lost but simply goes into some other part of the universe.

Relation Between Energy and Mass

A remarkable extension of the principle of energy conservation was discovered in the early twentieth century by Albert Einstein, whose theory of relativity made it clear that there is an intimate connection between the conservation of energy and mass. Einstein first reported this monumental finding in a short note entitled *Does the Inertia of a Body Depend on Its Energy Content?* which he published in 1905 as an afterthought to his famous first paper about relativity. In this note he analyzed the implications of relativity when applied to processes involving the emission of light by atoms. This analysis suggested that the measured mass of an atom must decrease when it loses energy. Thus, Einstein obtained what may well be the most famous equation in all of science.

$$\boxed{E = mc^2} \tag{1.8}$$

This states that mass, m, and energy, E, are not independent variables, but are related to each other by a constant of proportionality that is equal to the square of the velocity of light, c^2. Einstein actually wrote the equation as $m = E/c^2$, which better emphasizes the fact that the mass of an object depends on how much energy it has, and so *any* energy change must be accompanied by a change in mass.

For example, the combustion of methane, $CH_4 + 2\ O_2 \rightarrow CO_2 + 2\ H_2O$, is accompanied by the release of −604.5 kJ of heat (per mole of methane). Equation 1.8 implies that this change in energy must also be accompanied by a decrease in mass of about 6.7 ng (6.7×10^{-9} g). Although such a change in mass is too small to be readily measurable, it does clearly imply that mass is not strictly conserved during chemical reactions.

A more dramatic demonstration of the validity of Eq. 1.8 is the experimentally observed annihilation of an electron and a positron to form two high-energy (gamma ray) photons, $e^- + e^+ \rightarrow 2\gamma$, in which the entire mass of the electron and positron is converted into energy (in the form of two photons with no rest mass). This process also implies that the energy released in the nuclear reaction, $4H \rightarrow He + 2e^- + 2e^+$, is approximately equivalent to the difference in mass between one helium atom and four hydrogen atoms, which is 0.029 g or 2.6×10^9 kJ! (per mole of He).

Comparison of the previous chemical and nuclear reactions makes it clear why nuclear fusion might someday prove to be an attractive alternative to fossil fuels as a source of energy, if a practical means of performing such reactions in a safe and controlled way can be developed. Alternatively, future generations may decide that the safest place to carry out nuclear fusion reactions is in the Sun, where they already occur naturally, and thus focus research efforts on improving the efficiency with which the Sun's highly reliable and freely distributed supply of energy may best be harvested, stored, and transported.

1.3 Quantization Is Everywhere

Given that atoms and molecules are over 100,000 times smaller than the thickness of a piece of paper, it should not be too surprising that the way the world looks and behaves on such very short length scales is quite different from the macroscopic world of our everyday experience. In other words, some aspects of the atomic world can seem kind of strange because it is hard to project our macroscopic experiences onto the submicroscopic scale. Among the most persistently troubling examples of such difficulties are those associated with a blurring of the lines between what we perceive as the wave-like and particle-like properties of objects.

In our everyday experience, we have little trouble distinguishing the difference between a wave on the ocean and a ball bouncing on the beach. That is because our macroscopic experience tells us that waves and particles are very different sorts of objects. However, on the atomic scale such distinctions are not so clear, as the same object can behave both like a wave and like a particle. This phenomena is also closely related to the quantization of energy, as we shall see.[5]

Much of our everyday experience suggests that energy is a continuous function. For example, when we are driving a car, we are able to continuously accelerate from a state of zero kinetic energy up to a dangerously high kinetic energy. The same is true of the kinetic energy of a baseball or a billiard ball. Moreover, we expect a pendulum or a ball on a spring to be capable of oscillating over a continuous range of amplitudes, and thus to have a continuously variable potential energy.

However, on the atomic scale energies are usually quantized, in the sense that they have discrete rather than continuous values. The energy spacing between quantum states depends on the nature of the motion involved. When a given degree of freedom has a quantum state spacing that is small compared to the ambient thermal energy, then it will behave classically, while when the spacing is larger than the available thermal energy, it will behave nonclassically, as further discussed in Section 1.4.

The early development of quantum mechanics was marked by over two decades of bold speculation, aimed at repairing glaring disagreements between classical predictions and experimental measurements. The ensuing debate generated a fascinating plethora of trial-and-error attempts to resolve the discrepancies between theory and experiment.[6]

The Quantization of Light

One of the most famous failures of classical electrodynamics and thermodynamics pertains to the spectra of so-called *blackbodies*, which closely resemble coals glowing in a campfire and the light emitted by

---

[5] Even more generally, wave-particle duality is related to the quantization of *action*, which is defined as the product of momentum and position, whose units are the same as those of angular momentum and Planck's quantum of action h.

[6] This interesting early history of quantum theory is well described in a book entitled *The Conceptual Development of Quantum Mechanics* by Max Jammer, and admirably summarized in Chapter 10 of a book called *The World of Physical Chemistry* by Keith J. Laider.

stars overhead. Classical theory predicted that the intensity of the light radiated by such bodies should increase with increasing frequency, while experiments invariably showed intensities decreasing to zero at the highest frequencies. Max Planck resolved the discrepancy in 1900, by postulating that the energy emitted at each blackbody frequency, ν, is quantized in packets of $h\nu$, with a universal constant of proportionality, h, which now bears his name. However, it was initially far from clear whether the required quantization should be attributed to light or to the material from which the glowing body is composed, or both.

An important clarification of the above question was suggested by Einstein in the first of his three famous papers written in 1905, in which he presented various arguments, all leading to the conclusion that light itself is quantized in packets of energy $h\nu$, now known as photons.[7] The following are his own words (in translation) from the introduction to that paper.

> *It seems to me that the observations associated with blackbody radiation, fluorescence, the production of cathode rays by ultraviolet light, and other related phenomena connected with the emission or transformation of light are more readily understood if one assumes that the energy of light is discontinuously distributed in space. In accordance with the assumption to be considered here, the energy of a light ray spreading out from a point source is not continuously distributed over an increasing space but consists of a finite number of energy quanta which are localized at points in space, which move without dividing, and which can only be produced and absorbed as complete units. Reprinted with permission from Arons, A.B., Peppard, M.B., Am. J. Phys.* Einstein's Proposal of the Photon Concept — a Translation of the Annalen der Physik Paper of 1905, *1965, 33(367), 68. Copyright 1965, American Association of Physics Teachers.*

At the end of the above paper Einstein noted that the quantization of light could explain the so-called *photoelectric effect*, in which electrons are ejected when a metal surface is irradiated with light. The problematic feature of the experimental observations was that the kinetic energies of the ejected electrons were found to be proportional to the frequency of the light, rather than its intensity. Einstein pointed out that this apparently paradoxical observation can be readily understood if it is assumed that light is composed of particles (photons) with energy $h\nu$.

---

[7] Although Planck and Einstein developed our current understanding of photons, it is an interesting and little known fact that the term *photon* was first introduced in a short note submitted to the journal *Nature* in 1926 by a prominent physical chemist named Gilbert Newton Lewis – the same G.N. Lewis who created the Lewis dot-structure representation of chemical bonds, and the concept of Lewis acids and bases, as well as many other important ideas pertaining to chemical thermodynamics.

These speculations were not widely embraced for over a decade, until Robert Millikan reported the results of additional key experiments. The following extended quotation from the introduction of Millikan's 1916 paper, entitled *A Direct Photoelectric Determination of Planck's "h"*, provides an interesting glimpse into the prevailing view of Einstein's photon postulate.

> *It was in 1905 that Einstein made the first coupling of photo effects with any form of quantum theory by bringing forward the bold, not to say reckless, hypothesis of an electro-magnetic light corpuscle of energy $h\nu$, which energy was transferred upon absorption to an electron. This hypothesis may well be called reckless first because an electro-magnetic disturbance which remains localized in space seems a violation of the very conception of an electromagnetic disturbance, and second because it flies in the face of the thoroughly established facts of interference. The hypothesis was apparently made solely because it furnished a ready explanation of one of the most remarkable facts brought to light by recent investigations, viz., that the energy with which an electron is thrown out of a metal by ultra-violet light or X-rays is independent of the intensity of the light while it depends on its frequency. This fact alone seems to demand some modification of classical theory or, at any rate, it has not yet been interpreted satisfactorily in terms of classical theory.*
> *From Millikan, R.A.,* Physical Review *7, 355 (1916).*

Even after Millikan's paper, and after Einstein received a Nobel prize "for his services to theoretical physics, and especially for his discovery of the law of the photoelectric effect," the subject of photon quantization remained, and continues to be, an active and interesting area of research, all the results of which have so far been found to be entirely consistent with Einstein's original proposal. However, Einstein apparently retained some concerns about the photon concept, as illustrated by the following quotation from the end of his 1916 paper entitled *On the Quantum Theory of Radiation* (which is most famous for predicting stimulated emission, long before the development of lasers).

> *These properties of elementary processes... make the formulation of a proper quantum theory of radiation appear almost unavoidable. The weakness of the theory lies on the one hand in the fact that it does not get us any closer to making the connection with wave theory; on the other, that it leaves the duration and direction of the elementary processes to 'chance'. Nevertheless I am fully confident that the approach chosen here is a reliable one.*
> *From Einstein's 1916 paper translated in* Sources of Quantum Mechanics *Ed. B.L. van der Waerden (Dover, New York, 1967).*

Wave-Particles and Particle-Waves

The photoelectric effect is also closely related to the photoionization of atoms and molecules by light. In both cases the energy of the emitted electron is proportional to the frequency of the light. Also, in both cases no electrons are emitted when the photon energy $h\nu$ is too small. This makes sense, because some energy is required in order to overcome the binding energy of the electron to the material. So, for both the emission of photoelectrons from a metal and the photoionization of molecules one finds that

$$K = h\nu - \Phi \tag{1.9}$$

where K is the kinetic energy of the ejected electron and Φ is its binding energy, before it was ejected. The latter binding energy (or *work function*) is a positive number whose value depends on the nature of the metal or atom to which the electron is bound.

Einstein's explanation of the photoelectric effect, as expressed in Eq. 1.9, clearly implies that the energy of a photon is given by the following simple expression.

$$\boxed{E = h\nu} \tag{1.10}$$

If we combine Einstein's famous formula $E = mc^2$ (Eq. 1.8) with the definition of the momentum of a particle $p = mv$, we obtain the following relation between momentum and energy $p = Ev/c^2$. Recall that the velocity and frequency of light are $v = c$ and $\nu = c/\lambda$, respectively (where λ is the wavelength of light). These identities may be combined, as follows, to reveal a remarkably simple relation between the momentum of a photon and its wavelength.

$$p = \frac{Ev}{c^2} = \frac{h\nu c}{c^2} = \frac{h\nu}{c} = \frac{h(c/\lambda)}{c} = \frac{h}{\lambda}$$

The momentum of light can be observed experimentally by measuring the *radiation pressure* exerted by light as it reflects off the surface of a mirror. This pressure is exactly consistent with the particle-like properties of photons.[8]

The appearance of ν and λ in the equations for the energy and momentum of a photon indicates that the particle and wave properties of photons are inextricably linked. Such observations led a physics graduate student named Louis de Broglie to propose in his 1924 PhD thesis that particles such as electrons, protons, and atoms may also have wave-like properties. This astonishing prediction was beautifully confirmed in the following experiments, which clearly revealed that electrons and atoms do indeed have wave-like properties.

In the late 1920s, Otto Stern set out to systematically test de Broglie's hypothesis by conducting experiments in which beams of various kinds of atoms and molecules were directed at salt crystals. His results revealed, for example, that a beam of He atoms undergoes diffraction when it is scattered off of salt crystals. The observed diffraction fringe spacing was

---

[8] Quite remarkably, the pressure exerted by light can also be correctly predicted using purely classical theory in which light is described as an electromagnetic wave.

found to be related to the momentum of the He atoms (and the lattice spacing of the rock salt crystal), exactly as predicted by de Broglie.[9]

Thus, not only photons but all other particles appear to have a wavelength that is related to momentum.

$$\boxed{p = \frac{h}{\lambda}} \tag{1.11}$$

The apparently universal validity of this expression is one of the clearest pieces of evidence for the blurred distinction between particles and waves. All waves have particle-like properties and all particles have wave-like properties, but particles with very large momenta have very small wavelengths and waves with very long wavelengths have very small momenta. Objects with large momentum (such as macroscopic billiard balls) are more readily observable as particles while those with small momentum (such as photons) are more readily observable as waves.

Other examples of the particle-like properties of light include phenomena known as Compton scattering and Raman scattering. These both involve the inelastic scattering of light by a chemical substance. In other words, the energy of a photon is changed as it either gives up or gains energy from an object with which it collides. This is similar to what would happen if you threw a baseball at a spring mattress (or a trampoline), and the ball bounced back with a different velocity, because it exchanged energy with the springs.

Compton scattering is named after Arthur Compton, a professor of Physics at the University of Chicago, whose experiments performed in 1922 revealed the particle-like properties of photons in remarkable details. Compton's experiments showed that when an X-ray photon hits an electron, the two particles bounce off each other just like billiard balls on a pool table. The angles of the outgoing electron and photon are exactly those required in order to conserve both the energy and momentum of the two particles.[10]

---

[9] The diffraction of beams of He and H_2 molecules produced by crystals of NaCl and LiF were reported by Otto Stern and coworkers in 1930.

[10] More specifically, if the input photon energy is $h\nu_{in}$ and the initially stationary electron is kicked out with an energy of $\Delta\varepsilon = \frac{1}{2}mv^2 = h(\nu_{in} - \nu_{out})$ and a momentum of $\Delta p = mv = h\left(\frac{1}{\lambda_{in}} - \frac{1}{\lambda_{out}}\right)$, then the observed deflection angles of the outgoing electron, ϕ, and photon, θ, will be related as follows.

$$\cos\theta = \frac{\Delta\varepsilon}{c\Delta p}\cos\phi$$

where $\cos\theta$ is also related the change in the photon wavelength $\Delta\lambda = \lambda_{out} - \lambda_{in}$. These scattering predictions are obtained by equating the energy and momentum of the incoming and outgoing photon and electron, as described for example in Appendix XVII A in a book entitled *Light* by R.W. Ditchburn.

$$\Delta\lambda = \frac{h}{mc}(1 - \cos\theta)$$

Raman scattering, on the other hand, is a process in which a photon exchanges energy with a vibrating molecule.[11] More specifically, when a photon inelastically scatters from a molecule with a vibrational frequency of ν, then the change in the energy of the photon is observed to be exactly equal to $h\nu$. This remarkable result implies that not only photons but also molecular vibrations have quantized energies equal to $h\nu$ (as further described in the next subsection, as well as in Section 1.4 and Chapters 7 and 9).

Quantization of Atomic and Molecular Energies

The intimate connection between particles and waves implies that the internal energies of atoms and molecules must be quantized. This quantization is quite similar to that of musical instruments whose sound derives from the resonant acoustic frequencies and overtones that give each instrument its unique timbre. However, making the connection between particles, waves, and molecular quantization was a great challenge to the development of quantum theory in the first quarter of the twentieth century.

The Danish physicist and natural philosopher, Niels Bohr, took the first great leap in this direction by suggesting a remarkably simple semiclassical explanation for the line spectra of atoms. Bohr's theory proved to be stunningly successful, not only in accurately predicting the line spectrum of hydrogen but also the astronomical absorption lines of He^{+}, which had previously not been correctly assigned. Although Bohr's atomic theory is now viewed as only a tentative first step in the early development of quantum mechanics, the following comment made by Einstein reveals the high regard with which he still held Bohr's theory many years after it had fallen out of favor.

> *That this insecure and contradictory foundation was sufficient to enable a man of Bohr's unique instinct and perceptiveness to discover the major laws of the spectral lines and of the electron-shells of the atoms together with their significance for chemistry appeared to me like a miracle and appears to me as a miracle even today. This is the highest form of musicality in the sphere of thought. Source: Albert Einstein (author), Paul Arthur, Schilpp (editor).* Autobiographical Notes. A Centennial Edition. *Open Court Publishing Company. 1979. p. 45.*

---

[11] The discovery of Raman scattering was first reported by Chandrasekhara Raman in a news release he sent to a Calcutta newspaper on February 28, 1928, the day after he first observed the production of new colors when light from the Sun was scattered from various liquids. Interestingly, the same phenomenon was discovered entirely independently by Leonid Mandelshtam and Grigorii Landsberg in Moscow, as described in their notebook pages dated February 21, 1928, although their observations were not published until July of that year.

What Bohr suggested in his seminal 1913 paper was a deceptively simple hybrid of Einstein's proposal regarding the quantization of light and the classical atomic picture of an electron rotating around a nucleus. Bohr simply required that the rotating electron could only absorb or emit light in quanta of energy equal to $h\nu$. Moreover, Bohr postulated that the frequency of an absorbed photon must be equal to the arithmetic mean of the classical rotational frequencies of the electron before and after the optical transition has occurred. In other words, Bohr assumed that $\nu = (\nu_i + \nu_f)/2$, where ν is the photon frequency, while ν_i and ν_f are the initial and final (classical) orbital frequencies of the electron.

With no more than the above combination of elementary assumptions, Bohr was able to correctly predict the ground state energy and optical spectrum of a hydrogen atom, in exact agreement with experiment as well as with the predictions of current quantum theory (as described in Section 8.1 of Chapter 8).

Although some aspects of Bohr's atomic model have been superseded by later developments in quantum theory, many of his ideas continue to be of interest, particularly with regard to the boundaries between classical and quantum phenomena. Bohr's simple semiclassical procedure can be used to correctly predict the quantization of molecular vibrational, rotational, and translational energies (as further described below and in the following subsection).

For example, we may use Bohr's method to describe a particle that is confined to a *box* of length L. Such a *particle-in-a-box* may be used to understand the quantization of electrons confined within molecules or semiconductor *quantum-wells*, as well as the translational quantization of atoms and molecules confined in a macroscopic container. The following famous expression for the quantized energy levels of a particle of mass m confined to a one-dimensional box of length L may be obtained quite easily using Bohr's method (as further described in the following subsection).

$$\boxed{\varepsilon_n = \frac{n^2 h^2}{8mL^2}} \tag{1.12}$$

The quantum number $n = 1, 2, 3, \ldots$ can take on any positive integer values. When we delve more deeply into quantum theory, we will see that Eq. 1.12 is exactly the same as that obtained using more advanced quantum mechanical procedures (as described in Chapter 7).

If we assume that the box has a length of one meter ($L = 1$ m) and the mass of the particle is the same as that of a nitrogen molecule (N_2), then

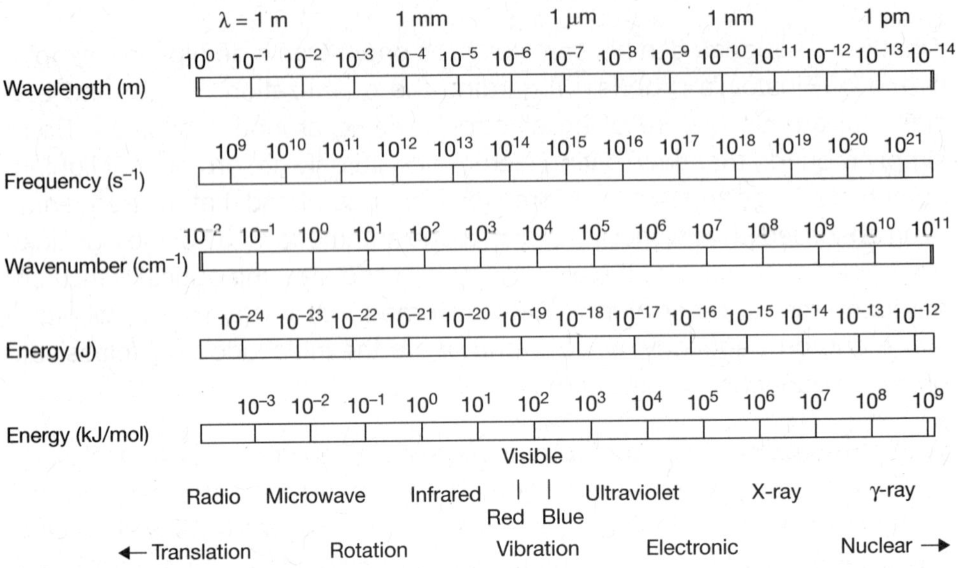

FIGURE 1.2 Relation between electromagnetic (photon) wavelengths, frequencies, wavenumbers (defined on page 33), and energies, and molecular quantum state spacings. Note that the values are expressed on a logarithmic scale, so each major division represents a different order of magnitude (factor of ten change).

Eq. 1.12 predicts that the spacing between the first two quantum states will be extremely small, $\Delta\varepsilon \approx 10^{-45}$ J (or in molar units, $\Delta\varepsilon \approx 10^{-25}$ kJ/mol). On the other hand, if the particle is an electron confined to an atom, which is like a box of Ångstrom size ($L \approx 1\text{Å} = 10^{-10}$ m), then the energy difference between the ground and first excited state becomes many orders of magnitude larger, $\Delta\varepsilon \approx 10^{-17}$ J ($\approx 10^4$ kJ/mol). In Section 1.4 we will see why quantum states with very small spacings, such as translational states, are expected to behave highly classically, while those with large quantum spacings, such as electronic states, are expected to behave strongly quantum mechanically.

Figure 1.2 illustrates the range of electromagnetic (photon) wavelengths, frequencies, and energies, and how these correspond to typical molecular quantum state spacings for translational, rotational, vibrational, electronic, and nuclear motions. Thermal energies may be equated with the average photon energy emitted by a blackbody radiator of temperature T, which is $h\nu = k_B T$ (where k_B is Boltzmann's constant). At room temperature ($T \approx 300$ K) this is in the infrared region of the electromagnetic spectrum, in the same frequency range as typical molecular vibrational motions, far below typical molecular electronic quantum spacings, and far above translational quantum spacings (for molecules in a macroscopic container).

More About Bohr's Quantization Method

The following is a more detailed description of the method that Bohr used in his original theoretical derivation of the hydrogen spectrum,[12] as well as illustrations of how his method may be applied to other simple quantum mechanical systems (in ways that Bohr had apparently not originally envisioned).

Contrary to presentations of Bohr's theory often encountered in other textbooks, Bohr did not assume that electrons have wave-like properties, and he also did not assume any angular momentum quantization condition. Rather, Bohr *predicted* angular momentum quantization by assuming only the Einstein photon energy relation $\varepsilon = h\nu$ and the mean-frequency condition $\nu = (\nu_i + \nu_f)/2$. However, he later decided that momentum quantization was a more universally applicable constraint, and so he elevated it to the status of a primary postulate in his "old" quantum theory. Moreover, although Bohr initially considered only circular electron orbitals, Sommerfeld later relaxed that restriction and in so doing correctly predicted the existence of orbitals with different magnetic quantum numbers.

In his original semiclassical quantum treatment of the hydrogen atom Bohr envisioned starting with a motionless electron that is far removed from the nucleus, so that it is no longer bound or rotating. He then imposed Einstein's relation $\varepsilon = h\nu$ (Eq. 1.10) to relate the energy and frequency of a photon. In other words, he required that an electronic transition can only occur if the energy of the photon matches the energy change of the electron $h\nu = \varepsilon_f - \varepsilon_i = \Delta\varepsilon$, in addition to the mean-frequency condition $\nu = (\nu_i + \nu_f)/2$. Bohr justified the latter mean-frequency assumption by showing that it is the only possible relationship between the photon and electron frequencies that is consistent both with the experimental Rydberg formula (Eq. 8.4 on page 266) and with classical predictions applied to transitions between the highest energy (most closely spaced) hydrogen orbitals.

Application of Bohr's method to a *particle-in-a-box* may be used to obtain Eq. 1.12. More specifically the box may be described by the following one-dimensional potential, which is equal to zero inside the box and infinity outside the box; $V(x) = 0$ when $0 < x < L$, and $V(x) = \infty$ when $x \leq 0$ or $x \geq L$. Since there is no potential energy inside the box, only kinetic energy distinguishes different states of the particle. A classical particle of mass m and velocity v has a kinetic energy of $\frac{1}{2}mv^2$. When confined to a box of length L, such a classical particle will bounce back and forth with a (non-sinusoidal) frequency of $\nu = v/(2L)$. Thus, the kinetic energy of such an electron is related in a simple way to its frequency, $2mL^2\nu^2$.

---

[12] As described in his paper entitled "On the Constitution of Atoms and Molecules," published in 1913 in the *Philosophical Magazine*, volume 26, pages 1–25.

We may now impose Einstein's photon quantization and Bohr's mean-frequency conditions, to obtain the following relation between the photon energy $h\nu$, and the change in frequency ν_j and kinetic energy ε_j of the particle: $h\nu = h(\nu_i + \nu_f/2)$ and $\varepsilon_f - \varepsilon_i = 2mL^2(\nu_f^2 - \nu_i^2)$. If we again assume that the particle is initially stationary, then its initial energy and frequency are both equal to zero. By applying the above expression to this zero-kinetic-energy state, we may readily obtain the particle round-trip frequency and kinetic energy for the next higher energy state, $\nu_1 = h/(4mL^2)$ and $\varepsilon_1 = h^2/(8mL^2)$. Repeated application of the above procedure yields Eq. 1.12.

It is also easy to apply Bohr's method to explain the quantization of molecular vibrations. These may be represented as harmonic oscillators, which are mathematically equivalent to the pendulum in an old-fashioned clock, or a ball rolling back and forth on a quadratic potential energy surface. Since the frequency of a harmonic vibrational oscillator, ν, is independent of its amplitude, the mean frequency associated with any two vibrational energies is simply equal to the harmonic vibrational frequency of the oscillator, $(\nu_i + \nu_f)/2 = \nu$. In other words, Bohr's semiclassical method predicts that if light is to be absorbed by a molecular vibrational mode, then its frequency ν must be equal to the classical vibrational frequency of the molecule. Note that this also implies that molecular vibrations have quantum states that are evenly spaced, with an energy difference of $\Delta\varepsilon = h\nu$ between neighboring states. The only exception is the very lowest energy state. In other words, if we assume that the molecule begins with absolutely no vibrational energy and so $\nu_i = 0$, then the photon frequency required to excite the molecule becomes $\nu = (0 + \nu)/2 = \nu/2$. So, the first nonzero quantum state of a harmonic oscillator is predicted to have a *zero-point* energy of $h\nu/2$, which is exactly the same as that obtained using modern quantum mechanical theory (as described in Chapter 6).

Similarly, the rotation of a particle in a circle of radius r may be represented as a particle confined to a ring of length $L = 2\pi r$, and a similar procedure may be used to predict that its quantum states will have energies of $\varepsilon_n = n^2\hbar^2/(2mr^2)$ where $n = 0, \pm 1, \pm 2, \ldots$ and $\hbar = h/(2\pi)$. Thus, Bohr's method may be used to correctly predict the quantization of molecular translational, rotational, vibrational, and electronic motions.

1.4 Thermal Energies and Populations

Relation Between Energy and Probability

Energy plays a key role in determining the probability of finding a system in a given state. Not surprisingly, states of lower energy have a higher probability than those of higher energy. This is, for example, why the density of the atmosphere decreases with increasing altitude (i.e., with increasing gravitational potential energy). It is also why the density of a vapor is lower than that of the liquid with which it is at equilibrium (because molecules in the liquid are stabilized by their intermolecular cohesive interaction energy). The quantitative connection between energy and probability was investigated by Maxwell, Boltzmann, and Gibbs, whose insights led to the following tremendously important, and yet remarkably simple proportionality.[13]

$$\boxed{P(\varepsilon) \propto e^{-\beta\varepsilon}} \tag{1.13}$$

$P(\varepsilon)$ is the probability of finding a system in a state of a given energy, ε, and $\beta = 1/k_BT$, where k_BT is a measure of thermal energy (equal to Boltzmann's constant, k_B, times the absolute temperature, T). In other words, the probability of observing any system in a state of energy ε is proportional to the *Boltzmann factor* $e^{-\beta\varepsilon}$, and this in turn only depends on the ratio $\varepsilon/k_BT = \beta\varepsilon$. So, the probability of occupying a state of a given energy not only decreases with increasing energy but also increases with temperature. This makes good physical sense, since our experience indicates that increasing the temperature of a liquid (which has a low energy) can convert it to a vapor (which has high energy).[14]

When Boltzmann's constant is multiplied by Avogadro's number it becomes equivalent to the gas constant, $R = \mathcal{N}_A k_B$. So, if we choose to express energies in molar units, then we should identify $\beta = 1/RT$. In other words, k_BT and RT are essentially equivalent, and so physical chemists tend to switch back and forth between expressing thermal energy as k_BT or RT, depending on whether the context calls for using molecular or molar units.

We can turn the proportionality in Eq. 1.13 into an equality by noting that the total probability of observing a system in any state must be equal to 1. In other words, we require that $\sum_i P(\varepsilon_i) = 1$, where the sum is carried out over all the energies (quantum states) of the system. This also implies that the

---

[13] The exponential relation between probability and energy is a consequence of the fact that the probability of observing statistically independent events is the product of their individual probabilities (see, for example, the paragraph preceding Eq. 1.31 on page 36). Equation 1.13 may be viewed as the one fundamental axiom from which all of statistical mechanics may be derived. Equivalently, one may take Boltzmann's famous expression for entropy $S = k_B \ln \Omega$ (see Eq. 2.9 on page 59) to be the fundamental axiom of statistical mechanics. In other words, Eq. 1.13 may be derived from $S = k_B \ln \Omega$ or conversely one may derive the latter equation from the former. The close connection between these two equations will become more evident when we revisit this subject in Section 5.4.

[14] The energy difference between a liquid and a vapor derives from the cohesive interactions between molecules. These interactions lower the energy of a liquid, while in a vapor the density is sufficiently low so that there are virtually no intermolecular interactions.

constant of proportionality that is missing in Eq. 1.13 must be $1/\sum_i P(\varepsilon_i)$, and so $P(\varepsilon)$ is exactly given by the following expression.

$$\boxed{P(\varepsilon_i) = \frac{e^{-\beta\varepsilon_i}}{\sum_i e^{-\beta\varepsilon_i}}} \tag{1.14}$$

Note that the sum of $P(\varepsilon_i)$ over all states is $\sum_i e^{-\beta\varepsilon_i}/\sum_i e^{-\beta\varepsilon_i} = 1$, as expected.

The denominator in Eq. 1.14 plays a surprisingly important role in chemical thermodynamics – it is called the *partition function*, and is often represented by the letter q.[15]

$$\boxed{q \equiv \sum_i^{\text{all states}} e^{-\beta\varepsilon_i}} \tag{1.15}$$

The value of q for a given system is related in a very simple way to the number of quantum states that are thermally populated at a given temperature. In order to see why this is the case, notice that Eq. 1.15 consists of a sum of terms, each of which are equal to a number between zero and one. The lowest energy terms in the series are each approximately equal to one (since $e^{-\beta\varepsilon_i} \approx 1$ whenever $\varepsilon_i < k_B T$), while the high-energy terms are approximately zero (since $e^{-\beta\varepsilon_i} \approx 0$ whenever $\varepsilon_i > k_B T$). Thus, q represents the average number of terms in the sum that have a value near one, and so *the value of q indicates the number of states that are thermally populated.*[16]

Some quantized systems have more than one state with the same energy. Such states are referred to as having *degenerate* energies. For example, if there are two states of the same energy, then the latter energy is said to have a "degeneracy of two," or to be "two-fold degenerate" (while a state with a degeneracy of one is referred to as "nondegenerate"). We may explicitly indicate the degeneracies of various energy levels by reexpressing the partition function (Eq. 1.15) in the following form.

$$\boxed{q \equiv \sum_j^{\text{all energies}} g_j e^{-\beta\varepsilon_j}} \tag{1.16}$$

Note that Eq. 1.16 is identical to Eq. 1.15, except that the Boltzmann factors for each of the ε_i states that have the same energy have been grouped

[15] When describing a macroscopic system composed of many molecules (maintained at constant temperature and volume), the partition function is often designated as $\mathcal{Q}$, or sometimes by other letters such as $\mathcal{Z}$, as further discussed in Section 5.3.

[16] The above analysis relies on the implicit assumption that the molecule's ground state energy is equal to zero, $\varepsilon_0 = 0$. If some other reference energy is used, then the value of q would change by a constant factor equal to $e^{-\beta\varepsilon_0}$.

together by including the degeneracies g_j of each ε_j energy level. In other words, the value of the partition function (q) predicted by Eqs. 1.15 and 1.16 is exactly the same; the only difference between the two expressions is that the summation index i extends over all *quantum states*, while the summation index j extends over all *energy levels* (each of which contains g_j individual quantum states). Similarly, the probability of finding a system at an energy of ε_j is necessarily g_j times greater than the probability of finding the system in any one of the g_j-fold degenerate quantum states.

$$P(\varepsilon_j) = \frac{g_j e^{-\beta\varepsilon_j}}{\sum_j g_j e^{-\beta\varepsilon_j}} \tag{1.17}$$

Chemical Reaction Equilibrium Constants

Equation 1.14 can be used to describe a wide variety of chemical processes. For example, the chemical equilibrium constant for a reaction of the form $A \rightleftharpoons B$ is $K = [B]/[A] = (n_B/V)/(n_A/V) = n_B/n_A = P(B)/P(A)$. In other words, the equilibrium constant for any such reaction is equal to the ratio of the probabilities of finding the system in the product or reactant states.[17]

If the reactants and products have different (nondegenerate) ground state energies, ε_A and ε_B, respectively (and if we assume that no other quantum states are significantly populated), then the equilibrium constant becomes simply $K = (e^{-\beta\varepsilon_B}/q)/(e^{-\beta\varepsilon_A}/q) = e^{-\beta\Delta\varepsilon}$, where $\Delta\varepsilon = \varepsilon_B - \varepsilon_A$.

We may further generalize the above result by allowing for the fact that each molecule may have a different degeneracy. If the reactant and product molecules have g_j states (or configurations) of the same energy, then $P(A) = g_A e^{-\beta\varepsilon_A}/q$ and $P(B) = g_B e^{-\beta\varepsilon_B}/q$. Thus, the equilibrium constant for such a reaction is predicted to have the following form.

$$K = \left(\frac{g_B}{g_A}\right) e^{-\beta\Delta\varepsilon} \tag{1.18}$$

It is also interesting to see what happens if we take the logarithm of both sides of Eq. 1.18, and then multiply by $-1/\beta = -RT$.

$$-RT \ln K = \Delta\varepsilon - RT \ln\left(\frac{g_B}{g_A}\right) \tag{1.19}$$

---

[17] Strictly speaking, when Eq. 1.14 or 1.17 is used to calculate $P(A)$ and $P(B)$, the above expression for the equilibrium constant pertains to reactions that take place between molecules in the gas phase in a container of constant volume and temperature, but the same idea can be extended to reactions carried out at constant pressure, as well as to reactions in nonideal gases, liquids, and supercritical fluids, as we will see in Chapters 5 and 10.

Notice that the left-hand side of Eq. 1.19 is identical to the commonly encountered expression for the standard Gibbs energy of reaction, $\Delta G = -RT \ln K$. Moreover, Eq. 1.19 is reminiscent of another well-known relation, $\Delta G = \Delta H - T\Delta S$, where ΔH and ΔS are enthalpy and entropy changes associated with the reaction. This suggests that there is a close connection between ΔH and the reaction energy $\Delta\varepsilon$ as well as between ΔS and the ratio of the number states that are available to the product and reactant molecules g_B/g_A, such that $\Delta S = R\ln(g_B/g_A) = R\ln g_B - R\ln g_A$.

We will revisit these profoundly important interconnections from several different perspectives as we gain a more complete understanding of thermodynamics and statistical mechanics in Chapters 2–5. Moreover, Section 4.2 and Chapter 10 describe more general treatments of chemical reactions.

Exercise 1.3

Consider a system that has two energy levels, where the lower energy (ground) state has an energy of $\varepsilon_0 = 0$ and is nondegenerate ($g_0 = 1$), and the upper energy (excited) state has an energy of ε_1 and is three-fold degenerate ($g_1 = 3$).

- What are the smallest and largest possible values of the partition function for this system, and at what temperatures will these values be achieved?

  Solution. The partition function of this two-level system is $q = g_0 e^{-\beta\varepsilon_0} + g_1 e^{-\beta\varepsilon_1} = 1 + 3e^{-\beta\varepsilon_1}$. At zero temperature, β approaches infinity and so $q = 1$ (which is its minimum value), while at infinite temperature, β approaches zero and so $q = 4$ (which is its maximum value). Note that these two values of q are also equivalent to the number of states that are thermally populated at the corresponding temperatures. In other words, at very low temperatures only the ground state is thermally populated and so $q = 1$, while at very high temperatures, both the ground state and the three excited states all become nearly equally populated and so $q = 1 + 3 = 4$.

- What is the value of the partition function when the temperature (times k_B or R) is exactly equivalent to ε_1?

  Solution. When $k_B T = \varepsilon_1$, then $\beta\varepsilon_1 = \varepsilon_1/(k_B T) = 1$, and so $q = 1 + 3e^{-1} \approx 2.1$. Note that exactly the same result would be obtained if we had expressed all energies in molar units and equated $RT = \varepsilon_1$, so $\beta\varepsilon_1 = \varepsilon_1/(RT) = 1$.

- Consider a chemical reaction in which the system is transformed from the lower energy (ε_0) reactant, which has a single conformation state ($g_0 = 1$), to the higher energy (ε_1) product, which has three conformational states of approximately the same energy ($g_1 = 3$). At what value of $\beta\Delta\varepsilon = \beta(\varepsilon_1 - \varepsilon_0)$ will the product and reactant populations be equal to each other?

Solution. Equation 1.18 implies that the equilibrium constant for such a reaction is $K = (g_1/g_0)e^{-\beta\Delta\varepsilon} = 3e^{-\beta\Delta\varepsilon}$. The reactant and product concentrations will be equal to each other when the equilibrium constant is equal to one. So, we may substitute $K = 1$ and solve the resulting expression to obtain $\beta\Delta\varepsilon = -\ln(1/3) = \ln(3) \approx 1.1$.

- Predict the value of the equilibrium constant for such a reaction at 298 K, if the energy difference between the reactants and products is 5 kJ/mol.

Solution. At 298K $RT \approx 8.3$ J/mol × 298 K $\times 10^{-3}$ kJ/J ≈ 2.5 kJ/mol, so $\beta\Delta\varepsilon = 5$ (kJ/mol)/2.5 (kJ/mol) ≈ 2. Thus, at 298 K the equilibrium constant is predicted to be $K = (g_1 e^{-\beta\varepsilon_1})/(g_0 e^{-\beta\varepsilon_0}) = (g_1/g_0)e^{-\beta\Delta\varepsilon} = 3e^{-2} \approx 0.4$.

Quantized Harmonic Oscillator

One of the simplest and most important applications of Eq. 1.14 pertains to a system with an evenly spaced ladder of quantum states, $\varepsilon_n = n\Delta\varepsilon$, where $\Delta\varepsilon$ is a constant energy spacing and n is a nonnegative integer (0, 1, 2, 3, ... so $\varepsilon_n = 0, \Delta\varepsilon, 2\Delta\varepsilon, 3\Delta\varepsilon, \ldots$). Such an energy level structure arises in many situations, including molecular vibrations (described as harmonic oscillators) as well as light (which also consists of harmonic electromagnetic oscillations). Both experimental observations and quantum mechanical predictions agree that all such systems have an evenly spaced ladder of energy quantum states.

Since light of a given frequency (color) is composed of photons of energy $h\nu$, a beam of light must have a total energy of $n\Delta\varepsilon = nh\nu$, where n is the number of photons in the beam. Similarly, molecular vibrations can also only have energies of $n\Delta\varepsilon = nh\nu$ where ν is the frequency of the molecular vibration.[18]

In applying Eq. 1.14 to such systems, one may use the following mathematical manipulation to express the infinite sum that appears in Eq. 1.14 (and Eq. 1.15) in a simple closed form. This begins by defining a new variable, $x = e^{-\beta\Delta\varepsilon}$, in terms of which the partition function reduces to the following simple power series.

$$q = \sum_{n=0}^{\infty} x^n = 1 + x + x^2 + x^3 + \cdots. \tag{1.20}$$

[18] A macroscopic crystal is like a very large molecule, for which $h\nu$ becomes the energy of each vibrational mode of the solid. These modes are also sometimes called *phonons* because of their similarity to photons.

Since $\beta\Delta\varepsilon$ is invariably positive, it follows that $0 < x < 1$. Under such conditions the above series converges exactly to $1/(1-x)$, and so the partition function reduces to the following expression.

$$q = \frac{1}{1 - e^{-\beta\Delta\varepsilon}} \tag{1.21}$$

Thus, the probability in Eq. 1.14 may be expressed as follows.

$$P(\varepsilon_n) = \frac{e^{-\beta n\Delta\varepsilon}}{q} = e^{-\beta n\Delta\varepsilon}\left(1 - e^{-\beta\Delta\varepsilon}\right) \tag{1.22}$$

This equation indicates that the probability of finding the system in a quantum state of energy ε_n decreases exponentially with increasing n (at any fixed temperature). In other words, the oscillator is less likely to occupy higher energy states (as expected). Moreover, as the temperature approaches absolute zero, only the lowest energy (ground) state will be populated, since $P(\varepsilon_0) = e^{-0} = 1$, while for all other states $P(\varepsilon_n) = e^{-\beta n\Delta\varepsilon} \approx 0$ whenever $k_BT << \Delta\varepsilon$ or $\beta\Delta\varepsilon >> 1$. At high temperature, on the other hand, all the states for which $\beta\varepsilon_n << 1$ will have a finite probability, while states for which $\beta\varepsilon_n >> 1$ will have essentially zero probability.[19] Thus, the temperature determines how many quantum states will be significantly populated. At low temperatures only the very lowest state will be populated, while at higher temperatures only states for which $\varepsilon_n < k_BT$ will be significantly populated.

Average Thermal Energies

The Boltzmann probabilities, $P(\varepsilon)$, may be used to calculate the average energy of the system by weighting each quantum state energy by the probability that it will be occupied, and then summing over all states.[20]

$$\boxed{\langle\varepsilon\rangle = \sum_i \varepsilon_i P(\varepsilon_i) = \frac{\sum \varepsilon_i e^{-\beta\varepsilon_i}}{\sum e^{-\beta\varepsilon_i}} = -\frac{1}{q}\left(\frac{dq}{d\beta}\right)} \tag{1.23}$$

---

[19] Note that in the high-temperature limit the harmonic oscillator partition function becomes

$$q \approx \frac{1}{1-(1-\beta\Delta\varepsilon)} = \frac{1}{\beta\Delta\varepsilon} = \frac{k_BT}{\Delta\varepsilon} = \frac{k_BT}{h\nu}$$

and so, in this limit, q represents the number of harmonic oscillator quantum of states whose energy is less than k_BT.

[20] The last equality in Eq. 1.23 is obtained by noting that the derivative of q with respect to β is $\sum -\varepsilon_n e^{-\beta\varepsilon_n}$, and so $\langle\epsilon\rangle = \frac{\sum \varepsilon_n e^{-\beta\varepsilon_n}}{q} = -\frac{1}{q}\left(\frac{dq}{d\beta}\right)$.

For a system with a ladder of evenly spaced energies, the above general expression reduces to the following result.[21]

$$\langle \varepsilon \rangle = \frac{\Delta\varepsilon\, e^{-\beta\Delta\varepsilon}}{1 - e^{-\beta\Delta\varepsilon}} = \frac{\Delta\varepsilon}{e^{+\beta\Delta\varepsilon} - 1} \tag{1.24}$$

At very low temperature, the denominator of Eq. 1.24 approaches infinity, and so $\langle \varepsilon_n \rangle = 0$ (which makes sense since in this case all the population goes to the ground state of energy $\varepsilon_0 = 0$). In the high temperature limit, $\Delta\varepsilon << k_B T$ and so $\beta\Delta\varepsilon << 1$ and $e^{\beta\Delta\varepsilon} \approx 1 + \beta\Delta\varepsilon$, which implies that $\langle \varepsilon_n \rangle \approx \Delta\varepsilon/(\beta\Delta\varepsilon) = 1/\beta = k_B T$. In other words, at high temperatures the ladder of states is populated up to an energy of approximately $k_B T$. Note that this simple result pertains not only to very high temperatures but also to a system of lower temperatures, which has very closely spaced energy levels. This is because $\beta\Delta\varepsilon$ will be small either at high temperatures or when the energy spacings are very small (compared to $k_B T$). Thus, *the high temperature limit is equivalent to the classical limit in which energies are essentially continuous*.

Equipartition of Energy

There is a deeper reason why the average energy of an oscillator approaches $k_B T$ at high temperature. This becomes evident when we investigate the properties of systems with different sorts of Hamiltonians (as further described in Section 1.5). Such an analysis reveals that the average energy of a system depends on the number of quadratic terms that appear in its Hamiltonian. So, for example, a free particle that is moving in the x-direction has only kinetic energy (and no potential energy), so its Hamiltonian is $H_x = \frac{1}{2} m v_x^2$, which is a quadratic function of the particle's velocity (in the x-direction), and has an average energy of $\frac{1}{2} k_B T$. Similarly, the Hamiltonian of a particle that is free to move in three dimensions has three quadratic terms, $H_{xyz} = \frac{1}{2} m v_x^2 + \frac{1}{2} m v_y^2 + \frac{1}{2} m v_z^2$, and so has an average energy that is three times larger, $3\left(\frac{1}{2} k_B T\right) = \frac{3}{2} k_B T$. In other words, *each quadratic term appearing in a molecule's Hamiltonian contributes $\frac{1}{2} k_B T$ to the average energy per molecule, or $\frac{1}{2} RT$ to the average energy per mole*.

The energy of a harmonic oscillator (such as a vibrating diatomic molecule) contains both kinetic and potential energy contributions, each of which contribute a quadratic term to its Hamiltonian, $H_{\text{vib}} = \frac{1}{2}\mu v^2 + \frac{1}{2} f \delta^2$. The two quadratic variables in this expression are the velocity v and displacement δ of the oscillator (while μ and f are the corresponding reduced mass

[21] Equation 1.24 follows from Eq. 1.23, since $dq/d\beta = -\Delta\varepsilon\, e^{-\beta\Delta\varepsilon}/(1 - e^{-\beta\Delta\varepsilon})^2$, which leads to Eq. 1.24 when divided by $-q$ (and the last identity in Eq. 1.24 is obtained by multiplying both the numerator and denominator by $e^{+\beta\Delta\varepsilon}$).

and harmonic force constant).[22] Each of the quadratic terms again produce an average energy of $\frac{1}{2}k_BT$, which explains why the average energy of a classical harmonic oscillator is equal to k_BT, since $\langle\varepsilon_{\text{vib}}\rangle = \frac{1}{2}k_BT + \frac{1}{2}k_BT = k_BT$.

Rotational degrees of freedom also contribute quadratically to the Hamiltonian. For example, a diatomic molecule rotating about a given axis has a rotational kinetic energy of $H_{\text{rot}} = \frac{1}{2}I\omega^2$, where $I = \mu r_b^2$ is the moment of inertia of the diatomic and ω is its angular frequency of rotation (in units of radians per second). Since a diatomic has two independent axes of rotation, there are two quadratic rotational terms in its Hamiltonian. So, we expect these to contribute an additional k_BT to the average energy.[23]

We may combine the above results to predict the total energy of a diatomic, including all seven of its quadratic Hamiltonian terms (two vibrational, two rotational, and three translational). Thus, we expect a diatomic molecule to have an average energy of $\frac{7}{2}k_BT$, in the classical limit. However, we only expect this classical prediction to hold at high temperature, when k_BT is larger than the molecule's quantum spacings (as further discussed in the following subsection).

Transition from Quantum to Classical Behavior

One of the first experimental clues that something was amiss with the purely classical view of physical chemistry came from experimental measurements of the heat capacities of solids and gases.[24] In both cases, experiments revealed that at low temperatures chemical systems have smaller heat capacities than expected based on classical predictions. For example, the anomalous low-temperature heat capacities of solids were

---

22 More specifically, for a diatomic harmonic oscillator $v = |\vec{v}_2 - \vec{v}_1| = dr/dt$ (where $\vec{v}_1$ and $\vec{v}_2$ are the velocity vectors of each atom) and $\delta = r - r_b$ (where δ is the diatomic's instantaneous bond length, and r_b is its average bond length). The reduced mass of a diatomic is defined as the following ratio of atomic masses $\mu = \frac{m_1 m_2}{m_1 + m_2}$ and vibrational force constant of a diatomic is $f = \mu\,(2\pi\nu)^2$ (where ν is its harmonic vibrational frequency).

23 In general, rotation about three independent axes is required to reorient an object in three-dimensional space. A diatomic has only two independent axes of rotation because rotation about the diatomic bond does not alter its orientation. More specifically, rotation about the bond axis corresponds to the rotation of electrons, whose quantum spacings are typically so large that they are not thermally active at ambient temperatures.

24 The heat capacity of a substance represents the amount of heat that is required to increase its temperature by one degree, as will be further discussed and quantified later.

not understood until Albert Einstein (1907) and later Peter Debye (1912) showed that these are consistent with the quantization of phonons (vibrational motions of solids). Similar nonclassical behavior of the heat capacity of diatomic gases was also recognized prior to 1901, as poignantly attested by Gibbs in the following quotation from the preface to his *Statistical Mechanics* (the first and still classic foundational book on this subject).

> *... we do not escape difficulties in as simple a matter as the number of degrees of freedom of a diatomic gas. It is well known that while theory would assign to the gas six degrees of freedom per molecule, in our experiments on specific heats we cannot account for more than five. Certainly one is building on an insecure foundation, who rests his work on hypotheses concerning the constitution of matter.*
>
> *Difficulties of this kind have deterred the author from attempting to explain the mysteries of nature, and have forced him to be contented with the more modest aim of deducing some of the more obvious propositions relating to the statistical branch of mechanics ...*[25]
>
> *From Gibbs, J.W.* Elementary principles in statistical mechanics, developed with especial reference to the rational foundation of thermodynamics. *(1902).*

Gibbs (who died in 1903) did not live to see these difficulties elegantly resolved, simply by introducing quantization into his own statistical mechanical formalism. As soon as quantization is introduced, the strange nonclassical behaviors of both molecular heat capacities and blackbody radiation become easy to explain. Both molecular and electromagnetic harmonic oscillators have an evenly spaced ladder of quantum states, and so are described by Eqs. 1.21–1.24.

In the classical limit, the energy of a mole of gas molecules is $U = \mathcal{N}_A \langle \varepsilon \rangle = \frac{\mathrm{D}}{2} \mathcal{N}_A k_B T = \frac{\mathrm{D}}{2} RT$, where D represents the number of quadratic terms in each molecule's Hamiltonian (and $R = k_B \mathcal{N}_A$ is the gas constant and $\mathcal{N}_A$ is Avogadro's number). In this limit, the molar heat capacity (at constant volume) is easy to calculate, since it is equal to the first

[25] The fact that Gibbs expected a diatomic molecule to have six degrees of freedom suggests that he viewed diatomics as rigid (nonvibrating) molecules with three translational and three rotational degrees of freedom. We now know that molecules have an additional vibrational degree of freedom. Moreover, the reason that we now consider a diatomic to have only two (rather than three) active rotational degrees of freedom is because rotation about the bond axis is of such a high frequency that it has an energy spacing that is typically much larger than $k_B T$. This latter degree of rotational freedom corresponds to the rotation of electrons, and so gives rise to electronic state quantization.

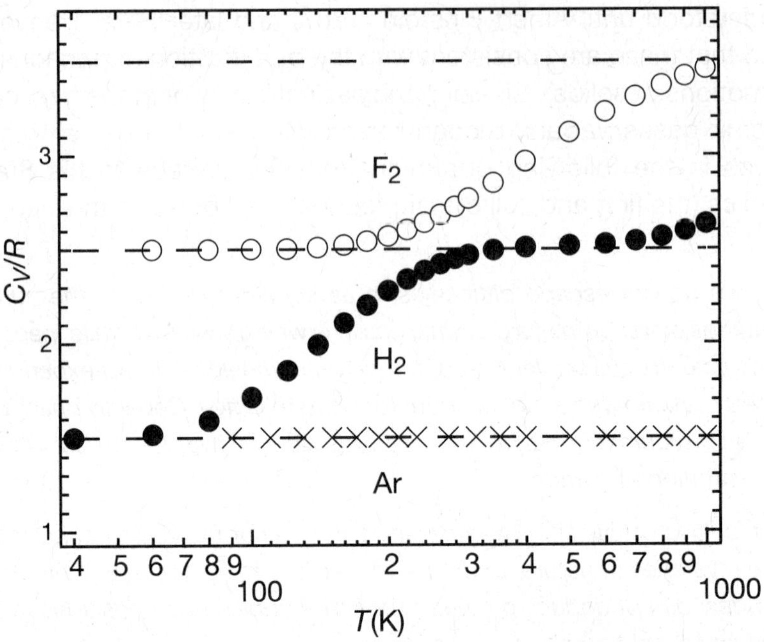

FIGURE 1.3 The experimental heat capacities (points) of argon and two diatomic molecules are plotted as a function of temperature. The horizontal dashed lines indicate the predicted heat capacity of a classical molecule with only translation (bottom), or only translation and rotation (middle), or translation, rotation, and vibration (top) degrees of freedom. E.W. Lemmon, M.O. McLinden, and D.G. Friend, "Thermophysical Properties of Fluid Systems" in NIST Chemistry WebBook, NIST Standard Reference Database Number 69, Eds. P.J. Linstrom and W.G. Mallard. National Institute of Standards and Technology, Gaithersburg MD, 20899, http://webbook.nist.gov. Gurvich, L.V., Veyts, I.V., and Alcock, C.B., Eds., Thermodynamic Properties of Individual Substances, 4th ed., Hemisphere Publishing Corp., New York, 1989.

derivative of the energy, $C_V = (\partial U/\partial T)_V = (\mathrm{D}/2)R$.[26] However, as temperature decreases, some degrees of freedom are expected to become inactive, because $k_B T$ becomes smaller than the corresponding quantum state spacing.

The experimental results in Figure 1.3 reveal the transition from quantum to classical behavior with increasing temperature for two diatomic molecules (H_2 and F_2), while argon behaves perfectly classically over this entire temperature range. Note that argon gas has three translational degrees

---

[26] C_V represents the amount of energy (heat) exchange that is required to produce a temperature change of 1 K, in a system that is held at a fixed volume. The heat capacity of a system held at constant pressure C_P may differ from C_V because some energy is expended (as work) when the volume of the system changes (under a fixed external pressure).

of freedom and so it is predicted to have a classical translational heat capacity of $\frac{3}{2}R$, which is exactly what is observed experimentally. This is consistent with the fact that the quantum spacing between translational states of argon are much smaller than k_BT (even at temperatures below 100 K), as implied by Figure 1.2.[27]

At low temperatures the heat capacity of a hydrogen molecule H_2 also indicates that it has three thermally active degrees of freedom. This implies that the rotational quantum states of H_2 are sufficiently large that they only become active at temperatures well above 100 K. This is quite unusual, as most molecules have rotational quantum state spacings that are so small that they would behave classically at 100 K. The reason that H_2 is exceptional is that it has an unusually high rotational frequency (because of its small mass). Hydrogen also has a very high vibrational frequency (again because of its small mass). Thus, both the rotational and vibrational quantum spacings of H_2 are significantly larger than those of other (heavier) diatomic molecules. So, only the three translations of H_2 are thermally active below about 100 K, and even at temperatures as high as 400 K, the vibrational motion of H_2 is evidently not yet active.

The heavier diatomic F_2 has significantly smaller rotational and vibrational quantum state spacings than H_2. Even at very low temperatures the two rotational degrees of freedom of F_2 are clearly already active. Moreover, the experimental heat capacity of F_2 implies that its vibrational quantum state spacing has a magnitude on the order of $\Delta\varepsilon \approx k_B \times 1000K \approx 10^{-20}$ J when expressed in molecular units, or $R \times 1000K \approx 10$ kJ/mol when expressed in molar units.

Recall that vibrational quantum spacings are equal to Planck's constant times the molecule's vibrational frequency, $\Delta\varepsilon = h\nu = hc\tilde{\nu} = hc/\lambda$, where $\tilde{\nu}$ is the frequency expressed in *wavenumber* units, which is equivalent to one over the wavelength of light with the same frequency $\tilde{\nu} = 1/\lambda$. In other words, λ is the wavelength of light whose frequency is resonant with that of the molecular vibration. Molecular vibrations typically have wavenumber frequencies somewhere between 100 cm^{-1} and 5000 cm^{-1} (which corresponds to wavelengths between 0.0002 cm and 0.01 cm, or between 2 μm and 100 μm) in the infrared region of the electromagnetic spectrum.

---

[27] In fact, translational quantum spacings are typically much smaller than k_BT even at 1 K, and so translational kinetic energy can almost always be treated classically. Recall that Eq. 1.12 implies that spacings between a molecule's translational quantum states depend on the size of the box that contains the molecule. Thus, in a macroscopic container this energy spacing is so small that translational energies invariably behave classically, but when a particle is confined to a container of nanometer (or smaller) dimensions, then its translational quantum state spacing becomes much larger.

The temperature dependence of the vibrational contribution to the heat capacity may be compared with theoretical predictions obtained using Eq. 1.24 (with $\Delta\varepsilon = hc\tilde{\nu}$). More specifically, the following vibrational contribution to the heat capacity of a diatomic is obtained by differentiating Eq. 1.24 with respect to temperature.[28]

$$C_V^{\text{vib}} = -\frac{1}{RT^2}\frac{d}{d\beta}\left[\frac{\Delta\varepsilon}{e^{\beta\Delta\varepsilon} - 1}\right] = R\frac{(\beta\Delta\varepsilon)^2 e^{\beta\Delta\varepsilon}}{\left[e^{\beta\Delta\varepsilon} - 1\right]^2} \tag{1.25}$$

Since translational and rotational energies typically have much smaller energy spacings, these degrees of freedom behave classically down to quite low temperatures. In other words, the sum of the translational and rotational contribution to the molar energy of a diatomic is $\frac{3}{2}RT + RT = \frac{5}{2}RT$, and so the corresponding contribution to the heat capacity is simply $\frac{5}{2}R$. Hence, the following expression is expected to accurately represent the total molar heat capacity of a diatomic gas such as F_2.

$$\frac{C_V}{R} = \frac{5}{2} + \frac{(\beta\Delta\varepsilon)^2 e^{\beta\Delta\varepsilon}}{\left[e^{\beta\Delta\varepsilon} - 1\right]^2} \tag{1.26}$$

Exercise 1.4

The diatomic molecule Br_2 has an experimental vibrational frequency of $\tilde{\nu} = 323\ \text{cm}^{-1}$, and is sufficiently heavy that its rotational (and translational) motion may be treated classically, while its vibrational motion may be treated as a quantized harmonic oscillator.

- Express the vibrational frequency in Hz = 1/s units, and determine the corresponding vibrational quantum state spacing in both J per molecule and kJ/mol units.

  Solution. The vibrational frequency of Br_2 is $\nu = c\tilde{\nu} \approx (3 \times 10^8$ m/s)(323 cm^{-1})(100 cm/m) $= 9.7 \times 10^{12}$ Hz. The corresponding quantum state spacing is $\Delta\varepsilon = h\nu \approx (6.6 \times 10^{-34}$ J s)$(9.7 \times 10^{12}$ s$^{-1}) \approx 6.4 \times 10^{-21}$ J. To convert this energy to kJ/mol units, we must multiply by Avogadro's number and divide by 1000 J/kJ, to obtain $\Delta\varepsilon = \mathcal{N}_A h\nu \approx (6 \times 10^{23}$ 1/mol)$(6.4 \times 10^{-21}$ J)$(10^{-3}$ kJ/J) ≈ 3.8 kJ/mol. Note that $\Delta\varepsilon$ is also exactly equal to the energy of photons whose frequencies are equal to the vibrational frequency of Br_2.

- Predict the value of the vibrational partition function q_{vib} and total heat capacity C_V of Br_2 when $\beta\Delta\varepsilon = \beta h\nu =$ 0.1, 1, and 10, and explain the physical implications of your predictions.

[28] More specifically, we may use the chain rule to first differentiate Eq. 1.24 with respect to β and then multiply by $d\beta/dT = -1/k_B T^2 = -k_B\beta^2$, which becomes $d\beta/dT = -1/RT^2$ when expressed in molar units.

Solution. The values of q_{vib} and C_V may be obtained using Eqs. 1.21 and 1.26. At the highest of the above three temperatures $\beta\Delta\varepsilon = 0.1$, $q_{vib} \approx 10.5$, and $C_V \approx R(\frac{5}{2} + 0.999) \approx \frac{7}{2}R \approx 29$ J/(K mol). At this high temperature the value of q_{vib} indicates that more that 10 vibrational quantum states are thermally populated, and the heat capacity is equivalent to that of a classical system whose Hamiltonian contains seven quadratic terms (three for the diatomic's translational motion, two for its rotational motion, and two for its vibrational motion). At the lowest temperature $\beta\Delta\varepsilon = 10$, $q_{vib} \approx 1$, and $C_V \approx R(\frac{5}{2} + 0.005) \approx \frac{5}{2}R$. So, at this temperature quantized vibrations are deactivated, which is consistent with the fact that the value of q_{vib} implies that only the ground vibrational state is thermally populated, while the value of C_V implies that there are only five thermally active quadratic (translational and rotational) terms in the Hamiltonian. When $\beta\Delta\varepsilon = 1$, we expect that the quantized vibrational motions will begin to be thermally activated, which is consistent with the predicted values of $q_{vib} \approx 1.6$ and $C_V \approx R(\frac{5}{2} + 0.92)$.

1.5 Classical Energy Hyperspheres

In order to calculate probabilities and average energies for degrees of freedom that are not quantized (or have very small quantum state spacings), it is convenient to replace sums by integrals. For example, for systems with a continuum of energies Eq. 1.14 becomes

$$P(\varepsilon) = \frac{e^{-\beta\varepsilon}}{\int e^{-\beta\varepsilon} d\tau} \tag{1.27}$$

where τ represents any variable(s) on which the energy depends (such as the positions and velocities of each of the molecules in the system).[29]

---

[29] In other words, the partition function in Eq. 1.14 is now expressed as an integral.

$$\int e^{-\beta\varepsilon} d\tau$$

However, this integral differs from the sum in Eq. 1.15 in that it has units, while the sum does not. For example, for an ideal gas particle translating in three-dimensional space $d\tau$ has units of position times momentum (or action) cubed. Thus, the above integral may be reduced to a dimensionless form by dividing it by h^3, where h has units of action. For a quantum mechanical particle, the latter dimensionless integral becomes identical to the corresponding quantum mechanical partition function when h is equal to Planck's constant. In other words, for such a particle, Eq. 1.15 becomes equivalent to $(\int e^{-\beta\varepsilon} d\tau)/h^3$.

When energy is treated as a continuous (classical) variable, then $P(\varepsilon)$ is called a *probability density*. The actual probability of finding the system within an infinitesimal range of τ values (near ε) is $P(\varepsilon)d\tau$.

The average energy of any such classical degree of freedom is obtained from the following integral of ε times the corresponding probability.

$$\langle\varepsilon\rangle = \int \varepsilon P(\varepsilon)d\tau = \frac{\int \varepsilon e^{-\beta\varepsilon}d\tau}{\int e^{-\beta\varepsilon}d\tau} \tag{1.28}$$

Since translational degrees of freedom typically behave classically (in any system of macroscopic volume), Eqs. 1.27 and 1.28 may be used to calculate the associated probabilities and average energies.

Let's first consider only the kinetic energy along the x-direction, for which the energy is $\varepsilon_x = \frac{1}{2}mv_x^2$. The variable τ in Eqs. 1.27–1.28 may, in this case, be taken to be equal to v_x. The integral in the denominator can be evaluated analytically.[30]

$$\int e^{-\beta\varepsilon_x}d\tau = \int_{-\infty}^{\infty} e^{-mv_x^2/2k_BT}dv_x = \sqrt{\frac{2\pi k_BT}{m}} \tag{1.29}$$

Thus, the probability of observing a molecule with a velocity between v_x and $v_x + dv_x$ is

$$P(v_x)dv_x = \sqrt{\frac{m}{2\pi k_BT}}e^{-mv_x^2/2k_BT}dv_x \tag{1.30}$$

The above expression may readily be extended to three dimensions by recalling that the probabilities of observing a combination of statistically independent events behave very much like the probabilities of tossing a sequence of coins, or rolling a pair of dice. For example, the probability of tossing two heads in a row is $\frac{1}{2}\frac{1}{2} = \frac{1}{4}$, while that of tossing three heads in a row is $\frac{1}{2}\frac{1}{2}\frac{1}{2} = \frac{1}{8}$. Similarly, the probability of throwing snake eyes (two ones) with a pair of dice is $\frac{1}{6}\frac{1}{6} = \frac{1}{36}$. In other words, *the combined probabilities of statistically independent events always multiply*. This simple fact may be used to convert the probability associated with the velocity in the x-direction, to the probability associated with the total velocity in three dimensions.

$$\begin{aligned}&[P(v_x)\,dv_x]\left[P(v_y)\,dv_y\right][P(v_z)\,dv_z]\\ &= \left(\frac{m}{2\pi k_BT}\right)^{3/2} e^{-m\left(v_x^2+v_y^2+v_z^2\right)/2k_BT}dv_xdv_ydv_z\end{aligned} \tag{1.31}$$

[30] The last identity in this expression is obtained using $\int_{-\infty}^{\infty} e^{-ax^2}dx = 2\int_0^{\infty} e^{-ax^2}dx = \sqrt{\frac{\pi}{a}}$ (see Appendix B).

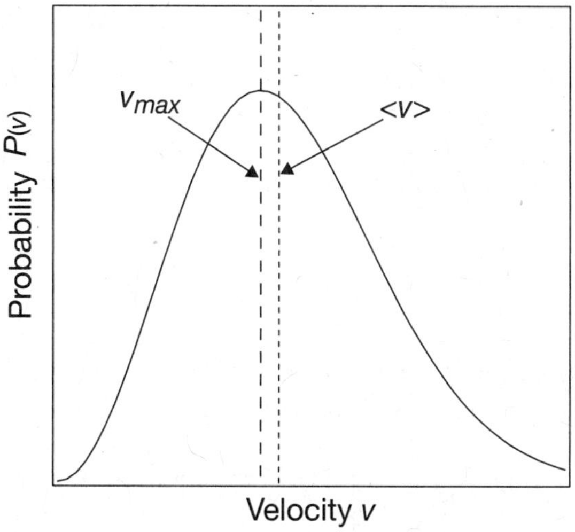

FIGURE 1.4 Equation 1.32 predicts the probability $P(v)$ that a molecule will have a given velocity v, when it is free to translate in three dimensions. Since $P(v)$ is not symmetrical in shape, the most probable velocity v_{max} is not the same as the average velocity $\langle v \rangle$.

We may also transform the above three-dimensional expression to a one-dimensional expression in terms of the total velocity, $v = \sqrt{v_x^2 + v_y^2 + v_z^2}$.

$$\boxed{P(v)dv = \left(\frac{m}{2\pi k_B T}\right)^{3/2} e^{-mv^2/2k_BT} 4\pi v^2 dv} \tag{1.32}$$

Notice that $4\pi v^2$ may be viewed as the "surface area" of a sphere of "radius" v, and so $4\pi v^2 dv$ is the "volume" of a shell of thickness dv. Thus, integration of $4\pi v^2 dv$ over a range of $0 \leq v \leq \infty$ is equivalent to integration of $dv_x dv_y dv_z$ over all velocities $-\infty \leq v_i \leq \infty$.[31]

The probability $P(v)$ in Eq. 1.32 (which is plotted in Fig. 1.4) is called the Maxwell-Boltzmann *velocity distribution function*. The most probable velocity is that at which $P(v)$ reaches its maximum value.

$$v_{max} = \sqrt{\frac{2k_BT}{m}} \tag{1.33}$$

The average velocity may be obtained by integrating v times $P(v)$.

$$\langle v \rangle = \int_0^\infty vP(v)dv = \sqrt{\frac{8k_BT}{\pi m}} \tag{1.34}$$

[31] In other words, the "volume" element $dv_x dv_y dv_z$ may be expressed as $v^2 \sin\phi d\phi d\theta dv$ in polar coordinates, and then integrated over all angles (i.e., from 0 to π for ϕ, and 0 to 2π for θ) to obtain $4\pi v^2 dv$.

Thus, the average velocity is $2/\sqrt{\pi} \approx 1.13$ times larger than v_{max}.

The total translational kinetic energy of a molecule (in three dimensions) is $\varepsilon = \frac{1}{2}mv^2$, so we may also change the independent variable in Eq. 1.32 from v to ε, to obtain the following expression for the probability as a function of energy, expressed in thermal units $\varepsilon^* = \varepsilon/k_BT = \beta\varepsilon$.[32]

$$\boxed{P(\varepsilon^*)\, d\varepsilon^* = \frac{2}{\sqrt{\pi}}\sqrt{\varepsilon^*}e^{-\varepsilon^*}\, d\varepsilon^*} \tag{1.35}$$

This *energy probability density* can be used to calculate the average translational kinetic energy of a molecule (in three dimensions).[33]

$$\langle\varepsilon\rangle = k_BT\frac{2}{\sqrt{\pi}}\int_0^\infty \varepsilon^*\sqrt{\varepsilon^*}e^{-\varepsilon^*}\, d\varepsilon^* = \frac{3}{2}k_BT \tag{1.36}$$

Thus, as expected, the average translational kinetic energy of a molecule in three dimensions is exactly three times its average energy in one dimension.

Exercise 1.5

Use the average translational kinetic energy of a molecule in three-dimensional space to obtain an expression for its root-mean-square velocity $\sqrt{\langle v^2\rangle}$, and determine whether that velocity is larger or smaller than the average velocity $\langle v\rangle$.

Solution. The average translational kinetic energy of a molecule is $\langle\frac{1}{2}mv^2\rangle = \frac{3}{2}k_BT$. Since $\frac{1}{2}m$ can be taken outside the average (because it is constant), we may rearrange the expression to obtain $\langle v^2\rangle = 3k_BT/m$ and so $\sqrt{\langle v^2\rangle} = \sqrt{3k_BT/m}$, which is slightly larger than $\langle v\rangle = \sqrt{8k_BT/\pi m}$.

The above results suggest that we may use a similar procedure to calculate probabilities and average energies for systems whose Hamiltonians contain any number of quadratic terms. In order to more easily see how we may generalize the above results, it is useful to note that the translational kinetic energy along each direction in space may be rescaled to produce a new variable $r_i \equiv v_i\sqrt{m/2}$, so that

$$\varepsilon = r_x^2 + r_y^2 + r_z^2$$

---

32 Note that $\varepsilon = \frac{1}{2}mv^2$ implies that $v = \sqrt{\frac{2\varepsilon}{m}}$, and so $dv = \frac{1}{\sqrt{2m\varepsilon}}d\varepsilon$.

33 The integral in Eq. 1.36 was evaluated using the fact that $\int_0^\infty x^b e^{-x}dx = \Gamma(b+1)$, for any real positive value of b. The gamma function, Γ, is further described in the paragraph following Eq. 1.37, and in Appendix B.

The right-hand side of the above expression looks just like the square of the radius of a sphere, $\mathtt{r}^2 = r_x^2 + r_y^2 + r_z^2$, and so we may equate $\varepsilon = \mathtt{r}^2$. We can do the same thing for systems with any number of translational degrees of freedom, or any other degrees of freedom that contribute quadratic terms to the energy.

For example, the two quadratic variables that contribute to the energy of a classical harmonic oscillator may be rescaled to express $\varepsilon_{\text{vib}} = r_1^2 + r_2^2 = \mathtt{r}^2$. If we add the two rotational degrees of freedom of a diatomic molecule, we may express the total vibrational plus rotational energy of the diatomic as the sum of four rescaled quadratic components, $\varepsilon_{\text{vib-rot}} = r_1^2 + r_2^2 + r_3^2 + r_4^2$.[34] Note that this is equivalent to the square of the radius of a four-dimensional hypersphere, $\mathtt{r}^2 = r_1^2 + r_2^2 + r_3^2 + r_4^2$ (or $\mathtt{r} = \sqrt{r_1^2 + r_2^2 + r_3^2 + r_4^2}$).

More generally, any system whose Hamiltonian contains $\mathtt{D}$ quadratic terms can be represented as a $\mathtt{D}$-dimensional hypersphere. In order to do this, we will need to make use of some interesting facts about hyperspheres. The volume of a hypersphere may be expressed as the following function of its radius $\mathtt{r}$ and dimension $\mathtt{D}$.

$$V_{\mathtt{D}} = \frac{\pi^{\mathtt{D}/2}}{\Gamma(\frac{\mathtt{D}}{2} + 1)} \mathtt{r}^{\mathtt{D}} \tag{1.37}$$

The gamma function, $\Gamma(b + 1)$, which appears in the numerator is closely related to the factorial function since $\Gamma(n + 1) = n! = n(n - 1)(n - 2) \ldots 1$, when n is a positive integer, and $\Gamma(1) = 0! = 1$. For half-integer values of b, the gamma function may be evaluated by noting that $\Gamma\left(\frac{1}{2}\right) = \sqrt{\pi}$ and $\Gamma(b + 1) = b\Gamma(b)$. So, it is not too hard to show that the above expression produces the correct "volumes" of $V_1 = 2\mathtt{r}$, $V_2 = \pi\mathtt{r}^2$, and $V_3 = \frac{4}{3}\pi\mathtt{r}^3$ for "spheres" in 1, 2, and 3 dimensions, respectively. The same formula also correctly predicts the volumes of hyperspheres in higher dimensions, each of which is enclosed by a surface that is equidistant from a single point.

The surface area of a hypersphere may be obtained by differentiating the volume with respect to $\mathtt{r}$.[35]

$$A_{\mathtt{D}} = \frac{dV_{\mathtt{D}}}{d\mathtt{r}} = \frac{\mathtt{D}\pi^{\mathtt{D}/2}}{\Gamma(\frac{\mathtt{D}}{2} + 1)} \mathtt{r}^{\mathtt{D}-1} \tag{1.38}$$

This surface area may be used to obtain the volume of a hyperspherical shell of radius $\mathtt{r}$ and thickness $d\mathtt{r}$. The volume of such a shell is simply

---

[34] More specifically, the rescaled variables are $r_1^2 = \frac{1}{2}\mu\Delta v^2$, $r_2^2 = \frac{1}{2}f\Delta r^2$, $r_3^2 = r_4^2 = \frac{1}{2}I\omega^2$.

[35] The derivative relation between volume and surface area follows from the fact that $V = \int_0^{\mathtt{r}} A\, d\mathtt{r}$, which implies that $A = dV/d\mathtt{r}$.

$A_{\mathbf{D}} d\mathbf{r}$, and so the probability of observing a state with a value of $\mathbf{r}$ between $\mathbf{r}$ and $\mathbf{r}+d\mathbf{r}$ in a system whose Hamiltonian contains $\mathbf{D}$ quadratic terms is

$$P_{\mathbf{D}}(\mathbf{r})d\mathbf{r} = [P_1(r_1)dr_1]^{\mathbf{D}} = \frac{\mathbf{D}}{\Gamma(\frac{\mathbf{D}}{2}+1)}\mathbf{r}^{*\mathbf{D}-1}e^{-\mathbf{r}^{*2}}d\mathbf{r}^* \tag{1.39}$$

where $P_1(r_1)$ is the probability associated with a single rescaled quadratic degree of freedom and $\mathbf{r}^* = \mathbf{r}/\sqrt{k_B T}$.[36] By equating $\varepsilon = \mathbf{r}^2$, we may express the above probability density in terms of $\varepsilon^* = \varepsilon/k_B T$.

$$P_{\mathbf{D}}(\varepsilon^*)\,d\varepsilon^* = \frac{\mathbf{D}}{2\Gamma(\frac{\mathbf{D}}{2}+1)}\varepsilon^{*\frac{\mathbf{D}}{2}-1}e^{-\varepsilon^*}d\varepsilon^* \tag{1.40}$$

The average value of any function of $\mathbf{r}$ may be obtained using $P_{\mathbf{D}}(\mathbf{r})$

$$\langle f(\mathbf{r})\rangle_{\mathbf{D}} = \int f(\mathbf{r})P_{\mathbf{D}}(\mathbf{r}) = \frac{\mathbf{D}}{\Gamma(\frac{\mathbf{D}}{2}+1)}\int_0^\infty f(\mathbf{r})\mathbf{r}^{*\mathbf{D}-1}e^{-\mathbf{r}^{*2}}d\mathbf{r}^* \tag{1.41}$$

while the corresponding average value of any function of ε may be obtained using $P_{\mathbf{D}}(\varepsilon^*)$.

$$\langle f(\varepsilon^*)\rangle_{\mathbf{D}} = \int f(\varepsilon^*)P_{\mathbf{D}}(\varepsilon^*)d\varepsilon^* = \frac{\mathbf{D}}{2\Gamma(\frac{\mathbf{D}}{2}+1)}\int_0^\infty f(\varepsilon^*)\varepsilon^{*\frac{\mathbf{D}}{2}-1}e^{-\varepsilon^*}d\varepsilon^* \tag{1.42}$$

Thus, for example, we may use Eq. 1.42 (or Eq. 1.41) to predict the average energy of any system whose Hamiltonian contains $\mathbf{D}$ quadratic terms, by replacing $f(\varepsilon^*)$ by $\varepsilon = \varepsilon^* k_B T$.[37]

$$\langle\varepsilon\rangle_{\mathbf{D}} = \frac{\mathbf{D}}{2}k_B T \tag{1.43}$$

In molar units, this becomes $U = (\mathbf{D}/2)RT$, from which we may easily obtain the expected classical result for the molar heat capacity.

$$C_V = \left(\frac{\partial U}{\partial T}\right)_V = \left(\frac{\mathbf{D}}{2}\right)R \tag{1.44}$$

---

[36] Equation 1.39 may be obtained by noting that in one dimension the partition function associated with a single rescaled coordinate r_1 is, $q = \int_{-\infty}^{\infty} e^{-r_1^2/k_BT}dr_1 = \sqrt{\pi k_B T}$, so $P_1(r_1)dr_1 = \frac{e^{-r_1^2/k_BT}dr_1}{\sqrt{\pi k_B T}}$, and thus $[P_1(\mathbf{r}_1)dr_1]^{\mathbf{D}} = \frac{e^{-\mathbf{r}^2/k_BT}}{(\pi k_B T)^{\frac{\mathbf{D}}{2}}}A_{\mathbf{D}}\,d\mathbf{r}$, where $\mathbf{r}^2 = \sum_{i=1}^{D} r_i^2$. Note that while r_1 extends over the entire real axis (including both positive and negative values), $\mathbf{r} = \sqrt{\mathbf{r}^2} > 0$ (and so extends only over positive values). In other words, $\mathbf{r}$ is the absolute "velocity" or "vibrational amplitude," in the appropriately rescaled units.

[37] The required integrals may be evaluated using the standard integrals given in Appendix B.

Notice that for a harmonic oscillator $\mathrm{D}=2$ and so Eq. 1.43 implies that the average vibrational energy of a diatomic molecule is $\langle\varepsilon\rangle = k_BT$, as expected. The above procedure may be readily used to calculate average values of any functions of the position, velocity, and/or energy for any system whose Hamiltonian is composed of a sum of quadratic terms.

Beyond the Quadratic Approximation

The above results are all restricted to systems whose Hamiltonians may be expressed as the sum of quadratic position and/or velocity (or momentum) variables. For translational kinetic energy this is exactly the case, while for vibrational and rotational motions the corresponding quadratic terms are approximations that ignore anharmonic and centrifugal effects that give rise to additional nonquadratic contributions to the Hamiltonian. However, the above procedure may be readily extended to treat such nonquadratic contributions.

For example, the average energy of any term in a classical Hamiltonian that has an n-th order power law form $\varepsilon = cs^n$ (where c is a constant and s is either a position or a momentum coordinate) may be obtained from Eq. 1.28.[38]

$$\langle\varepsilon\rangle = \frac{\int cs^n e^{-\beta cs^n} ds}{\int e^{-\beta cs^n} ds} = \frac{1}{n\beta} = \frac{1}{n}k_BT \tag{1.45}$$

The puckering vibrations of some four-membered ring compounds have potential energies that scale approximately as the fourth power (rather than square), of the corresponding displacement, $\varepsilon = ax^4$. Thus, in the classical (high-temperature) limit, such nonharmonic vibrations are expected to have an average potential energy of $\frac{1}{4}k_BT$, which is smaller (by a factor of two) than that of harmonic (quadratic) vibrations.

More generally, a particle in D-dimensions that is confined by a central force potential of the form $V(r) = cr^n$ has an average kinetic energy of $\langle K\rangle = \frac{\mathrm{D}}{2}k_BT$ and an average potential energy of $\langle V\rangle = \frac{\mathrm{D}}{n}k_BT$ (as further explained in Exercise 1.6). Thus, the ratio of the potential and kinetic energies of such a particle is predicted to depend in the following simple way on the value of n (and is independent of D).

$$\boxed{\frac{\langle V\rangle}{\langle K\rangle} = \frac{2}{n}} \tag{1.46}$$

This important result is equivalent to the virial theorem discovered in 1870 by Rudolf Clausius (who also played a major role in the development of

[38] The integrals in the numerator and denominator of Eq. 1.45 may both be obtained from the integral $\int_0^\infty x^b e^{-cx} dx$ in Appendix B by substituting the variable $cs^n = x$, as illustrated in Exercise 1.6.

thermodynamics). Note that for a harmonic oscillator Eq. 1.46 predicts that $\langle V\rangle/\langle K\rangle = 1$, as expected. On the other hand, for a particle confined by a potential that scales as $1/r$ (such as an electron bound to a proton or a planet rotating around the Sun), the ratio becomes $\langle V\rangle/\langle K\rangle = -2$, which is again consistent with experimental observations.

Exercise 1.6

The above results imply that a classical system with an nth-order power-law potential has an average potential energy that depends on both the power-law exponent n and the dimension of the system D.

- Show that a one-dimensional system with a potential of the form $V(x) = a|x|^n$ has an average potential energy of $\frac{1}{n}k_BT$.

  Solution. The desired average energy is equal to the following ratio of integrals.

$$\langle V\rangle = \frac{\int_{-\infty}^{\infty} Ve^{-\beta V}dx}{\int_{-\infty}^{\infty} e^{-\beta V}dx} = \frac{2\int_0^{\infty} ax^n e^{-\beta ax^n}dx}{2\int_0^{\infty} e^{-\beta ax^n}dx}$$

  If we equate $z = ax^n$, then the integrands of the upper and lower integrals become $ze^{-\beta z}$ and $e^{-\beta z}$, respectively. In order to express both integrals in terms of z, we must also relate dx to dz by making use of the fact that $x = (z/a)^{\frac{1}{n}}$ and so $dx/dz = (1/n)(1/a^{\frac{1}{n}})z^{(\frac{1}{n}-1)}$, which implies that $dx = (1/na^{\frac{1}{n}})z^{(\frac{1}{n}-1)}dz$. Thus, the above ratio of integrals may be expressed as follows, after canceling the constants $1/na^{\frac{1}{n}}$ that appear in both integrals and equating $zz^{(\frac{1}{n}-1)} = z^{\frac{1}{n}}$.

$$\langle V\rangle = \frac{\int_0^{\infty} z^{\frac{1}{n}}e^{-\beta z}dz}{\int_0^{\infty} z^{(\frac{1}{n}-1)}e^{-\beta z}dz}$$

  We may now make use of the integral $\int_0^{\infty} x^b e^{-cx}dx$ in Appendix B to obtain the desired result.

$$\langle V\rangle = \frac{\Gamma(\frac{1}{n}+1)/\beta^{(\frac{1}{n}+1)}}{\Gamma(\frac{1}{n}-1+1)/\beta^{(\frac{1}{n}-1+1)}} = \frac{\beta^{\frac{1}{n}}}{\beta^{(\frac{1}{n}+1)}}\left[\frac{\Gamma(\frac{1}{n}+1)}{\Gamma(\frac{1}{n})}\right] = \frac{1}{\beta}\left(\frac{1}{n}\right) = \frac{1}{n}k_BT$$

  Note that the second-to-last identity was obtained using $\Gamma(s+1) = s\Gamma(s)$ (see Appendix B), which implies that $\Gamma(s+1)/\Gamma(s) = s$.

- Extend the above result to a three-dimensional system with a potential of the form $V(r) = ar^n$, to show that $\langle V\rangle = \frac{3}{n}k_BT = \frac{\mathrm{D}}{n}k_BT$.

  Solution. In this case, the expression for $\langle V\rangle$ may be obtained by recalling that integration over three dimensions is equivalent to integrating the spherical shell $4\pi r^2 dr$ over the range $0 \le r \le \infty$.

$$\langle V\rangle = \frac{\int_0^{\infty} Ve^{-\beta V}4\pi r^2 dr}{\int_0^{\infty} e^{-\beta V}4\pi r^2 dr} = \frac{\int_0^{\infty} ar^{n+2}e^{-\beta ar^n}dr}{\int_0^{\infty} r^2 e^{-\beta ar^n}dr}$$

By equating $z = ar^n$, we again obtain $r = (z/a)^{\frac{1}{n}}$ and $dr = (1/na^{\frac{1}{n}})z^{(\frac{1}{n}-1)}dz$ which, after substitution in the above integrals (and again making use of the integral $\int_0^\infty x^b e^{-cx} dx$ in Appendix B), yields the desired result.

$$\langle V \rangle = \frac{\int_0^\infty z^{\frac{3}{n}} e^{-\beta z} dz}{\int_0^\infty z^{(\frac{3}{n}-1)} e^{-\beta z} dz} = \frac{\Gamma(\frac{3}{n}+1)/\beta^{(\frac{3}{n}+1)}}{\Gamma(\frac{3}{n})/\beta^{(\frac{3}{n})}} = \frac{3}{n} k_B T$$

Thus, we have verified that the average potential energy of an *n*th-order power-law potential in three dimensions is three times what it was in one dimension, which is consistent with the more general prediction that $\langle V \rangle = \frac{D}{n} k_B T$.

HOMEWORK PROBLEMS

Problems That Illustrate Core Concepts

1. Make a rough sketch showing how the following properties of a baseball depend on time, from the moment it is hit high into left field to the time it is caught by the outfielder. You may neglect the effects of friction in parts (a)–(c).
   (a) Kinetic energy (K)
   (b) Potential energy (V)
   (c) Total energy (E)
   (d) How would friction change the above results? Answer this question in the form of a typed (not handwritten) paragraph composed of complete sentences, which could be understood by an average baseball fan.

2. Calculate the kinetic energies of each of the following objects. Express your answers in joule (J) units. [Note that a joule is the SI unit of energy, as further described in Appendix B, and 1 mile = 1.609 km, 1 oz = 28.35 g, and 1 lb = 0.4536 kg.]
   (a) A baseball weighing 5.25 oz moving at 20 mph
   (b) A bicyclist weighing 175 lb moving at 20 mph
   (c) What would the ratio of the velocities of the baseball and the bicyclist have to be if they have the same kinetic energy?

3. Consider a car that weighs about 2000 kg (≈4400 lb) and is parked at an elevation of 30 m (≈100 ft) above the bottom of a hill with a constant slope equal to −0.5. [Recall that the acceleration of gravity is $g \approx 9.8$ m/s^2.]
   (a) What is the force on the car along the slope of the hill (in SI units)?
   (b) Estimate the final kinetic energy the car would have if its brakes gave way and it freely rolled down to the bottom of the hill (in SI units).
   (c) Estimate the maximum speed of the runaway car in both m/s and mph units.

4. Consider a process in which two gamma-ray photons are released when an electron and a positron are annihilated. [Note that the mass of a positron is equal to that of an electron, and the gamma-ray photon has no mass.]
   (a) What is the final energy of each photon (in J units)?
   (b) What is the final energy of each photon (in kJ/mol units)?
   (c) What is the final energy of each photon (in eV units)?

5. Consider an object weighing m (kg) that is initially at rest and experiences a constant

acceleration of a (m/s^2) for a duration of t (s). Express your answer to each of the following questions in the form of an equation involving only the variables m, a, and t (neglecting the effects of friction).

(a) What is the force on the object?

(b) Integrate the acceleration to obtain the final velocity of the object.

(c) Integrate the velocity to obtain the final distance that the object has traveled.

(d) Show that the total work performed on the object is exactly equal to its final kinetic energy.

6. In 1916 Millikan measured the kinetic energy of photoelectrons to determine Planck's constant. In order to understand how he did this, use Einstein's photoelectic formula, Eq. 1.9, to determine the slope of the photoelectron kinetic energy, K, with respect to the photon frequency.

7. Electron kinetic energies are often measured in units of electron-volts (1 eV $\approx 1.6 \times 10^{-19}$ J), which is the kinetic energy of an electron that is accelerated through a 1 volt potential. When an aluminum plate is irradiated with UV light of 253.5 nm wavelength, the ejected electrons are observed to have an average kinetic energy of about 0.8 eV. Use these results to determine the electron binding energy (or *work function*) Φ of aluminum (in eV units).

8. The energy of sunlight hitting the Earth is about 1400 W/m^2 (where 1 W = 1 J/s) in mid-day.

(a) How much work is equivalent to the energy of light shining on 1 m^2 area of the Earth in 1 s?

(b) If you assume that all the photons hitting the Earth have a wavelength of $\lambda \approx 500$ nm, estimate the number of photons striking a 1 m^2 area of the Earth in 1 s.

(c) Given that the Sun is about 150 million km from the Earth, estimate the total power output of the Sun (in W).

9. Consider the probability of finding a molecule in its ground state, $P(\varepsilon_0)$, or first excited state, $P(\varepsilon_1)$, given that energy separation between the two states is $\varepsilon_1 - \varepsilon_0 = \Delta\varepsilon = 2.5$ kJ/mol. You may assume that the states are nondegenerate ($g_0 = g_1 = 1$), unless stated otherwise.

(a) What is the ratio of the excited and ground state populations, $P(\varepsilon_1)/P(\varepsilon_0)$, at room temperature, $T \approx 300$ K? [Note that $RT = N_A k_B T \approx 2.5$ kJ/mol at 300 K.]

(b) What would happen to the ratio, $P(\varepsilon_1)/P(\varepsilon_0)$ if the temperature were increased by a factor of two? Why does this make sense?

(c) Given that diatomic molecule Cl_2 has a vibrational frequency (in wavenumber units) of 565 cm^{-1}, calculate the ratio of its excited to ground state vibrational population $P(\varepsilon_1)/P(\varepsilon_0)$ at $T \approx 300$ K. [Note that the energy of the transition is the same as that of a photon of the same frequency, ν, and $k_B T/hc \approx 209$ cm^{-1} at 300 K.]

(d) Given that the first rotational transition of Cl_2 has a frequency of ≈ 0.5 cm^{-1}, and a degeneracy of $g_0 = 1$ for the ground state and $g_1 = 3$ for the first excited state, calculate the ratio of the excited to ground state rotational populations $P(\varepsilon_1)/P(\varepsilon_0)$ at $T \approx 300$ K.

10. Experimental studies of a particular protein have determined the ratio of the unfolded (u) to the folded (f) state concentrations at equilibrium is $K = [u]/[f] = 0.4$ at 20°C. As a first approximation, one may represent the protein unfolding process as a simple statistical mechanical two-level system with energies, ε_f and ε_u, and degeneracies, g_f and g_u.

(a) Write an expression for K in terms of the parameters, ε_f, ε_u, g_f, g_u, and the temperature of the system, T.

(b) If you assume that the folded and unfolded states have the same effective degeneracy, use the experimental value of K at 20°C to estimate $\Delta\varepsilon = \varepsilon_u - \varepsilon_f$ (in kJ/mol units).

(c) If the experimental energy of unfolding is $\Delta\varepsilon \approx 20$ kJ/mol, then what does the experimental value of K at 20°C tell you about the ratio of the effective degeneracies of the unfolded and folded states?

11. Consider a system with two energy levels ε_0 and ε_1 whose degeneracies are g_0 and g_1, respectively. You may assume that $\varepsilon_0 = 0$ so that $\Delta\varepsilon = \varepsilon_1 - \varepsilon_0 = \varepsilon_1$.

(a) What will the partition function of this system reduce to at very low temperatures (when $k_B T << \Delta\varepsilon$)?

(b) What will the partition function of this system reduce to at very high temperatures (when $k_B T >> \Delta\varepsilon$)?

(c) Obtain an expression for the average energy of this system, expressed in terms of g_0, g_1, and $\beta\Delta\varepsilon$.

12. Given that translational and rotational degrees of freedom typically behave classically, but vibrations behave quantum mechanically, the total (average) energy of a diatomic is predicted to be $U = (5/2)RT + \langle\varepsilon\rangle$, where $\langle\varepsilon\rangle$ is obtained from Eq. 1.24.

(a) Verify the first equality in Eq. 1.25 using the chain rule, and then verify the second equality in Eq. 1.25 by evaluating the derivative with respect to β (and then manipulating the result to show that it is equivalent to the final expression).

(b) Use Eq. 1.26 to predict the molar heat capacity of Cl_2 gas at 300 K, using the information provided in problem 9, above, and compare your result with the experimental molar heat capacity of Cl_2, $C_V \approx 26$ J/(K mol) at 300 K.

13. Consider a particle that is confined by a one-dimensional quadratic (harmonic) potential of the form $U(x) = Ax^2$ (where A is a positive real number).

(a) What is the Hamiltonian of the particle (expressed as a function of velocity v and x)?

(b) What is the average kinetic energy of the particle (expressed as a function of T)?

(c) Use the virial theorem (Eq. 1.46) to obtain the average potential energy of the particle.

(d) What would the average kinetic and potential energies be if $U(x) = Ax^6$?

14. Equation 1.32 predicts the probability $P(v)$ that a molecule will have a given total velocity, or more specifically $P(v)dv$ is the probability that a molecule will have a velocity between v and $v + dv$.

(a) Use Eq. 1.32 to show that the following identities may be used to calculate a molecule's average velocity.

$$\langle v\rangle = \sqrt{8k_B T/\pi m} = \sqrt{8RT/\pi M}$$

(b) Use Eq. 1.32 to show that the following identities may be used to calculate a molecule's most probable velocity.

$$v_{max} = \sqrt{2k_B T/m} = \sqrt{2RT/M}$$

(c) Use the above results to predict both the average and the most probable velocities of N_2 and H_2 at 300 K.

15. The following questions pertain to the magnitude of the x-component of the speed (absolute value of the velocity) $c_x = |v_x|$ of atoms in a gas, whose probability distribution has the form $P(c_x) = ae^{-bc_x^2}$.

(a) Write down the integral you would need to evaluate in order to calculate the average value of c_x^2.

(b) Evaluate the above integral in terms of the constants a and b.

(c) Express b in terms of the mass m of each atom and the temperature T of the gas (and other constants), given that the

Boltzmann factor in $P(c_x)$ pertains to the x-component of the kinetic energy of each atom in the gas.

(d) How should the value of a differ from that in Eq. 1.30, given that speed c_x includes both the corresponding positive and negative velocities, and so the probability of observing a speed c_x is twice that of observing the corresponding velocity v_x?

(e) Use the above results to express the average value of c_x^2 in terms of m and k_BT, and then show that the average kinetic energy in the x-direction $\langle \varepsilon_x \rangle = \frac{1}{2}m\langle c_x^2 \rangle$ has the expected value.

Problems That Test Your Understanding

16. Consider a system at 300 K that has an infinite ladder of evenly spaced quantum states with an energy spacing of 0.25 kJ/mol. You may assume that the ground state has an energy of $\varepsilon_0 = 0$ and that all the states are nondegenerate ($g = 1$).

(a) What is the value of $\beta\Delta\varepsilon$ for this system?

(b) Estimate the number of quantum states that are thermally populated.

17. The diatomic iodine anion (I_2^-) has a vibrational frequency of approximately 110 cm^{-1}. Predict the total heat capacity of I_2^- in the gas phase at 300 K.

18. Consider a reaction of the general form $A \rightleftarrows B$, which has an equilibrium constant of 0.54 at 300 K. You may assume that this reaction can be approximated by a two-level system with an energy difference of $\Delta\varepsilon = \varepsilon_1 - \varepsilon_0 = 5$ kJ/mol.

(a) Determine the ratio of the degeneracies of the product and reactant molecules.

(b) Predict the equilibrium constant of the above reaction at 600 K.

19. The potential energy of an electron in a hydrogen atom has the following form, $V(r) = ar^{-1}$ (where a is a constant and r is the distance between the electron and the proton).

(a) Use the virial theorem to predict the average potential energy of a hydrogen 1s electron, given that its average kinetic energy is +1313 kJ/mol.

(b) Use the virial theorem to predict the kinetic energy of a hydrogen 2s electron, given that its total energy is −328 kJ/mol.

20. Consider a classical ball that is initially stationary, and located at $x = 0$, on a potential energy surface of the form $V(x) = x^2 - x + 1$ (in J units). You may assume that the ball does not experience any friction.

(a) What is the initial force on the ball along the x-direction (in SI units)?

(b) What is the total energy of the ball (in SI units)?

(c) What is the maximum value of the kinetic energy that the ball will attain after it is released?

21. Calculate the average velocity of an Ar atom at 300 K (in m/s units), given that its molar mass is 40 g/mol.

22. Obtain an expression for the root-mean-squared translational velocity $\sqrt{\langle v^2 \rangle}$ of a molecule of molar mass M, expressed in terms M, T, and other constants.

23. Obtain an expression for the average energy of a system that has exactly three quantum states of different energies, $\varepsilon_0 = 0$, ε_1, and ε_2 (all of which are nondegenerate, $g = 1$).

24. An object with a mass of 2 kg is found to be moving in the x-direction in such a way that its position has the following time dependence: $x(t) = 5t^2$ (where x and t are in SI units of m and s, respectively).

(a) Obtain an expression for the velocity of the object (as a function of time), $v(t)$.

(b) What is the kinetic energy of the object at time $t = 5$ s (in J units)?

(c) How much work would be required to slow the object back down to the velocity it had at $t = 0$ s?

(d) If the initial potential energy of the object (at $t = 0$ s) is 10 J, what is the value of its potential energy at $t = 5$ s?

(e) Write an expression for the Hamiltonian of the object, and determine its value.

25. The oxygen molecules in this room each have a mass of about 32 g/mol and are moving in three-dimensional space at a temperature of approximately 298 K.

(a) What is the average translational kinetic energy of each oxygen molecule (in kJ/mol units)?

(b) What is the average velocity of each oxygen molecule (in m/s units)?

26. The diatomic molecule CO has a vibrational frequency of $\tilde{\nu} = \nu/c = 2170\ \text{cm}^{-1}$ and may be treated as a quantized harmonic oscillator.

(a) What is the energy of one photon of light that has the same frequency as CO (in J units)?

(b) What is the value of the vibrational partition function of CO at 300 K?

(c) At what temperature would approximately five vibrational quantum states of CO be thermally populated?

27. The isomerization of butane from a trans to the gauche conformational state has an equilibrium constant of approximately $K = [gauche]/[trans] = 0.5$ at 300 K. Since there are two gauche conformers but only one trans conformer, we may approximate this isomerization reaction as a two-level system in which the trans state has an energy of ε_t and a degeneracy of g_t=1, and the gauche state has an energy of ε_g and a degeneracy of $g_g = 2$.

(a) Use the above information to estimate the energy difference between the gauche and trans states $\Delta\varepsilon = \varepsilon_g - \varepsilon_t$.

(b) At what temperature do you expect the equilibrium constant to be equal to 1?

(c) What value will the equilibrium constant approach at very low temperature?

(d) What value will the equilibrium constant approach at very high temperature?

Introduction to Chemical Thermodynamics

CHAPTER 2

2.1 What Is Thermodynamics Good For?

Thermodynamics is a beautifully logical, elegant, and versatile language for describing the properties of systems that are in a state of equilibrium. Before embarking on a description of the axiomatic foundations and formal structure of this language, it is interesting to consider why one might want to develop such a language, and what sorts of things one might hope to be able to say once it is developed.

An equilibrium system is defined as one whose macroscopically observable properties do not change with time. However, even the most apparently stable parts of the world around us clearly do undergo time-dependent changes. So, one may wonder why thermodynamics is of any practical use at all? One of the surprising realizations that emerge from studying this subject is that, even though thermodynamics pertains to systems at equilibrium, it can also explain why all things in the world tend to change with time.

A hint about why thermodynamics could be relevant to understanding our nonequilibrium world can be inferred by noting that the tendency of things to change is related to the fact that they are not yet in a state of equilibrium. So, understanding the properties of equilibrium systems may help us understand how and why things change. Even more importantly, we would like to know how much useful work one might hope to obtain by harnessing the tendency of a given chemical system to approach equilibrium. The desire to achieve such a fundamental and practical understanding is the driving force that led to the development of thermodynamics.

Another useful (and surprising) thing about thermodynamics is that it can be used to make quantitative statements about transformations involving intermediate states that stray arbitrarily far from equilibrium conditions. In other words, one may use thermodynamics to relate processes that follow equilibrium paths to those that follow nonequilibrium paths. Establishing such connections was a crowning achievement of nineteenth-century scientists and engineers. In particular, the French engineer, Sadi Carnot, and German physicist, Rudolf Clausius, first formulated the concept of Entropy. Subsequently the little known American scientist from Connecticut, J. Willard Gibbs,[1] extended and significantly generalized thermodynamics to demonstrate that all chemical processes are driven by chemical potential gradients (as we will see). Moreover, the chemical potential changes associated with a given chemical reaction determine the useful work that could be extracted theoretically from the reaction, as well as the excess entropy that would be produced if the reaction were allowed to take place without obtaining any useful work.

More recently, in the final years of the twentieth century, Chris Jarzynski derived an important new irreversible work theorem. Jarzynski's theorem

---

[1] Gibbs's contributions to science set the stage for the modern disciplines of engineering, physics, and chemistry. In many respects, Gibbs may be viewed as the founder of what we now call Physical Chemistry. Gibbs lived his entire life in the same house, a few blocks from Yale University, where he spent his career as a professor of mathematical physics. For most of his adult life he shared his family home with his two sisters until his death at the age of 64 (in 1903). After receiving a PhD in Engineering from Yale in 1863 – the first such degree granted in the United States – he spent a few years as a tutor at Yale and then spent three years in Europe (accompanied by his two sisters) studying the latest developments in mathematical physics. After returning to New England, he was hired by Yale as a professor (with no salary), and subsequently published two papers that introduced an elegant and powerful geometric view of thermodynamics, as well as his monumental 300-page paper on chemical thermodynamics entitled, *On the Equilibrium of Heterogeneous Substances*, whose publication was delayed because the *Connecticut Journal of Science* had to raise funds to cover the significant cost associated with its publication. Although Gibbs is today appreciated for his ground breaking and definitive contributions to thermodynamics (and statistical mechanics), at the time, very few people were able to understand his sophisticated mathematical arguments, besides Ludwig Boltzmann in Germany and James Maxwell in England (to whom Gibbs personally sent copies of his papers). Another (younger) German physicist, Max Planck, who was not aware of Gibbs's work, spent the first ten years of his academic career deriving thermodynamic results, all of which, he later realized, were previously published by Gibbs. Even more remarkably, as the introduction to Gibbs's seminal book on *Statistical Mechanics* (1902) makes poignantly clear (see page 31), the one element lacking in Gibbs's work on macroscopic and statistical thermodynamics was an account of the quantum mechanical properties of atoms, molecules, and light, whose discovery was initiated by Planck, in a paper published in 1900 (although Gibbs was apparently not aware of Planck's work).

can be used to predict the maximum amount of useful work that can theoretically be obtained from a given molecular process entirely from measurements of the actual work performed along *nonequilibrium paths*. The remarkable predictions of this theorem have since been verified by beautiful experiments (such as those involving the unfolding of a single RNA molecule) performed along *both* equilibrium and nonequilibrium paths (as further discussed in Section 5.5).

Thermodynamics can also be used to understand the interesting, and often surprising, ways in which chemical processes involving interactions between pairs of molecules are related to macroscopic *self-assembly* processes, such as those that give rise to biological structures (as further described in Section 4.2). Similarly, thermodynamics can be used to relate reactions taking place in the gas phase to those taking place in liquids, or more complex materials of biological or industrial importance. Moreover, thermodynamics may be used to describe the effects of extreme temperatures and pressures on chemical processes, such as those found in hydrothermal vents deep under the ocean, where life on Earth may have begun. The computational strategies required to apply thermodynamics to such complex systems and extreme conditions are introduced in Chapter 5.

However, it is important to keep in mind that one of the greatest strengths of thermodynamics is also one of its greatest limitations. Thermodynamics predicts exact mathematical relationships between macroscopically observable properties of a system. These relationships are sufficiently general in that they apply to any system whatsoever. In other words, one need not specify, or even know, the detailed molecular structure of a system in order to use thermodynamics to predict relationships between a given set of macroscopically observable properties and other properties that may not be so easy to directly measure. On the other hand, the generality of thermodynamics also implies that it cannot be used to determine the internal molecular structure of any chemical system.

In spite of the limitations of thermodynamics, its central importance and lasting value is attested by the following interesting statement made by Albert Einstein:

> *A theory is the more impressive, the greater the simplicity of its premises, the more different kinds of things it relates, and the more extended is its area of applicability. Therefore the deep impression that classical thermodynamics made upon me. It is the only physical theory of universal content concerning which I am convinced that, within the framework of applicability of its basic concepts, it will never be overthrown.*
>
> *Source: Albert Einstein (author), Paul Arthur Schilpp (editor).* Autobiographical Notes. A Centennial Edition. *Open Court Publishing Company. 1979. p. 31.*

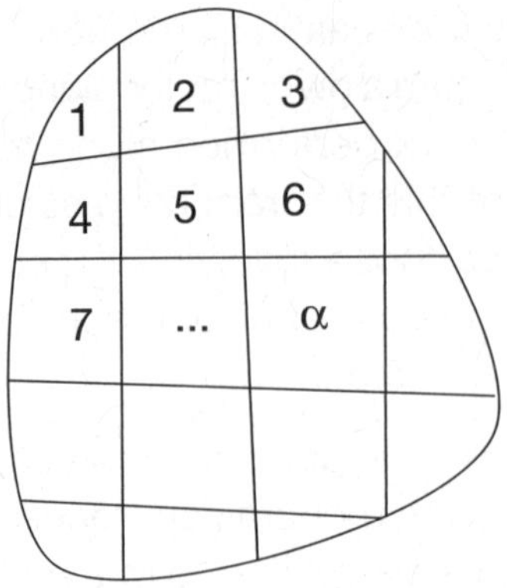

FIGURE 2.1 A large system may be divided into smaller subsystems.

Some Thermodynamic Vocabulary

Since thermodynamics is an exact mathematical language, the terms used to describe thermodynamic quantities must each be precisely defined. Establishing such definitions is required in order to relate the predictions of thermodynamics to measurable properties of the world around us.

The variables in thermodynamic equations are all macroscopically observable quantities, such as pressure (P), temperature (T), energy (U), and entropy (S). These quantities each represent averages over some macroscopic amount of time, or over a large number of identically prepared systems. Thermodynamics only requires knowing such average values.[2] All thermodynamic variables may also be classified into two broad categories depending on whether they are *intensive* or *extensive*.

Intensive variables are observables, such as T and P, which are independent of the size of a system. In other words if we label T_α as the temperature of a subsystem within a larger equilibrium system (see Fig. 2.1), then this temperature will be independent of the size of the subsystem, and will be the same as the temperature of all the subsystems that are in thermal equilibrium with each other.

$$T = T_1 = T_2 = T_3 = \ldots T_\alpha \tag{2.1}$$

Extensive variables are observables, such as U and S, whose values scale with the size of a subsystem. In other words, the energy of each subsystem depends on its size, and the energy of the total system is the sum of the energies of all the subsystems from which it is composed.

$$U = \sum_\alpha U_\alpha \tag{2.2}$$

[2] The way in which these averages are related to the underlying microscopic (molecular) components is the subject of statistical mechanics combined with quantum mechanics (since all atoms and molecules are quantum mechanical in nature), and so is beyond the scope of macroscopic thermodynamics.

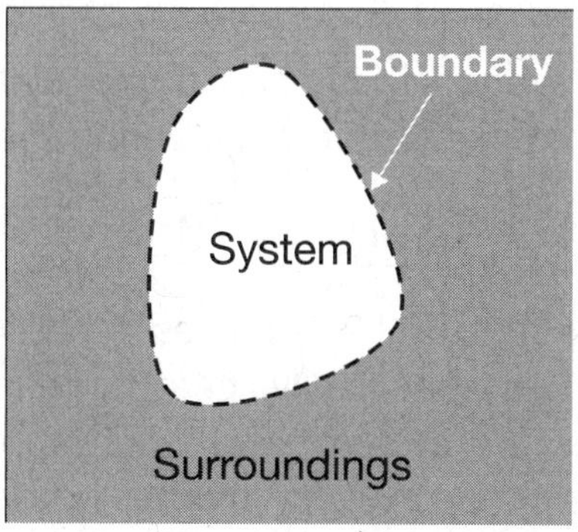

FIGURE 2.2 A thermodynamic analysis must begin by identifying the system and specifying the nature of the boundary that separates the system from its surroundings. For example, it is important to know whether the boundary is rigid or flexible, or thermally conducting or insulating, or is permeable to particular chemical species.

Another very important part of the vocabulary of thermodynamics is the distinction between a *system* and it *surroundings* (as shown in Fig. 2.2). Although we are perfectly free to define any part of the world as a system, it is important to specify what system we are talking about in any particular situation. For example, if we want to measure the temperature or volume of a system, we must obviously know where to perform those measurements. Similarly, identifying the system is required in order to quantify exchanges of energy or matter between a system and its surroundings.

The boundary between a system and its surroundings plays an important role in thermodynamics. If the boundary is thermally conductive, then the temperature of the system and surroundings will become equal to each other at equilibrium. If the boundary is flexible, then the pressures of the system and surroundings will become equal to each other at equilibrium. Similarly, if the boundary is permeable to a particular chemical species, then, at equilibrium, the chemical potential of that species will be the same inside and outside of the system. These relationships are consequences of the first and second laws of thermodynamics (as demonstrated in Chapter 3).

2.2 The Laws of Thermodynamics

Introduction to the First Law

The first law of thermodynamics is very closely related (but not identical) to the principle of energy conservation. More specifically, the first law states that the energy change of any system, ΔU, must result from exchanges of energy between the system and its surroundings. Moreover, the first law specifies that all energy exchanges can be classified as either thermodynamic *work* W or *heat* Q, exchanges.[3] Thus, the total energy change of any system is equal to the sum of the heat and work exchanged between the system and its surroundings during the process of interest.

$$\boxed{\Delta U = Q + W} \tag{2.3}$$

It is important to note that Eq. 2.3 implies a particular sign convention for heat and work. More specifically, heat is defined as being positive when heat is *added to the system*, and work is defined as being positive when work is *done on the system*. In other words, a positive exchange of heat or work will increase the energy of the system, while a negative heat or work exchange will decrease the energy of the system.[4]

As we have seen in Section 1.2, work arises from the action of some force along some displacement coordinate. The force and displacement can be associated either with macroscopic mechanical forces, or with chemical forces arising from chemical potential gradients (as further discussed in Section 3.1). Although heat exchanges can be more difficult to directly measure than work exchanges, Eq. 2.3 implies that heat may be identified as any energy exchange other than work.

$$Q \equiv \Delta U - W \tag{2.4}$$

Although this expression is mathematically unambiguous, it does not provide a physical explanation of what heat actually is.[5] One of the best ways

---

[3] The term *work* in thermodynamics has a somewhat more restricted definition than it does in other branches of physics. Although thermodynamic work is described by Eq. 1.1, not all work described by the latter equation is thermodynamic work, as further explained in footnote 2 on page 6, as well as in the discussion of the fundamental equation of thermodynamics in Section 3.1.

[4] Some textbooks, particularly those used by engineers, as well as some older physical chemistry books, define work done *by the system* as being positive work. You can easily tell what sign convention is assumed in a given book by noting how the first law is stated. For example, if work done *by* the system is defined as positive work, then the energy change of the system becomes $\Delta U = Q - W$, rather than Eq. 2.3.

[5] For example, Eq. 2.4 does not make it clear whether there is any relationship between Q and the term "heat" as it is used in our everyday language, or between Q and other forms of energy, such as kinetic, K, or potential, V, energy. In general, a heat exchange between the system and its surroundings may change both the potential and/or the kinetic energies of the system (and the same goes for a work exchange). Thus, there is (unfortunately) not a simple one-to-one relation between K or V and either W or Q. Moreover, the distinction between heat and work is further complicated by the fact that heat exchanges may be attributed to mechanical work done on molecular (thermally equilibrated) degrees of freedom, as further discussed in Section 3.1, on page 86 (and footnote 7).

to gain a better physical understanding of the relationship between energy, work, and heat is to consider simple ideal gas processes, such as those described in Section 2.3.

The challenges associated with understanding heat have a long and interesting history. In the eighteenth century, there was a raging debate as to whether or not heat is a substance (like a fluid) called caloric, which literally passes from one object to another. It was, in fact, not very easy to disprove the caloric theory of heat. However, Count Rumford's brilliantly quantitative cannon boring experiments clearly demonstrated that the exchange of heat between a system and its surroundings does not induce any measurable change in the mass of a system,[6] thus proving that, if heat is a "substance," then it must be composed of a material that has no mass. Moreover, James Prescott Joule's subsequent measurements of the temperature increase produced by mechanically stirring a fluid further demonstrated that there is a direct equivalence between the amount of mechanical energy dissipated and the amount of heat generated in such processes. The SI unit of energy, which is called a joule, was thus originally referred to as the mechanical equivalent of heat.

---

[6] Count Rumford was born in 1753, under the name Benjamin Thompson, in the town of Woburn, Massachusetts, which was then still a British colony. At the age of eighteen he obtained a job as a school teacher in Concord, New Hampshire (a city formerly known as Rumford). Thompson left New England for Europe during the days leading up to the American Revolution, after he was arrested and released several times by the "Sons of Liberty" as a suspected British spy. He soon became a well-known statesman in England and later moved to Bavaria, where he was appointed the Minister of War, Minister of Police, and Grand Chamberlain – the second most powerful person in the country, after the king. In return for his services he was named a Count of the Holy Roman Empire and chose the name Count Rumford (in memory of his origins in New England). Not long after that, while still in his 40s, Rumford returned to England, became a well-known scientist and founded the Royal Institution of London, where Humphry Davy and Michael Faraday later became prominent researchers. In 1798 Rumford wrote a paper entitled *An Inquiry Concerning the Source of Heat Which is Excited by Friction*. This paper was inspired by his earlier observations of cannon manufacturing in Munich, which clearly revealed how the work performed in drilling-out cannon barrels inevitably released a great quantity of heat. His carefully designed experiments included one that employed a team of horses to drill a cannon cylinder, while the cylinder was immersed in a large water tank. After the water came to a boil, he weighed the cannon and iron filings to show that no mass was lost, thus proving that heat exchange was not associated with a decrease in mass. His experiments also provided early evidence of the close relationship between heat and work.

Heat and Work Are Not State Functions

Energy differs in a very important way from heat and work, as U is a *state function*, but Q and W are not. That means that, while the energy difference between two states of a system is independent of the path taken in converting the system from one state to the other, the amount of heat and/or work exchanged between the system and surroundings is *path-dependent*. The following analogies may be helpful in clarifying the distinction between state functions and quantities such as heat and work, which are not state functions.

Consider a process involving hiking between two campsites in the mountains. The difference between the altitudes of the two campsites (like ΔU) certainly does not depend on what path one follows in going from one campsite to the other. However, the distance that is traveled, or the time that passes, while hiking along different trails may obviously be quite different. Thus, in this example altitude is a path-independent state function, while distance and time are path-dependent and so are not state functions (as they cannot be predicted simply by knowing the initial and final states of the process).

As another example, consider a pool of water that may be filled either using a faucet or by rain, and may be emptied either using a drain or by evaporation. The change in the water level of the pool is analogous to ΔU, and so is a path-independent state function, while the means by which water is added or removed are not state functions. More specifically, adding water to the pool from either faucet or rain is analogous to $W > 0$ and $Q > 0$, while removing water from the pool through the drain or by evaporation is analogous to $W < 0$ and $Q < 0$.

As yet another example, we may consider a system consisting of a metal block whose temperature (and energy) can be increased either by mechanically rubbing against its surface or by immersing it in a high-temperature oil bath. Both of these processes could produce exactly the same increase in the energy of the block (as indicated by its temperature change). Thus, either work or heat exchanges may be used to produce the same change in the state of a system.

Introduction to the Second Law

The second law of thermodynamics is arguably the most enigmatic and provocative fundamental principle underlying all processes in engineering, physics, chemistry, and biology. The discovery of this profound principle of nature originally emerged from the practical concerns of nineteenth-century engineers who desired to increase the efficiency of steam engines.

In essence, the second law introduces a new *extensive* thermodynamic *state function* called *entropy*, S, which dictates the direction in which systems may spontaneously evolve. We will see that S also dictates the maximum amount of work that can be obtained from a given chemical transformation, as well as the maximum efficiency with which work may be extracted from a heat engine.

Experimental evidence suggests that in order to operate most efficiently, a process must be carried out in such a way that the system remains arbitrarily close to equilibrium with its surroundings at all times. Under these conditions, a process is also fully *reversible*, in the sense that both the system and surroundings will return exactly to their initial states if the process is run in reverse. Less efficient processes do not have this property, and so are referred to as *irreversible*. Consideration of reversible and irreversible processes led Sadi Carnot and Rudolf Clausius to develop the concept of entropy.[7] This ultimately led Clausius to define the (infinitesimal) change in the entropy of a system dS in terms of the quantity of heat δQ that is *reversibly* added to the system at a given temperature T.[8]

$$\boxed{dS \equiv \frac{\delta Q}{T}} \tag{2.5}$$

We will revisit this important result in Section 3.1, where we will see how it may be derived from more general postulates of thermodynamics. These postulates also imply that the entropy of any isolated system (such as the entire universe) must increase as the result of any spontaneous (irreversible) processes.

If a given process produces an entropy change of dS in the system, and an entropy change of dS_{Surr} in the surroundings, then the total entropy change of the universe is $dS_{\text{Univ}} = dS + dS_{\text{Surr}}$. The second law further

---

[7] The German physicist Rudolf Julius Emanuel Clausius (1822–1888) first stated the second law of thermodynamics in his 1850 memoir entitled *On the Mechanical Theory of Heat*, although he did not formulate the second law explicitly in terms of entropy for another 15 years. In 1888 Willard Gibbs wrote a eulogy to Clausius, in which he summarized his contributions to thermodynamics as follows: "If we say, in the words used by Maxwell some years ago, that thermodynamics is 'a science with secure foundations, clear definitions, and distinct boundaries,' and ask when those foundations were laid, those definitions fixed, and those boundaries traced, there can be but one answer. Certainly not before the publication of that [1850] memoir. . . ."

[8] As suggested by the above discussion, a reversible process is one in which the system and its surroundings remain infinitesimally close to equilibrium with each other throughout the process, as further described in Section 2.3, as well as in Section 4.4 (in Chapter 4).

implies that this total entropy change must be nonnegative, as expressed by the following Clausius inequality.

$$\boxed{dS_{\text{Univ}} = dS + dS_{\text{Surr}} \geq 0} \tag{2.6}$$

The equality only holds when the process is reversible, while the inequality applies to irreversible processes. In other words, *the entropy of the universe is unaltered by a reversible process, while any irreversible process must produce an increase in the entropy of the universe*. The same is true for any other isolated system.

It is easy to show that the equality on the right-hand side of Eq. 2.6 applies to a reversible process. For any such process, Eq. 2.5 indicates that $dS = \delta Q/T$. Moreover, the heat added to the surroundings is necessarily opposite in sign to the heat added to the system, $\delta Q_{\text{Surr}} = -\delta Q$ (because any heat flowing out of the system must flow into the surroundings). Thus, $dS_{\text{Univ}} = dS + dS_{\text{Surr}} = \delta Q/T + \delta Q_{\text{Surr}}/T = \delta Q/T - \delta Q/T = 0$.

If we designate $\eth Q$ as the heat exchanged between the system and surroundings in a general process, which may be irreversible, then Eq. 2.6 requires that $dS + dS_{\text{Surr}} = \delta Q/T - \eth Q/T \geq 0$, which leads immediately to the following alternative form of the Clausius inequality.[9]

$$\delta Q \geq \eth Q \tag{2.7}$$

Let's now consider what would happen if a system is transformed from an initial state A to a final state B as a result of some *irreversible* process. Equation 2.5, combined with the above inequality, implies that $\Delta S = S_B - S_A = \int_A^B \delta Q/T > \int_A^B \eth Q/T$ for any such process. This also implies that the following alternative form of the Clausius inequality must

[9] We can confirm that latter inequality most easily by assuming that the surroundings behave like an ideal heat bath, in which case the heat exchange $\eth Q$ produces an entropy change of $dS_{\text{Surr}} = -\eth Q/T$. In other words, an ideal heat bath is *defined* as one that responds to any heat exchange in the same way, regardless of whether that heat exchange resulted from a reversible or irreversible process. The entropy change of the system may be obtained using Eq. 2.5, which requires knowing the heat exchange δQ obtained when following a reversible path (between the same two states). The heat exchange δQ associated with the latter (reversible) process will not in general be the same as the heat exchange $\eth Q$ associated with the former (irreversible) process. If the temperature of the system and bath are assumed to be the same, then Eq. 2.6 leads immediately to Eq. 2.7. If the temperature of the bath is not the same as that of the system, then Eq. 2.7 must be expressed in the more general form $\delta Q/T \geq \eth Q/T_{\text{Surr}}$ (Eq. 4.54), as further explained in the discussion of irreversible heat transfer processes in Section 4.4.

hold for any *cyclic process*, defined as a process in which the system begins and ends at the same state, so that $S_B - S_A = 0$.[10]

$$\boxed{\oint \frac{\delta Q}{T} \leq 0} \tag{2.8}$$

The following quite different expression for entropy was obtained by Ludwig Boltzmann, and is etched on his gravestone. This expression relates the entropy of any system to the phase space volume Ω that is available to the system.[11]

$$\boxed{S = k_B \ln \Omega} \tag{2.9}$$

Although Boltzmann derived Eq. 2.9 before the development of quantum mechanics, it turns out that Ω is also equivalent to the number of quantum mechanical states that are thermally accessible to the system. In other words, Ω is equivalent to the partition function of the whole system.[12]

Although the laws of thermodynamics may be stated in various ways, in the introduction to his famous paper *On the Equilibrium of Heterogeneous Substances*, Gibbs chose to simply quote Clausius's remarkably concise statements of the first and second laws:

- The energy of the universe is constant.
- The entropy of the universe seeks a maximum.

Introduction to the Third Law

The last of the three laws of thermodynamics was discovered by Walther Hermann Nernst in the early twentieth century. Nernst began his career as a physicist, but soon turned his attention to chemical thermodynamics, and is perhaps best known for his development of the electrochemical equation, which now bears his name. Nernst, along with his famous senior colleague at Leipzig University, Wilhelm Ostwald, as well as Jacobus Henricus van't Hoff and Svante August Arrhenius, all played a key role

---

[10] The symbol $\oint$ is used to designate any integral that follows a cycling path. Equation 2.8 only strictly applies to a process in which the system is continuously coupled to a bath of temperature T (although T may vary along the path).

[11] The phase space volume that is available to a system is defined as the region of position and momentum space that it explores when it is in a particular thermodynamic state (i.e., at a give temperature, pressure, and chemical composition). Equation 2.9 is also closely linked to the Boltzmann probability formula Eq. 1.13, as further discussed in footnote 13 on page 23.

[12] The connection Ω and the quantum mechanical partition function of a system is further discussed in Chapter 5, and particularly Sections 5.3 and 5.4.

in developing the discipline we now call *physical chemistry* (and all four eventually received Nobel prizes for their contributions).

In 1906 Nernst first formulated what he referred to as his *heat theorem*, which later became known as the third law of thermodynamics. Nernst was led to this law by experiments suggesting that the maximum work obtainable from a process could be calculated from the associated heat exchanged at temperatures close to absolute zero. In addition to its fundamental theoretical implications, the theorem was soon applied to industrial problems, including the commercial production of ammonia.

The third law is now most often described as stating that the entropy of any single component system approaches zero as its absolute temperature approaches zero. Thus, the third law implies that one may determine the total entropy of any substance by integrating the entropy change associated with producing it from its pure components at zero temperature.

The third law also played an important role in the development of quantum mechanics. In particular, experimental measurements of the low-temperature heat capacities of solids (like those of gases) were known to be inconsistent with classical predictions. Einstein was the first to demonstrate (in 1907) that the experimentally observed temperature dependence of the heat capacity of diamond could be explained simply by assuming that the vibrations (phonons) of the diamond lattice behaved like quantized harmonic oscillators. The success of Einstein's theory, and its later refinement by Peter Debye and others, provided a dramatic confirmation of the third law, as it indicated that at zero temperature any material should relax to its quantum mechanical ground state. The fact that this ground state has zero entropy may be inferred directly from Boltzmann's expression for entropy (Eq. 2.9), which requires that when only one state is populated, then $\Omega = 1$ and so $S = k_B \ln 1 = 0$.

It is quite remarkable that both Nernst's third law and Boltzmann's expression for entropy were developed using arguments that were independent of quantum mechanics, and yet both proved to be entirely consistent with quantum mechanical predictions.

2.3 Important Ideal Gas Examples

An ideal gas may be envisioned as a system composed of infinitesimally small (noninteracting) molecules. Real gases closely resemble ideal gases over a wide range of pressures and temperatures, up to and including the ambient conditions of the air around us. In this section we will investigate the experimental and theoretical properties of such gases and apply the first and second laws to various sorts of ideal gas processes. Doing so is both practically important, as the results are applicable to real gases, and conceptually rewarding, as the results may be used to better understand

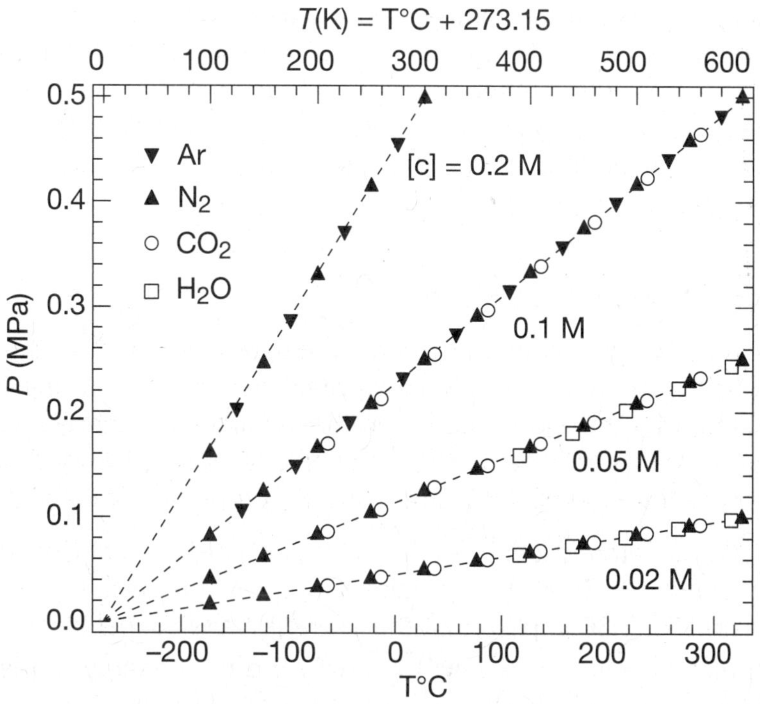

FIGURE 2.3 The experimental pressures of real gases all approach zero at the same temperature 0 K = −273.15°C. The data points are from E.W. Lemmon, M.O. McLinden, and D.G. Friend, "Thermophysical Properties of Fluid Systems" in NIST Chemistry WebBook, NIST Standard Reference Database Number 69, Eds. P.J. Linstrom and W.G. Mallard. National Institute of Standards and Technology, Gaithersburg MD, 20899, http://webbook.nist.gov.

the relationship between energy, entropy, work, and heat in both reversible and irreversible thermodynamic processes.

Gas Pressure and Absolute Temperature

Experimental measurements of the pressure of gases at various temperatures and densities make it clear why temperature has a natural (absolute) scale, with 0 K as the lowest possible temperature. This is clearly revealed in Figure 2.3, which shows how the pressures of various gases (measured in MPa, mega-Pascal, units)[13] depend on temperature (measured either in degrees Celsius °C or Kelvin K units), at various fixed gas densities or concentrations $[c] = n/V$ (in moles per liter M units). Note that all the *isochores* (lines)[14] converge to a common point when the pressure

[13] A pressure of one MPa is equivalent to 10^6 Pa $= 10^6$ J/m^3 $\approx$ 10 Atm.

[14] An isochore is a line, or curve, of constant volume, just as an isotherm is a curve of constant temperature, and an isobar is a curve of constant pressure. Note that a gas of constant volume (and a fixed number of moles) also has a constant concentration [*c*].

approaches zero and the temperature approaches $-273.15°C = 0$ K. In other words, the experimental pressures of real gases all point towards the origin of the absolute temperature scale. These results also suggest that the pressures of all gases depend only on their temperature and concentration, not on their chemical structure (and we will shortly see why that is the case).

Ideal Gas Equation of State

The energy of a monatomic ideal gas is equal to the sum of the kinetic energies of all its molecules, while its pressure results from the average force exerted by the molecules when they collide with the walls of their container. Here we will see how these simple facts may be used to derive the ideal gas law, and all the thermodynamic properties of ideal gases.

Consider a *monatomic* ideal gas consisting of N atoms enclosed in a container of volume V. Because this gas has only three (translational) degrees of freedom, Eq. 1.36 implies that the average translational kinetic energy of each molecule is $\langle\varepsilon\rangle = \left\langle\frac{1}{2}mv^2\right\rangle = \frac{3}{2}k_BT$. The total energy of the gas is simply N times the average energy of each molecule.

$$U = N\langle\varepsilon\rangle = N\left\langle\frac{1}{2}mv^2\right\rangle = \frac{3}{2}Nk_BT = \frac{3}{2}nRT \tag{2.10}$$

The energy density of the gas is defined as its energy per unit volume.

$$\frac{U}{V} = \frac{N}{V}\left\langle\frac{1}{2}mv^2\right\rangle = \frac{3}{2}\left(\frac{nRT}{V}\right) \tag{2.11}$$

More generally, we may consider an ideal gas composed of either diatomic or polyatomic molecules. If all of the internal degrees of freedom of such a gas behaved classically, then each quadratic term in its Hamiltonian would contribute an additional $\frac{1}{2}nRT$ to the energy. Thus, for example, a classical diatomic with two rotational degrees of freedom and a harmonic vibrational potential would have a classical energy density of $(7/2)(nRT/V)$.[15] Thus, any *classical* system composed of molecules whose Hamiltonian contains D quadratic terms is predicted to have the following classical energy density.

$$\boxed{\frac{U}{V} = \frac{\mathrm{D}}{2}\left(\frac{nRT}{V}\right)} \tag{2.12}$$

---

[15] Recall that a diatomic harmonic oscillator has a total of seven quadratic terms in its Hamiltonian (see Section 1.4).

Note that the units of energy density are the same as those of pressure, because pressure is defined as the force exerted on a unit area, and thus has SI units of $N/m^2 = (J/m)/m^2 = J/m^3$. So, we expect the pressure of any ideal gas to be proportional to its energy density.

The constant of proportionality between pressure and energy density may be determined by evaluating the average force exerted on a wall that is perpendicular to the x-direction (which is obviously also equivalent to the pressure on a wall that is facing in any other direction). Newton's second law, Eq. 1.3, implies that force F may be expressed as the time derivative of momentum $F = dp/dt = d(mv)/dt = m(dv/dt) = ma$. Every time a molecule of absolute velocity v_x collides with the wall it undergoes a change in momentum of $2mv_x$. Many such collisions produce a nearly constant average force on the wall. The rate at which molecules of velocity v_x collide with a unit area of the wall is $\frac{1}{2}(N/V)\langle v_x \rangle$ (where the factor of $\frac{1}{2}$ comes from the fact that only half of the molecules of velocity v_x are moving toward the wall). So, the average force per unit area induced by these collisions is $\frac{1}{2}(N/V)\langle v_x(2mv_x)\rangle = (N/V)\langle mv_x^2 \rangle$. We may also relate v_x^2 to the total velocity of each molecule by noting that $v^2 = v_x^2 + v_y^2 + v_z^2$. Since all three directions are equivalent, $\langle v_x^2 \rangle = \frac{1}{3}\langle v^2 \rangle$, and so the force per unit area becomes $(N/V)\langle mv_x^2 \rangle = \frac{1}{3}(N/V)\langle mv^2 \rangle = \frac{2}{3}(N/V)\left\langle \frac{1}{2}mv^2 \right\rangle$. The equation of state of an ideal gas may be obtained by combining the latter expression with Eq. 2.11.

$$P = \frac{2}{3}\left(\frac{N}{V}\right)\left\langle \frac{1}{2}mv^2 \right\rangle = \frac{2}{3}\left(\frac{U}{V}\right) = \frac{nRT}{V} \tag{2.13}$$

Note that this implies that the pressure of an ideal gas is proportional to both its temperature T and its molar concentration $[c] = n/V$ (or molecular number density $\rho = N/V$).

$$\boxed{P = \frac{nRT}{V} = [c]RT = \frac{Nk_BT}{V} = \rho k_B T} \tag{2.14}$$

Since only the *translational* kinetic energy of molecules contribute to the pressure of an ideal gas, we expect Eq. 2.14 to hold for gases composed of molecules with any number of internal degrees of freedom (even if those degrees of freedom are quantized and so do not behave classically). This expectation is confirmed by experimental measurements performed on real gases, such as those shown in Figure 2.4.

Pressure-Volume Work

Since pressure is defined as the force per unit surface area, the force used to compress the gas is simply related to the product of the external pressure (applied to the piston) and area of the piston, $F = PA$. As a result, the work

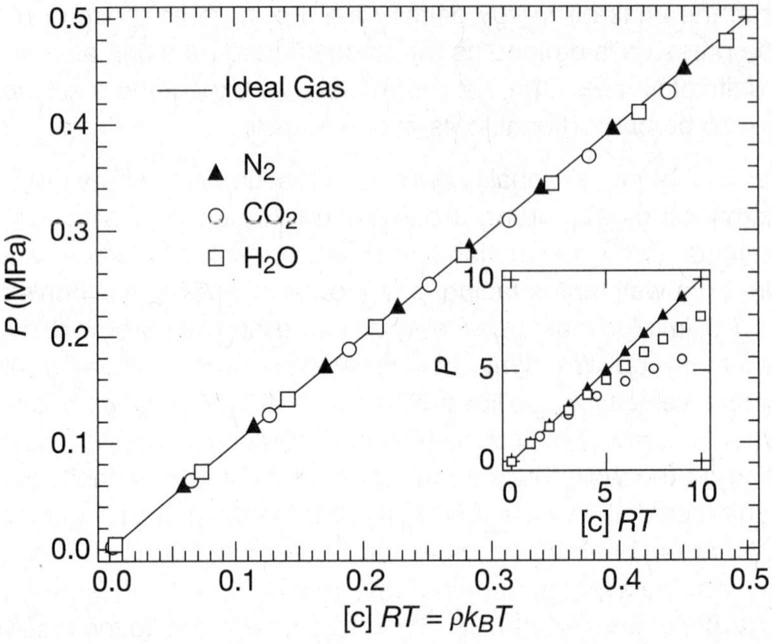

FIGURE 2.4 The experimental pressures of real gases agree essentially perfectly with the ideal gas law (Eq. 2.14) at sufficiently low pressures (up to a few Atm). Significant deviations from the ideal gas law only become evident above 1 MPa ($\approx$10 atm), even for strongly interacting molecules such as water or carbon dioxide (as shown in the inset graph). The experimental results for N_2 and CO_2 were obtained at 300 K, while those for H_2O pertain to water vapor (steam) at 600 K. Note that 1 MPa = 1 kJ/L so when [c] is expressed in M units and *RT* is expressed in kJ/mol, then [c]*RT* has units of MPa. The data points are from E.W. Lemmon, M.O. McLinden, and D.G. Friend, "Thermophysical Properties of Fluid Systems" in NIST Chemistry WebBook, NIST Standard Reference Database Number 69, Eds. P.J. Linstrom and W.G. Mallard. National Institute of Standards and Technology, Gaithersburg MD, 20899, http://webbook.nist.gov.

δW required to change the volume of the gas by an infinitesimal amount dV is proportional to the externally applied pressure.[16]

$$\delta W = -Fdh = -PAdh = -P\,dV \tag{2.15}$$

The negative sign in Eq. 2.15 is required since we define work to be positive $\delta W > 0$ when work is done *on the system*. In other words, when a gas

---

[16] Since work is not a state function, and so is not an exact differential function (like V, U, and S, etc.), we use the notation δW rather than dW to indicate an infinitesimal exchange of work. As in Eq. 2.7, the symbol δ is used to designate any infinitesimal exchange of work or heat, whether reversible or irreversible, and the symbol δ designates *reversible* exchanges of work or heat.

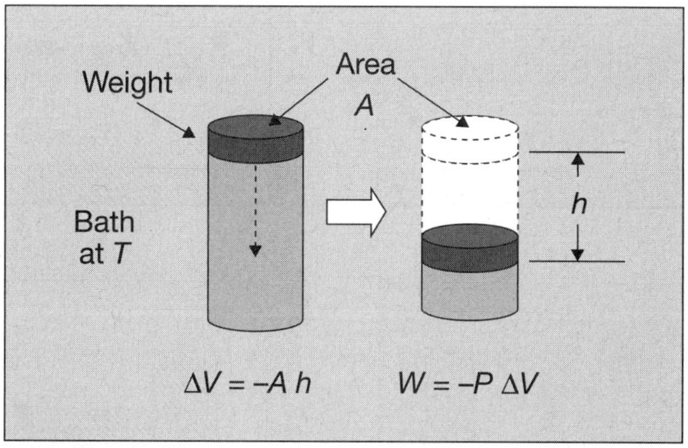

FIGURE 2.5 Illustration of the pressure-volume (*PV*) work that is done when a weight compresses a gas under isothermal conditions (at temperature *T*).

is *compressed* $dV < 0$, work is done on the system $\delta W > 0$, so we must include a negative sign in Eq. 2.15 to get $\delta W = -P\,dV > 0$.

Figure 2.5 illustrates what happens when work is done by putting a weight on top of the gas. This is equivalent to imposing a constant force (or constant external pressure) on the system. It is easy to calculate the total work exchanged as the gas is compressed from its initial volume V_1 to its final volume V_2 with a constant external pressure P.

$$W = \int_{V_1}^{V_2} -P\,dV = -P(V_2 - V_2) = -P\Delta V \qquad (2.16)$$

In other words, the work done on the gas is determined by the magnitude of the *external* pressure applied to the gas multiplied by the volume change *of the gas*.

Notice that when the weight is initially placed on the gas cylinder, there is an imbalance between the internal and external pressure of the gas; it is this imbalance that induces the gas to compress. Once the gas settles down to its final (equilibrium) state, the internal and external pressures will both be exactly the same. The final volume of the gas may be determined using the ideal gas law (Eq. 2.14). In other words, the final volume of the gas ($V = V_2$) is determined by its temperature, pressure, and number of moles, $V = nRT/P$.

Reversible and Irreversible Isothermal Processes

The difference between reversible and irreversible processes may be illustrated by considering what happens when an ideal gas expands *isothermally* (at constant T) along three different paths, as illustrated in Figure 2.6. Note that each of the processes in Figure 2.6 starts at the

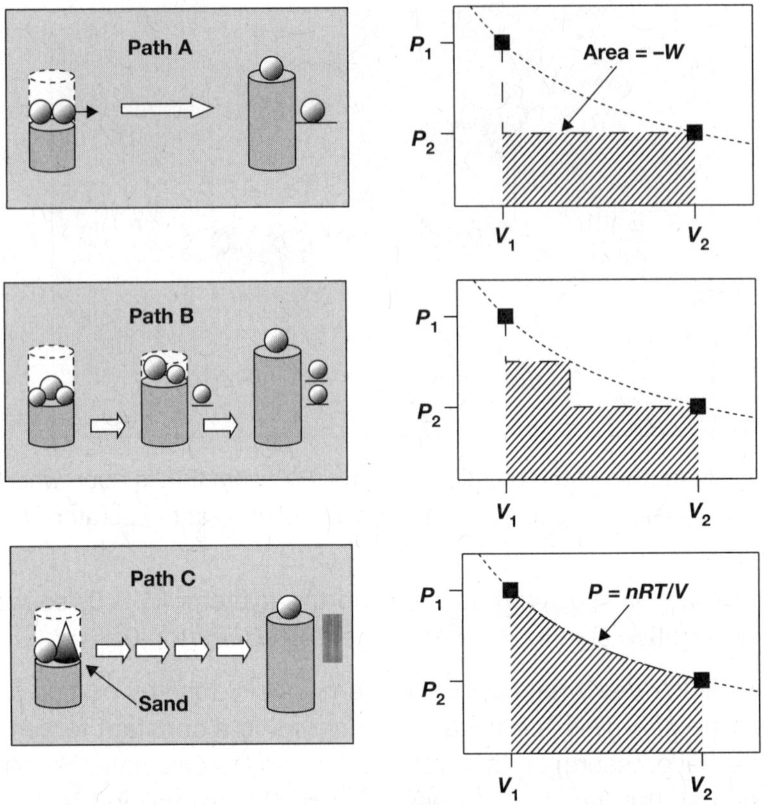

FIGURE 2.6 The amount of work performed by an expanding isothermal ideal gas depends on the path that is followed. In path A, a single large weight is removed, thus instantaneously decreasing the external pressure from P_1 to P_2. In path B, two smaller weights are removed one at a time, and so more work is performed. In path C, arbitrarily small grains of sand are removed one at a time, so that the internal and external pressures remain very nearly in equilibrium with each other at every point along the path, and thus the maximum possible amount of work is performed (by the system on the surroundings).

same initial pressure $P_1 = nRT/V_1$, and ends at the same final pressure $P_2 = nRT/V_2$, so they represent three different paths between the same initial and final states. Although the initial and final weights on the gas cylinder are the same in all three processes, we will see that the amount of work exchanged along the three paths is not the same, and only one of the three paths is reversible.

Path A in Figure 2.6 illustrates a process in which the gas is initially equilibrated in a cylinder held down by two large weights. Then one of the weights is rolled off, allowing the gas to expand to a new (larger volume) equilibrium state. The graph on the right-hand side indicates that the work exchanged along this path has a magnitude that is exactly equal to the shaded area, $|W| = P|\Delta V|$. The sign of the work exchange is negative because work is done by the gas (system) on the surroundings (by lifting the weight). This is also consistent with Eqs. 2.15 and 2.16, because in this case $V_2 > V_1$ and so $\Delta V = V_2 - V_1 > 0$ and $W = -P\Delta V < 0$.

Path B in Figure 2.6 illustrates a process in which two smaller weights are removed one at a time. The magnitude of work exchanged at each step is again equivalent to the area of the rectangles shown in the graph to the right. Notice that a greater total amount of work is exchanged along this path. In other words, when smaller weights are removed, there is a smaller imbalance between the internal and external pressure, and more work is performed.

Path C in Figure 2.6 pertains to the limit in which weight is removed from the cylinder in very small increments (like grains of sand). When the size of the grains of sand becomes arbitrarily small, the internal and external pressures remain nearly perfectly balanced throughout the process. In this limit, the path followed by the system is determined by the ideal gas law, Eq. 2.14, and so the total work exchanged may be calculated by integrating $-P\,dV$ over the curved path traced by the ideal gas equation of state.

$$
\begin{aligned}
W &= \int_{V_1}^{V_2} -P\,dV = -\int_{V_1}^{V_2} \frac{nRT}{V} dV = -nRT \int_{V_1}^{V_2} \frac{1}{V} dV \\
&= -nRT \ln(V)\Big|_{V_1}^{V_2} = -nRT\,[\ln(V_2) - \ln(V_1)] \\
&= -nRT \ln\left(\frac{V_2}{V_1}\right) = nRT \ln\left(\frac{V_1}{V_2}\right) \qquad (2.17)
\end{aligned}
$$

The different sizes of the shaded areas in the three graphs on the right-hand side of Figure 2.6 clearly imply that the maximum amount of work that can be done by an expanding gas is obtained when weight is removed in very small steps, in such a way that the external and internal pressures remain almost exactly balanced throughout the expansion process.

The ideal gas expansion process described by path C is one example of a much more general principle, which states that the maximum amount of work that can be obtained from any process is achieved when the process is carried out along a *reversible* path. A reversible process is defined as one in which the internal and external intensive variables of the system and surroundings remain essentially perfectly balanced throughout the process. An irreversible process, on the other hand, is one that follows a path along which there is a significant imbalance between the internal and external intensive variables of the system and surroundings.[17]

---

[17] A more general description of reversible and irreversible processes, and their relationship to the second law, is contained in Section 4.4.

The heat exchanged in each of the above isothermal ideal gas expansions may be determined by combining the first law, Eq. 2.3, with what we know about the energy of an ideal gas, Eq. 2.12. Note that the latter equation implies that the energy of any ideal gas only depends on its temperature (and is independent of its volume or pressure). So, in any *isothermal* ideal gas expansion process, the energy of the gas necessarily remains constant, $\Delta U = 0$. Moreover, since the energy of the system is constant, the first law requires that $Q = -W$. In other words, since the work exchanged in all of the above ideal gas expansion processes is negative, there must also be an equal and opposite exchange of heat between the system and the surrounding isothermal bath. For example, we may use Eq. 2.17 to obtain the heat exchanged during the *reversible* isothermal expansion of an ideal gas.

$$Q = -W = nRT \ln\left(\frac{V_2}{V_1}\right) \tag{2.18}$$

Since $V_2/V_1 > 1$ for an expansion process, the heat exchange is positive, which implies that heat is transferred from the surroundings to the system.

Adiabatic Processes

In order to understand why heat must be exchanged in isothermal processes, such as those shown in Figure 2.6, it is instructive to consider what would happen if the gas were thermally insulated, so it could not exchange any heat with the surroundings $Q = 0$. A process in which no heat is exchanged is called an *adiabatic* process. For any such process the first law implies that $\Delta U = W$. If a weight is lifted adiabatically, then work is done by the system and so $W = \Delta U < 0$. In other words, the energy of a gas must decrease whenever it lifts a weight adiabatically. This also implies that the temperature of the gas must decrease in any such process.[18] In other words, the gas must necessarily cool down as it expends energy by lifting a weight under adiabatic conditions. So, in order to maintain the gas at a constant temperature (in an isothermal expansion process), heat must be absorbed from the surrounding bath.

In order to quantitatively predict the temperature change associated with any adiabatic ideal gas process, we must know the heat capacity of the gas. More specifically, Eq. 2.12 implies that when an ideal gas experiences an infinitesimal temperature change dT its energy will change by $dU = (\mathrm{D}/2)nRdT = C_V dT$, where C_V is the heat capacity of an ideal gas

[18] The amount of temperature decrease is determined by the heat capacity of the gas, which in turn depends on its molecular structure. However, regardless of the nature of the gas, the first law requires that the energy expended by doing work is exactly equivalent to the decrease in the energy of the system.

composed of molecules whose Hamiltonian contains D thermally active (classical) terms. Since the energy change in any adiabatic process is also exactly equal to the associated work exchange, the temperature change of the gas is directly related to the amount of work that is exchanged, $dT = \delta W / C_V$. More generally, for any non-infinitesimal adiabatic ideal gas process $\Delta U = W$ and so $\Delta T = W / C_V$, as long as the heat capacity of the gas remains essentially constant over the corresponding temperature range.

A particularly important example of an adiabatic process is the *reversible adiabatic expansion of an ideal gas*. This process is quite similar to path C in Figure 2.6, except that the gas remains thermally insulated as the grains of sand are removed, so no heat exchange can take place between the system and its surroundings. The work exchanged when each (infinitesimal) grain of sand is removed is $\delta W = -P\,dV = -[(nRT)/V]dV = (\mathrm{D}/2)\,nRdT$. In order to relate the temperature, volume, and pressure changes in any such process, we may rearrange the latter identity to separate the variable T and V and then, after dividing by nR, integrate both sides as follows.

$$
\begin{aligned}
-\int_{V_1}^{V_2} \frac{1}{V} dV &= \frac{\mathrm{D}}{2} \int_{T_1}^{T_2} \frac{1}{T} dT \\
-\ln(V)\Big|_{V_1}^{V_2} &= \frac{\mathrm{D}}{2} \ln(T)\Big|_{T_1}^{T_2} \\
-\ln\left(\frac{V_2}{V_1}\right) &= \frac{\mathrm{D}}{2} \ln\left(\frac{T_2}{T_1}\right) \\
\ln\left(\frac{V_1}{V_2}\right) &= \ln\left(\frac{T_2}{T_1}\right)^{\frac{\mathrm{D}}{2}} \\
\left(\frac{V_1}{V_2}\right) &= \left(\frac{T_2}{T_1}\right)^{\frac{\mathrm{D}}{2}}
\end{aligned}
\tag{2.19}
$$

Notice that the last equality may be rearranged to obtain $T_2/T_1 = (V_1/V_2)^{2/\mathrm{D}} = (V_2/V_1)^{-2/\mathrm{D}}$. We may use these expressions to determine how much the temperature of a classical ideal gas will change when it undergoes an adiabatic reversible expansion from an initial volume V_1 to a final volume V_2, or how much its volume will change if its temperature decreases from T_1 to T_2. The above expressions can also be combined with the ideal gas equation of state (Eq. 2.14) to obtain the following relations between the pressure, volume, and temperature for an adiabatic reversible ideal gas process that begins at T_1, V_1, P_1 and ends at T_2, V_2, P_2.

$$
\boxed{\frac{P_2}{P_1} = \left(\frac{V_1}{V_2}\right)^{\frac{\mathrm{D}+2}{\mathrm{D}}} = \left(\frac{T_2}{T_1}\right)^{\frac{\mathrm{D}+2}{2}}}
\tag{2.20}
$$

This equation can be used, for example, to obtain the dependence of pressure on volume, $P = P_0(V_0/V)^{(D+2)/D}$, or the dependence of temperature on pressure, $T = T_0(P/P_0)^{2/(D+2)}$, for any *adiabatic reversible* ideal gas process that starts at T_0, V_0, P_0.

Recall that a monatomic ideal gas has only three translational degrees of freedom ($D = 3$), and these degrees of freedom invariably behave classically. A classical diatomic molecule, on the other hand, has $D = 7$ quadratic terms in its Hamiltonian. However, we also know that molecular vibrations often behave nonclassically, since vibrational quantum states typically have energy spacings that are comparable to (or greater than) $k_B T$. So, Eq. 2.19 is only expected to accurately describe a monatomic ideal gas (with $D = 3$), or a gas composed of diatomic or polyatomic molecules in the classical limit, at very high temperatures. At lower temperatures one must take molecular rotational and vibrational (and electronic) quantization into account in order to accurately predict the temperature dependence of the average energy U of a gas as we did, for example, in obtaining Eqs. 1.24 and 1.26.

Ideal Gas Entropy

Ideal gas processes may also be used to gain a better physical understanding of entropy. The Clausius entropy formula (Eq. 2.5) implies that one may determine the entropy change of a system by measuring the heat exchanged when the system is *reversibly* transformed from the initial to the final state of interest. In general, one must integrate the right-hand side of Eq. 2.5 over such a reversible path.

$$\Delta S = \int \frac{\delta Q}{T} \tag{2.21}$$

It is simplest to first consider *isothermal* processes, and in particular the isothermal expansion of an ideal gas. In order to calculate ΔS, we must perform the process reversibly, as illustrated by path C in Figure 2.6. For any such isothermal reversible process we may take $1/T$ out from under the integral to obtain the following expression for ΔS.[19]

$$\Delta S = \left(\frac{1}{T}\right) \int \delta Q = \frac{Q_{rev}}{T} \tag{2.22}$$

[19] Note that Eq. 2.22 is applicable to any isothermal reversible process, whether or not the system is an ideal gas.

The last equality follows simply from the fact that the total reversible heat exchange Q_{rev} is the integral of all the infinitesimal heat exchanges δQ that take place along the reversible path.

In other words, Eq. 2.22 indicates that the entropy change of an ideal gas that undergoes an isothermal expansion from an initial volume V_1 to a final volume V_2 may be determined by carrying out the expansion reversibly, and then dividing the net heat exchange by the temperature of the system. The total heat exchanged in such a reversible isothermal ideal gas expansion is given by Eq. 2.18. So, we may combine Eqs. 2.18 and 2.22 to obtain the following important equation for the entropy change in any isothermal ideal gas process.

$$\boxed{\Delta S = nR \ln\left(\frac{V_2}{V_1}\right)} \tag{2.23}$$

Notice that this implies that the entropy change per molecule is $\Delta S = k_B \ln(V_2/V_1) = k_B \ln V_2 - k_B \ln V_1 = S_2 - S_1$,[20] which clearly suggests that $S = k_B \ln V$ and looks very similar to Boltzmann's famous entropy formula, Eq. 2.9.[21] So it appears that Boltzmann's parameter Ω, which represents the number of different configurations (or states) in which we are likely to find the system, is proportional to the volume accessible to an ideal gas molecule. This seems quite reasonable since the number of places a gas molecule can go is proportional to the volume of its container. Notice that we have obtained this result by applying the Clausius entropy formula (Eq. 2.5) to derive an expression for entropy that is consistent with Boltzmann's seemingly quite different formula (Eq. 2.9).

Recall that an *adiabatic* process is one in which no heat is exchanged, and so $\delta Q = 0$. Thus, Eq. 2.21 implies that $\Delta S = 0$ *for any adiabatic reversible process*. But this appears to contradict Eq. 2.23, since the volume of an ideal gas certainly changes when it undergoes an adiabatic expansion – how could it be that there is no entropy change even though the gas expands? The key to resolving this apparent paradox is to recall that Eq. 2.23 was obtained by considering a reversible *isothermal* expansion, while in a reversible *adiabatic* expansion the temperature of the system does not remain constant.

Since $\Delta S = 0$ for any adiabatic reversible process, it must also be true that during an adiabatic expansion the influence of temperature and volume on

---

[20] Recall that $N = \mathcal{N}_A n$ is the number of molecules in the gas and $R = k_B \mathcal{N}_A$, so $nR = Nk_B$.

[21] More specifically, comparison of Eqs. 2.9 and 2.23 implies that $S = k_B \ln \Omega = k_B \ln(cV) = k_B \ln V + k_B \ln c$, (where c is independent of volume, but may depend on temperature).

entropy exactly cancel each other. Equation 2.19 relates the volume and temperature changes occurring during an adiabatic ideal gas process. So, we may combine Eqs. 2.19 and 2.23 (with the fact that $\Delta S = 0$), to infer that changing the temperature of an ideal gas produces an entropy change of $\Delta S = nR\ln(T_2/T_1)^{\mathrm{D}/2}$. We will later learn how to derive this important result in a more general way, to show that the entropy change associated with any ideal gas process – involving any change in volume and/or temperature – is given by the following expression.

$$\boxed{\Delta S = nR\left[\ln\left(\frac{V_2}{V_1}\right) + \ln\left(\frac{T_2}{T_1}\right)^{\frac{\mathrm{D}}{2}}\right]} \tag{2.24}$$

Comparison of the last term in Eq. 2.24 with Boltzmann's formula (Eq. 2.9) also suggests that the number of quantum states that are thermally accessible to a classical ideal gas molecule (or any classical system whose Hamiltonian contains D quadratic terms) is proportional to $T^{\mathrm{D}/2}$, which again turns out to be exactly right (as we will see). Also, note that $\ln(T_2/T_1)^{\mathrm{D}/2} = (\mathrm{D}/2)\ln(T_2/T_1)$, and so we can draw on Eq. 1.14 (and set $n = 1$) to reexpress Eq. 2.24 in the following commonly encountered molar form.

$$\Delta S = R\ln\left(\frac{V_2}{V_1}\right) + C_V\ln\left(\frac{T_2}{T_1}\right) \tag{2.25}$$

It is important to keep in mind that Eqs. 2.24 and 2.25 are only strictly true for a classical ideal gas (whose heat capacity C_V is temperature-independent).

Isothermal ideal gas processes, such as those illustrated in Figure 2.6, may be further related to the Clausius inequality (Eq. 2.7). Since entropy is a state function, and so is path-independent, the entropy of an isothermal ideal gas will be given by Eq. 2.23 regardless of whether the path connecting the initial and final states is reversible or irreversible. We have also seen that less work is performed along the irreversible paths (A and B) than along the reversible path (C), and so irreversible paths must produce a smaller heat exchange (because $Q = -W$ for any isothermal ideal gas process). Since the heat exchanged along paths A and B is positive (as heat is absorbed by the system), it is also clear that $Q_A/T < Q_B/T < Q_C/T$, thus confirming that Eq. 2.7 holds for these ideal gas processes.[22]

---

[22] Although the processes in Figure 2.6 all involved a positive heat exchange, one may carry out a similar analysis of the corresponding compression processes (for which $Q < 0$) to show that Eq. 2.7 holds equally well for processes involving either positive or negative heat exchanges.

An Irreversible Free Expansion

A *free expansion* is an irreversible process in which a gas expands from volume V_1 to volume V_2 *without lifting any weight at all*. If no weight is lifted, then $W=0$. If we further assume that the system is an ideal monatomic gas, and the free expansion is carried out isothermally, then $Q=\Delta U=\frac{3}{2}nR\Delta T=0$. Since neither work nor heat are exchanged between the system and its surroundings, the entropy of the surroundings must remain unchanged, while the entropy change of the system will again be $\Delta S=nR\ln(V_2/V_1)$ (because entropy is a state function). So, the entropy change of the universe resulting from such a highly irreversible free expansion is $\Delta S_{Univ}=\Delta S+0=nR\ln(V_2/V_1)>0$. Thus, the entropy of the universe does indeed increase as the result of this irreversible process, as predicted by the Clausius inequality (Eq. 2.6). Similar calculations can be performed to show that $\Delta S_{Univ}>0$ for *any* irreversible ideal gas process, while $\Delta S_{Univ}=0$ is true *only* when the process is carried out reversibly (as demonstrated in the following exercise).

Exercise 2.1

The following questions pertain to one mole of a monatomic ideal gas system, which is initially equilibrated at $T=300$ K and $V=0.2\,\text{m}^3$.

- What is the initial pressure of the gas (expressed in MPa units)?

  Solution. The ideal gas equation of state (Eq. 2.13) may be used to calculate the pressure of the gas, $P=nRT/V\approx(1\text{ mol})(8.3\text{ J/K mol})(300\text{ K})/(0.2\text{ m}^3)\approx 1.2\times10^4\text{ Pa}=0.012$ MPa. Note that 1 Atm is approximately equivalent to 0.1 MPa, so this gas has a pressure of about 0.12 Atm.

- How would the energy of the gas change if it either expanded isothermally to a volume of 2 m^3 or was heated to 400 K while keeping the volume fixed?

  Solution. The energy of an ideal gas is independent of its volume (or pressure), as it depends only on temperature. For one mole of a monatomic ideal gas $\Delta U=\frac{3}{2}R\Delta T$. So, in any isothermal ideal gas process $\Delta U=0$, while changing the temperature of a monatomic ideal gas by 100 K would produce an energy change of $\Delta U=\frac{3}{2}R\Delta T=1.5(8.3\text{ J/K mol})(100\text{ K})(0.001\text{ kJ/J})\approx 1.2$ kJ/mol.

- How much work and heat would be exchanged between the system and surroundings if the gas expanded *isothermally* and *reversibly* to a volume of 2 m^3?

  Solution. The pressure-volume work exchange in any process may be calculated by integrating the external pressure times the volume change of the gas $W = -\int P dV$. In a reversible ideal gas process, the internal and external pressures remain infinitesimally close to each other throughout the process and so we may replace the external pressure by the internal pressure of the gas $P = nRT/V$. For this isothermal process, $W = -RT \int_{V_1}^{V_2} (1/V) dV = -RT \ln(V_2/V_1) \approx -(8.3$ J/K mol)(300 K)(ln 10)(0.001 kJ/J) ≈ -5.7 kJ/mol. Moreover, since $\Delta U = 0$ for an isothermal ideal gas process, the first law (Eq. 2.3) implies that $Q = -W \approx 5.7$ kJ/mol. Note that the signs of W and Q indicate that work was done by the gas on the surroundings, and heat was absorbed from the surroundings by the system.

- What is the entropy change of the gas in the above reversible isothermal process?

  Solution. The above reversible heat exchange, combined with Eq. 2.5, may be used to calculate the entropy change of the gas, $\Delta S = \int \delta Q/T = Q_{rev}/T = R \ln(V_2/V_1) \approx 19$ J/(K mol). Note that since entropy is a state function, it is path-independent, and so the entropy of the gas would change by exactly the same amount in any process that started and ended at the same two equilibrium states. However, only in a reversible process would the entropy change of the surroundings be exactly equal and opposite to that of the system.

- If the gas expanded *adiabatically* and *reversibly* to a final volume of 2 m^3, calculate the final temperature of the gas, the amount of work exchanged between the gas and the surroundings, and the entropy change of the gas.

  Solution. We may calculate the final temperature of the ideal monatomic gas using Eq. 2.19, $T_2 = T_1(V_1/V_2)^{2/3} = (300\ \text{K})\ 0.1^{\ 2/3} \approx 65$ K. For any adiabatic process $\Delta U = W$, and so $W = \frac{3}{2} R \Delta T = 1.5$ (8.3 J/K mol)(0.001 kJ/J) (65 K–300 K) ≈ -2.9 kJ/mol. The entropy change of the gas may be obtained using Eq. 2.24, $\Delta S = R\{\ln(2/0.2) + \ln[(65/300)^{3/2}]\} = R[\ln(10) - \ln(10)] = 0$ J/(K mol).

A Reversible Carnot Cycle

A reversible Carnot engine operates in a cycle, such as that shown in Figure 2.7, consisting of adiabatic ($Q = 0$) and isothermal ($T =$ constant) branches. Since a cyclic process is one that returns the system to its initial state at the end of each cycle, it is necessarily the case that $\Delta U = 0$, and so $-W = Q$. When a Carnot cycle is carried out in a clockwise direction, it operates as a *heat engine* that converts some of the heat $Q_2 > 0$ absorbed from the high-temperature reservoir into work $-W = Q = Q_1 + Q_2 > 0$ with an efficiency of $-W/Q_2 = (T_2 - T_1)/T_2$, as shown in the following exercise.

Exercise 2.2

Use the first and second laws to prove that the efficiency $\eta_{\text{Carnot}} \equiv -W/Q_2$ of a Carnot engine is equal to $(T_2 - T_1)/T_2$, and then use this result to calculate the amount of work performed by the engine in Figure 2.7 for every joule of heat that is absorbed from the high-temperature reservoir, as well as the amount of heat that is wasted by being dumped into the low-temperature reservoir.

Solution. Since energy and entropy are state functions, the first and second laws imply that for any cyclic process $\Delta U = Q + W = \oint dU = 0$ and $\Delta S = \oint dS = \oint \delta Q/T = 0$. Thus, the first law requires that $-W = Q = Q_2 + Q_1$, while the second law requires that $\oint \delta Q/T = Q_2/T_2 + Q_1/T_1 = 0$ or $Q_1/Q_2 = -T_1/T_2$. These results may be combined to obtain the desired expression for the efficiency of a Carnot engine.

$$\eta_{\text{Carnot}} = \frac{-W}{Q_2} = \frac{Q_2 + Q_1}{Q_2} = 1 + \frac{Q_1}{Q_2} = 1 - \frac{T_1}{T_2} = \frac{T_2 - T_1}{T_2} \tag{2.26}$$

When applied to the Carnot engine represented in Figure 2.7, this result implies that for every joule of heat absorbed from the high-temperature reservoir only $-W = (400 - 350)/400 = 0.125$ J of work will be done by the engine, while $Q_1 = -W - Q_2 = -0.875$ J of heat will be wasted.

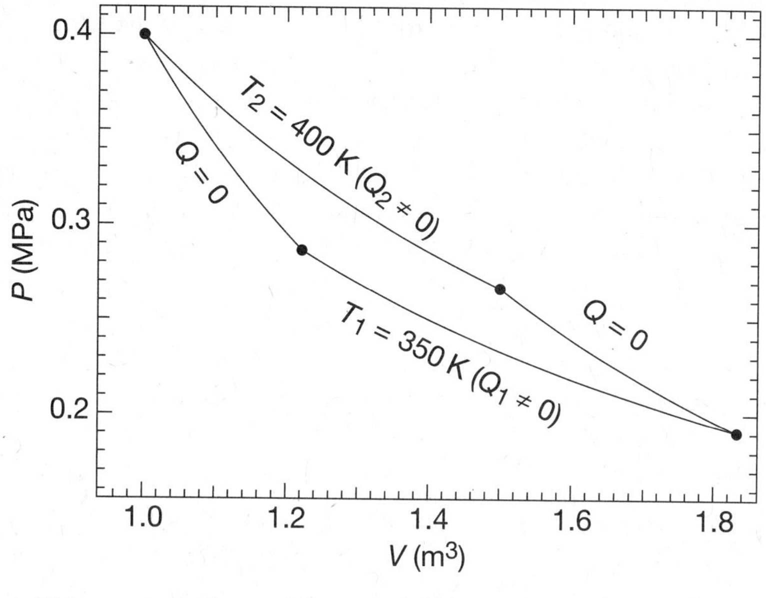

FIGURE 2.7 This graph depicts a Carnot cycle that, when traversed in the clockwise direction, functions as a heat engine that performs work by extracting heat ($Q_2 > 0$) from a 400 K reservoir and depositing heat ($Q_1 < 0$) in a 350 K reservoir. On the other hand, when a Carnot cycle is run in the counterclockwise direction, then it functions like a heat pump (or air conditioner), which transfers heat from the cold to the hot reservoir.

When a Carnot engine is run in the counterclockwise direction then it operates as a *heat pump* (or air conditioner) in which mechanical work $W < 0$ is done in order to pump heat out of the low-temperature reservoir and into the high-temperature reservoir. The efficiency of refrigerators or air conditioners can be quantified in terms of their *coefficient of performance* $\mathrm{COP} \equiv Q_1/W$, which relates the amount of heat extracted from the low-temperature reservoir to the amount of work performed. For an ideal Carnot refrigerator $\mathrm{COP} = T_1/(T_2 - T_1)$, which indicates that it is hardest to pump heat out of a very cold reservoir into a very hot one. This expression for COP also implies that an ideal (reversible) Carnot air conditioner could pump heat out of a house at 75°F (297 K) to an outside temperature of 100°F (311 K) with a COP of approximately 21, while commercial air-conditioning systems typically have COP values of 2 to 5.

HOMEWORK PROBLEMS

Problems That Illustrate Core Concepts

1. A given system of unknown internal structure performs the task of lifting a 1 lb (0.454 kg) weight by one 1 ft (0.305 m), in Earth's gravitational field ($g \approx 9.8$ m/s^2), with a steady lifting rate of 0.5 ft/s, so it takes the system 2 s to complete the process. All of the energy exchanged between the system and the surroundings comes from a chemical process (battery) inside the system, which produces a steady power of 1 W (and is only turned on during the lifting process). The system is connected to a heat-sink, which keeps it at a constant temperature. You may assume that both the weight and the heat-sink are part of the surroundings.
   (a) How much work was exchanged in the above process? (Express your answer in J, and be careful to include the correct sign.)
   (b) How much energy is dissipated as heat, which is exchanged between the system and the heat-sink? (Express your answer in J, and be careful to include the correct sign.)
   (c) What is ΔU for the above process? (Express your answer in J, and be careful to include the correct sign.)
   (d) If the system were not connected to a heat-sink, would you expect its temperature to increase or decrease during the weight-lifting process? (Explain in words.)

2. When one mole of an ideal gas undergoes a reversible isothermal (constant temperature) expansion from a volume V_1 to a larger volume V_2 the amount of heat exchanged between the gas (system) and its surrounding temperature bath is $Q = \mathcal{N}_A k_B T \ln(V_2/V_1) = RT \ln(V_2/V_1)$. However, when the same gas undergoes a spontaneous, and highly *irreversible*, free expansion between the same two volumes (at the same temperature), then no heat and no work are exchanged between the system and surroundings (so the state of the surroundings remains unchanged).
   (a) What is the entropy change of the gas and why is it the same for both processes?
   (b) What are the entropy changes of the surroundings in the two processes?
   (c) What are the entropy changes of the universe in the two processes?

3. Boltzmann's entropy formula, Eq. 2.9, can also be used to predict the entropy of an ideal gas which expands isothermally, by making use of the fact that the number of translation

quantum states that are accessible to a single gas atom is proportional to the volume of the gas, $\Omega = aV$ (where a is a constant of proportionality that depends on temperature, but not on volume).

(a) Use Eq. 2.9 to show that the entropy change of a single gas atom, which expands isothermally from a volume V_1 to a volume V_2, can be expressed as $k_B \ln(V_2/V_1)$.

(b) If you assume that entropies of gas atoms are additive, use the above result to show that the entropy change of a mole of gas that expands isothermally from a volume V_1 to a volume V_2 can be expressed as $R \ln(V_2/V_1)$.

4. The experimental density of CO_2 gas at 0.1 MPa and 300 K is 0.001773 g/cm^3.

(a) What is the experimental concentration (in M units) and number density (molecules/m^3 units) of CO_2 gas at 0.1 MPa and 300 K?

(b) Compare the above experimental concentration and number density values to those obtained from the ideal gas law.

5. Consider the following two isothermal monatomic ideal gas expansion processes.

- Path A: Sudden (irreversible) decrease in pressure from P_1 to P_2 (where $P_2 < P_1$).
- Path B: Gradual (reversible) decrease in pressure from P_1 to P_2, such that the internal and external pressures remain in equilibrium at every step along the path.

(a) Draw a graph of P vs. V illustrating the path followed in each of the above processes.

(b) Obtain expressions for ΔU, W, and Q for each of the above processes (express your results as functions of n, T, P_1, and/or P_2).

6. Calculate the work exchanged (in J) along each of the paths described in problem 5, assuming that $n = 1$, $P_1 = 1$ MPa, $P_2 = 0.1$ MPa, and $T_1 = T_2 = 300$ K (and confirm that the signs and relative magnitudes of your answers make physical sense).

7. Where did the energy required to maintain the system at a constant temperature in the two isothermal processes described in problem 5 come from, and how is this energy related to the amount of work that was performed along each path?

8. A *reversible adiabatic* expansion of one mole of a monatomic ideal gas is carried out by starting at $P_1 = 1$ Pa and $T_1 = 300$ K, and ending at $T_2 = 200$ K.

(a) What are the initial and final volumes of the gas?

(b) What is the final pressure of the gas?

(c) How much work was exchanged during the process (include the correct sign)?

9. Obtain expressions for the entropy change of the gas ΔS, the surroundings (assumed to be an ideal isothermal bath) ΔS_{Surr}, and the universe ΔS_{Univ}, for each of the following processes.

(a) The *isothermal irreversible* process described by path A in problem 5.

(b) The *isothermal reversible* process described by path B in problem 5.

(c) Is the ΔS_{Univ} that you obtained in (a) greater than or less than that in a free expansion (as described in problem 2)? Why does your answer makes sense?

Problems That Test Your Understanding

10. Consider a system consisting of *one mole* of a single-component monatomic ideal gas that expands *isothermally* from a volume of 10 liters to 100 liters.

(a) What is the entropy change of the gas?

(b) Combine your result in (a) with Boltzmann's expression for entropy, Eq. 2.9, to show that the translational partition function of an ideal gas must be proportional to volume (at constant temperature).

11. The entropy of a harmonic oscillator may be predicted by combining Boltzmann's entropy formula with the partition function of a harmonic oscillator (and recalling that both expressions pertain to the number of thermally accessible quantum states). In answering the following questions, you may assume that the frequency of the harmonic oscillator is $\tilde{\nu} = \nu/c = 100\ \text{cm}^{-1}$.
   (a) What is the vibrational entropy of a harmonic oscillator at 0 K?
   (b) At what temperature will $k_B T$ be exactly equal to the quantum spacing $\Delta\varepsilon$ of the harmonic oscillator?
   (c) When $k_B T$ is much larger than the energy spacing, what is the simplest way of expressing how the number of thermally accessible states of a harmonic oscillator depends on $k_B T$ and $\Delta\varepsilon$? [Hint: Recall that when x is small, $e^x \approx 1 + x$.]
   (d) Use the result in (c) to obtain an expression for the absolute (vibrational) entropy of a harmonic oscillator at high temperature.

12. Consider 0.1 mole of a monotomic ideal gas (system) which is initially equilibrated at a temperature of 500 K in a container with a volume of 1 m^3, and then undergoes a *reversible isothermal* expansion to a volume of 2 m^3.
   (a) What is the energy change of the gas (in J units)?
   (b) How much work is exchanged in the above process (in J units)?
   (c) What is the entropy change of the gas (in J/K units)?
   (d) What is the entropy change of the universe (in J/K units)?

13. Consider 0.1 mole of a monotomic ideal gas (system) that is initially equilibrated at a temperature of 500 K in a container with a volume of 1 m^3, and then a stopcock is opened to allow the gas to undergo a *free expansion* to a volume of 2 m^3.
   (a) How much work is exchanged in this process (in J units)?
   (b) How much heat is exchanged in this process (in J units)?
   (c) What is the entropy change of the gas (in J/K units)?
   (d) What is the entropy change of the universe (in J/K units)?

14. Consider 0.1 mole of a monotomic ideal gas (system) that is initially equilibrated at a temperature of 500 K in a container with a volume of 1 m^3, and then undergoes a *reversible* expansion to a volume of 2 m^3 in which *no heat is exchanged*.
   (a) What is the temperature change of the gas (in K units)?
   (b) What is the energy change of the gas (in J units)?
   (c) How much work is exchanged in this process (in J units)?
   (d) What is the entropy change of the gas (in J/K units)?

15. Consider a system consisting of one mole of a monatomic ideal gas that expands from a volume of $V_1 = 0.5$ m^3 to a volume of $V_2 = 1$ m^3, and is maintained at a constant temperature of $T = T_1 = T_2 = 300$ K by a surrounding ideal isothermal bath. In answering each of the following questions make sure your answers all have physically reasonable signs.
   (a) What is the entropy change of the system (in J/K units)?
   (b) Briefly (and clearly) explain, in words, how the entropy change of the system would depend on the nature of the path that is followed in converting the system from the above initial state to the above final state.

(c) If the process were carried out reversibly, what would be the entropy change of the surroundings (in J/K units)?

(d) If the process were carried out reversibly, how much heat exchange would take place between the system and surroundings?

(e) If the process were carried out as a free expansion (with no work or heat exchange), how much would the entropy of the universe (system plus surroundings) change (in J/K units)?

16. Consider a system composed of one mole of a diatomic ideal gas that is initially contained in a cylinder of volume $V_1 = 0.1$ m^3 at a temperature of $T_1 = 200$ K, and is then heated to temperature of 300 K, at constant pressure. You may assume that all of the translational, rotational, and vibrational degrees of freedom are thermally active (so they behave classically). Make sure all your answers have physically reasonable signs.

(a) What is the pressure of the system (in Pa = J/m^3 units)?

(b) What is the energy change of the system (in J units)?

(c) What is the volume change of the system (in m^3 units)?

(d) How much work is exchanged between the system and surroundings (in J units)?

(e) What is the heat exchange between the system and surroundings (in J units)?

(f) What is the entropy change of the system (in J/K units)?

17. The following questions pertain to one mole of a monatomic ideal gas system, which is initially equilibrated at $T = 300$ K and $V = 1$ m^3, and then is allowed to expand irreversibly by removing a weight so that the external pressure suddenly decreases by a factor of 10, while maintaining contact with an isothermal bath at 300 K.

(a) What is the initial pressure of the gas (expressed as a number with units)?

(b) What is the energy change of the gas?

(c) What is the work exchanged between the gas and its surroundings?

(d) What is the heat exchanged between the gas and its surroundings?

(e) What is the entropy change of the gas?

(f) What is the entropy change of the universe for this process?

18. Consider a system consisting of one mole of a monatomic ideal gas that begins at an initial temperature of 1000 K and a pressure of 0.1 MPa, and then expands adiabatically and reversibly until its temperature has decreased to 100 K.

(a) What is the energy change of the gas?

(b) How much work is exchanged between the gas and its surroundings?

(c) What is the entropy change of the gas?

(d) What is the entropy change of the universe for this process?

19. Consider a gas composed of one mole of diatomic molecules that is initially equilibrated at a temperature of 300 K and is then heated to 301 K at constant volume.

(a) If the translation, rotation, and vibrational degrees of freedom of the gas all behaved classically, how much heat would be required to produce the above temperature change?

(b) Estimate the amount of heat required to produce the above temperature change if the vibrational quantum spacing of the gas is approximately equal to $k_B T$.

20. A reversible Carnot engine that operates in a cycle consisting of two isothermal and two adiabatic branches converts the heat absorbed from the high-temperature reservoir into work done by the system (per cycle) with an efficiency that is given by Eq. 2.26. Consider a Carnot engine that contains 0.1 mole of an ideal gas and expands from a volume of 1 m^3 to a volume of 2 m^3 at 500 K, and then deposits heat into the surroundings at 250 K. First, calculate the heat absorbed by the system from the high-temperature reservoir, and then predict the total amount work done by the system during each complete cycle.

Axiomatic Foundations of Thermodynamics

CHAPTER 3

3.1 Fundamental Equation and Postulates

Functions and Partial Derivatives

Thermodynamic state functions, like other mathematical functions, may in general depend on one or more independent variables. For example, a simple one-dimensional function such as $y = f(x)$ has a single independent variable, x. Such a function may be represented as a curve in the x-y plane, in which each value of x is associated with a single value of y. The slope, $s = dy/dx$, represents the derivative of the function at each point. This slope determines how much the function y will be affected by a small change in the independent variable x.

$$dy = sdx = \left(\frac{dy}{dx}\right) dx \tag{3.1}$$

The derivative dy/dx represents the *susceptibility* of the function y to a small change in x.

More generally, a function of two variables, $z = f(x, y)$, may be represented as a surface suspended above the x-y plane. In this case, the following equation expresses the influence of small changes in x and y on the function, z.

$$dz = \left(\frac{\partial z}{\partial x}\right)_y dx + \left(\frac{\partial z}{\partial y}\right)_x dy \tag{3.2}$$

Each of the slopes $(\partial z/\partial x)_y$ and $(\partial z/\partial y)_x$ again represents the susceptibilities of the function z to small changes in x and y. The slope $(\partial z/\partial x)_y$ is called the *partial derivative* of z with respect to x, because it is performed while holding y constant. Similarly, the partial derivative of z with respect to y is performed while holding x constant.

We may extend the above procedure to functions with any number of independent variables such as $Y = F(x_1, x_2, x_3 \ldots)$, in which $x_1, x_2, x_3 \ldots$ represent any set of independent variables over which the function Y is defined. Thus, a change in Y may again be expressed in terms of partial derivatives with respect to each of the variables x_i, while holding all of the other variables constant.

$$dY = \left(\frac{\partial Y}{\partial x_1}\right)_{x_i \neq x_1} dx_1 + \left(\frac{\partial Y}{\partial x_2}\right)_{x_i \neq x_2} dx_2 + \left(\frac{\partial Y}{\partial x_3}\right)_{x_i \neq x_3} dx_3 + \ldots \tag{3.3}$$

It can sometimes also be convenient to choose some other set of independent variables $x'_1, x'_2, x'_3 \ldots$ to describe a given function. For example, these new variables may represent the coordinates along a new set of axes, which are rotated with respect to the original axes. Thus, we may represent the same function either as $Y(x_1, x_2, x_3 \ldots)$ or as $Y(x'_1, x'_2, x'_3 \ldots)$. Having defined this new set of variables, we may again represent the effect of a small change in these new variables on Y.

$$dY = \left(\frac{\partial Y}{\partial x'_1}\right)_{x'_i \neq x'_1} dx'_1 + \left(\frac{\partial Y}{\partial x'_2}\right)_{x'_i \neq x'_2} dx'_2 + \left(\frac{\partial Y}{\partial x'_3}\right)_{x'_i \neq x'_3} dx'_3 + \ldots \tag{3.4}$$

The above properties of functions play a central role in thermodynamics. For example, we may express the energy of a given system as a function of temperature and pressure, $U(T, P)$. Thus, the partial derivatives of U determine how it will change as the result of small changes in T and P, since $dU = (\partial U/\partial T)_P dT + (\partial U/\partial P)_T dP$. Alternatively, we may transform to a new set of independent variables by expressing the energy of the system as a function of its entropy and volume $U(S, V)$, and thus $dU = (\partial U/\partial S)_V dS + (\partial U/\partial V)_S dV$. It is sometimes also convenient to express S as a function of energy and volume, $S(U, V)$, in which case $dS = (\partial S/\partial U)_V dU + (\partial S/\partial V)_U dV$.[1]

[1] Notice that in writing all of the thermodynamic expressions in this paragraph, we have implicitly assumed that n (or N) is held fixed, and so we have not treated n as one of the variables of either U or S.

Exercise 3.1

The energy and entropy of a system containing n moles of a monatomic ideal gas can be expressed as a function of temperature, pressure, and/or volume. We may use what we know about ideal gas thermodynamics to obtain expressions for the total differential of $U(T, P)$ and $S(T, V)$, and various associated partial derivatives.

- Write an expression for the total differential dU, by treating T and P as the independent variables, and obtain equations for the two partial derivatives that appear in the resulting expression.

  Solution. The total differential of the energy may be represented as a sum of two terms that indicate how U changes in response to an infinitesimal changes in T and P.

  $$dU = \left(\frac{\partial U}{\partial T}\right)_P dT + \left(\frac{\partial U}{\partial P}\right)_T dP$$

  The first partial derivative may be evaluated using the fact that $U = \frac{3}{2}nRT$, and so $(\partial U/\partial T)_p = \frac{3}{2}nR$. The second partial derivative is equal to zero, $(\partial U/\partial P)_T = 0$, because U depends only on T and is independent of pressure (or volume). So, the total differential of the energy of a monatomic ideal gas reduces to simply $dU = \frac{3}{2}nRdT$.

- Use what you know about $S(T, V)$ to obtain an expression for dS and the two associated partial derivatives.

  Solution. The total differential of $S(T, V)$ is given by the following general expression.

  $$dS = \left(\frac{\partial S}{\partial T}\right)_V dT + \left(\frac{\partial S}{\partial V}\right)_T dV$$

  The partial derivatives may be evaluated by making use of the fact that Eq. 2.24 implies that $S = S_0 + nR\ln(V/V_0) + \frac{3}{2}nR\ln(T/T_0)$. So, the first of the above partial derivatives is $(\partial S/\partial T)_V = \frac{3}{2}nR/T$ and the second is $(\partial S/\partial V)_T = nR/V$. Note that the latter derivatives were obtained by recalling that $d\ln x/dx = 1/x$, and treating S_0, V_0, and T_0 as constants (so the only variables on which S depends are T and V). Thus, we may express the entropy change of an ideal gas as $dS = (3nR/2T)dT + (nR/V)dV$.[2]

---

[2] Alternatively, we could have obtained the expression for dS by noting that $V/V_0 = (V_0/V_0) + (V - V_0)/V_0 = 1 + (1/V_0)\Delta V$, which becomes $V/V_0 = 1 + (1/V)dV$ when V is infinitesimally close to V_0 [and similarly $T/T_0 = 1 + (1/T)dT$ when T is infinitesimally close to T_0], and then made use of the fact that $\ln(1 + x) \approx x$ (when $x << 1$).

- Use the above results, and the chain rule, to confirm that $(\partial U/\partial S)_V = T$.

Solution. The chain rule may be used to express $(\partial U/\partial S)_V$ as a product, or ratio, of two other derivatives (both evaluated at constant volume), $(\partial U/\partial S)_V = (\partial U/\partial T)_V(\partial T/\partial S)_V = (\partial U/\partial T)_V/(\partial S/\partial T)_V$. Moreover, $(\partial U/\partial T)_V = (\partial U/\partial T)_P$, because the energy of an ideal gas is independent of both volume and pressure. So, we may now use the previously obtained expressions for $(\partial U/\partial T)_V$ and $(\partial S/\partial T)_V$ to verify that $(\partial U/\partial S)_V = (\frac{3}{2}nR)/[\frac{3}{2}(nR/T)] = T$.

The Fundamental Equation of Thermodynamics

The connection between thermodynamics and partial differential equations was developed in a beautifully general way by Willard Gibbs. He did this by recognizing that functions such as $P(T, V, \ldots)$ or $U(S, V, \ldots)$ may be envisioned as surfaces, like mountain ranges, which undulate up and down as one moves to different locations (defined by the values of the variables $T, V, \ldots$ or $S, V, \ldots$). Such single-valued functions, or surfaces, are also called exact differential functions, and thus Gibbs recognized that one may use the mathematical tools of partial differential equations to investigate how thermodynamic functions will respond to changes in the values of the associated variables.

Gibbs also recognized that one can only change the energy of a system by applying some sort of force in order to change one or more of the variables that define the state of the system.[3] Thus, one may express any differential change in the energy of the system as the following sum of terms, each of which is a product of a generalized applied force $\mathcal{F}_i$ times a generalized displacement $d\chi_i$ of the corresponding system variable.

$$dU = \sum_i \mathcal{F}_i d\chi_i \tag{3.5}$$

In other words, the above expression implies that the energy of the system, $U(\chi_1, \chi_2, \chi_3 \ldots)$, is a function of the variables χ_i, and that the

[3] The variables of a thermodynamic system can be either extensive or intensive thermodynamics functions, and they may also include externally applied gravitational, electrical, or magnetic fields, as well as other sorts of constraints on the system, such as partitions or semipermeable membranes that may be used to separate various components of the system from each other.

generalized forces $\mathcal{F}_i$ are equivalent to the partial derivatives of U with respect to χ_i.[4]

$$\mathcal{F}_i = \left(\frac{\partial U}{\partial \chi_i}\right)_{\chi_j \neq \chi_i} \tag{3.6}$$

As a simple example, consider a system consisting of a ball on a hill, such as that illustrated in Figure 1.1 (on page 7). In order to increase the potential energy of the ball, one must push it up to a higher elevation. In other words, $dU = (\partial U/\partial h)dh = mg(\partial h/\partial h)dh = mgdh$. So, the work that must be performed to elevate the ball is exactly equal to the product of $(\partial U/\partial h)$ and dh.

As another example, consider a system whose energy is changed by performing work to compress it to a smaller volume. In this case, $dU = -P\,dV = (\partial U/\partial V)dV$.[5] This clearly implies that $P = -(\partial U/\partial V)$, although we still need to specify exactly what other variables are held constant when performing the above partial derivative (as we will see).

More generally, one may consider the effects of other variables, such as gravitational, magnetic, and electric potentials, on the energy of a system (as further discussed in Section 3.6). However, since many chemical processes of practical interest do not involve significant variations in gravitational, magnetic, or electrical fields, it is often appropriate to neglect these contributions to the energy of the system. Moreover, any work exchanges associated with these variables are entirely analogous to pressure-volume work, and so we may consider $P\,dV$ to represent other sorts of mechanical work exchanges.

The variables of greatest relevance to chemical processes are those that specify the chemical composition of the system, expressed in terms of the numbers of molecules of different types N_i. Even when all other variables are held constant, we may change the energy of a system by either adding or removing molecules, $dU = (\partial U/\partial N_i)dN_i$. The latter partial derivative is called the *chemical potential*, $\mu_i \equiv (\partial U/\partial N_i)$, whose central significance to

---

[4] The identification of the generalized force $\mathcal{F}_i$ with $(\partial U/\partial \chi_i)$ appears to be missing the negative sign that is present in 3.6. This difference comes from the fact that Eq. 3.6 represents the force that must be *applied* to the system in order to change its energy, while Eq. 1.4 is the equal and opposite force experienced by the system. This connection is also implicit in Newton's third law, which states that if two bodies exert forces on each other, then these forces are equal in magnitude and opposite in direction.

[5] Recall that the negative sign must be present since the energy of the system is increased ($dU > 0$) when work is performed on the system by applying pressure ($P > 0$) to produce a decrease in volume ($dV < 0$) of the gas.

chemical processes was first recognized by Willard Gibbs.[6] Thus, μ_i is the coefficient that determines how much the energy of a system will change when a single molecule (of type i) is added to the system. Again, we will have to be careful to specify exactly which variables must be held constant when a molecule is added, as explained below.

All of the above ways of changing the energy of the system represent work exchanges of different sorts. Mechanical pressure-volume work is $\delta W = -PdV$, while the chemical work associated with adding a molecule to the system is $\delta W = \mu dN$. In other words, the generalized force and displacement associated with pressure-volume work are $\mathcal{F} = -P$ and $d\chi = dV$, while for a chemical exchange $\mathcal{F} = \mu$ and $d\chi = dN$.

What is not yet clear is how $\mathcal{F} d\chi$ can be used to describe heat exchanges. In other words, what generalized force $\mathcal{F}$ and displacement $d\chi$ are associated with depositing energy in a system's thermal degrees of freedom? An important clue is provided by the second law expression $dS = \delta Q/T$ (Eq. 2.5), which can be rearranged to $\delta Q = TdS$. This suggests that $TdS = \mathcal{F} d\chi$ is equivalent to an infinitesimal change in a system's thermal energy.

In other words, quite remarkably, depositing energy in the thermal degrees of freedom of a system[7] is equivalent to displacing the system's entropy dS under the influence of a generalized force equal to the system's temperature T. Moreover, since the effect of dS on dU may be expressed as $dU = (\partial U/\partial S)dS$, the temperature of the system is evidently equivalent to the partial derivative $(\partial U/\partial S)$, in the same way that the pressure of the system is associated with the partial derivative $(\partial U/\partial V)$.

In summary, the first law of thermodynamics, Eq. 2.3, indicates that all energy changes are due to either heat or work exchanges $dU = \delta Q + \delta W$. The energy change associated with infinitesimal changes in the system's volume V or the chemical composition N_i are equivalent to reversible work exchanges.

$$\delta W = -PdV + \sum_i \mu_i dN_i$$

[6] Gibbs actually expressed μ_i in terms of a change in the mass of a particular component, rather than the number of molecules of that component, and he called the associated partial derivative of the energy simply the "potential" for the substance i. It is now more common to express μ_i as a derivative with respect to the number of molecules (or moles) of type i and to call it the chemical potential.

[7] A system's thermal degrees of freedom consist of the kinetic and potential energies of the molecules in the system, whose populations are determined by the associated Boltzmann factors (as expressed by Eq. 1.14 or Eq. 1.27).

On the other hand, adding an infinitesimal amount of energy to the thermal degrees of freedom of the system is equivalent to a reversible heat exchange.[8]

$$\delta Q = TdS \tag{3.7}$$

This also implies that S, V, and N_i are the *natural variables* of $U(S, V, N_i)$, in the sense that δQ and δW may be expressed directly in terms of displacements of the variables S, V, and N_i (as further explained in Section 4.4). So, we can now write the complete expression for the influence of these variables on U.

$$dU = \left(\frac{\partial U}{\partial S}\right)_{V,N_i} dS + \left(\frac{\partial U}{\partial V}\right)_{S,N_i} dV + \sum_i \left(\frac{\partial U}{\partial N_i}\right)_{S,V,N_j \neq N_i} dN_i$$

$$\boxed{dU = TdS - PdV + \sum_i \mu_i dN_i} \tag{3.8}$$

Equation 3.8 is called the *fundamental equation* of thermodynamics (in the energy representation). Note that a comparison of the coefficients in the above expressions for dU indicates that we can now be more specific in identifying the partial differential expressions for T, P, and μ_i.

$$T \equiv \left(\frac{\partial U}{\partial S}\right)_{V,N_i} \tag{3.9}$$

$$P \equiv -\left(\frac{\partial U}{\partial V}\right)_{S,N_i} \tag{3.10}$$

$$\mu_i \equiv \left(\frac{\partial U}{\partial N_i}\right)_{S,V,N_j \neq N_i} \tag{3.11}$$

It is sometimes convenient to treat S (rather than U) as the dependent variable, so that $S(U, V, N_i)$ leads to the following *entropy representation* of the fundamental equation.

$$dS = \left(\frac{\partial S}{\partial U}\right)_{V,N_i} dU + \left(\frac{\partial S}{\partial V}\right)_{U,N_i} dV + \sum_i \left(\frac{\partial S}{\partial N_i}\right)_{U,V,N_j \neq N_i} dN_i \tag{3.12}$$

---

[8] What is not yet so obvious is why the identity $dS = \delta Q/T$ only holds when the heat exchange is performed *reversibly* (which also implies that the heat exchange is performed while maintaining the temperature of the system infinitesimally close to that of the surroundings). The reasons for these peculiar restrictions will become clear when we look more closely into what exactly is meant by *reversible* and *irreversible* processes, in Section 4.4.

We may also obtain the following expression for dS by isolating dS in Eq. 3.8.[9]

$$dS = \left(\frac{1}{T}\right) dU + \left(\frac{P}{T}\right) dV - \sum_i \left(\frac{\mu_i}{T}\right) dN_i \tag{3.13}$$

This entropy representation of the fundamental equation is sometimes very useful in facilitating thermodynamic derivations, as we will see. Comparison of Eqs. 3.12 and 3.13 implies the following additional identities.

$$\frac{1}{T} = \left(\frac{\partial S}{\partial U}\right)_{V,N_i} \tag{3.14}$$

$$\frac{P}{T} = \left(\frac{\partial S}{\partial V}\right)_{U,N_i} \tag{3.15}$$

$$\frac{\mu_i}{T} = -\left(\frac{\partial S}{\partial N_i}\right)_{U,V,N_j \neq N_i} \tag{3.16}$$

Generalization of the Laws of Thermodynamics

The first and second laws of thermodynamics grew out of an effort to describe the properties of mechanical devices such as steam engines. From a more modern perspective, thermodynamics is viewed as a mathematical system built upon a few fundamental postulates, or axioms. These axioms can be expressed in forms that are both simpler and more abstract than the laws of thermodynamics, and yet contain the same essential physical information.

The first postulate of thermodynamics essentially states that equilibrium states are like mathematical functions in the sense that there can't be two different equilibrium states for a given system.

> **Postulate I.** *For any system there exists one and only one equilibrium state with a given total energy (U), volume (V), and composition (N_1, N_2, N_3, ...), in the absence of any additional internal constraints.*

The term *constraints*, as it is used in thermodynamics, refers to the parameters that describe the state of the system. These include external constraints imposed by the boundaries of the system, as well as the

---

[9] Performing such an inversion is only strictly possible when S is a single-valued function of U. As we will see in the next section, one of the fundamental postulates of thermodynamics implies that this is the case, as Postulate III requires that S is a monotonically increasing function of U.

total energy and chemical composition of the system.[10] In addition to such external constraints, a system may also have internal constraints that further restrict the state of the system, as we will see.

The restriction regarding additional internal constraints is an important one, as imposing an additional constraint can move a system to a different equilibrium state. For example, one could imagine introducing an internal constraint consisting of a partition that prevents molecules of different types from freely mixing. Such a system would clearly have a different equilibrium state before and after removing the partition. Similarly, we could envision a partition that maintained a different pressure in two parts of a system. Such a system would again have different equilibrium states with and without the partition. There are many other types of internal constraints that one could imagine imposing. These all have the effect of forcing the system into a new equilibrium state. Imposing any such constraint on an unconstrained equilibrium system also invariably requires doing work on the system.

A central question of thermodynamics is what determines the equilibrium state of any system? In other words, if you initially have a system of some composition that is not in equilibrium, what drives it toward its equilibrium state? This is clearly a deep question, and one whose answer is far from obvious. Even Carnot and Clausius, who identified entropy as an important new state function, did not initially suspect that this function also determines the unique state to which any system will evolve as it approaches equilibrium.

In order to describe the evolution of a system within the realm of thermodynamics, we must consider how a system that is initially at equilibrium, under a particular set of constraints, evolves to a different equilibrium state when the constraints on the system are changed. As a specific example,

---

[10] Note that some care must be taken in clarifying what is meant by the composition of a system. If the system is composed of nonreactive molecules then the number of each of these chemical components ($N_1, N_2, N_3 \ldots$) completely specifies the composition of the system. However, in systems that may undergo chemical reactions we must use some other method to specify the composition of the system in a way that is invariant to chemical transformations. For example, we may view the components of the system as the number of atoms of each type, or even more generally we may take $N_1, N_2, N_3 \ldots$ to specify the number of fundamental particles in the system (neutrons, protons, and electrons, etc., whose numbers we may safely assume to be conserved under typical laboratory conditions). Thus, under the most general conditions the specification of a fixed set of values of $N_1, N_2, N_3 \ldots$ essentially amounts to stating that the system is *closed*, in the sense that there is no exchange of material between the system and its surroundings.

consider a gas that is initially constrained within a container of volume V_1, and then the constraint is removed by opening a stopcock so that the gas can expand into a larger volume V_2. Our experience tells us that the gas will spontaneously evolve so as to fill the larger volume uniformly. We also know that such an expansion increases the entropy of the system. In other words, a gas will never spontaneously evolve to a state of lower entropy (unless it is forced to do so by performing work on the system).

The second postulate of thermodynamics essentially states that there exists some function whose value is maximized at equilibrium.

> ***Postulate II.*** *For any equilibrium system there exists an extensive function called entropy, S, whose value is maximized when removing all internal constraints while maintaining the system isolated from its surroundings.*

The term *isolated* implies that there is no exchange of either heat or work between the system and its surroundings (which also implies that no chemical substances can enter or leave the system). Postulate II is sometimes more loosely paraphrased by stating that the entropy of any isolated system is maximized at equilibrium. Such a statement is problematic because it implies that we can know the entropy of a nonequilibrium state. That is why it is important to invoke internal constraints in talking about the maximization of entropy. In other words, if we want to talk about the entropy of a system changing, we must specify the initial and final *equilibrium* states of the system. The only way we can move a system to a new equilibrium state is by imposing (or releasing) a constraint. So, when we say that the entropy of an isolated system is maximized at equilibrium, what we really mean is that entropy is maximized when an internal constraint is removed, without exchanging any work or heat with the surroundings.

The third postulate essentially states that there is a one-to-one relationship between the entropy S and energy U.

> ***Postulate III.*** *The entropy of any system is a continuous, continuously differentiable, and monotonically increasing function of its energy at constant volume and composition.*

In other words, this postulate is equivalent to the following inequality.

$$\boxed{\left(\frac{\partial S}{\partial U}\right)_{V,N_i} > 0} \tag{3.17}$$

Note that the subscripts V, N_i indicate that the derivative is evaluated while holding the variables V and N_i constant. This postulate is intimately linked to the second law of thermodynamics, since Eq. 2.5 is a direct consequence of Postulates II and III (as we will see).

If S is a continuous monotonically increasing function of U, that also implies that U is a continuous monotonically increasing function of S.

$$\boxed{\left(\frac{\partial U}{\partial S}\right)_{V,N_i} > 0} \tag{3.18}$$

We have already seen that the latter partial derivative appears in the fundamental equation of thermodynamics (Eq. 3.8) and is equivalent to the absolute temperature of the system T (see Eq. 3.9).

$$T \equiv \left(\frac{\partial U}{\partial S}\right)_{V,N_i}$$

Although this may seem like a rather strange way to define the absolute temperature, we will see that this definition does in fact exactly conform with our experience regarding temperature.[11]

The above three postulates are sufficient to obtain almost all of the results of thermodynamics. In other words, these postulates may be used to establish formal relations, not only between U and S, but also between all other thermodynamic functions that may be useful in describing processes carried out under various conditions, including those involving chemical reactions as well as various sorts of exchanges of work and/or heat between a system and its surroundings (as we will see).

The following final postulate is equivalent to the third law of thermodynamics.

> ***Postulate IV.*** *The entropy of any single-component equilibrium system vanishes when its absolute temperature approaches zero.*

Note that this postulate implies that entropy, like T, is invariably positive and has a natural reference value with respect to which it may be measured. Strictly speaking, the above postulate only applies to systems that have a single (nondegenerate) quantum mechanical lowest energy (ground) state. However, for all practical purposes, Postulate IV may be

---

[11] Another surprising consequence of the above postulates is that they also imply that S is a concave function of U (and/or V and/or N_i). In other words, the second derivative of S with respect to U (or V or N_i) is necessarily negative. This may be demonstrated by again considering the process of removing an internal constraint from the system. Since removal of such a constraint must increase the system's entropy (or leave it unchanged), it follows that S cannot be a convex function of U (or its other extensive variables), and so U must be a convex function of S. Although the details of this proof are not very difficult, its demonstration is left as a challenge to the reader.

considered to apply to all systems (since any physically reasonable degree of ground-state degeneracy, combined with Eq. 2.9, would produce an entropy that is vanishingly small in comparison with that of the system at high temperature).

3.2 Temperature and Thermal Equilibrium

The fundamental equation and postulates of thermodynamics may be used to better understand what drives all systems toward a state of equilibrium. For example, Figure 3.1 illustrates a situation in which two subsystems, A and B, which may initially have different temperatures, are separated by a thermally conducting partition (such as a copper plate). We know from experience that the two subsystems will tend to transfer heat from the hotter side to the colder side, until the two sides equilibrate to the same temperature. The following analysis shows how this behavior arises as a natural consequence of the second law (or Postulate II).

For simplicity, we assume that each subsystem has a constant volume and number of moles, and that the outside of the container is sealed and thermally insulated. In other words, everything inside the big box (consisting of A and B) is *isolated*, and so Postulate II requires that the total entropy of A and B must approach a maximum at equilibrium. For bookkeeping purposes, we will treat subsystem A as the *system*, and subsystem B as the *surroundings*, so that a positive exchange of energy corresponds to energy flowing from B to A.

Because entropy is an extensive property, the total entropy is the sum of the entropy of A and B.

$$S = S_A + S_B$$

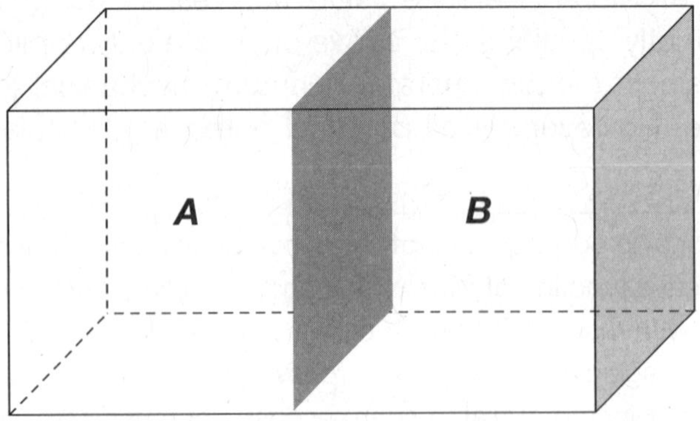

FIGURE 3.1 Two closed, constant-volume subsystems are separated by a thermally conducting partition. The second law requires that at thermal equilibrium the temperatures of the two subsystems will necessarily become equal ($T_A = T_B$).

We may use the fundamental equation, Eq. 3.12, to relate an infinitesimal change in the entropy of each side to the corresponding change in energy.[12]

$$dS_A = \left(\frac{\partial S_A}{\partial U_A}\right)_{V_A,N_A} dU_A$$

$$dS_B = \left(\frac{\partial S_B}{\partial U_B}\right)_{V_B,N_B} dU_B$$

Thus, the total entropy change is the sum of the above two terms.

$$dS = dS_A + dS_B = \left(\frac{\partial S_A}{\partial U_A}\right)_{V_A,N_A} dU_A + \left(\frac{\partial S_B}{\partial U_B}\right)_{V_B,N_B} dU_B$$

Since energy is conserved, we may equate $dU_A = -dU_B = dU$, where dU is an infinitesimal amount of energy that is transferred from B (the surroundings) to A (the system). We may also use Eq. 3.14 (or Eq. 3.9) to replace the two partial derivatives in the above equation by $1/T_A$ and $1/T_B$, respectively. Thus, the above equation can be reexpressed in the following form.

$$dS - \left(\frac{1}{T_A}\right) dU - \left(\frac{1}{T_B}\right) dU - \left[\left(\frac{1}{T_A}\right) - \left(\frac{1}{T_B}\right)\right] dU$$

Moreover, since V_A, N_A, V_B, N_B are all constants, there can be no mechanical or chemical work exchange between the system and its surroundings, and so $dU = \delta Q$.

$$dS = \left[\left(\frac{1}{T_A}\right) - \left(\frac{1}{T_B}\right)\right] \delta Q \tag{3.19}$$

We may now invoke Postulate II, which requires that the entropy of an isolated system is maximized at thermal equilibrium. This implies that at equilibrium the derivative of S will be equal to zero, and so an infinitesimal (but nonzero) exchange of heat $\delta Q \neq 0$ will produce no change in entropy $dS = 0$. In other words, at thermal equilibrium $dS/\delta Q = 0$ or

$$dS = 0 \times \delta Q = \left[\left(\frac{1}{T_A}\right) - \left(\frac{1}{T_B}\right)\right] \delta Q \tag{3.20}$$

which can only be true if

$$\left(\frac{1}{T_A}\right) - \left(\frac{1}{T_B}\right) = 0$$

or

$$T_A = T_B$$

---

[12] Since the volume and number of molecules on each side are both constant, $dV_i = 0$ and $dN_i = 0$, and so only the first term on the right-hand side of Eq. 3.12 contributes to dS_i.

Thus, we have proven that Postulate II implies that the temperature of the two subsystems in Figure 3.1 must be equal to each other at thermal equilibrium. However, it is important to keep in mind that the above derivation required that the partition between the two subsystems is thermally conductive. If the partition were not thermally conductive (if it were composed of a thermally insulating material), then the two subsystems need not have the same temperature at equilibrium.

A more detailed description of the above heat transfer process may be performed by assuming that the two subsystems are initially separated by an insulating partition and have *different equilibrium temperatures*. Once the insulating partition is replaced by a thermally conductive partition, then heat will flow between the two subsystems, and will continue to do so until the temperatures of the two subsystems become equal.

We can use the second law (Postulate II) to verify that heat will flow from the hotter to the colder subsystem. For example, if A (the system) is initial at a slightly higher temperature than B (the surroundings), then $T_A > T_B$ and $(1/T_A) - (1/T_B) < 0$. Since the total entropy can only increase during any spontaneous process, we know that $dS > 0$. The latter two inequalities, combined with Eq. 3.19, clearly require that $\delta Q < 0$. Such a negative heat exchange implies that heat leaves the system. In other words, if $T_A > T_B$, then heat will flow out of the higher-temperature system (A) and into the lower-temperature surroundings (B).[13]

More generally, in any system with a nonuniform temperature, heat will always spontaneously tend to flow from higher to lower temperatures. The only way we could force heat to flow against a thermal gradient is by doing work. For example, we could use an electrical refrigerator, heat pump, or air conditioner to pump heat out of a cold system into its warmer surroundings (as described when discussing the Carnot cycle on pages 74–76).

3.3 Chemical and Phase Equilibria

Chemical Equilibria and the Chemical Potential

The same procedure we used to show that temperature must become uniform at thermal equilibrium can also be used to obtain relationships between other intensive variables. The most important intensive variable governing chemical equilibria is the chemical potential (defined by Eqs. 3.11 or 3.16). Now we will see that the second law predicts that chemical reactions are driven toward a state in which the chemical potentials of the reactant and product species are equal to each other.

---

[13] A similar argument may be used to show that if $T_A < T_B$, then heat would flow into the lower-temperature system (A) from the higher-temperature surroundings (B).

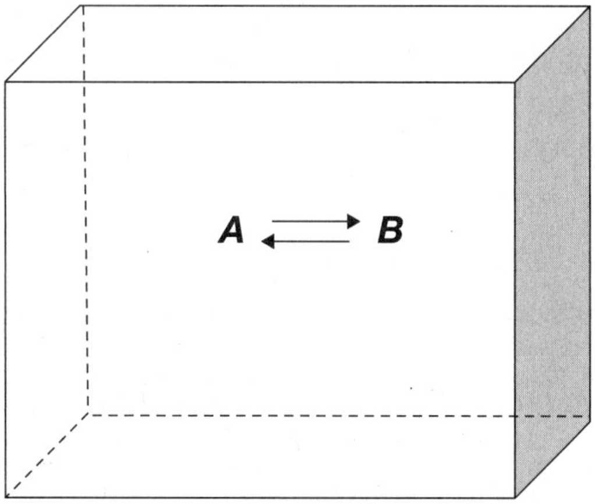

FIGURE 3.2 A closed, insulated, constant-volume system contains molecules that can interconvert between two different chemical states (A and B). The second law requires that the chemical potentials of the two compounds must become equal ($\mu_A = \mu_B$) at equilibrium.

Figure 3.2 illustrates a system in which one chemical component (A) may react to form another component (B). For example, A and B could be the gauche and trans conformational isomers of n-butane, or they could be the native and denatured states of a protein. Although the system may in general contain other nonreacting species (such as a solvent), we will focus attention only on the two components of interest. We will also assume that the container that holds the system is insulated and sealed (so neither heat nor matter can enter or leave), and that the volume of the container is constant. Thus, no work or heat exchange can take place between the system and its surroundings (so it is isolated).

Now imagine that we initially put some amounts of A and B into the system that are not equal to their equilibrium concentrations. The entropy representation of the fundamental equation (Eq. 3.12) implies that the total entropy change may be expressed as follows.

$$dS = \left(\frac{\partial S}{\partial U}\right)_{V,N_A,N_B} dU + \left(\frac{\partial S}{\partial V}\right)_{U,N_A,N_B} dV + \left(\frac{\partial S}{\partial N_A}\right)_{U,V,N_B} dN_A + \left(\frac{\partial S}{\partial N_B}\right)_{U,V,N_A} dN_B$$

Since the energy and volume of the system are constant $dU = dV = 0$, the first two terms are necessarily equal to zero. Equation 3.16 may be used to replace the last two partial derivatives by $-\mu_A/T_A$ and $-\mu_B/T_B$, and since the temperature of the two components must be equal at thermal equilibrium, we may substitute $T_A = T_B = T$. Moreover, stoichiometric balance requires that $dN_B = -dN_A = dN$ (where dN represents the number of molecules of type B that are formed at the expense of molecules of type A). Thus, the fundamental equation in this case reduces to

$$dS = \left[\left(\frac{\mu_A}{T}\right) - \left(\frac{\mu_B}{T}\right)\right] dN$$

At equilibrium, entropy will be maximized, and so $dS/dN = 0$ or $dS = 0 \times dN$, which implies that $\mu_A/T - \mu_B/T = 0$. Thus, *the chemical potentials of the reactant and product must be exactly equal at equilibrium*.

$$\mu_A = \mu_B \tag{3.21}$$

Equation 3.21 is an example of a much more general result, which holds for any chemical process carried out in any system. For example, we will later see that Eq. 3.21 also holds in a system held in a container of constant pressure, rather than constant volume, or even in an open system in which the exchange of molecules may take place between the system and the surroundings. Even more generally, chemical potentials will necessarily balance at equilibrium in systems of arbitrary complexity, whose chemical constituents may reside in different phases (such as a gas, liquid, or solid), or in different locations on a surface, or inside of cavities in a porous material, or within the binding pocket of a protein. In any such circumstances the chemical potentials of each chemical species will be uniform throughout the system at equilibrium (just as the temperature is uniform throughout a thermally equilibrated system).

The above result may also be extended to chemical reactions involving more complex stoichiometry,

$$\sum_i^{\text{Reactants}} a_i A_i \rightleftharpoons \sum_i^{\text{Products}} b_i B_i \tag{3.22}$$

where a_i and b_i are stoichiometric coefficients for the reactant A_i and product B_i species, respectively. For any such reaction one may use the following equation to express the relationship between the chemical potentials of the reacting species at equilibrium.

$$\sum_i^{\text{Reactants}} a_i \mu_i = \sum_i^{\text{Products}} b_i \mu_i \tag{3.23}$$

It is sometimes convenient to express Eq. 3.23 even more compactly by defining all the stoichiometric coefficients as ν_i, some of which are positive and equal to b_i, and others of which are all negative and have magnitudes that are equal to a_i. With this definition we may rewrite Eq. 3.23

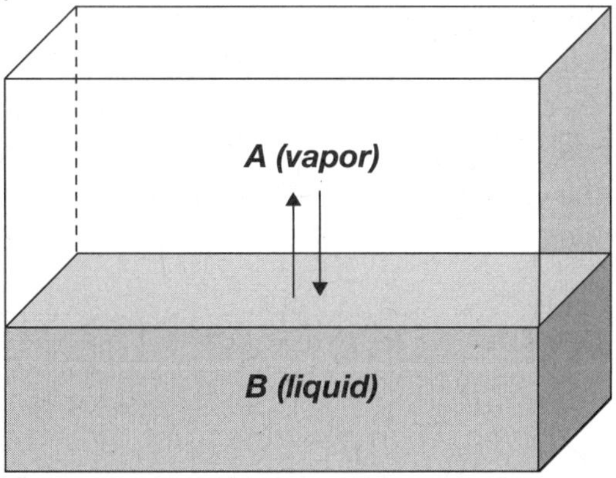

FIGURE 3.3 A closed, insulated, constant-volume system contains two phases (A and B). The second law requires that at equilibrium the temperature, pressure, and chemical potentials of the two phases will necessarily become equal ($T_A = T_B$, $P_A = P_B$, and $\mu_A = \mu_B$).

as $\sum b_i\mu_i - \sum a_i\mu_i = \sum \nu_i\mu_i = 0$ and so the equilibrium condition may be expressed in the following compact form.

$$\boxed{\sum_i \nu_i\mu_i = 0} \quad \text{at equilibrium} \tag{3.24}$$

Phase Equilibrium

Consider a system composed of two subsystems whose boundary allows the exchange of molecules, as well as heat and mechanical work. For example, one subsystem could be a vapor phase A and the other a liquid phase B, as shown in Figure 3.3. In such a situation, we may again invoke the second law to obtain general relations between all of the intensive variables of the two subsystems at equilibrium.

For simplicity we will assume there is only one chemical component in the system (such as water). In other words, we assume that the number of water molecules in the vapor phase is N_A and in the liquid phase is N_B. Note that when molecules are exchanged between the two phases, the volumes V_A and V_B, as well as the energies U_A and U_B, of each phase will also change. We further assume that both phases are enclosed by a sealed and insulated container of fixed total volume (so the entire system is isolated from its surroundings). When we apply the fundamental equation to each phase, we obtain the following expressions.

$$dS_A = \left(\frac{1}{T_A}\right) dU_A + \left(\frac{P_A}{T_A}\right) dV_A - \left(\frac{\mu_A}{T_A}\right) dN_A \tag{3.25}$$

$$dS_B = \left(\frac{1}{T_B}\right) dU_B + \left(\frac{P_B}{T_B}\right) dV_B - \left(\frac{\mu_A}{T_A}\right) dN_B \tag{3.26}$$

The entropy change of the entire system is again $dS = dS_A + dS_B$. Since the total energy, volume, and number of molecules are all conserved, we may equate $dU_A = -dU_B = dU$, $dV_A = -dV_B = dV$, and $dN_A = -dN_B = dN$. Thus, an infinitesimal entropy change of the whole system, resulting from the transfer of some molecules between phase A and phase B, may be expressed as follows.

$$dS = \left(\frac{1}{T_A} - \frac{1}{T_B}\right) dU + \left(\frac{P_A}{T_A} - \frac{P_B}{T_B}\right) dV - \left(\frac{\mu_A}{T_A} - \frac{\mu_B}{T_B}\right) dN \tag{3.27}$$

At equilibrium $dS = 0$ even when $dU \neq 0$ and/or $dV \neq 0$ and/or $dN \neq 0$, so all three quantities in parentheses must be equal to zero. This implies that all three of the following intensive variables of the two phases must be equal at equilibrium.

$$T_A = T_B$$
$$P_A = P_B$$
$$\mu_A = \mu_B$$

In other words, in any two-phase equilibrium, all of the intensive variables of the two phases – temperatures, pressures, and chemical potentials – must be in perfect balance with each other.

If the system contained more than one chemical species, then similar expressions would hold for each chemical species. Moreover, if the system contained molecules that can chemically react with each other, then Eq. 3.24 would impose an additional equilibrium condition that must hold for any such reactive species.

Gibbs Phase Rule

One of the elegant (and surprising) thermodynamic results that Willard Gibbs obtained in his 300-page paper entitled *On the Equilibrium of Heterogeneous Substances* came to be known as Gibbs phase rule. This could well have been called a law (rather than a rule) as it is an exact result that applies to *any* equilibrium system composed of any number of phases and chemical components. More specifically, Gibbs phase rule provides a remarkably simple expression for the total number of thermodynamic degrees of freedom f available to any chemical system, where Gibbs defined f as the number of intensive variables that one may independently vary in a system containing p separate phases and c different chemical components.[14]

[14] Note the term *degrees of freedom* when applied to f is not the same as the number of quadratic terms D in a system's Hamiltonian.

For example, if the system is a one-component monatomic ideal gas, then we know that its equation of state is specified by two intensive variables, such as the temperature T and pressure P. These two intensive variables may be used to determine any other intensive variables of the system, such as its number density $\rho = N/V = P/(k_B T)$ or energy density $U/V = (3/2)P$.[15] More generally, the equilibrium state of any single-component system is determined by specifying the values of any two of its intensive variables.[16]

A system that contains p phases and c chemical components has $p \times c$ additional variables, since one must specify the chemical composition of each phase. This suggests that such a system may have a total of as many as $pc + 2$ thermodynamic degrees of freedom. However, more careful consideration reveals that $pc + 2$ significantly overestimates the actual value of f. This is because not all of the pc variables are independent. First of all, the composition of each phase is determined by the mole fraction of each component in that phase $\chi_i \equiv N_i/N_T$ (where $N_T = \sum_i N_i$ is the total number of molecules in each phase). Since the mole fractions in each phase must add up to one, $\sum_i \chi_i = 1$, we only need to know $c - 1$ mole fractions in order to determine the mole fraction of the last component. This reduces the number of degrees of freedom to $p(c - 1) + 2$, but we are not done yet.

We also know that the chemical potentials of each component must be the same in all the p phases that are in equilibrium with each other. For a two-phase system (A and B) this would introduce one equation $\mu_A = \mu_B$, while for p phases there would be $p - 1$ such equations. Since the same is true for each component, there are a total of $c(p - 1)$ equations that interrelate the chemical potential of all the components in all the phases. Thus, the remaining number of independent degrees of freedom in a system with p phases and c components is equal to $[p(c - 1) + 2] - [c(p - 1)] = c - p + 2$.

---

[15] So, we could have equally well chosen N/V and U/V (or any other combination of two intensive variables) to determine all of the other intensive variables of the gas.

[16] The fundamental equation also makes it clear why other single-component systems can have no more than two degrees of freedom. Since we are concerned with intensive properties, the size of the system is arbitrary, so we may pick N to have any fixed value ($dN = 0$), and so Eq. 3.8 becomes $dU = TdS - P\,dV$. This expression indicates that the energy of the system depends on two variables (S and V), which implies that any one-component system cannot have more than two degrees of freedom. We could also have divided the entire equation by N, which is a constant and so $dU/N = d(U/N)$, etc., and thus obtain $d(U/N) = Td(S/N) - Pd(V/N)$, where S/N and $V/N = 1/\rho$ are another pair of intensive variables whose values would be sufficient to specify the state of the system.

This is how Gibbs obtained the phase rule, which determines the number of independent thermodynamic degrees of freedom f in any system with c components and p phases.[17]

$$\boxed{f = c - p + 2} \tag{3.28}$$

Exercise 3.2

How many thermodynamic degrees of freedom are available to a single-component, single-phase system such as pure argon gas (or pure liquid water)?

Solution. Since in any such system $c = 1$ and $p = 1$, Eq. 3.28 indicates that there are $f = 1 - 1 + 2 = 2$ thermodynamic degrees of freedom. This implies that one may independently vary both the pressure and the temperature of the system (over some range) without undergoing a phase transition (so that argon remains a gas or the liquid water remains a liquid).

The number of thermodynamic degrees of freedom is equivalent to the number of thermodynamic variables that may be independently varied. So, if a system has only one thermodynamic degree of freedom, then one may not independently vary both the temperature T and the pressure P of the system while maintaining the system in a state with the same number of phases (and components). For example, a single-component ($c = 1$), two-phase ($p = 2$) system has only one thermodynamic degree of freedom ($f = 1$), and so one may not independently vary both T and P. In other words, in any such system there must be a one-to-one functional relationship between T and P. This functional relationship may be represented by a curve in the $P - T$ plane, which is called a two-phase *coexistence curve*.

Figure 3.4 shows several such coexistence curves in the phase diagrams of water and carbon dioxide. These curves trace the locus of P and T

[17] Recall that Eq. 3.8 was obtained assuming that the only kind of mechanical work exchange is pressure-volume work. If we had allowed for additional kinds of mechanical work, such as work against magnetic, electric, or gravitational fields, then each such additional term in the fundamental equation would have introduced an additional degree of freedom. So, for example, if we introduce an electric field as a new external variable, then a single-component (single-phase) system would have three rather than two degrees of freedom, and thus the number 2 in Eq. 3.28 would be replaced by 3. Similarly, if we had confined ourselves to processes that occur at constant pressure, so only temperature variations are considered, then the number 2 in Eq. 3.28 should be replaced by the number 1.

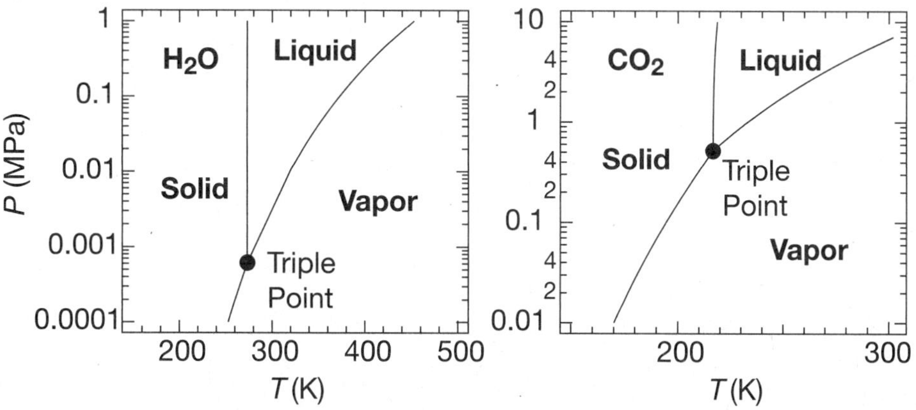

FIGURE 3.4 These temperature-pressure phase diagrams of water and carbon dioxide show several two-phase coexistence curves (liquid-vapor, solid-liquid, and solid-vapor) and triple points, at which the three phases simultaneously coexist. Note that although the solid-liquid boundary in the phase diagram of water appears to be perfectly vertical, it in fact has a small negative slope (and so it is possible to melt ice under high pressure at a constant temperature near 273 K). From E.W. Lemmon, M.O. McLinden, and D.G. Friend, "Thermophysical Properties of Fluid Systems" in NIST Chemistry WebBook, NIST Standard Reference Database Number 69, Eds. P.J. Linstrom and W.G. Mallard. National Institute of Standards and Technology, Gaithersburg MD, 20899, http://webbook.nist.gov.

values along which each of these single-component systems can coexist in two separate phases: vapor-liquid, solid-liquid, or solid-vapor. Note that at ambient pressure (0.1 MPa), the phase diagram of water crosses two coexistence curves as the temperature of the system is increased (as water melts and then boils), while carbon dioxide only crosses one coexistence curve (corresponding to the sublimation of dry ice to CO_2 vapor). The solid phase of water that appears in Figure 3.4 is hexagonal ice (Ih). Many other forms of ice (with different crystal structures) exist at higher pressures (and lower temperatures). Each of these different solid phases of ice are separated from neighboring phases by coexistence curves.

The phase diagrams in Figure 3.4 also indicate the locations of triple points, at which the solid, liquid, and vapor phase all simultaneously coexist. The reason that this can only happen at a single isolated point in the phase diagram (rather than along a curve) is again explained by Gibbs phase rule (Eq. 3.28). In other words, when a single-component system has three phases simultaneously in equilibrium, then $f = 1 - 3 + 2 = 0$. The fact that there are no degrees of freedom remaining means that one cannot vary either the temperature or the pressure and maintain three coexisting phases. In other words, three phases of H_2O (or CO_2) can only coexist at isolated points in the temperature-pressure phase diagram.

The phase rule can also be used to understand what sorts of structures may be found in the phase diagrams of systems containing several chemical components. For example, the phase diagram of a two-component system may be represented as a three-dimensional function of P, T, and the mole fraction χ (of either one of the components). The phases in such a system will be separated by coexistence surfaces (rather than curves), and these surfaces may intersect to describe curves at which three phases can coexist, and points at which four phases can coexist.

Chemically reactive multicomponent chemical systems require special consideration. Each chemical equilibrium imposes an additional relation (Eq. 3.24), which further reduces the number of degrees of freedom of the system. As a result, the actual number of *independent* components in a system that can undergo r different equilibrium reactions is $c-r$. So, applying Eq. 3.28 to chemically reactive systems requires replacing c by the number of *independent* components, $c-r$, and so the phase rule for such a system becomes $f=c-r-p+2$.

Gibbs phase rule is essentially geometrical in origin, as it expresses the fact that the equilibrium phase diagram of any material must contain single-phase regions separated by boundaries between neighboring phases, and these boundaries may intersect to form lower-dimensional shapes. The eighteenth-century mathematician Leonhard Euler[18] (whom we will shortly reencounter), obtained a famous relation pertaining to any polyhedron,[19] which is remarkably reminiscent of Gibbs phase rule. If we designate the number of vertices, edges, and faces in a given polyhedron by the letters f, c, and p, respectively, then Euler's formula may be expressed as $f=c-p+2$, as illustrated in Figure 3.5.

---

[18] Euler (pronounced "oiler") was born in Switzerland, where he obtained a PhD in the mathematics of sound propagation at the age of 19, but spent much of his subsequent academic career in Russia and Germany. In his early twenties, Euler wrote an article entitled *Meditations Upon Experiments Made Recently on the Firing of Canon*, in which he first suggested that the letter e be used to represent the natural logarithm base (2.718..., an irrational number, like π). When he was 29 years old, he demonstrated that $e^{i\theta}=\cos\theta+i\sin\theta$ (Euler's formula), which leads to the remarkable identity $e^{i\pi}=-1$ linking e, π, i, and -1 (where $i\equiv\sqrt{-1}$). In 1778, when Euler was 68 years old and nearly blind, he produced an average of one mathematical paper every week. His influence on the later development of mathematics was so extensive that a bewildering array of theorems, equations, identities, formulas, functions, and conjectures are associated with his name. A generation later, the great French mathematician Pierre-Simon Laplace is reported to have said "read Euler, read Euler, he is the master of us all."

[19] A polyhedron is a three-dimensional solid enclosed by flat faces.

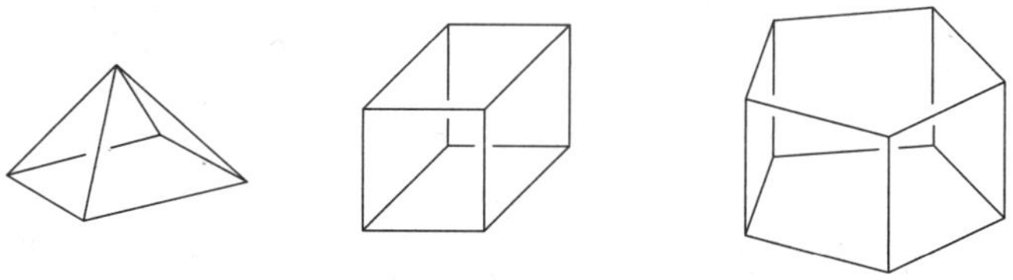

FIGURE 3.5 The number of vertices (f), edges (c), and faces (p) in each of the above polyhedra are related by Euler's formula, $f = c - p + 2$, which has exactly the same form as Gibbs phase rule (reflecting the fact that both formulas are essentially geometrical in origin).

3.4 Euler and Gibbs-Duhem Relations

The fact that the energy of any system can be expressed as a function of the extensive variable S, V, and N_i implies that the function U scales with the size of the system. In other words, if S, V, and N_i are all multiplied by some factor λ, then U will also change by the same factor.

$$U(\lambda S, \lambda V, \lambda N_i) = \lambda U(S, V, N_i) \tag{3.29}$$

Any mathematical function with the above property is called a homogeneous function (of first order). Leonhard Euler[20] proved that any such functions must also have the following special property.[21]

$$U = \left(\frac{\partial U}{\partial S}\right)_{V,N_i} S + \left(\frac{\partial U}{\partial V}\right)_{S,N_i} V + \sum_i \left(\frac{\partial U}{\partial N_i}\right)_{S,V,N_j \neq N_i} N_i \tag{3.30}$$

Note that this equation looks somewhat similar to the fundamental equation, except that dU, dS, dV, and dN_i are replaced by U, S, V, and N_i. Also, note that we may make use of the definitions in Eqs. 3.9 – 3.11 to write the above equation as follows.

$$\boxed{U = TS - PV + \sum_i \mu_i N_i} \tag{3.31}$$

[20] See footnote 18 on page 102.

[21] The proof of Eq. 3.30 follows directly from taking the derivative of Eq. 3.29 with respect to λ. Since $[d(\lambda U)/d\lambda] = U = [\partial(\lambda U)/\partial(\lambda S)][\partial(\lambda S)/\partial\lambda] + [\partial(\lambda U)/\partial(\lambda V)][\partial(\lambda V)/\partial\lambda] + \ldots = (\partial U/\partial S)S + (\partial U/\partial V)V + \ldots$, where the partial derivatives are each performed with all other variables held constant.

This important thermodynamic identity, known as the *Euler relation*, may also be expressed in the following entropy representation.

$$\boxed{S = \frac{U}{T} + \frac{PV}{T} - \sum_i \frac{\mu_i N_i}{T}} \tag{3.32}$$

If we evaluate the differential of Eq. 3.31, using the usual rule for calculating the derivatives of sums and products, we obtain the following additional identity.

$$dU = TdS + SdT - PdV - VdP + \sum_i \mu_i N_i + \sum_i N_i d\mu_i \tag{3.33}$$

This expression appears at first glance to be inconsistent with the fundamental equation of thermodynamics, Eq. 3.8. However, Eq. 3.33 is a necessary consequence of Eq. 3.29 and so must also be correct. The only way out of this apparent dilemma is to recognize that both Eqs. 3.8 and 3.33 can be simultaneously correct, if the following is also true.

$$\boxed{SdT - VdP + \sum_i N_i d\mu_i = 0} \tag{3.34}$$

This surprising identity, known as the *Gibbs-Duhem relation*, may be used to derive other important thermodynamic results.[22]

Consider, for example, a single-component system, for which the Gibbs-Duhem relation may be rearranged to the following form.

$$d\mu = \left(\frac{V}{N}\right) dP - \left(\frac{S}{N}\right) dT \tag{3.35}$$

Notice that this expression implies that $V/N = (\partial\mu/\partial P)_T$ and $S/N = -(\partial\mu/\partial T)_P$ for any single-component system.[23] If we further assume that the temperature is held constant (so $dT = 0$), then we may integrate both

[22] Pierre Maurice Marie Duhem (1861–1916) was a French physicist, mathematician, and philosopher of science. His doctoral thesis focused on the Gibbs's chemical potential and its relation to chemical reaction equilibria. His results contradicted those of a more senior scientist, Marcellin Berthelot. Although Duhem was right, Duhem's thesis was rejected, apparently as a result of Berthelot's influence.

[23] We will also later see that very similar relations may be obtained for multicomponent mixtures if we replace V/N by the partial molar volume, $\bar{v}_i \equiv (\partial V/\partial N_i)_{T,P,N_j \neq N_i} = (\partial\mu_i/\partial P)_T$ and replace S/N by the partial molar entropy, $\bar{s}_i \equiv (\partial S/\partial N_i)_{T,P,N_j \neq N_i} = -(\partial\mu_i/\partial T)_P$.

sides of Eq. 3.35 to determine how the chemical potential depends on pressure.

$$\int d\mu = \mu - \mu^0 = \int_{P_0}^{P} \left(\frac{V}{N}\right) dP \qquad (3.36)$$

Note that μ^0 is the chemical potential at the reference pressure P_0, while μ is the chemical potential at the pressure P.[24]

Exercise 3.3

Use the Gibbs-Duhem relation to verify that $\Delta\mu = -T\Delta S$ for any isothermal ideal gas process.

Solution. The ideal gas law Eq. 2.13 may be used to replace $V/N = k_B T/P$ in Eq. 3.36 and then integrate from P_0 to P (at constant temperature) to obtain $\Delta\mu = k_B T[\ln P - \ln P_0] = k_B T \ln(P/P_0) = k_B T \ln(V_0/V) = -k_B T \ln(V/V_0) = -T\Delta S$. Note that we used the ideal gas law (again) to equate $P/P_0 = [nRT/V]/[nRT/V_0] = V_0/V$, and then used Eq. 2.23 to equate $k_B \ln(V/V_0) = \Delta S$.

The intermediate result obtained in the above exercise implies that $\Delta\mu = \mu - \mu^0 = -k_B T \ln(V/V_0)$, or equivalently $\mu = \mu^0 - k_B T \ln(V/V_0)$. When combined with the ideal gas law (Eq. 2.14), this leads to the following important expressions.

$$\boxed{\mu = \mu^0 + k_B T \ln\left(\frac{P}{P_0}\right)} \qquad (3.37)$$

$$\boxed{\mu = \mu^0 + k_B T \ln\left(\frac{\rho}{\rho_0}\right)} \qquad (3.38)$$

$$\boxed{\mu = \mu^0 + k_B T \ln\left(\frac{[c]}{[c]_0}\right)} \qquad (3.39)$$

The latter two identities (where $\rho = N/V$ and $[c] = n/V = \rho/\mathcal{N}_A$) are more generally applicable, as they also describe the dependence of a solute's

---

[24] This equation is the basis of the Maxwell equal area method, which Boltzmann used to determine the liquid-vapor coexistence curve from the mathematical equation of state that had recently been proposed by van der Waals. More specifically, since the chemical potential of the vapor and liquid molecules must be equal at equilibrium, Eq. 3.36 implies that $\mu_{vap} - \mu_{liq} = \int_{vap}^{liq} (V/N) dP = 0$. This, combined with the fact that the pressures of the two phases must also be equal when they are in equilibrium, may be used to determine not only how P depends on T along the coexistence curve, but also the vapor and liquid densities, ρ_{vap} and ρ_{liq}.

chemical potential on concentration in any real gas, liquid, or solid solution that is sufficiently dilute so that interactions between solute molecules may be neglected.[25]

3.5 Transformed Potential Functions

Legendre Transformations

Although energy plays a special role in thermodynamics, other thermodynamic energy functions (or potentials) play equally important roles in various thermodynamic applications. These other thermodynamic potentials may be derived from $U(S, V, N_i)$ using a mathematical procedure called Legendre transformation.[26] This procedure is used to exchange the natural variables of U (S, V, and N_i) for other natural variables (such as T, P, and/or μ_i). We will later see that there is a close connection between the natural variables of a thermodynamic potential and the driving force that pushes chemical processes toward their equilibrium state.

---

[25] Note that if both μ and μ^0 were expressed in molar units, then the first of the above expressions would become $\mu = \mu^0 + RT \ln\left(\frac{P}{P_0}\right)$. Similarly, the other expressions would also look identical when expressed in molar units, except that $k_B T$ would be replaced by RT. One also often encounters Eqs. 3.37–3.39 written without explicitly including the reference state pressure P_0, number density ρ_0, or concentration $[c]_0$. When this is done, then it is understood that the reference state has a value of one (in whatever units are used to describe P, ρ, or $[c]$). In other words, if it is assumed that $P_0 = 1$ when describing gas-phase chemical processes, then Eq. 3.37 may be expressed as $\mu = \mu^0 + k_B T \ln P$. Similarly, if it is assumed that $[c]_0 = 1$ when describing solution-phase chemical processes, then Eq. 3.39 may be expressed as $\mu = \mu^0 + k_B T \ln [c]$. However, it is always important to keep in mind that μ^0 represents the chemical potential at the reference state. So, when P_0 does not appear in Eq. 3.37, then μ^0 implicitly pertains to the state at which $P_0 = 1$. Similarly, when $[c]_0$ does not appear in Eq. 3.39, then μ^0 implicitly pertains to the state at which $[c]_0 = 1$.

[26] Adrien-Marie Legendre was a French mathematician (1752–1833) who was born into a wealthy family and was elected to the French Academy of Sciences when he was 30 years old. A few years later he was elected to the Royal Society of London. But then came the French Revolution. Legendre described his experience during those years as follows, in a letter to a young mathematician friend, Karl Gustav Jacob Jacobi: "I married following a bloody revolution that had destroyed my small fortune; we had great problems and some very difficult moments, but my wife staunchly helped me to put my affairs in order little by little and gave me the tranquility necessary for my customary work and for writing new works which have steadily increased my reputation."

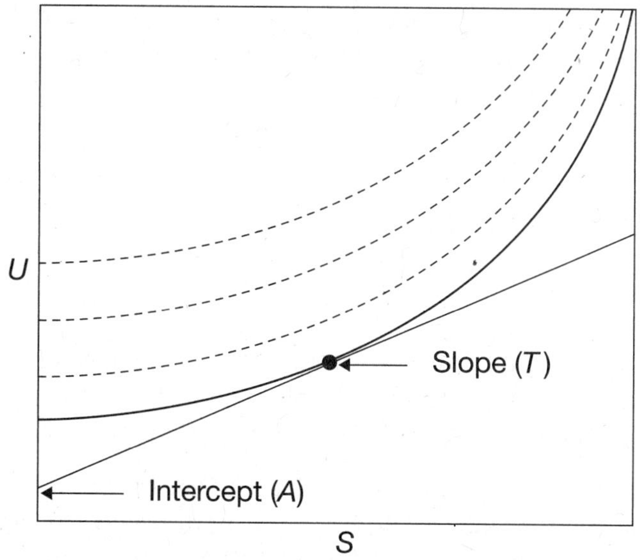

FIGURE 3.6 Transforming the function $U(S)$ into a new function in which the variable S is replaced by the variable $T = (\partial U/\partial S)_{V,N_i}$ requires eliminating the redundancy inherent in the fact that an infinite family of functions (dashed curves) have the same slope as $U(S)$. However, if both the slope and the intercept of U are incorporated into the transformation, then the new function will retain all of the information contained in U.

The difficulty associated with transforming from a variable such as S to the variable $T = (\partial U/\partial S)_{V,N_i}$ is that T only specifies the slope of U with respect to S, as illustrated in Figure 3.6. However, we may produce a unique transformation of U if we define a new function that describes how the intercept (A) shown in Figure 3.6 depends on the corresponding slope (T). Since every point along U is associated with a unique intercept and slope, the way that A depends on T may be used to uniquely describe the transformation of $U(S)$ to a new potential function $A(T)$.

More specifically, a simple geometric analysis, guided by Figure 3.6, may be used to show that the following *Legendre transformation* correctly describes how $A(T)$ is related to $U(S)$, or more generally how $A(T, V, N_i)$ is related to $U(S, V, N_i)$.

$$A(T, V, N_i) = U(S, V, N_i) - \left(\frac{\partial U}{\partial S}\right)_{V,N_i} S$$

$$A = U - TS \tag{3.40}$$

This new state function, which is called the Helmholtz energy, is also referred to as the Legendre transform of U whose natural variables are T, V, and N_i.

The same procedure can be used to generate many other thermodynamic potentials. One important example is the enthalpy $H(S,P,N_i)$, which is obtained by performing a Legendre transformation of U to replace the variable V by $(\partial U/\partial V)_{S,N_i} = -P$ (see Eq. 3.10).

$$H(S,P,N_i) = U(S,V,N_i) - \left(\frac{\partial U}{\partial V}\right)_{S,N_i} V$$

$$H = U - (-P)V = U + PV \tag{3.41}$$

Note that performing a Legendre transformation always requires subtracting a quantity that is equal to the partial derivative of the old function (with respect to the old variable) multiplied by the old variable.

Gibbs energy $G(T,P,N_i)$ may be obtained either by performing two Legendre transformations of $U(S,V,N_i)$ or by performing a single Legendre transformation of either $A(T,V,N_i)$ or $H(S,P,N_i)$. For example, if we start with H, we may obtain G by transforming the variable S to $(\partial U/\partial S)_{V,N_i} = T$.

$$G(T,P,N_i) = H(S,P,N_i) - \left(\frac{\partial U}{\partial S}\right)_{V,N_i} S$$

$$\begin{aligned} G &= H - TS \\ &= A + PV \\ &= U + PV - TS \end{aligned} \tag{3.42}$$

It is easy to show that the same result can also be obtained by starting with A and transforming the variable V to $(\partial U/\partial V)_{S,N_i} = -P$.

Exercise 3.4

Legendre transformations can be performed equally well in the opposite direction. Show that the Legendre transformation of $A(T,V,N_i)$ to $U(S,V,N_i)$ implies that $(\partial A/\partial T)_{V,N_i} = -S$.

Solution. Changing the independent variable from T to S is equivalent to performing a Legendre transformation of T to $(\partial A/\partial T)_{V,N_i}$.

$$U(S,V,N_i) = A(T,V,N_i) - \left(\frac{\partial A}{\partial T}\right)_{V,N_i} T$$

Notice that we may also rearrange Eq. 3.40 to obtain $U = A + TS$. Comparison of these two identities implies that $(\partial A/\partial T)_{V,N_i} = -S$.

Legendre transformation may also be used to change one or more of the independent variables N_i to $(\partial U/\partial N_i)_{S,V,N_j \neq N_i} = \mu_i$. This can be useful in correctly describing biochemical processes in which the pH is held constant by buffering agents. This is because using a buffer to maintain a constant pH is equivalent to holding the chemical potential of dissolved protons constant, rather than holding the total number of protons constant. More generally, one may also use Legendre transformations to obtain new functions from $S(U, V, N_i)$ in which one or more of the variables are replaced by the corresponding derivative of S.

Special Features of Potential Functions

The potential functions generated by performing various Legendre transformations of $U(S, V, N_i)$ each have special features that make them uniquely useful. For example, we will see that there is an intimate connection between enthalpy and heat, as well as between Helmholtz energy and work. Moreover, Gibbs energy is related in a special way to chemical potentials and the amount of useful (chemical) work that may be extracted from chemical reactions. More generally, each of the thermodynamic potentials obtained by performing a Legendre transformation on $U(S, V, N_i)$ are particularly useful when treating situations in which the natural variables of the potential are also those that are experimentally held fixed. Thus, $A(T, V)$ is useful for describing chemical processes that occur at constant T and V. Similarly, $G(T, P)$ is useful for describing chemical processes that take place under constant T and P conditions.

In order to understand the connection between enthalpy and heat, we must first obtain the fundamental equation for dH. Since, $H \equiv U + PV$, the total differential of the enthalpy is $dH = dU + PdV + VdP$. This may be combined with $dU = TdS - PdV + \sum_i \mu_i dN_i$ to obtain the following enthalpy representation of the fundamental equation.

$$\boxed{dH = TdS + VdP + \sum_i \mu_i dN_i} \tag{3.43}$$

Notice that the second equality implies that $H(S, P, N_i)$, so Eq. 3.43 is the Legendre transform of the fundamental equation in which the variable V is replaced by P.

In a *closed system* at *constant pressure*, the last two terms in Eq. 3.43 are equal to zero and so

$$dH = TdS = \delta Q \tag{3.44}$$

In other words, the enthalpy change associated with any process that is carried out under these conditions is exactly equal to the heat exchange that would occur if the process were carried out reversibly.

A similar procedure may be used to obtain the fundamental equation pertaining to the Helmholz energy. More specifically, from $A \equiv U - TS$ we obtained $dA = dU - TdS - SdT$, and then substitute

$dU = TdS - PdV + \sum_i \mu_i dN_i$ to obtain the following Helmholtz representation of the fundamental equation.

$$\boxed{dA = -SdT - PdV + \sum_i \mu_i dN_i} \tag{3.45}$$

In a *closed system* at *constant temperature*, the first and last terms on the right-hand side are equal to zero, which indicates that dA is equivalent to the amount of mechanical (pressure-volume) work that would be exchanged if such a process were carried out reversibly.

$$dA = -PdV = \delta W_{PV} \tag{3.46}$$

For a more general *isothermal* process in an *open system*, there may be an additional exchange of chemical work between the system and surroundings $\sum_i \mu_i dN_i = \delta W_N$.

$$dA = -PdV + \sum_i \mu_i dN_i = \delta W_{PV} + \delta W_N = \delta W \tag{3.47}$$

In other words, *the Helmholtz energy change for any isothermal process is equivalent to the associated total reversible work exchange*. Thus, A is sometimes referred to as the *reversible work function*.

The same strategy may be used to understand why Gibbs energy plays such a central role in chemistry. Since $G \equiv U - TS + PV$, its total differential is $dG = dU - TdS - SdT + PdV + VdP$. If we again introduce the fundamental equation for dU, we obtain the following Gibbs representation of the fundamental equation (which is the Legendre transform of Eq. 3.8 expressed in terms of the variables T, P, and N_i).

$$\boxed{dG = -SdT + VdP + \sum_i \mu_i dN_i} \tag{3.48}$$

At *constant temperature and pressure*, $dG = \sum_i \mu_i dN_i$, which implies that Gibbs energy determines the amount of chemical work that would be obtained from any such chemical process if it were carried out reversibly.

$$dG = \sum_i \mu_i dN_i = \delta W_N \tag{3.49}$$

We may also combine $G = U - TS + PV$ directly with the Euler relation $U = TS - PV + \sum_i \mu_i N_i$, to obtain the following important identity.

$$\boxed{G = \sum_i \mu_i N_i} \tag{3.50}$$

Notice that we have not placed any restriction (such as constant temperature or pressure, etc.) in obtaining Eq. 3.50, and so it is applicable to any and all systems. In other words, *Gibbs energy is equal to the sum of all the chemical potentials of the molecules in any system*.

Partial Molar Thermodynamic Functions

Each of the terms on the left-hand side of Eq. 3.50 may be expressed as $\mu_i N_i = (\partial G/\partial N_i)_{T,P,N_j \neq N_i} N_i = (\partial G/\partial n_i)_{T,P,n_j \neq n_i} n_i$ (where the last identity is obtained by replacing N_i by $n_i N_A$). The latter partial derivative $(\partial G/\partial n_i)_{T,P,n_j \neq n_i} \equiv \overline{G}_i$ is called the *partial molar Gibbs energy* of component i. Thus, Eq. 3.50 may be reexpressed in the following equivalent form.

$$\boxed{G = \sum_i n_i \overline{G}_i} \tag{3.51}$$

In other words, the Gibbs energy of any system is equal to the sum of the partial molar Gibbs energies of each of the chemical components of the system. The following alternative derivation of Eq. 3.51 makes it clear why similar expressions may also be obtained for other partial molar quantities.

Notice that since T and P are intensive variables, $G(T, P, n_i)$ is a homogeneous function of only the variables n_i. In other words, if we multiply the number of moles of each component in the system by any factor λ, while holding T and P constant, then the total Gibbs energy of the system will increase by exactly the same factor, $G(T, P, \lambda n_i) = \lambda G(T, P, n_i)$. Thus, the same mathematical identity that led to Eq. 3.30 also leads to Eq. 3.51.

$$G = \left(\frac{\partial G}{\partial n_1}\right)_{T,P,n_j \neq n_1} n_1 + \left(\frac{\partial G}{\partial n_2}\right)_{T,P,n_j \neq n_2} n_2 + \cdots = \sum_i n_i \overline{G}_i$$

The total volume of any system is also necessarily a homogeneous function of n_i, when the system is maintained at constant temperature and pressure, since the volume must double when the number of moles of all the components are doubled (at constant T and P). In other words, this implies that $V(T, P, \lambda n_i) = \lambda V(T, P, n_i)$, which leads to the following identity.

$$V = \left(\frac{\partial V}{\partial n_1}\right)_{T,P,n_j \neq n_1} n_1 + \left(\frac{\partial V}{\partial n_2}\right)_{T,P,n_j \neq n_2} n_2 + \cdots = \sum_i n_i \overline{V}_i \tag{3.52}$$

Thus, the volume of any system is the sum of the partial molar volumes of all of its chemical components. The same is true for any other extensive function X (where X could be U, H, S, etc.).

$$X = \sum_i \left(\frac{\partial X}{\partial n_i}\right)_{T,P,n_j \neq n_i} n_i = \sum_i n_i \overline{X}_i \tag{3.53}$$

Notice that such identities arise because T and P are *intensive* variables. The same would not be the case if the variables that were held constant were *extensive*.[27]

3.6 Other Sorts of Thermodynamic Work

The above fundamental equations pertain to systems in which only pressure-volume work $\delta W_{PV} = -P\,dV$ or chemical work $\delta W_N = \mu dN$ is considered. More generally, one may treat processes that involved other sorts of mechanical and electrochemical work by adding additional terms to the fundamental equation (as described in Section 3.1). In each case, the corresponding term in the fundamental equation has the form $\mathcal{F}d\chi$, where $\mathcal{F}$ is a generalized force and $d\chi$ is the associated generalized displacement.

For example, consider a system consisting of a rubber band that is stretched by applying a force (line tension) f, which increases the length of the rubber band by $d\ell$. The associated work could be included in the fundamental equation (Eq. 3.8) by adding an additional term equal to $fd\ell$. If we further assume that both the total volume and the chemical composition of the rubber band remain the same during the stretching process (so $dV = 0$ and $dN_i = 0$), then the fundamental equation reduces to the following form.

$$dU = TdS + fd\ell \tag{3.54}$$

Recall that $TdS = \delta Q$ is the reversible heat exchange between the system and surroundings, so if we stretch the rubber band reversibly and adiabatically (so $TdS = \delta Q = 0$), then the total energy change of the rubber band will be equal to the work that is done on the rubber band when it is stretched $dU = \delta W = fd\ell$.

If you try stretching a rubber band while holding it against your lip, you will notice that its temperature increases slightly when it is stretched (and decreases when it is relaxed). Thus, if we were to stretch the rubber band isothermally (while it was immersed in a constant temperature bath), then it would give off heat to the bath $\delta Q = TdS < 0$. This implies that the entropy of a rubber band *decreases* $dS < 0$ when it is stretched isothermally. In other words, the polymeric molecules (elastomers) in the rubber band have a lower entropy when they are all stretched in the same direction than they do when they are relaxed.

As another example, consider a film whose surface area increases by $d\mathcal{A}$ under the influence of a surface tension γ. We may include

[27] For example, the Helmholtz energy of a system is not an extensive function of n_i at constant temperature and volume and so $A \neq \sum_i (\partial A/\partial n_i)_{T,V,n_j \neq n_i} n_i$, but A is an extensive function of n_i at constant temperature and pressure, so $A = \sum_i (\partial A/\partial n_i)_{T,P,n_j \neq n_i} n_i = \sum_i n_i \overline{A}_i$.

the associated work in the fundamental equation for any such process.

$$dU = TdS + \gamma d\mathcal{A} \tag{3.55}$$

It is easy to see how we could use the same procedure to describe situations in which work is performed by lifting a system (of mass m) in a gravitational field, in which case $\delta W = mgdh$. We could also similarly describe work performed against a magnetic or electric field. The latter work plays an important role in electrochemistry.

Electrochemical Processes

Electrical work $\mathcal{E}d\mathcal{Q}$ is exchanged when a voltage $\mathcal{E}$ displaces a charge $d\mathcal{Q}$. Since the smallest unit of electric charge is the charge of one electron e, we may equate $\mathcal{Q} = Ne$, where N is the total number of fundamental charges in $\mathcal{Q}$. We may also express the charge in molar units as $\mathcal{Q} = n\mathcal{F}$, where $n = N/\mathcal{N}_A$ and $\mathcal{F} = \mathcal{N}_A e$, and $\mathcal{F}$ is Faraday's constant (which is equal to the magnitude of the charge of one mole of electrons).

An electrochemical system may contain various neutral and charged molecules. If we designate the charge (oxidation state) of a given molecule as Z, then $Ze\mathcal{E}dN$ is the electrical work associated with applying a voltage $\mathcal{E}$ to dN such molecules. We may incorporate this electrical work into the fundamental equation by summing over all the molecules in the system.

$$dU = TdS - PdV + \sum_i \mu_i dN_i + \sum_i Z_i e\mathcal{E}dN_i \tag{3.56}$$

Note that μ is the chemical potential of a given molecule before a voltage is applied and $Ze\mathcal{E}$ is the work associated with applying the voltage to that molecule. The sum of these two terms is referred to as the molecule's *electrochemical potential* $\tilde{\mu}$.

$$\boxed{\tilde{\mu} \equiv \mu + Ze\mathcal{E}} \tag{3.57}$$

We may now also combine the two sums in Eq. 3.56 to express the fundamental equation as follows.

$$\boxed{dU = TdS - PdV + \sum_i \tilde{\mu}_i dN_i} \tag{3.58}$$

Note the chemical and electrochemical potential are identical for any neutral molecule ($Z = 0$) since $\tilde{\mu} = \mu + 0e\mathcal{E} = \mu$. Moreover, the chemical and electrochemical potentials of any molecule or ion are equivalent to each other when no voltage is applied to the system ($\mathcal{E} = 0$) since $\tilde{\mu} = \mu + Ze\,0 = \mu$.

Various Legendre transformations may again be performed in order to change the independent variables S and V to the corresponding conjugate variables. For example, we may transform the independent variable S to T and V to P to obtain the following fundamental equation for the Gibbs energy of an electrochemical system.

$$dG = -SdT + VdP + \sum_i \tilde{\mu}_i dN_i \tag{3.59}$$

This indicates that the Gibbs energy change associated with any electrochemical process performed at constant T and P is equivalent to the corresponding electrochemical work $dG = \sum_i \tilde{\mu}_i dN_i$.

HOMEWORK PROBLEMS

Problems That Illustrate Core Concepts

1. Use Eqs. 2.12, 2.13, and 2.24 to obtain expressions for the following partial derivative for a *diatomic* ideal gas, assuming that the temperature of the system is sufficiently low that *only translational and rotational motions are thermally active*. Note that you may rewrite Eq. 2.24 by assuming that $V_1 = V_0$ and $T_1 = T_0$ are both constants, while $V_2 = V$ and $T_2 = T$ are variables.

(a) $\left(\frac{\partial U}{\partial T}\right)_V$ (b) $\left(\frac{\partial U}{\partial V}\right)_T$

(c) $\left(\frac{\partial V}{\partial P}\right)_T$ (d) $\left(\frac{\partial P}{\partial T}\right)_V$

(e) $\left(\frac{\partial S}{\partial V}\right)_T$

(f) Use the above results, combined with the chain rule, to obtain expressions for $(\partial U/\partial P)_V$, $(\partial U/\partial P)_T$, and $(\partial S/\partial P)_T$. Note that the chain rule can only be applied when the same variable is held constant in all of the partial derivatives.

2. Obtain expressions for the following partial derivatives for a diatomic ideal gas, assuming that its translational, rotational, and vibrational motions all behave classically.

(a) $\left(\frac{\partial S}{\partial V}\right)_T$ (b) $\left(\frac{\partial S}{\partial T}\right)_V$

3. Express the results in problems 1(a) and 2(b) in a form that is appropriate for a system whose Hamiltonian contains D thermally active (classical) quadratic terms (and note that the heat capacity of such a system is given by Eq. 1.44). Show that your result implies that $C_V = (\partial U/\partial T)_V$ can also be expressed as $T(\partial S/\partial T)_V$, and then use Eq. 2.5 to show that the latter identity is consistent with the fact that the constant-volume heat capacity of a system is defined as the amount of heat required to increase the temperature of a system by one degree, $C_V = (\delta Q/\partial T)_V$, in a system that is maintained at constant volume.

4. The heat capacity of a constant-pressure system is defined as $C_P = (\delta Q/\partial T)_P = T(\partial S/\partial T)_P$. Use Eq. 2.13 to convert Eq. 2.25 for $S(T, V)$ to the corresponding equation for $S(T, P)$, and use the latter equation to show that $C_P = C_V + R$ for an ideal gas (where both C_P and C_V are expressed in molar units). Note that $C_P = C_V + R$ will be true for any gas whose C_V is temperature-independent. In other words, the relation between C_P and C_V does not depend on the value of D (although Eq. 1.44 indicates that C_V itself does depend on D).

5. Use what you know about the partial derivatives of a single-component monatomic ideal gas with a fixed number of molecules ($dN = 0$) to fill in the blanks in the following equation with expressions that depend only on T, V, and N, $dS = \_\_\_\_ dT + \_\_\_\_ dV$, and show that your result is consistent with the fundamental equation in the entropy representation $dS = (1/T)dU + (P/T)dV$ (see Eq. 3.13).

6. Consider one mole of a monatomic ideal gas at 300 K that undergoes a free expansion from a volume of 1 m^3 to 10 m^3.

(a) How much does the chemical potential of each molecule in the gas change?

(b) Multiply the above result by $\mathcal{N}_A$ to obtain the chemical potential change in molar units.

(c) How much reversible work would be required to return the gas to its original volume (by following a reversible isothermal path, as in problem 2 of Chapter 2)?

7. How many thermodynamic degrees of freedom are available to an equilibrium system consisting of an ice cube floating in a glass of water, and how does that influence the possible temperature and pressure variations of the system?

8. Is it possible for the solid, liquid, and vapor phases of water to all be in equilibrium with each other at 0.1 MPa (see Fig. 3.4)? Explain your answer in terms of Gibbs phase rule.

9. Would it be possible to vary the temperature of a system consisting of water and ethanol (CH_3CH_2OH) while maintaining an equilibrium between solid, liquid, and vapor phases? Would the same be true if the system were held at *constant pressure*?

10. Show how one could obtain $G(T, P, N_i)$ by performing a Legendre transformation of $A(T, V, N_i)$, and make use of Eq. 3.42 ($G = A + PV$) to identify the partial derivative of A, which is equal to P.

11. The fact that $A = U - TS$ implies that the total differential of A may be expressed as $dA = dU - TdS - SdT$.

(a) Combine the latter expression with the fundamental equation (Eq. 3.8) to obtain an expression for dA in terms of dT, dV, and dN_i.

(b) Use the results you obtained in (a) to determine how the partial derivatives of A with respect to its natural variable are related to S, P, and μ_i.

(c) Confirm that the result you obtained for P is the same as that obtained in problem 10, and the result you obtained for S is consistent with the fact that $(\partial A/\partial T)_{V,N_i} = -S$ (which was demonstrated in exercise 3.4 on page 108).

12. Fill in the missing variables, and identify the correct sign, in each of the following identities.

(a) $\left(\dfrac{\partial H}{\partial S}\right)_{\_,\_} = \pm T$

(b) $\left(\dfrac{\partial H}{\partial P}\right)_{S,N_i} = \pm \_$

(c) $\left(\dfrac{\partial G}{\partial T}\right)_{\_,\_} = \pm S$

(d) $\left(\dfrac{\partial G}{\partial N_i}\right)_{\_,\_} = \pm \mu_i$

13. Consider one mole of a monatomic ideal gas at 300 K that expands isothermally from a volume of 1 m^3 to 10 m^3.

(a) What is the energy change of the gas?

(b) What is the entropy change of the gas?

(c) What is the Helmholtz energy change of the gas?

(d) What is the enthalpy change of the gas?

(e) What is the Gibbs energy change of the gas?

14. Obtain expressions for each of the following standard derivatives, for n moles of a monatomic ideal gas (express your results in terms of n, T, P, and/or R):

(a) $C_V \equiv T(\partial S/\partial T)_V$ (b) $C_P \equiv T(\partial S/\partial T)_P$

(c) $\alpha_P \equiv (1/V)(\partial V/\partial T)_P$

(d) $\kappa_T \equiv -(1/V)(\partial V/\partial P)_T$

15. Consider a processes in which ethanol is added to water at $T = 298$ K and $P = 0.1$ MPa. Pure water (MW = 18.015 g/mol) has a density of 0.997 g/cm^3 and pure ethanol (MW = 46.07 g/mol) has a density of 0.789 g/cm^3.

(a) What is the partial molar volume of pure water (in cm^3/mol=ml/mol units)?

(b) When 0.1 mole of ethanol is added to 10 moles of water, the total volume of the system is found to increase by 5.51 cm^3. Use this experimental result to estimate the partial molar volume ethanol in water (in cm^3/mol units)?

(c) Note that the partial molar volume of a solute is expected to depend on the solvent in which it is dissolved. How does the partial molar volume of ethanol in water [which you obtained in (b) above] compare with the partial molar volume of pure ethanol (obtained from the density of pure ethanol)?

16. Write a general expression that relates the total entropy S of a multicomponent mixture to the numbers of moles n_i and partial molar entropies $\overline{S}_i$ of the components in the system.

Problems That Test Your Understanding

17. Obtain expressions for the following partial derivatives for a system consisting of n moles of a monatomic ideal gas:

(a) $\left(\frac{\partial V}{\partial T}\right)_P$ (b) $\left(\frac{\partial P}{\partial V}\right)_T$

(c) Confirm that both of the following expressions for C_V are equivalent to each other by obtaining monatomic ideal gas expressions for the partial derivatives on both sides of the equation:

$$T\left(\frac{\partial S}{\partial T}\right)_V = \left(\frac{\partial U}{\partial T}\right)_V$$

18. Obtain equations for the following derivatives for N_2 gas, expressed in terms of the variables n, T, V, and/or P (and other constants). You may assume that the temperature of the system is sufficiently low that only the translational and rotational degrees of freedom N_2 are thermally active (and the pressure is sufficiently low that N_2 behaves like an ideal gas).

(a) $\left(\frac{\partial P}{\partial T}\right)_V$ (b) $\left(\frac{\partial U}{\partial T}\right)_V$

(c) $\left(\frac{\partial S}{\partial T}\right)_V$ (d) $\left(\frac{\partial S}{\partial P}\right)_V$

19. A given gas has a (molar) heat capacity at constant pressure of $C_P \sim 50$ J/(K mol). Estimate the number of thermally active quadratic terms in the Hamiltonian of this gas.

20. Perform a Legendre transform of $G(T, P, N_i)$ in order to obtain a new potential function whose natural variables are T, V, and N_i (show your work by indicating the partial derivative that you used in order to transform the variable P to V).

21. Fill in the missing variables and identify the correct sign in each of the following identities (for a system of constant composition, so all N_i are implicitly held constant).

(a) $\left(\frac{\partial H}{\partial \_}\right)_P = \pm T$ (b) $\left(\frac{\partial G}{\partial \_}\right)_{\_} = V$

(c) $\left(\frac{\partial A}{\partial \_}\right)_{\_} = -P$ (d) $\left(\frac{\partial U}{\partial V}\right)_S = \pm \_$

(e) $\left(\frac{\partial S}{\partial U}\right)_V = \pm \_$

22. Consider a process in which one mole of a monatomic ideal gas is compressed from a volume of $V_1 = 1.459$ m^3 to $V_2 = 1$ m^3 at a constant temperature of $T = 353.7$ K.

(a) What is the entropy change of the gas (in J/K units)?

(b) What is the change in the value of PV for the gas (in J units)?

(c) What is the energy change of the gas (in J units)?

(d) What is the enthalpy change of the gas (in J units)?

(e) What is the Helmholtz energy change of the gas (in J units)?

(f) What is the Gibbs energy change of the gas (in J units)?

23. Perform a Legendre transform of $U(S, V, N_i)$ in order to obtain a new potential function whose natural variables are S, P, and N_i (and make sure to indicate the partial derivative that is used in order to transform the variable V to P).

24. Fill in the missing variables, and identify the correct sign, in each of the following identities (for a system of constant composition, so all N_i are implicitly held constant).

(a) $\left(\frac{\partial U}{\partial S}\right)_{\_} = \pm T$ (b) $\left(\frac{\partial G}{\partial P}\right)_T = \pm \_$

(c) $\left(\frac{\partial A}{\partial V}\right)_{\_} = \pm P$ (d) $\left(\frac{\partial H}{\partial P}\right)_S = \pm \_$

(e) $\left(\frac{\partial S}{\partial V}\right)_{\_} = \pm \frac{P}{T}$

25. Consider a single-component system of constant composition ($dN = 0$), whose energy U is expressed as a function of T and P (rather than S and V).

(a) Fill in the blanks in the following partial derivatives.

$$dU = \left(\frac{\partial \_}{\partial \_}\right)_{\_} dT + \left(\frac{\partial \_}{\partial \_}\right)_{\_} dP$$

(b) Obtain expressions for the above partial derivatives if the system is a monatomic ideal gas.

Thermodynamic Calculation Strategies and Applications

CHAPTER 4

Practical thermodynamics calculations often require manipulating basic identities in order to obtain expressions that are best suited to a particular application. In this chapter we will learn how to perform various sorts of practical thermodynamic manipulations pertaining to systems composed of either nonreactive or chemically reactive components.

More specifically, in Section 4.1 we will see how to transform any thermodynamic partial derivative into a form that is compatible with practical experimental measurements. In Sections 4.2 and 4.3 we will see how to obtain reaction thermodynamic functions from experimental equilibrium constants. These methods will be applied both to binary reactions (such as protein-ligand binding) and to molecular self-assembly processes (such as those that give rise to living systems).

The second law implies that all spontaneous (or living) processes are ultimately driven by an increase in the entropy of the universe. In Section 4.4 we will see how the Clausius inequality may be used to quantify the driving forces, and useful energy content, of chemical reactions carried out under various experimental conditions.

4.1 Reduction of Thermodynamic Derivatives

Thermodynamic calculations often require transforming partial derivatives into a form that is more useful for a particular calculation or measurement. The values of some thermodynamic partial derivatives are extensively tabulated, or can readily be experimentally measured. If you know how to translate derivatives from one form to another, then you can express derivatives that would be very difficult to measure in terms of derivatives that have already been measured (or could easily be measured). Performing such manipulations is referred to as the reduction of thermodynamic derivatives.

For example, the derivative $(\partial S/\partial P)_T$ would be quite difficult to directly measure experimentally. However, we will see that this derivative is also exactly equivalent to $-(\partial V/\partial T)_P$, which is quite easy to measure (as it only requires determining how much the volume of the system changes with temperature, at constant pressure).

Other commonly tabulated thermodynamic quantities include the constant-pressure heat capacity C_P, the isobaric thermal expansion coefficient α_P, and the isothermal compressibility κ_T, each of which are defined as follows.

$$C_P \equiv T\left(\frac{\partial S}{\partial T}\right)_P = \left(\frac{\partial H}{\partial T}\right)_P \tag{4.1}$$

$$\alpha_P \equiv \frac{1}{V}\left(\frac{\partial V}{\partial T}\right)_P = -\frac{1}{\rho}\left(\frac{\partial \rho}{\partial T}\right)_P \tag{4.2}$$

$$\kappa_T \equiv -\frac{1}{V}\left(\frac{\partial V}{\partial P}\right)_T = \frac{1}{\rho}\left(\frac{\partial \rho}{\partial P}\right)_T \tag{4.3}$$

The constant-volume heat capacity C_V may be obtained from C_P, α_P, and κ_T, by making use of the following identity (whose derivation is given in Exercise 4.4).

$$C_V \equiv T\left(\frac{\partial S}{\partial T}\right)_V = \left(\frac{\partial U}{\partial T}\right)_V = C_P - \frac{TV\alpha_P^2}{\kappa_T} \tag{4.4}$$

Notice that C_P and C_V are total (not molar) heat capacities. For a single-component system, the molar heat capacities are C_V/n and C_P/n. For a multicomponent system, Eq. 3.53 implies that we may express C_P as the sum of the partial molar heat capacities of each component, $C_P = \sum_i n_i(\partial C_P/\partial n_i)_{T,P,n_j \neq n_i} = \sum_i n_i \overline{C}_{P,i}$. The following is a selection of other important thermodynamic identities, all of which we will soon learn how to derive.

The temperature derivative of the pressure at constant volume is frequently encountered in thermodynamic manipulations.

$$\left(\frac{\partial P}{\partial T}\right)_V = \left(\frac{\partial S}{\partial V}\right)_T = \frac{\alpha_P}{\kappa_T} \tag{4.5}$$

The coefficients α_P and κ_T are related in the following way.

$$\left(\frac{\partial \alpha_P}{\partial P}\right)_T = -\left(\frac{\partial \kappa_T}{\partial T}\right)_P \tag{4.6}$$

The isentropic compressibility κ_S (which is also referred to as the adiabatic compressibility) may be obtained from κ_T, α_P, and C_P using the following identity.

$$\kappa_S \equiv -\frac{1}{V}\left(\frac{\partial V}{\partial P}\right)_S = \frac{1}{\rho}\left(\frac{\partial \rho}{\partial P}\right)_S = \kappa_T - \frac{TV\alpha_P^2}{C_P} \tag{4.7}$$

Moreover, κ_S can be obtained directly from experimental sound velocity measurements, and so one may rearrange Eq. 4.7 to determine κ_T from experimentally measured values of κ_S, α_P, and C_P.

Partial Derivative Relations

The following relations between partial derivatives are the basis of all derivative reduction procedures. We will first express each relation in a general form that applies to any function $F(x, y, z)$, and then illustrate its application using thermodynamic examples.

Given any function $F(x, y, z)$ we know that we may express dF in terms of dx, dy, and dz, as follows.

$$dF = \left(\frac{\partial F}{\partial x}\right)_{y,z} dx + \left(\frac{\partial F}{\partial y}\right)_{x,z} dy + \left(\frac{\partial F}{\partial z}\right)_{x,y} dz \tag{4.8}$$

The following two cross-derivatives are necessarily equal to each other (as are many other such pairs of cross-derivatives).

$$\left[\frac{\partial}{\partial y}\left(\frac{\partial F}{\partial x}\right)_{y,z}\right]_{x,z} = \left[\frac{\partial}{\partial x}\left(\frac{\partial F}{\partial y}\right)_{x,z}\right]_{y,z}$$

Notice that the variable z is held constant in all of the above partial derivatives. Leaving this constant variable out makes it easier to see that the above identity arises simply from the fact that we may switch the order in which partial derivatives are evaluated.

$$\boxed{\left[\frac{\partial}{\partial y}\left(\frac{\partial F}{\partial x}\right)_{y}\right]_{x} = \left[\frac{\partial}{\partial x}\left(\frac{\partial F}{\partial y}\right)_{x}\right]_{y}} \tag{4.9}$$

It is important to note that the above identity requires that variables that are held constant (x and y) are the same as the variables in the derivatives.

Two other cross-derivative relations may be obtained from Eq. 4.8 by changing the order in which the derivatives with respect to x and z, or y and z, are performed. Such cross-derivative relations—which are also referred to as *Maxwell relations* — are often useful in thermodynamic derivations.

Exercise 4.1

Use the fundamental equations for dU, dG, and dA (with the variables N_i held constant) to derive the following thermodynamic identities.

$$\left(\frac{\partial T}{\partial V}\right)_S = -\left(\frac{\partial P}{\partial S}\right)_V \quad \left(\frac{\partial S}{\partial P}\right)_T = -\left(\frac{\partial V}{\partial T}\right)_P \quad \left(\frac{\partial P}{\partial T}\right)_V = \left(\frac{\partial S}{\partial V}\right)_T$$

Solution. For a system of fixed composition, the fundamental equation becomes $dU = (\partial U/\partial S)_V dS + (\partial U/\partial V)_S dV = TdS - PdV$. Thus, Eq. 4.9 implies that the corresponding cross-derivatives must be equal to each other.

$$\left[\frac{\partial}{\partial V}\left(\frac{\partial U}{\partial S}\right)_V\right]_S = \left[\frac{\partial}{\partial S}\left(\frac{\partial U}{\partial V}\right)_S\right]_V$$

$$\left(\frac{\partial T}{\partial V}\right)_S = -\left(\frac{\partial P}{\partial S}\right)_V$$

Similarly, using $dG = -SdT + VdP$ and $dA = -SdT - PdV$ we may obtain the following two identities.

$$-\left(\frac{\partial S}{\partial P}\right)_T = \left(\frac{\partial V}{\partial T}\right)_P \quad \left(\frac{\partial P}{\partial T}\right)_V = \left(\frac{\partial S}{\partial V}\right)_T$$

Note that the second of the above identities is the same as the first equality in Eq. 4.5.

For any function of two variables $F(x, y)$ we know that $dF = (\partial F/\partial x)_y dx + (\partial F/\partial y)_x dy$. We may divide the latter expression through by dx, while holding the function F itself constant, to obtain the following identity.

$$\left(\frac{\partial F}{\partial x}\right)_F = \left(\frac{\partial F}{\partial x}\right)_y \left(\frac{\partial x}{\partial x}\right)_F + \left(\frac{\partial F}{\partial y}\right)_x \left(\frac{\partial y}{\partial x}\right)_F$$

$$0 = \left(\frac{\partial F}{\partial x}\right)_y + \left(\frac{\partial F}{\partial y}\right)_x \left(\frac{\partial y}{\partial x}\right)_F$$

The above expression may be rearranged to obtain the following very useful partial derivative relation.

$$\boxed{\left(\frac{\partial y}{\partial x}\right)_F = -\frac{\left(\frac{\partial F}{\partial x}\right)_y}{\left(\frac{\partial F}{\partial y}\right)_x}} \tag{4.10}$$

More generally, for any three variables x, y, and z (any of which could be either intensive or extensive thermodynamic functions), the above relation may be reexpressed in the following form.

$$\left(\frac{\partial x}{\partial y}\right)_z \left(\frac{\partial y}{\partial z}\right)_x \left(\frac{\partial z}{\partial x}\right)_y = -1 \tag{4.11}$$

Exercise 4.2

Use Eq. 4.10 (or Eq. 4.11) to verify that $(\partial S/\partial V)_U = P/T$ and $(\partial P/\partial T)_V = \alpha_P/\kappa_T$. Note that these identities are equivalent to Eq. 3.15 and the second equality on the right-hand side of Eq. 4.5.

Solution. We may use Eq. 4.10 to transform $(\partial S/\partial V)_U$ by bringing the energy U up into the numerators of the two partial derivatives on the right-hand side, and then use Eqs. 3.9 and 3.10 to obtain Eq. 3.15.

$$\left(\frac{\partial S}{\partial V}\right)_U = -\frac{\left(\frac{\partial U}{\partial V}\right)_S}{\left(\frac{\partial U}{\partial S}\right)_V} = \frac{P}{T}$$

The identity $(\partial P/\partial T)_V = \alpha_P/\kappa_T$ may be verified by first applying Eq. 4.10 to the derivative $(\partial P/\partial T)_V$ and then making use of the definitions of α_P and κ_T (Eqs. 4.2 and 4.3).

$$\left(\frac{\partial P}{\partial T}\right)_V = \frac{-\left(\frac{\partial V}{\partial T}\right)_P}{\left(\frac{\partial V}{\partial P}\right)_T} = \frac{-V\alpha_P}{-V\kappa_T} = \frac{\alpha_P}{\kappa_T}$$

Equations 4.10 and 4.11 look somewhat similar to relations that may be obtained using the chain rule, except for the negative sign and the fact that different variables are held constant. More specifically, if all the partial derivatives had the same variables held constant, then the chain rule could be used to obtain the following identities.

$$\left(\frac{\partial y}{\partial x}\right)_F = \left(\frac{\partial y}{\partial z}\right)_F \left(\frac{\partial z}{\partial x}\right)_F = \frac{\left(\frac{\partial z}{\partial x}\right)_F}{\left(\frac{\partial z}{\partial y}\right)_F} \tag{4.12}$$

$$\left(\frac{\partial x}{\partial y}\right)_F \left(\frac{\partial y}{\partial z}\right)_F \left(\frac{\partial z}{\partial x}\right)_F = 1 \tag{4.13}$$

Exercise 4.3

Use the chain rule to prove that $(\partial A/\partial P)_T = PV\kappa_T$, and to confirm that $(\partial U/\partial T)_V = C_V$ and $(\partial H/\partial T)_P = C_P$.

Solution. The first identity may be obtained by multiplying and dividing by dV (at constant T).

$$\left(\frac{\partial A}{\partial P}\right)_T = \left(\frac{\partial A}{\partial V}\right)_T \left(\frac{\partial V}{\partial P}\right)_T = PV\kappa_T$$

The second equality is obtained using the definition of κ_T (Eq. 4.3) and the fact that $(\partial A/\partial V)_T = -P$ (which is one of the two partial derivatives that appear in the fundamental equation for the Helmholtz energy $dA = -SdT - PdV$). The same method can be used to confirm the two heat capacity relations, by first multiplying and dividing by dS, either at constant V or at constant P, and then using the fact that $TdS = \delta Q$.

$$\left(\frac{\partial U}{\partial T}\right)_V = \left(\frac{\partial U}{\partial S}\right)_V \left(\frac{\partial S}{\partial T}\right)_V = T\left(\frac{\partial S}{\partial T}\right)_V = \left.\frac{\delta Q}{dT}\right|_V = C_V$$

and

$$\left(\frac{\partial H}{\partial T}\right)_P = \left(\frac{\partial U}{\partial S}\right)_P \left(\frac{\partial S}{\partial T}\right)_P = T\left(\frac{\partial S}{\partial T}\right)_P = \left.\frac{\delta Q}{dT}\right|_P = C_P$$

The following partial derivative relation is useful for changing a variable that is held constant (z) to a different variable (y). It is obtained by starting with $dF = (\partial F/\partial x)_y dx + (\partial F/\partial y)_x dy$ and then dividing through by dx, while holding z constant.

$$\left(\frac{\partial F}{\partial x}\right)_z = \left(\frac{\partial F}{\partial x}\right)_y + \left(\frac{\partial F}{\partial y}\right)_x \left(\frac{\partial y}{\partial x}\right)_z \tag{4.14}$$

Exercise 4.4

Use Eq. 4.14 to manipulate the partial derivatives $(\partial S/\partial T)_V$ and $(\partial V/\partial P)_S$ in order to relate C_V to C_P (Eq. 4.4) and κ_S to κ_T (Eq. 4.7).

Solution. To relate $C_V = T(\partial S/\partial T)_V$ to $C_P = T(\partial S/\partial T)_P$, we need to change the variable that is held constant from V to P, using Eq. 4.14.

$$\left(\frac{\partial S}{\partial T}\right)_V = \left(\frac{\partial S}{\partial T}\right)_P + \left(\frac{\partial S}{\partial P}\right)_T \left(\frac{\partial P}{\partial T}\right)_V$$

$$\frac{1}{T}C_V = \frac{1}{T}C_P - \left(\frac{\partial V}{\partial T}\right)_P \frac{\alpha_P}{\kappa_T}$$

$$\frac{1}{T}C_V = \frac{1}{T}C_P - V\frac{\alpha_P^2}{\kappa_T}$$

$$C_V = C_P - \frac{TV\alpha_P^2}{\kappa_T}$$

Similarly, relating $\kappa_S = -(1/V)(\partial V/\partial P)_S$ to $\kappa_T = -(1/V)(\partial V/\partial P)_T$ requires changing the constant variable from S to T.

$$\left(\frac{\partial V}{\partial P}\right)_S = \left(\frac{\partial V}{\partial P}\right)_T + \left(\frac{\partial V}{\partial T}\right)_P \left(\frac{\partial T}{\partial P}\right)_S$$

$$-V\kappa_S = -V\kappa_T + V\alpha_P\left[\frac{-(\partial S/\partial P)_T}{(\partial S/\partial T)_P}\right]$$

$$-\kappa_S = -\kappa_T + \alpha_P\left[\frac{(\partial V/\partial T)_P}{(C_P/T)}\right]$$

$$-\kappa_S = -\kappa_T + \alpha_P\left[\frac{V\alpha_P}{(C_P/T)}\right]$$

$$\kappa_S = \kappa_T - \frac{TV\alpha_P^2}{C_P}$$

Derivative Reduction Strategy

The following sequence of steps can serve as a useful guide in reducing thermodynamic derivatives to a combination of C_P, C_V, α_P, and κ_T or κ_S (along with other thermodynamic variables and constants).

1. Make use of fundamental equation derivatives and/or Maxwell (cross-derivative) relations.
2. Bring potentials (U, H, A, or G) to the numerator (then return to 1).
3. Bring S to the numerator and use C_P and C_V (or return to 1).
4. Bring V to the numerator and use α_P or κ_T (or return to 1).

Although these rules provide a useful starting point, it is not uncommon to encounter examples that require extending this general procedure. One such strategy is to use Legendre transform relations, or the corresponding fundamental equations, to replace a potential function (after step 2 above), as illustrated by the second and third examples in the following exercise.

Exercise 4.5

Verify that $(\partial U/\partial V)_T = -P + T(\alpha_P/\kappa_T)$ using three different derivative reduction strategies.

- We could begin by noting that the natural variables of U are S and V, which suggests that it may be helpful to change the constant T in $(\partial U/\partial V)_T$ to a constant S, using Eq. 4.14.

$$\left(\frac{\partial U}{\partial V}\right)_T = \left(\frac{\partial U}{\partial V}\right)_S + \left(\frac{\partial U}{\partial S}\right)_V\left(\frac{\partial S}{\partial V}\right)_T$$

  Now we may use the partial derivative identities implicit in the fundamental equation $dU = TdS - PdV$ to identify two of the derivatives on the right-hand side.

$$\left(\frac{\partial U}{\partial V}\right)_T = -P + T\left(\frac{\partial S}{\partial V}\right)_T$$

  Next we may reduce $(\partial S/\partial V)_T$ as described in Exercise 4.1.

$$\left(\frac{\partial U}{\partial V}\right)_T = -P + T\left(\frac{\partial P}{\partial T}\right)_V = -P + T\left(\frac{\alpha_P}{\kappa_T}\right)$$

- A second strategy for obtaining the same result may begin by substituting $U = A + TS$, and then differentiating each term.

$$\left(\frac{\partial U}{\partial V}\right)_T = \left[\frac{\partial(A+TS)}{\partial V}\right]_T = \left(\frac{\partial A}{\partial V}\right)_T + T\left(\frac{\partial S}{\partial V}\right)_T + S\left(\frac{\partial T}{\partial V}\right)_T$$

  Note that the last partial derivative on the right-hand side is necessarily equal to zero because T is held constant so $dT|_T = 0$ and thus $(\partial T/\partial V)_T = 0$. Moreover, the first derivative on the right-hand side is implicit in the Helmholtz energy fundamental equation $dA = -SdT - PdV$, which leads to the following expression that can then be further reduced as described above.

$$\left(\frac{\partial U}{\partial V}\right)_T = -P + T\left(\frac{\partial P}{\partial T}\right)_V + 0$$

- A third strategy could begin by equating $dU = TdS - PdV$, and then again proceeding as above.

$$\left(\frac{\partial U}{\partial V}\right)_T = T\left(\frac{\partial S}{\partial V}\right)_T - P\left(\frac{\partial V}{\partial V}\right)_T = T\left(\frac{\partial P}{\partial T}\right)_V - P = -P + T\left(\frac{\alpha_P}{\kappa_T}\right)$$

The identity obtained in the above exercise can be rearranged to the following form, which is also known as the *thermodynamic equation of state*, as it represents a general thermodynamic equation for the pressure of any system.

$$\boxed{P = T\left(\frac{\partial P}{\partial T}\right)_V - \left(\frac{\partial U}{\partial V}\right)_T} \tag{4.15}$$

The derivatives on the right-hand side $(\partial P/\partial T)_V$ and $(\partial U/\partial V)_T$ are also known as the *thermal pressure coefficient* and *internal pressure*, respectively. Notice that for an ideal gas $(\partial P/\partial T)_V = P/T = nR/V = [c]R$ and $(\partial U/\partial V)_T$ is equal to zero. Thus, any deviation from the ideal gas equation leads to a nonzero internal pressure, which arises from the influence of intermolecular interactions on the energy of real gases and fluids (as further described in Chapter 5).

Exercise 4.6

Another important example of the influence of gas nonideality is quantified by the Joule-Thompson coefficient $\mu_{JT} \equiv (\partial T/\partial P)_H$, which describes the temperature change experienced by a gas when it expands adiabatically.

- Transform the above Joule-Thompson partial derivative to show that $(\partial H/\partial P)_T = -\mu_{JT} C_P$.

  Solution. Equation 4.10 can be used to accomplish this in a single step.

$$\mu_{JT} = \left(\frac{\partial T}{\partial P}\right)_H = \frac{-\left(\frac{\partial H}{\partial P}\right)_T}{\left(\frac{\partial H}{\partial T}\right)_P} = \frac{-\left(\frac{\partial H}{\partial P}\right)_T}{C_P}$$

- Further reduce $(\partial H/\partial P)_T$ to show that it is equal to $V(1 - T\alpha_P)$.

  Solution. Since the natural variables of H are S and P, it is useful to start by changing the constant variable from T to S.

$$\left(\frac{\partial H}{\partial P}\right)_T = \left(\frac{\partial H}{\partial P}\right)_S + \left(\frac{\partial H}{\partial S}\right)_P \left(\frac{\partial S}{\partial P}\right)_T$$
$$= V + T\left[-\left(\frac{\partial V}{\partial T}\right)_P\right] = V - TV\alpha_P = V(1 - T\alpha_P)$$

- Show that the same result may also be obtained by replacing H in $(\partial H/\partial P)_T$.

  Solution. Since P and T are the natural variables of G, it may be useful to use $H = G + TS$ and thus equate $dH = dG + TdS + SdT$.

$$\left(\frac{\partial H}{\partial P}\right)_T = \left(\frac{\partial G}{\partial P}\right)_T + T\left(\frac{\partial S}{\partial P}\right)_T + S\left(\frac{\partial T}{\partial P}\right)_T$$
$$= V - TV\alpha_P + 0 = V(1 - T\alpha_P)$$

The results obtained in the above exercise may be used to relate the Joule-Thompson coefficient of a single-component gas to its molar volume $\overline{V} = V/n$, molar heat capacity, $\overline{C}_P = C_P/n$, and thermal expansion coefficient α_P.

$$\boxed{\mu_{JT} \equiv \left(\frac{\partial T}{\partial P}\right)_H = \frac{\overline{V}(T\alpha_P - 1)}{\overline{C}_P}} \tag{4.16}$$

Recall that for an ideal gas $\alpha_P = 1/T$ so $\mu_{JT} = (V/C_P)(1 - 1) = 0$, and so the Joule-Thompson coefficient of an ideal gas is equal to zero. Thus, the temperature of an ideal gas will not change in a Joule-Thompson expansion experiment. On the other hand, if a nonideal gas experiences a pressure drop of ΔP, then its temperature will change by $\Delta T \approx \mu_{JT} \Delta P$.

4.2 Chemical Reaction Thermodynamics

Chemical reactions may be viewed as examples of more general self-assembly processes, in which molecules combine with each other to form various sorts of aggregates and higher-order structures. This section describes general relations between reaction thermodynamic functions and equilibrium constants. The results are then applied to binary reactions (such as the formation of molecular dimers and antibody-antigen complexes). Section 4.3 describes the thermodynamics of higher-order self-assembly processes (such as the condensation of rain drops and the formation of surfactant micelles). The latter processes have much in common with the biological self-assembly of phospholipids, amino acids, and nucleic acids to create cell membranes, proteins, genes, and living systems.

The Gibbs energy change associated with a chemical reaction is defined as the difference between the product and reactant partial molar Gibbs energies $\Delta G_{rxn} \equiv \sum_i \nu_i \overline{G}_i$, where the constants ν_i represent the stoichiometric coefficients of each reactant and product species (see Eq. 3.24).

$$\boxed{\Delta G_{rxn} \equiv \sum_i \nu_i \left(\frac{\partial G}{\partial n_i}\right)_{T,P,n_j \neq n_i} = \sum_i \nu_i \overline{G}_i} \tag{4.17}$$

At equilibrium Eq. 3.24 implies that $\Delta G_{rxn} = 0$, while if $\Delta G_{rxn} \neq 0$ the reaction may proceed spontaneously, either in the forward direction (when $\Delta G_{rxn} < 0$) or in the backward direction (when $\Delta G_{rxn} > 0$).[1]

We may determine the partial molar entropy difference between the products and reactants ΔS_{rxn} by differentiating ΔG_{rxn} with respect to T (at constant P).

$$\boxed{\Delta S_{rxn} \equiv \sum_i \nu_i \overline{S}_i = -\sum_i \nu_i \left(\frac{\partial \overline{G}_i}{\partial T}\right)_P = -\left(\frac{\partial \Delta G_{rxn}}{\partial T}\right)_P} \tag{4.18}$$

The second equality follows from the fact that $S = -(\partial G / \partial T)_P$.[2]

---

[1] In Section 4.4 we will see that these inequalities pertaining to chemical reaction free energies are general consequences of the Clausius inequality. More specifically, we will see that for any chemical reaction, carried out under any conditions, the sum $\sum_i \nu_i \mu_i$ will be equal to zero at equilibrium (Eq. 4.52) and will be less than zero for a reaction that can occur spontaneously in the forward direction (Eq. 4.53).

[2] The partial molar entropy $\overline{S}_i$ is defined as the derivative of S with respect to n_i (at constant P, T, and $n_j \neq n_i$). If we change the order in which we apply this derivative

Similarly, the partial molar volume difference between the reactants and products may be obtained by differentiating ΔG_{rxn} with respect to P (at constant T), recalling that $V = (\partial G/\partial P)_T$.

$$\boxed{\Delta V_{\text{rxn}} \equiv \sum_i \nu_i \overline{V}_i = \sum_i \nu_i \left(\frac{\partial \overline{G}_i}{\partial P}\right)_T = \left(\frac{\partial \Delta G_{\text{rxn}}}{\partial P}\right)_T} \tag{4.19}$$

We may combine Eq. 4.17 with the fact that $G = H - TS$ to obtain the following well-known relation.

$$\boxed{\Delta G_{\text{rxn}} = \Delta H_{\text{rxn}} - T\Delta S_{\text{rxn}}} \tag{4.20}$$

The enthalpy of reaction may be obtained by rearranging this equation $\Delta H_{\text{rxn}} = \Delta G_{\text{rxn}} + T\Delta S_{\text{rxn}}$. Alternatively, we may obtain ΔH_{rxn} using the following expression.

$$\boxed{\Delta H_{\text{rxn}} \equiv \sum_i \nu_i \overline{H}_i = \sum_i \nu_i \left[\frac{\partial(\overline{G}_i/T)}{\partial(1/T)}\right]_P = \left[\frac{\partial(\Delta G_{\text{rxn}}/T)}{\partial(1/T)}\right]_P} \tag{4.21}$$

This identity (called the van't Hoff equation) is a consequence of the Gibbs-Helmholtz relation $[\partial(G/T)/\partial(1/T)]_P = H$, which may be derived as follows.

$$\begin{aligned}\left[\frac{\partial(G/T)}{\partial(1/T)}\right]_P &= G\left[\frac{\partial(1/T)}{\partial(1/T)}\right]_P + \frac{1}{T}\left[\frac{\partial G}{\partial(1/T)}\right]_P = G + \frac{1}{T}\left[-T^2\left(\frac{\partial G}{\partial T}\right)_P\right] \\ &= G - T\left(\frac{\partial G}{\partial T}\right)_P = G - T(-S) = G + TS = H \end{aligned} \tag{4.22}$$

The connection between ΔG_{rxn} and chemical equilibrium constants may be obtained by combining Eqs. 4.17 and 3.39. More specifically, Eq. 3.39 implies that we may express the partial molar Gibbs energy of each component in a reaction in terms of its value at a given standard state concentration $[c_i]_0$.[3]

$$\overline{G}_i = \overline{G}_i^0 + RT\ln\left(\frac{[c_i]}{[c_i]_0}\right) \tag{4.23}$$

---

and the derivative with respect to T (at constant P and constant composition), we find that $\overline{S}_i = -(\partial\overline{G}_i/\partial T)_P$ (where the latter derivative is implicitly performed while holding the composition of the system constant).

[3] Recall that Eq. 3.39 was derived from the entropy change associated with isothermal ideal gas processes. However, the same result is also applicable to the chemical potential change associated with changing the concentration of a solute dissolved in a liquid (as long as the solute concentration is sufficiently low that we may neglect solute-solute interactions). In other words, when the concentration of a solute is changed from $[c]_0$ to $[c]$, then its chemical potential changes by $\mu - \mu^0 = k_B T\ln([c]/[c]_0)$, which is equivalent to Eq. 4.23, when expressed in molar units.

Thus, Eq. 4.17 may be used to express ΔG_{rxn} as follows.[4]

$$\Delta G_{rxn} = \sum_i \overline{G}_i^0 + RT \sum_i \nu_i \ln\left(\frac{[c_i]}{[c_i]_0}\right) = \Delta G^0_{rxn} + RT \sum_i \ln\left(\frac{[c_i]}{[c_i]_0}\right)^{\nu_i}$$

$$= \Delta G^0_{rxn} + RT \ln\left\{\prod_i \left(\frac{[c_i]}{[c_i]_0}\right)^{\nu_i}\right\} = \Delta G^0_{rxn} + RT \ln Q_{rxn} \qquad (4.24)$$

The quantity $RT \ln Q_{rxn}$ represents the change in Gibbs energy associated with changing each reactant and product concentration from its standard state value $[c_i]_0$ to its actual concentration $[c_i]$. When the reaction has reached equilibrium, then Eq. 3.24 implies that $\Delta G_{rxn} = 0$, and the reaction quotient Q_{rxn} becomes equivalent to the equilibrium constant.

$$\boxed{K_c \equiv Q^{eq}_{rxn} = \prod_i \left(\frac{[c_i]^{eq}}{[c_i]_0}\right)^{\nu_i}} \qquad (4.25)$$

Thus, at equilibrium Eq. 4.24 reduces to the following well-known expression.

$$\boxed{\Delta G^0_{rxn} = -RT \ln K_c} \qquad (4.26)$$

Implicit Standard State Conditions

It is important to keep in mind that ΔG^0_{rxn} pertains to the Gibbs energy of reaction when the reactant and product species are all at the standard state concentration. If the standard state concentration is taken to be $[c_i]_0 = 1$ M, then we may express the equilibrium constant as $K_c = \prod_i [c_i]^{\nu_i}$, where $[c_i]$ are equilibrium concentrations of the corresponding chemical species (each measured in molar units). However, even though $[c_i]_0$ does not appear in the latter expression, it is important to remember that ΔG^0_{rxn} implicitly pertains to a standard state concentration of 1 (in whatever units are used to measure the reactant and product concentrations).

One may express gas phase equilibrium constants in pressure units $K_P = \prod_i (P_i)^{\nu_i}$, in which case the implicit standard state is that at which all of the product and reactant species have a partial pressure of 1, in whatever units are used to measure the pressures P_i. A similar procedure may

[4] The symbol $\prod_i f_i$ represents a product over all the terms indexed by f_i (in the same way that $\sum_i$ is the sum of all such terms). The third equality in Eq. 4.24 makes use of the fact that $\ln(A) + \ln(B) = \ln(A \cdot B)$, and so $\sum_i \ln f_i = \ln(\prod_i f_i)$.

be used to express equilibrium constants in terms of other composition variables and implicit standard state conditions.[5]

Binary Reaction Equilibria

The above results may be used to describe simple chemical processes, such as the reaction of two monomers M to form a dimer A, or the reaction of an antigen A_g with an antibody A_b to form an antibody-antigen complex A_gA_b. The association equilibrium constants K_A for all such binary reactions may be obtained using Eq. 4.25 and related to the corresponding reaction free energy ΔG^0 using Eq. 4.26.

For a simple dimerization reaction of the form $M + M \rightleftharpoons A$, Eqs. 4.25 and 4.26 imply that $\Delta G^0 = -RT \ln K_A = -RT \ln([A]/[M]^2)$, which can be rearranged to the following form.

$$K_A = \frac{[A]}{[M]^2} = e^{-\Delta G^0/RT} \tag{4.27}$$

Note that when $[A] = [M]$, this equilibrium relation implies that $K_A = [A]/[M]^2 = 1/[M]$, or $[M] = 1/K_A$. Thus, the value of $1/K_A$ is equivalent to the monomer concentration at which the dimer and monomer concentrations become equal to each other. In other words, $1/K_A$ is a measure of the concentration at which a significant degree of dimerization begins to occur. At lower concentrations the system will consist primarily of monomers, and at higher concentrations it will consist primarily of dimers.

An antigen-antibody binding reaction of the form $A_g + A_b \rightleftharpoons A_gA_b$ has an association equilibrium constant of $K_A = [A_gA_b]/([A_g][A_b])$. In this case, $1/K_A$ represents the antigen concentration at which half of the antibodies are complexed with an antigen, since $[A_g] = 1/K_A$ when $[A_gA_b] = [A_b]$. Biochemists often refer to K_A as the affinity constant (or $1/K_A$ as the affinity) for such antibody-binding reactions. The sensitivity with which an antibody responds to an antigen increases as the affinity constant becomes larger (and the affinity concentration becomes smaller). In other words, the affinity $1/K_A$, expressed in molar units, represents the antigen concentration required in order to induce a significant antibody response.

---

[5] For example, it is sometimes convenient to express equilibrium constants as a ratio of mole fractions $K_\chi = \prod_i (\chi_i)^{\nu_i}$, where $\chi_i \equiv n_i/n$ (and $n = \sum_i n_i$ is the total number of moles in the system). The standard state that is implicit in K_χ pertains to reactant and product species, each of which has a mole fraction of $\chi_i = 1$. The latter standard state can easily lead to confusion, as it pertains to solutes that are sufficiently dilute that solute-solute interactions may be neglected, but the mole fraction of each solute species is nevertheless scaled up to a hypothetical value of $\chi_i = 1$.

Equation 4.27 may be solved to predict the concentrations of monomers $[M]$ and dimers $[A]$ as a function of the total (input) monomer concentration $[M_T] = [M] + 2[A]$. The latter mass balance relation may be rearranged to $[A] = \frac{1}{2}([M_T] - [M])$ and combined with Eq. 4.27 to express the equilibrium constant as $K_A = ([M_T] - [M])/(2[M]^2)$, which is equivalent to the following equation.

$$2K_A[M]^2 + [M] - [M_T] = 0 \tag{4.28}$$

We may solve this quadratic equation to obtain $[M]$ and $[A] = ([M_T] - [M])/2$ as a function of $[M_T]$, as illustrated in Figure 4.1(a).[6]

4.3 Self-Assembly Thermodynamics

Higher-order self-assembly processes, in which n monomers combine to form large aggregates, may be described in much the same way as simpler dimerization reactions. However, we will see that when n is large, such self-assembly processes behave quite differently than when $n = 2$. More specifically, while dimer formation occurs gradually over a broad range of monomer concentrations, higher-order aggregates tend to form abruptly when the monomer concentration reaches a particular value, called the *critical aggregation concentration* (or *critical micelle concentration*). This is illustrated in Figure 4.1, which shows how the equilibrium concentrations of free monomers $[M]$ and aggregates $[A]$ depend on the total monomer concentration $[M_T]$, for a dimerization ($n = 2$) and a higher-order ($n = 100$) aggregation process.

The aggregation equilibrium constant K_A for a self-assembly reaction of the form $nM \rightleftharpoons A$ may again be obtained from Eqs. 4.25 and 4.26.

$$K_A = \frac{[A]}{[M]^n} = e^{-\Delta G^0/RT} \tag{4.29}$$

Thus, the total monomer concentration $[M_T]$ is the sum of free $[M]$ and bound $n[A]$ monomer concentrations.

$$[M_T] = [M] + n[A] = [M] + nK_A[M]^n \tag{4.30}$$

Note that although Eq. 4.30 cannot in general be analytically inverted to obtain an expression for $[M]$ as a function of $[M_T]$, it can readily be graphically inverted by using Eq. 4.30 to calculate $[M_T]$ at various values of $[M]$

[6] In order to solve Eq. 4.28 we treat K_A and $[M_T]$ as input constants, and $[M]$ as the variable. Recall that the two roots of a quadratic equation of the form $ax^2 + bx + c = 0$ are equal to $\left(-b \pm \sqrt{b^2 - 4ac}\right)/(2a)$. When applied to Eq. 4.28, we find that only one of the two roots corresponds to the physically relevant (positive) free monomer concentration.

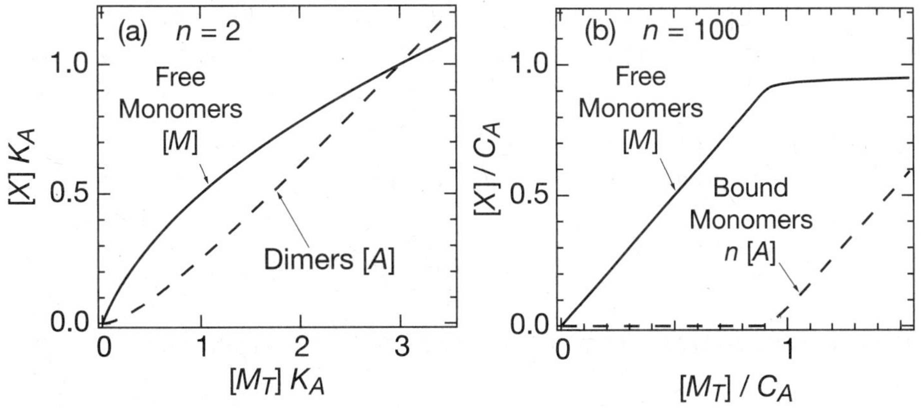

FIGURE 4.1 The way in which aggregate formation depends on the total monomer concentration $[M_T]$ is qualitatively different for a dimerization reaction (a) and a high-order aggregation process (b). In a dimerization process, the concentration of dimers $[A]$ becomes equal to the concentration of free monomers $[M]$, when $[M] = 1/K_A$. In a higher-order aggregation process, the formation of aggregates begins abruptly when the total monomer concentration approaches C_A (after which the concentration of free monomers also remains approximately equal to C_A).

and then plotting $[M]$ as a function of $[M_T]$. This is how the curves in Figure 4.1(b) were generated.

At equilibrium the chemical potentials of the free monomers and aggregates come into perfect balance with each other. More specifically, an aggregation reaction of the form $nM \rightleftharpoons A$ will come to equilibrium when $n\mu_M = \mu_A$. Equation 4.23 (and the associated footnote 3 on page 129) may be used to express the latter chemical potentials in terms of their values under standard state conditions.[7]

$$n\,\mu_M = n(\mu_M^0 + k_B T \ln[M]) = \mu_A^0 + k_B T \ln[A] = \mu_A \tag{4.31}$$

In order to treat self-assembly processes involving aggregates of various sizes, it is useful to introduce a notation that expresses all chemical potentials and concentrations in monomeric form. More specifically, we may define μ_n and C_n as the chemical potential and concentration of monomers contained in aggregates of size n. For example, a free monomer has an aggregate of size of 1, and so $\mu_1 = \mu_M$ and $C_1 = [M]$. Similarly, aggregates of size n have chemical potentials of $\mu_n = \mu_A/n$ and concentrations of $C_n = n[A]$. In other words, the total chemical potential of an aggregate

[7] Each of the concentrations in Eq. 4.31 is expressed in the same units, and is implicitly divided by the standard state concentration, and thus the chemical potentials μ_i^0 pertain to that standard state concentration.

of size n is equal to the sum of the chemical potentials of the corresponding monomers, $\mu_A = n\mu_n$, and the concentration of monomers in such an aggregate is n times larger than that of the corresponding aggregate $C_n = n[A]$.

Using the above monomeric notation we may express Eq. 4.31 as follows.

$$n\,\mu_1 = n\left(\mu_1^0 + k_B T \ln C_1\right) = n\mu_n^0 + k_B T \ln\left(\frac{C_n}{n}\right) = n\mu_n \qquad (4.32)$$

Dividing through by n yields a remarkably simple equilibrium condition.

$$\boxed{\mu_1 = \mu_n} \qquad (4.33)$$

This indicates that, at equilibrium, *the chemical potentials of the free and bound monomers are equal to each other*. Since the same is true for aggregates of any size, this relation implies that the chemical potentials of all the monomers in aggregates of any size must all be equal to each other at equilibrium.

Equation 4.32 may be rearranged to obtain the following expression for the equilibrium concentration of monomers in aggregates of size n in terms of the difference between the chemical potentials of the aggregated and free monomers under standard state concentration conditions $\Delta\mu_n^0 = \mu_n^0 - \mu_1^0$.[8]

$$\boxed{C_n = n\left(C_1 e^{-\beta\Delta\mu_n^0}\right)^n} \qquad (4.34)$$

The *critical aggregate concentration* C_A may be defined by requiring that the free monomer and aggregate concentrations will be equal to each other $[M] = C_1 = \frac{1}{n}C_n = [A]$ when $C_1 = C_A$. This definition, combined with Eq. 4.34, results in the following expression for C_A.[9]

$$\boxed{C_A = \left(e^{+\beta\Delta\mu_n^0}\right)^{\frac{n}{n-1}} = \left(\frac{1}{K_A}\right)^{\frac{1}{n-1}}} \qquad (4.35)$$

Note that for a binary ($n = 2$) reaction Eq. 4.35 indicates that $C_A = 1/K_A$, thus confirming that C_A is equivalent to the concentration at which $[M] = [A]$ (as further discussed on page 131). Moreover, when n is large, then

---

[8] Equation 4.34 is obtained by first dividing Eq. 4.32 through by n to get $\Delta\mu_n^0 = \mu_n^0 - \mu_1^0 = -k_B T \ln[(1/C_1)(C_n/n)^{1/n}]$ and then solving for C_n.

[9] The first equality in Eq. 4.35 is obtained from Eq. 4.34 by setting $C_n = nC_1$ and then replacing C_1 by C_A and isolating C_A. The second equality in Eq. 4.35 is obtained using Eq. 4.29, and noting that $\Delta G^0/RT = n\Delta\mu_n^0/k_B T = \beta n\Delta\mu_n^0$, so $K_A = e^{-\Delta G^0/RT} = e^{-\beta n\Delta\mu_n^0}$ and thus $e^{+\beta n\Delta\mu_n^0} = 1/K_A$.

$n/(n-1) \approx 1$, and so $C_A \approx e^{+\beta\Delta\mu_n^0}$. The latter approximate equality is sometimes taken as the definition of the critical aggregation concentration, but Eq. 4.35 is preferable, as it is applicable to aggregates of any size and assures that C_A is always equivalent to the value of C_1 at which $C_1 = [M] = [A] = \frac{1}{n}C_n$.

Notice that Eq. 4.35 implies that C_A and K_A each pertain to a particular aggregate size n. Thus, both C_A and K_A are, in general, functions of n.

Equations 4.34 and 4.35 may be combined to obtain the following relation between the concentrations of aggregated C_n and free C_1 monomers at equilibrium.

$$\boxed{C_n = nC_A\left(\frac{C_1}{C_A}\right)^n} \tag{4.36}$$

Note that when $C_1 = C_A$ then Eq. 4.36 implies that $C_n = nC_A = nC_1$, or equivalently $\frac{1}{n}C_n = C_1$ (again confirming that $[A] = [M]$).

Equations 4.34 through 4.36 are quite general, as they may be applied to situations in which aggregates of various sizes are simultaneously in equilibrium with each other. For example, if we know how $\Delta\mu_n^0$ depends on n, then we can use Eq. 4.34 to predict the resulting aggregate-size distribution C_n. Conversely, an experimentally measured aggregate-size distribution C_n may be used to determine $\Delta\mu_n^0$.

As a simple example, it is interesting to consider what will happen if we assume that $\Delta\mu_n^0 = \mu_n^0 - \mu_1^0$ has the same value for all aggregates. Under such conditions Eq. 4.35 implies that C_A would also be approximately independent of n. In other words, this corresponds to a system in which monomers have little preference for one aggregate size over another. At low concentrations, when $C_1 < C_A$ (so that $C_1/C_A < 1$), Eq. 4.36 predicts that C_n will be a rapidly decreasing function of n. In other words, no large aggregates will form when $C_1 < C_A$. However, when C_1 approaches C_A (so that $C_1/C_A \approx 1$), then Eq. 4.36 predicts that aggregates of all sizes will abruptly begin to form. This is consistent with our experience of what happens when the relative humidity of the atmosphere approaches 100%. In other words no condensation takes place until the relative humidity reaches 100%, at which point water droplets of various sizes abruptly begin to form. Thus, a relative humidity of 100% is equivalent to the point at which the water vapor concentration C_1 reaches C_A.

More generally, we may apply Eqs. 4.34 through 4.36 to situations in which $\Delta\mu_n^0$ is not constant, and depends in some more interesting way on n. For example, the formation of micelles from surfactants (such as soap molecules), or the formation of biological vesicles from phospholipids, often produces aggregates with a relatively narrow size distribution, centered around some particular aggregate size n^*. This behavior suggests that $\Delta\mu_n^0$ has a minimum value near n^*, so that a chemical potential gradient drives the system towards the formation of aggregates of this special size. We might also expect that the width of the aggregate-size distribution should be related to how steeply $\Delta\mu_n^0$ drives the system towards n^*.

A wide variety of such aggregate formation processes may be represented by considering what happens if $\Delta\mu_n^0$ is a quadratic function of n, with a minimum at $n = n^*$. More specifically, let's assume that $\mu_n^0 - \mu_{n^*}^0 = \kappa^*(n - n^*)^2 = \kappa^*\Delta n^2$, where κ^* is a positive number that dictates the steepness of the quadratic chemical potential well (centered around $n = n^*$). Since $\Delta\mu_n^0 = \mu_n^0 - \mu_1^0 = (\mu_n^0 - \mu_{n^*}^0) + (\mu_{n^*}^0 - \mu_1^0)$ we may express $\Delta\mu_n^0$ as follows, in terms of $\kappa^*\Delta n^2 = \kappa^*(n - n^*)^2$ and $\Delta\mu_{n^*}^0 = \mu_{n^*}^0 - \mu_1^0$.

$$\Delta\mu_n^0 = \kappa^*\Delta n^2 + \Delta\mu_{n^*}^0 \tag{4.37}$$

When Eq. 4.37 is combined with Eq. 4.34, we obtain the following equation for the aggregate-size distribution.

$$C_n = n\left[C_1 e^{-\beta\Delta\mu_{n^*}^0} e^{-\beta\kappa^*\Delta n^2}\right]^n \tag{4.38}$$

Notice that this expression depends on temperature ($\beta = 1/k_BT$), as well as on the quadratic chemical potential parameters $\Delta\mu_{n^*}^0$, κ^*, and n^*. The free monomer concentration C_1 establishes the chemical potentials of all the mutually equilibrated aggregates (and depends on the total number of monomers that have been added to the system).

We may further manipulate Eq. 4.38 to obtain the following expression for the aggregate concentration $[A] = \frac{1}{n}C_n$ in terms of the aggregation parameters C_A^*, κ^*, and n^*, and the variables n and C_1.[10]

$$[A] = \frac{C_n}{n} = \left[C_A^*\left(\frac{C_1}{C_A^*}\right)^{n^*} e^{-\beta n^*\kappa^*\Delta n^2}\right]^{n/n^*} \tag{4.39}$$

Equation 4.39 predicts that the aggregate-size distributions should have an approximately Gaussian shape. More specifically, notice that when $n \approx n^*$ (and so $n/n^* \approx 1$), Eq. 4.39 predicts that C_n should be approximately proportional to $e^{-\beta n^*\kappa^*\Delta n^2} = e^{-\Delta n^2/(2\sigma^2)}$, which is a Gaussian function peaked at $n = n^*$, with a standard deviation (width) of $\sigma = \sqrt{k_BT/2n^*\kappa^*}$.

Figure 4.2 shows examples of aggregate-size distributions predicted using Eq. 4.39, for a system with $n^* = 100$, $\sigma = 10$, and $C_A^* = 0.01$ M. Notice that the formation of aggregates is again quite abrupt, as virtually no aggregation takes place when $C_1 < C_A$, while when C_1 approaches C_A^*

[10] Equation 4.39 was obtained from Eq. 4.38 using the fact that $X^n = (X^{n^*})^{n/n^*}$ and noting that Eq. 4.35 implies that $e^{-\beta\Delta\mu_{n^*}^0} = (1/C_A^*)^{(n^*-1)/n^*}$, and so $(e^{-\beta\Delta\mu_{n^*}^0})^{n^*} = e^{-\beta n^*\Delta\mu_{n^*}^0} = (1/C_A^*)^{(n^*-1)} = C_A^*(1/C_A^*)^{n^*}$ and $(e^{-\beta\kappa^*\Delta n^2})^{n^*} = e^{-\beta n^*\kappa^*\Delta n^2}$.

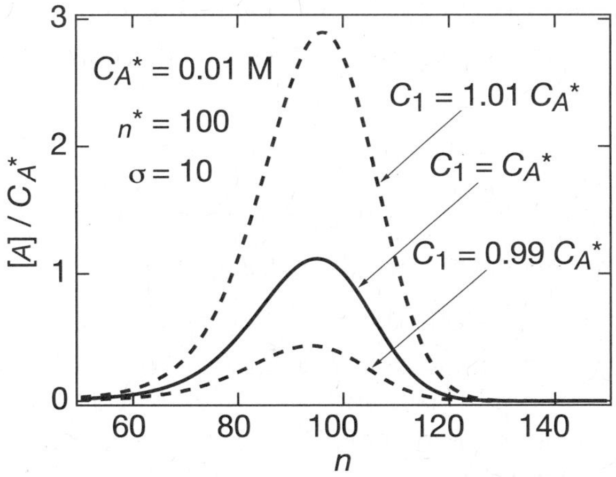

FIGURE 4.2 The solid curve represents the aggregate-size distribution predicted using Eq. 4.39 when $C_1 = C_A^* = 0.01$ M, in a self-assembling system with $n^* = 100$ and $\sigma = 10$. The dashed curves illustrate the abrupt onset of aggregation at C_A^*, as they reveal the large change in aggregate concentration induced by a 1% change in the free monomer concentration.

the aggregate concentration grows dramatically, with an aggregate-size distribution that is determined by the values of n^* and σ.

The aggregate-size distributions predicted by Eq. 4.39, and shown in Figure 4.2, do not have a perfectly Gaussian shape (and are not centered at exactly $n = n^*$). This is because the n/n^* exponent in Eq. 4.39 is a function of n (and is only exactly equal to one when $n = n^*$).

4.4 Spontaneous Consequences

The second law (or Postulate II on page 90) requires that any spontaneous process must produce an increase in the entropy of the universe (or any other isolated system). Thus, entropy may be viewed as the driving force for all spontaneous processes in the universe. But, we would also like to know what drives spontaneous processes in various (non-isolated) subsystems within the universe, such as systems at constant temperature and pressure. As we will see, the answer to this question is intimately linked to the Legendre transforms of U. In order to uncover this connection, we must first revisit the first and second laws, as applied to both reversible and irreversible processes.

Reversibility and the Fundamental Equation

A reversible process is *defined* as one in which the system and surroundings remain arbitrarily close to equilibrium *with each other* at all times. We also know that at equilibrium the intensive variables of a system and its surroundings must be equal to each other. (In both cases, the constraints imposed by the boundary between the system and surroundings determine which intensive variables may equilibrate.) This implies that a reversible process has the following characteristics:

- $T = T_{Surr}$ if heat (Q) can be exchanged.
- $P = P_{Surr}$ if mechanical work (W_{PV}) can be exchanged.
- $\mu = \mu_{Surr}$ if chemical work (W_N) can be exchanged.[11]

An irreversible process is one that does not meet one or more of the above criteria, so either $T \neq T_{Surr}$ and/or $P \neq P_{Surr}$ and/or $\mu \neq \mu_{Surr}$. Under such conditions, the amount of heat and work exchanged between the system and surroundings depends on how both the system and the surroundings evolve during the process.

For example, when a closed system undergoes an infinitesimal *irreversible* change, the resulting energy change may be expressed as follows.

$$dU = \delta Q + \delta W_{PV} = \delta Q - P_{Surr} dV \tag{4.40}$$

This indicates that the irreversible work δW_{PV} is proportional to the external pressure (P_{Surr}) times the volume change of the system (dV). In other words, δW_{PV} depends on properties of both the system and its surroundings.

On the other hand, if the system were to undergo the same transformation along a *reversible* path, then $dS = \delta Q/T$ so $\delta Q = TdS$ and $P = P_{Surr}$ so $\delta W_{PV} = -PdV$, and thus the first law implies that

$$dU = \delta Q + \delta W_{PV} = TdS - PdV \tag{4.41}$$

Notice that the heat and work exchanges in this reversible process depend only on properties of the system.

Now consider a more general *irreversible* process, involving both chemical and mechanical work, which leads to the following energy change.

$$dU = \delta Q + \delta W_{PV} + \delta W_N = \delta Q - P_{Surr} dV - \sum_i \mu_{i,Surr}\, dN_i \tag{4.42}$$

If the system were to undergo the same transformation along a *reversible* path, then $\delta Q = \delta Q = TdS$, $P = P_{Surr}$ and $\mu_i = \mu_{i,Surr}$, and so

$$dU = \delta Q + \delta W_{PV} + \delta W_N = TdS - PdV - \sum_i \mu_i dN_i \tag{4.43}$$

Again, the latter expression depends only on properties of the system. Note that Eq. 4.43 is identical to the fundamental equation (Eq. 3.8). In other words, the terms on the right-hand side of the fundamental equation

[11] In a multicomponent system this restriction applies to each component.

represent the amount of heat and work that would be exchanged if the system were transformed along a reversible path.[12]

Rectification of Second Law Inequalities

The Clausius inequality may be used to better understand the physical significance of the second law, when applied to both reversible and irreversible processes. We will see that it is easier to trace the energetic and entropic consequences of irreversibility if we first *rectify* the Clausius inequality by converting it to an equality.

The Clausius inequality Eq. 2.7 implies that $\eth Q \leq \delta Q$, where $\eth Q$ is the heat exchanged in the processes of interest and δQ is the heat that would have been exchanged if the process had been carried out reversibly.[13] We may rectify this inequality by expressing it in the following form.

$$\eth Q = \delta Q - \eth \epsilon \tag{4.44}$$

In other words, the new parameter $\eth \epsilon$ (with units of energy) quantifies the irreversibility of a given process. Note that the Clausius inequality requires that $\eth \epsilon$ *is always positive*, and can only be equal to zero for a reversible process.

$$\eth \epsilon \geq 0 \tag{4.45}$$

We will also see that $\eth \epsilon$ is exactly equal to the magnitude of the additional work that could have been performed by the system (on the surroundings) if the process had been carried out reversibly. Equivalently, $\eth \epsilon$ is equal to the amount of work that one would have to perform on the system in order to undo the effects of an irreversible (spontaneous) transformation.

---

[12] However, it is important to keep in mind that the fundamental equation is not restricted to reversible processes, as it may always be used to determine the energy change of a system from its equation of state. For example, one may carry out a process along what is called a quasi-static irreversible path, in which the system's energy changes exactly as indicated by Eq. 4.43, but no heat or work is exchanged. A simple example of such a process is one in which an ideal gas is allowed to slowly expand without doing any work or absorbing any heat. If the same transformation had been carried out reversibly, then both heat and work would necessarily be exchanged, and the amounts of each could be determined by integrating TdS and $-PdV$ in the fundamental equation.

[13] This form of the Clausius inequality applies only to processes in which the system is coupled to a bath so that $T = T_{Surr}$, although T and T_{Surr} may in general vary throughout the process. The Clausius inequality and second law lead to additional interesting consequences of the second law when T and T_{Surr} differ by more than an infinitesimal amount, as discussed in the *Irreversible Heat Transfer* subsection on pages 143–147.

Since energy is a state function, the first law implies that the sum of the heat and work exchange is independent of whether the process is carried out reversibly or irreversibly.

$$dU = \delta Q + \delta W = \eth Q + \eth W \tag{4.46}$$

The latter equality can be rearranged (and combined with Eq. 4.44) to produce the following two equivalent rectified forms of the Clausius inequality.

$$\delta Q - \eth Q = \eth W - \delta W = \eth\epsilon \tag{4.47}$$

In other words, a spontaneous process (for which $\eth\epsilon > 0$) will invariably absorb less heat from the surroundings, and perform less work on the surroundings, than a reversible process (for which $\eth\epsilon = 0$).[14]

The physical significance of $\eth\epsilon$ may be further revealed by considering a simple process in which an ideal gas expands from a smaller volume V_1 to a larger volume V_2, along either an irreversible or a reversible path, as illustrated in the following exercise.

Exercise 4.7

Compare the value of $\eth\epsilon$ associated with a highly irreversible (and spontaneous) free expansion of an ideal gas to the work that would have been exchanged if the gas had undergone a reversible isothermal expansion (at the same temperature).

Solution. If the initial volume of the gas is V_1 and it undergoes a free expansion to a final volume of $V_2 > V_1$ (at temperature T), then no work would be exchanged between the system and surroundings, and so $W = \int \eth W = 0$. Moreover, since the expansion is isothermal (and the gas is ideal), $\Delta U = Q + W = 0$, and so

$$\textit{Irreversible free expansion:}\quad W = \int \eth W = -\int \eth Q = -Q = 0$$

On the other hand, if the process were carried out reversibly (and isothermally), then we could calculate the work exchange by integrating the equation of state of the gas to obtain

$$\textit{Reversible isothermal expansion:}\quad W = -\int P\,dV = -RT\ln(V_2/V_1) = -Q$$

where the last equality again follows from the fact that the process is isothermal and so $Q = -W$. Thus, Eq. 4.47 implies that

$$\int \delta W - \int \eth W = -RT\ln(V_2/V_1) - 0 = -\int \eth\epsilon < 0$$

which is exactly equal to the amount of (negative) work that would have been done by the system (on the surroundings) if the free expansion had been carried out along a reversible (isothermal) path.

[14] Recall that work done *by the system on the surroundings* is negative, so an irreversible process is one in which $\eth W$ is less negative than δW, which means that less work is done by the system on the surroundings.

Note that the above exercise implies that $\int \delta\epsilon > 0$ is equal to the amount of (positive) work that one would have to perform on the gas in order to undo the effects of its irreversible expansion, by reversibly compressing it from V_2 back to V_1. Moreover, although the initial and final states of the system are the same in both the reversible and irreversible expansions, the surroundings are not left in the same final state after the reversible and irreversible processes. The irreversible (free) expansion produces no entropy change in the surroundings (because $Q = 0$), so $\Delta S_{\text{Univ}} = \Delta S = R\ln(V_2/V_1)$. In the reversible expansion, on the other hand, the entropy increase of the system is exactly compensated by an entropy decrease in the surroundings, so $\Delta S_{\text{Univ}} = \Delta S + \Delta S_{\text{Surr}} = R\ln(V_2/V_1) - R\ln(V_2/V_1) = 0$.

Spontaneous Processes in Non-Isolated Systems

Processes that are carried out at constant temperature or pressure are in general not isolated, as they may involve the exchange of heat (from a constant temperature bath) or the exchange of mechanical (pressure-volume) work between the system and surroundings. Here we will see how the rectification procedure described in the previous section may be used to identify the thermodynamic functions that determine the direction in which spontaneous processes will proceed in non-isolated systems.

For an isothermal process, Eqs. 3.47 and 4.47 may be combined to obtain the following relationship between the Helmholtz energy and work along either a reversible or an irreversible path.

$$dA = \delta W = \delta W - \delta\epsilon \tag{4.48}$$

Any process that is carried out at constant temperature and volume in a closed system cannot involve the exchange of either mechanical or chemical work between the system and surroundings, so $\delta W = \delta W_{PV} + \delta W_N = 0$. For any such process, the above expression implies that dA can only be negative (or zero).

$$dA|_{T,V} = -\delta\epsilon \leq 0 \tag{4.49}$$

In other words, *any spontaneous process taking place in a closed system at constant T and V must produce a decrease in Helmholtz energy.*[15]

A similar procedure may be used to show that Gibbs energy determines the direction in which spontaneous transformations will proceed in any closed system at constant temperature and pressure. Recall that Eq. 3.49 requires that Gibbs energy is equivalent to the amount of reversible chemical work that may be obtained from any process at constant temperature

[15] It is important to note that a closed system is not the same as a nonreactive system. In other words, spontaneous chemical reaction can still take place within a closed system. The maximum amount of work that could be theoretically extracted from such spontaneous reactions is $-\int \delta\epsilon$, as we will see.

and pressure $dG|_{T,P} = \delta W_N = \delta W_N - \delta\epsilon$. Thus, in a closed system (for which $\delta W_N = 0$) we immediately obtain the following important result.

$$dG|_{T,P} = -\delta\epsilon \leq 0 \tag{4.50}$$

In other words, *any spontaneous process taking place in a closed system at constant T and P must produce a decrease in Gibbs energy.*

The above procedure may be readily extended to many other sorts of processes. One interesting example is a process carried out in a closed system at constant S and V. Since no mechanical or chemical work exchange can occur between a closed, constant-volume system and its surroundings, the energy change of the system is $dU = \delta Q = \delta Q - \delta\epsilon = TdS - \delta\epsilon$. Moreover, if we assume that S is also constant (so $TdS = 0$), we obtain the following result.

$$dU|_{S,V} = -\delta\epsilon \leq 0 \tag{4.51}$$

This implies that *in any isentropic (constant S) process in which no mechanical or chemical work is exchanged, the systems will be driven spontaneously toward states of lower energy.*

The above result is related to our everyday experience of the fact that objects tend to spontaneously fall downhill, and will only go uphill if work is performed to lift them. More specifically, consider a river flowing gently toward the sea. If the river has an approximately constant temperature and velocity (and if we neglect evaporation), then we may reasonably assume that the entropy and volume of the river remain essentially constant as it rolls along. Thus, such a river closely approximates a closed system of constant S and V, and so the fact that it flows downhill (to lower potential energy) is consistent with the second law prediction expressed in Eq. 4.51.

In summary, the following consequences of the Clausius inequality pertain to closed systems in which different variables are held constant.

Spontaneity and Equilibrium Conditions		
At constant	Spontaneous	Equilibrium
S, V	$dU\|_{S,V} < 0$	$dU\|_{S,V} = 0$
S, P	$dH\|_{S,P} < 0$	$dH\|_{S,P} = 0$
T, V	$dA\|_{T,V} < 0$	$dA\|_{T,V} = 0$
T, P	$dG\|_{T,P} < 0$	$dG\|_{T,P} = 0$

The above results also imply that spontaneous chemical reactions can only proceed in a direction that lowers the sum of all the chemical potentials of the system's components $\sum_i \mu_i dN_i < 0$. For example, at constant S and V the fundamental equation (Eq. 3.8 on page 87) implies that $dU = \sum_i \mu_i dN_i$, and so the second column in the above table implies that $\sum_i \mu_i dN_i < 0$ whenever a spontaneous chemical reaction occurs within such a system. Similarly, at constant T and P the fundamental equation for Gibbs energy (Eq. 3.48 on page 110) implies that $dG = \sum_i \mu_i dN_i$. Thus, the second column in the above table requires that $\sum_i \mu_i dN_i < 0$ for any spontaneous chemical reaction that occurs at constant T and P. Clearly the same argument can be repeated for dH and dA to show that $\sum_i \mu_i dN_i < 0$ for any spontaneous chemical processes. Furthermore, the third column implies that $\sum_i \mu_i dN_i = 0$ once all chemical reactions have reached equilibrium.

If we introduce the *reaction variable* $\tilde{N}$, defined such that $d\tilde{N} \equiv dN_i/\nu_i$, where ν_i represent the corresponding stoichiometric coefficients (as defined in Eq. 3.24), then $\sum_i \mu_i dN_i = \sum_i \mu_i \nu_i d\tilde{N}$. Since the latter sum must be equal to zero at equilibrium (for an infinitesimal nonzero value of $d\tilde{N}$), the following identity applies to all chemical equilibria.[16]

$$\boxed{\sum_i \nu_i \mu_i = 0} \text{ for a reaction at equilibrium} \tag{4.52}$$

Moreover, all spontaneous chemical processes will proceed in the direction that leads to a decrease in the stoichiometric sum of the reactant and product chemical potentials.[17]

$$\boxed{\sum_i \nu_i \mu_i < 0} \text{ for a spontaneous reaction} \tag{4.53}$$

Irreversible Heat Transfer

All of the results obtained in the previous two subsections were derived from the Clausius inequality as expressed in Eq. 2.7. This form of the Clausius inequality is restricted to systems that remain in thermal contact with a temperature bath. In other words, Eq. 2.7 implies that any heat that is transferred out of the system at temperature T is absorbed by the surrounding bath that is at essentially the same temperature.

In this section we consider what would happen if heat were transferred from a system at temperature T to a bath at a significantly different temperature T_{Surr}. For example, this would be the case if we allowed a hot metal block to slowly cool while surrounded by a gas of a much lower temperature (assuming that the gas volume is very large, and that it is sufficiently well

---

[16] Equation 4.52 is the same as Eq. 3.24, but now we have proven that it holds for all chemical equilibria, regardless of the nature of the constraints imposed by the boundary between the system and surroundings.

[17] The connection between Eqs. 4.52 and 4.53 and the thermodynamics of chemical reactions and molecular self-assembly processes are described in Sections 4.2 and 4.3.

circulated that it remains at essentially the same low temperature while the metal block slowly cools). Such a process is an example of a thermally irreversible heat exchange. For any such process Eq. 2.7 must be replaced by the following more generally applicable form of the Clausius inequality.

$$\frac{\delta Q}{T} \geq \frac{\delta Q}{T_{\text{Surr}}} \tag{4.54}$$

Note that this inequality is obtained directly from Eq. 2.6, which implies that $dS_{\text{Univ}} = dS + dS_{\text{Surr}} = \delta Q/T + \delta Q_{\text{Surr}}/T_{\text{Surr}} = \delta Q/T - \delta Q/T_{\text{Surr}} \geq 0$.[18]

Thermodynamic derivations based on Eq. 4.54 lack some of the elegant simplicity of those based on the more restricted form of the Clausius inequality Eq. 2.7. For example, Eq. 4.46 does not apply under the more general conditions associated with Eq. 4.54. In such cases the excess entropy produced in the universe (or any other isolated system) is related in the following way to the associated heat and work exchanges.

$$= \frac{\delta Q}{T} - \frac{\delta Q}{T_{\text{Surr}}} = \frac{dU - \delta W}{T} - \frac{dU - \delta W}{T_{\text{Surr}}} \tag{4.55}$$

Consider, for example, a hot metal block surrounded by a cooler gas, as described above. In particular, assume that the temperature of the hot metal block (system) decreases from T_1 to T_2 as the result of the slow exchange of heat with the surrounding gas of temperature T_{Surr}. For simplicity, we may also assume that the system remains at constant volume and composition, so no mechanical or chemical work is exchanged during the cooling process ($\delta W = \delta W = 0$), and that the heat capacity C_V of the system is effectively constant over the temperature range of interest. This implies that a total heat exchange of $Q = C_V(T_2 - T_1) < 0$ occurs during the cooling process.

Now consider the following two different paths, both of which lead to the same heat exchange. One is an irreversible path in which the initially hot system at $T = T_1$ is suddenly exposed to the cold gas of temperature $T_{\text{Surr}} = T_2$. The second path is a reversible one in which the bath temperature closely tracks the temperature of the system $T \approx T_{\text{Surr}}$ as it slowly cools. In other words, in this case the bath temperature changes so that it continuously stays infinitesimally close to the temperature of the

[18] In replacing $\delta Q_{\text{Surr}}/T_{\text{Surr}}$ by $-\delta Q/T_{\text{Surr}}$ we have assumed that the surroundings behave like an ideal heat bath, in the sense that any heat added to the bath has the same effect on its entropy as the same quantity of heat exchanged in a reversible process. It is also important to note that even this more general form of the Clausius inequality requires that both the system and bath are each internally thermally equilibrated (such that they each have a well-defined temperature).

system. Approximating such a reversible cooling process experimentally would require exposing the hot metal block sequentially to a series of baths, each of which has a temperature just below that of the block. So, once the block cools to the temperature of the first bath, then it is moved into a bath of a slightly lower temperature, and so on, until it finally reaches the last bath of temperature T_2.

Notice that the amount of heat required to decrease the temperature of the block from T_1 to T_2 is exactly the same along both paths, $Q = C_V(T_2 - T_1)$. However, the second law predicts that the reversible process will produce no excess entropy, while the irreversible process produces an increase in the entropy in the universe. The magnitude of the latter excess entropy may be obtained by noting that the heat exchange required to cool the system by an infinitesimal amount is $\delta Q = \delta Q = C_V dT$, and so the total entropy change of the system is

$$\Delta S = \int_{T_1}^{T_2} \frac{\delta Q}{T} = \int_{T_1}^{T_2} \frac{C_V}{T} dT = C_V \ln\left(\frac{T_2}{T_1}\right)$$

The entropy change of the bath is

$$\Delta S_{\text{Surr}} = -\frac{\int_{T_1}^{T_2} \delta Q}{T_{\text{Surr}}} = -\frac{\int_{T_1}^{T_2} C_V dT}{T_2} = -\frac{C_V(T_2 - T_1)}{T_2}$$

So, the total excess entropy produced as the result of the irreversible heat exchange is given by the following expression.

$$\Delta S_{\text{Univ}} = C_V \left[\ln\left(\frac{T_2}{T_1}\right) - \left(1 - \frac{T_1}{T_2}\right)\right] \tag{4.56}$$

In order to better understand the physical significance of this result, it is convenient to reexpress it in terms of the unitless variable $\tau \equiv (T_2 - T_1)/T_1$, which measures the initial temperature difference between the bath at T_2 and the system at T_1 (see the following exercise for further details).

$$\Delta S_{\text{Univ}} = C_V \left[\ln(1 + \tau) - \left(\frac{\tau}{1 + \tau}\right)\right] \tag{4.57}$$

If the temperature difference between the system and bath is small ($|\tau| << 1$), we may expand the above expression to second order in τ to obtain the following interesting result.[19]

$$\Delta S_{\text{Univ}} \approx C_V \left[\left(\tau - \frac{\tau^2}{2}\right) - \left(\tau - \tau^2\right)\right] = \left(\frac{C_V}{2}\right)\tau^2 \tag{4.58}$$

[19] The Taylor series expansions of the following two functions (when $x << 1$) are $\ln(1 + x) = x - (x^2/2) + \mathcal{O}(x^3)$ and $x/(1 + x) = x[1/(1 + x)] = x - x^2 + \mathcal{O}(x^3)$, where the symbol $\mathcal{O}(x^3)$ represents all terms of cubic or higher order.

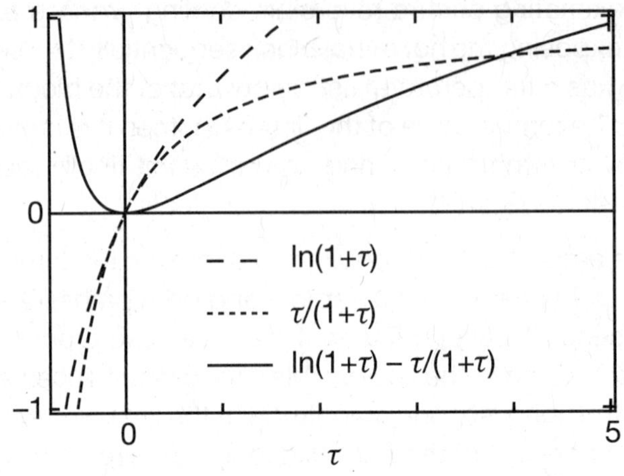

FIGURE 4.3 The solid curve is proportional to the entropy produced in the universe as a result of an irreversible heat exchange between a metal block that is initially at temperature T_1 and then cools by exchanging heat with an ideal gas bath at temperature T_2, where $\tau = (T_2 - T_1)/T_2$ (see Eq. 4.57). The dashed curves show that $\ln(1+\tau)$ is invariably larger than $\tau/(1+\tau)$, and so their difference is always positive.

Note that the terms that are linear in τ exactly cancel, so $\Delta S_{\text{Univ}} = 0$ to first order in τ. *This implies that no excess entropy is produced as long as the difference in temperature between T and T_{Surr} remains infinitesimal.* The remaining second-order term ($\propto \tau^2$) is invariably positive. Thus, Eqs. 4.57 and 4.58 apply equally well to both cooling ($\tau < 0$) and heating ($\tau > 0$) processes, and so a positive excess entropy is produced regardless of the sign of the heat exchange. Figure 4.3 shows a plot of each of the terms that contribute to Eq. 4.57. The fact that the solid curve remains positive at all finite (positive or negative) values of τ confirms the second law prediction that a positive excess entropy is invariably produced regardless of the magnitude of the temperature difference between the system and surroundings.

Exercise 4.8

Consider an irreversible heat transfer process between a system at temperature T_1 and a bath at temperature T_2 (assuming that the system's heat capacity C_V is temperature-independent).

- Show that Eqs. 4.56 and 4.57 are equivalent to each other.

Solution. If we replace τ by $(T_2 - T_1)/T_1$ in Eq. 4.57, we obtain

$$\Delta S_{\text{Univ}} = C_V \left[\ln\left(1 + \frac{T_2 - T_1}{T_1}\right) - \left(\frac{\frac{T_2-T_1}{T_1}}{1 + \frac{T_2-T_1}{T_1}}\right)\right]$$

$$= C_V\left[\ln\left(\frac{T_1+T_2-T_1}{T_1}\right)-\left(\frac{\frac{T_2-T_1}{T_1}}{\frac{T_1+T_2-T_1}{T_1}}\right)\right]$$

$$= C_V\left[\ln\left(\frac{T_2}{T_1}\right)-\left(\frac{T_2-T_1}{T_2}\right)\right]$$

$$= C_V\left[\ln\left(\frac{T_2}{T_1}\right)-\left(1-\frac{T_1}{T_2}\right)\right]$$

- Explain why the physically relevant range of τ values is $-1<\tau<\infty$.

Solution. Since $\tau=(T_2-T_1)/T_1$, the value of τ will approach ∞ when the *bath* temperature becomes arbitrarily large ($T_2 >> T_1$), while when the *system* temperature becomes arbitrarily large ($T_1 >> T_2$), then τ will approach $-T_1/T_1=-1$.

HOMEWORK PROBLEMS

Problems That Illustrate Core Concepts

1. Verify the following identity: $\left(\frac{\partial P}{\partial T}\right)_V=\frac{\alpha_P}{\kappa_T}$.

2. Verify the following identity: $\left(\frac{\partial S}{\partial V}\right)_U=\frac{P}{T}$.

3. Reduce the following partial derivatives to an expression containing only α_P, κ_T, C_P, C_V and/or other thermodynamic variables:

(a) $\left(\frac{\partial S}{\partial P}\right)_T$ (b) $\left(\frac{\partial P}{\partial S}\right)_V$ (c) $\left(\frac{\partial P}{\partial S}\right)_H$

(d) $\left(\frac{\partial V}{\partial T}\right)_U$ (e) $\left(\frac{\partial U}{\partial P}\right)_T$ (f) $\left(\frac{\partial U}{\partial P}\right)_V$

(g) $\left(\frac{\partial H}{\partial P}\right)_T$ (h) $\left(\frac{\partial U}{\partial T}\right)_P$

4. Manipulate the derivative $(\partial H/\partial T)_P$ to show that it is equal to $T(\partial S/\partial T)_P$, without making use of the fact that both derivatives are equal to C_P.

5. The following are two expressions for the internal pressure $(\partial U/\partial V)_T$:

$$\left(\frac{\partial U}{\partial V}\right)_T=-\left[\frac{\partial(P/T)}{\partial(1/T)}\right]_V=T^2\left[\frac{\partial(P/T)}{\partial T}\right]_V$$

(a) Manipulate the derivative on the far left to show that it is equal to

$$T\left(\frac{\partial P}{\partial T}\right)_V-P$$

(b) Manipulate the derivative on the far right to show that it is also equal to the above expression.

(c) Show that the derivative in the middle is equal to the one on the far right.

6. Verify the following identities:

(a) $\left(\frac{\partial S}{\partial N}\right)_{U,V}=-\frac{\mu}{T}$ (b) $\left(\frac{\partial P}{\partial T}\right)_S=\frac{C_P}{TV\alpha_P}$

(c) $\frac{C_P}{C_V}=\frac{\kappa_T}{\kappa_S}$ (d) $\left(\frac{\partial \alpha_P}{\partial P}\right)_T=-\left(\frac{\partial \kappa_T}{\partial T}\right)_P$

(e) $\frac{1}{V}\left(\frac{\partial V}{\partial T}\right)_S=-\frac{C_V\kappa_T}{TV\alpha_P}$

7. The gas phase dimerization reaction $2NO_2 \leftrightharpoons N_2O_4$ has an equilibrium constant of $K_c=4.72$ at 373 K (when all concentrations are measured in M units).

(a) What is ΔG° for this reaction?

(b) At what reactant and product concentrations is the reaction free energy equal to $\Delta G°$?

(c) Use the quadratic equation to obtain an expression for $[NO_2]$ as a function of the total nitrogen concentration $[N]_T = [NO_2] + 2[N_2O_4]$, and plot both $[NO_2]$ and $[N_2O_4]$ as functions of $[N]_T$ (over a range of $0 < [N]_T < 1$).

(d) How is $[NO_2]$ related to K_c when $[NO_2] = [N_2O_4]$, and what is the value $[N]_T$ at this point?

(e) The reaction $2\ H \rightleftharpoons H_2$ has an equilibrium constant of $K_c \approx 5 \times 10^{14}$ at 1000 K. At what total hydrogen concentration $[H]_T$ will the equilibrium concentration of atomic hydrogen [H] be equal to that of diatomic hydrogen $[H_2]$?

8. A particular surfactant solution has a critical micelle concentration of 0.01 M and forms aggregates (micelles) with an approximately Gaussian aggregate-size distribution containing 120 ± 12 surfactant molecules at 300 K.

(a) Make a plot of [M] vs. $[M]_T = [M] + 120[A]$ for this aggregation reaction, and mark the location of critical micelle concentration [cmc] on both the horizontal and vertical axes of the graph. Note: The easiest way to make such graphs is to start by calculating $[M]_T$ as a function of [M] (over a range of $0 < [M] < 0.97$ [cmc], with at least 100 points over that range), and then invert the axes to display [M] vs. $[M]_T$ (and set the ranges of both the horizontal and vertical axes to 0–0.011 M).

(b) If the standard chemical potential (per monomer) is assumed to be a quadratic function of the aggregate size n, such that $\mu_n - \mu_{n^*} = \kappa^*(n - n^*)^2$, then the aggregates are predicted to have an approximately Gaussian distribution, $C \exp[-(n - n^*)^2/(2\sigma^2)]$, with $\sigma = \sqrt{k_B T/(2\kappa^* n^*)}$. Use this approximation, and the experimental information given above to estimate the values of n^* and κ^*. Hint: Note that the experimental aggregate-size distribution implies that $\langle n \rangle \pm \sigma = 120 \pm 12$, and start by considering how $\langle n \rangle$ should be related to n^*.

(c) How would you expect the width of the distribution to change if T were increased?

(d) How would you expect the width of the distribution to change if κ^* were larger?

9. Show that a spontaneous chemical process that is carried out in a closed system at constant S and P is driven by a decrease in enthalpy. Hint: Start with the fundamental equation for dH and note that $\sum \mu_i dN_i = \delta W_N$, then use the rectified form of the Clausius inequality.

Problems That Test Your Understanding

10. Reduce the following partial derivatives to an expression containing only α_P, κ_T, C_P, C_V, and/or other thermodynamic variables.

(a) $\left(\frac{\partial S}{\partial P}\right)_T =$ (b) $\left(\frac{\partial P}{\partial T}\right)_G =$

11. Reduce the following partial derivatives to an expression containing only α_P, κ_T, C_P, C_V, and/or other thermodynamic variables.

(a) $\left(\frac{\partial P}{\partial T}\right)_V =$ (b) $\left(\frac{\partial U}{\partial V}\right)_T =$

12. Reduce the following partial derivatives to an expression containing only α_P, κ_T, C_P, C_V, and/or other thermodynamic variables.

(a) $\left(\frac{\partial T}{\partial P}\right)_S =$ (b) $\left(\frac{\partial G}{\partial V}\right)_T =$

13. Consider one mole of a monatomic ideal gas at a pressure of 0.5 MPa and a temperature

of 500 K. Obtain a numerical value, including units, for each of the following partial derivatives for this gas.

(a) $\left(\frac{\partial P}{\partial T}\right)_V =$ (b) $\left(\frac{\partial G}{\partial P}\right)_T =$

14. Consider one mole of a monatomic ideal gas that undergoes a free expansion from an initial volume $V_1 = 1\,\text{m}^3$ to a final volume of $V_2 = 5\,\text{m}^3$ at a constant temperature of 400 K.

(a) What is the Gibbs energy change of the gas (in kJ/mol units)?

(b) How much reversible work exchange would be required to return the gas to its initial state (in kJ/mol units)?

15. Reduce the following partial derivatives to an expression containing only the derivatives α_P, κ_T, C_P, C_V, and/or other thermodynamic variables, and indicate the SI units of the expression.

(a) $\left(\frac{\partial S}{\partial V}\right)_T =$ (b) $\left(\frac{\partial T}{\partial V}\right)_S =$

(c) $\left(\frac{\partial A}{\partial P}\right)_T =$ (d) $\left(\frac{\partial U}{\partial V}\right)_T =$

16. Reduce the derivative $(\partial S/\partial T)_A$ to an expression containing only α_P, κ_T, C_P, C_V, and/or other thermodynamic variables.

17. If n_1 moles of a solute were added to n_0 moles of solvent, while maintaining the system at constant volume and temperature, then the pressure of the system would change by an amount that is determined by the partial derivative, $(\partial P/\partial n_1)_{T,V,n_0}$, which is not very easy to directly measure experimentally. Show that this partial derivative can be reexpressed in terms of the solute partial molar volume, as well as α_P, κ_T, C_P, C_V, and/or other thermodynamic variables.

18. A particular antibody-antigen complex has a dissociation equilibrium constant of $\sim 10^{-10}$ (M). Estimate the antigen concentration at which the concentration of the antigen-antibody complex will be equal to the concentration of the unbound antibody.

19. The isomerization of glucose-6-phosphate (G6P) to fructose-6-phosphate (F6P) has an equilibrium constant of approximately 0.5 (in water at 298 K and 0.1 MPa).

(a) Use the above experimental equilibrium constant to determine the standard Gibbs energy for the isomerization reaction ΔG^0.

(b) Combine your results in (a) with the experimental value of $\Delta S^0 \sim 24$ J/(K mol) for the above isomerization reaction to determine the corresponding standard enthalpy change ΔH^0.

(c) What is the experimental value of $(\partial \Delta G^0/\partial T)_{P,N_i}$?

20. Use the graphs below to fill in the blanks in the following sentences with either a letter (such as A, a, or b, etc.) or a number with units (you may assume that the numbers on the x- and y-axes of both graphs are in M units).

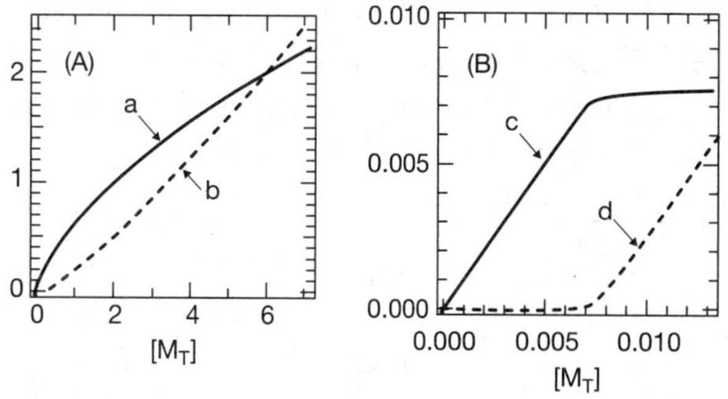

(a) Graph ______ pertains to a low-order aggregation reaction (such as a dimerization equilibrium).

(b) Curve ______ represents the concentration of free monomers in a micelle forming solution.

(c) The critical aggregation concentration for the above micelle formation reaction is approximately ______.

(d) The equilibrium constant (affinity constant) for the above dimerization reaction is approximately ______.

21. The following graphs contain aggregate-size distributions pertaining to four different self-assembly processes.

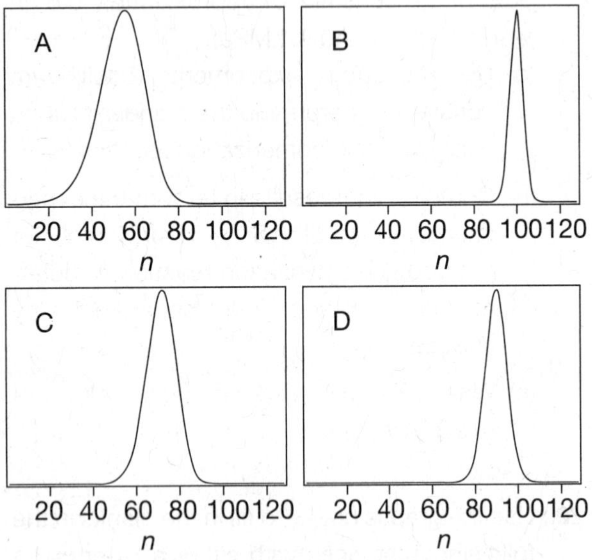

(a) Which of the above self-assembly processes has the largest average aggregate size?

(b) Which of the above self-assembly processes has the weakest dependence of the monomer chemical potential on aggregate size?

(c) What do you expect to happen to the above distributions as the temperature is increased?

(d) What do you expect to happen to the above distributions as the monomer concentration is decreased below the critical aggregation concentration?

22. The following graph shows the relative concentrations of micelles (aggregates) that begin to form in a system as soon as the free monomer concentration increases to about $[M] = C_1 = 0.02$ M. The temperature of the system is 300 K, and you may assume that the chemical potential (per monomer) has the following n-dependence: $\mu_n^0 \approx a(n - b)^2$.

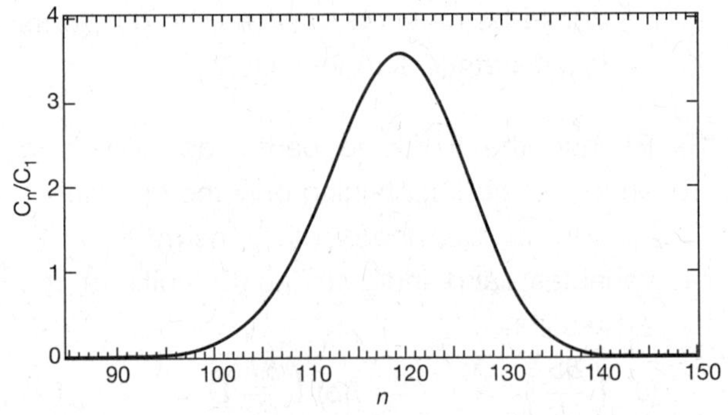

(a) Estimate the critical micelle concentration of this system.

(b) Estimate the value of the parameter b for this system.

(c) Estimate the standard deviation σ for this aggregate-size distribution.

(d) Estimate the value of the parameter a for this system.

Nonideal Systems and Computer Simulations

CHAPTER 5

In Chapter 2 we learned how the energy, entropy, and equation of state of a low-density gas depend on its volume and temperature (as well as the number D of thermally active quadratic terms in its Hamiltonian), as given by Eqs. 2.12, 2.13, and 2.24. The latter equations, combined with the general thermodynamic identities described in Chapter 4, may be used to predict all of the other thermodynamic functions and partial derivatives of any such gas.

This chapter is focused on the more challenging task of predicting the thermodynamic properties of nonideal system in which intermolecular interactions play a significant role. In Sections 5.1 and 5.2 we will learn how to experimentally quantify and theoretically approximate the properties of nonideal gases and liquids. In Sections 5.3 and 5.4 we will see how the statistical mechanics formalism developed by Boltzmann and Gibbs, which we first encountered in Chapter 1, may be extended to produce a general theoretical framework for relating the molecular and macroscopic properties of nonideal systems. In Section 5.5 we will see how the latter results led to the discovery of three remarkable statistical mechanical identities that facilitated the development of computer simulation strategies for predicting the thermodynamic properties of complex fluids and biological systems.

5.1 Quantifying Nonidealities

Recall that the equations of state of real gases are described by the ideal gas law $P = [c]RT = \rho k_B T$ (see Eq. 2.14 and Fig. 2.4). This also implies that the chemical potential of each component within such a gas depends logarithmically on its partial pressure $\mu = \mu^0 + k_B T \ln(P/P_0)$ (see Eq. 3.37).[1] More specifically, μ^0 is the chemical potential of the component of interest when it has a partial pressure of P_0, while μ is its chemical potential at a partial pressure of P. This also implies that the chemical potential of a solute in either a real gas or real liquid solution has a similar logarithmic concentration dependence $\mu = \mu^0 + k_B T \ln([c]/[c]_0)$ (see Eq. 3.39), as long as the solute is sufficiently dilute that solute-solute interactions may be neglected.

In order to describe the experimental properties of nonideal gases and liquids, G.N. Lewis[2] suggested extending the above expressions by replacing P by the *fugacity* f of a dense gas, and $[c]$ by the *activity* a of a concentrated solution. In other words, Lewis suggested expressing the chemical potentials in any chemical system using a form that looks identical to that for ideal gases and solutions.[3]

$$\mu = \mu^0 + k_B T \ln\left(\frac{f}{f_0}\right) \tag{5.1}$$

$$\mu = \mu^0 + k_B T \ln\left(\frac{a}{a_0}\right) \tag{5.2}$$

If the systems behave perfectly ideally, then $f/f_0 = P/P_0$ and $a/a_0 = [c]/[c]_0$, and so we would recover Eqs. 3.37 and 3.39. If the systems are not ideal, then Eq. 5.1 implies that $f/f_0 = e^{\beta \Delta\mu}$ (where $\beta = 1/k_B T$). Thus, given experimental data for the difference between the chemical potential

---

[1] In a multicomponent system a similar expression relates the μ_i to its partial pressure P_i. If the chemical potential were expressed in molar units, then $k_B T$ would be replaced by RT in the above expression. Since the molar chemical potential of component i is equivalent to its partial molar Gibbs energy, $\mathcal{N}_A \mu_i = (\partial G/\partial n_i)_{P,T,n_j \neq n_i} = \overline{G}_i$. If we had expressed the chemical potentials in molar, rather than molecular, units, then the above expression would become $\overline{G}_i = (\partial G/\partial n_i)_{P,T,N_j \neq N_i} = \overline{G}_i^0 + RT \ln(P/P_0)$.

[2] Gilbert Newton Lewis was a leading early twentieth-century physical chemist who presided over the rise of the University of California at Berkeley to the preeminent position it still holds today as a leading Department of Chemistry, and particularly Physical Chemistry. G.N. Lewis also created the Lewis dot-structure representation of chemical bonds, and the concept of Lewis acids and bases. When G.N. Lewis was once asked to define physical chemistry, he famously replied that "physical chemistry is anything that is interesting."

[3] Although the subscripts i have not been included in Eqs. 5.1 and 5.2, it is understood that these expressions apply to each component in the system, each of which will in general have different fugacity or activity values.

(or partial molar Gibbs energy) of a gas at two different pressures, we could determine f/f_0. Similarly, given experimental data for $\Delta\mu$ of a solute at two different concentrations, we could use Eq. 5.2 to obtain $a/a_0 = e^{\beta\Delta\mu}$.

If the reference states in Eqs. 5.1 and 5.2 pertain to a gas or solution that is dilute enough that it behaves ideally, then we may equate $f_0 = P_0$ and $a_0 = [c]_0$. For example, the standard reference states for gases and solutions may be taken to be $P = 1$ atm and $[c]_0 = 1$ M, respectively. Since gases and solutions are often very nearly ideal under these standard state conditions, we may equate $f/f_0 = f/1 = f$ and $a/a_0 = a/1 = a$. Thus, Eqs. 5.1 and 5.2 are sometimes expressed in the following form.

$$\mu = \mu^0 + k_B T \ln f$$

$$\mu = \mu^0 + k_B T \ln a$$

However, it is again important to keep in mind that such expressions contain an implicit reference state of $P_0 = 1$ atm or $[c]_0 = 1$ M, and that μ^0 corresponds to the chemical potential of the gas phase solute when it has a partial pressure of 1 atm or of the liquid phase solute when it has a concentration of 1 M. As the pressure or concentration of the solute increases, we expect f and a to deviate from the solute's actual partial pressure P and concentration $[c]$, respectively.

Since $f = P$ at low pressure and $a = [c]$ at low concentration, the fugacity and activity may be expressed as $f = \phi P$ and $a = \gamma[c]$, where ϕ and γ are the dimensionless fugacity and activity coefficients, both of which approach 1 under ideal (low-pressure or dilute solution) conditions.

Equations 5.1 and 5.2 can provide a convenient means of representing the experimentally measured nonideal properties of dense gases and concentrated solutions. In the absence of experimental data, one might hope to predict f and a using a molecularly detailed theoretical description of the system of interest. More generally, the thermodynamic properties of any chemical system could be obtained from a theoretical prediction of the chemical potential μ_i or *absolute activity* λ_i of each component in the system of interest.

$$\boxed{\lambda_i \equiv e^{\beta\mu_i}} \tag{5.3}$$

Thus, if we wish to predict (rather than measure) μ_i or λ_i, we must make use of a theory that relates the chemical structures of the components in a system to its thermodynamic properties. The remainder of this chapter describes various theoretical and computer simulation strategies for obtaining such predictions.

Exercise 5.1

Use the Gibbs-Duhem equation for a single-component system (Eq. 3.35 on page 104) to obtain an expression that may be used to determine the fugacity of a single-component nonideal gas from experimental pressure and volume measurements (at constant temperature).

Solution. Under isothermal conditions, Eq. 3.35 becomes $d\overline{G} = \overline{V}dP$. Recall that when $\mu = (\partial G/\partial N)_{P,T}$ is expressed in molar units, it is equivalent to $\overline{G} = \mathcal{N}_A\mu = (\partial G/\partial n)_{P,T} = G/n$, and $\overline{V} = (\partial V/\partial n)_{P,T} = V/n$ is the partial molar volume of the gas. Thus, by measuring $\overline{V}$ as a function of P we may determine the chemical potential change of the gas.

$$\mathcal{N}_A \int_{\mu_0}^{\mu} d\mu = \int_{\overline{G}_0}^{\overline{G}} d\overline{G} = \int_{P_0}^{P} \overline{V}dP$$

$$\Delta\mu = \mu - \mu_0 = \frac{1}{\mathcal{N}_A}\int_{P_0}^{P} \overline{V}dP$$

If the reference pressure is $P_0 = 0.1$ MPa (and the gas behaves ideally at this pressure), then $f/f_0 = f/P_0 = f/0.1 = e^{\beta\Delta\mu}$ or $f = 0.1\, e^{\beta\Delta\mu}$ MPa. Note that, if the gas was ideal then we would obtain $e^{\beta\Delta\mu} = P/P_0 = P/0.1$, and thus $f = 0.1P/0.1 = P$, as expected.

5.2 Simple Models of Molecular Fluids

The van der Waals Equation

The first attempt to develop a molecular theory of real (nonideal) fluids was proposed in 1873 by the Dutch physical chemist, Johannes van der Waals.[4] His molecular description of nonideal gases set the stage for over 100 years of theoretical investigations aimed at predicting and understanding the properties of condensed phase (liquid and solid) chemical systems.

[4] Johannes van der Waals was born in in 1837 and began his career as a school teacher. He later went on to obtain degrees in mathematics and physics, and eventually, at the age of 36, obtained a PhD. His now famous van der Waals equation of state was first introduced in his PhD thesis entitled *On the Continuity of the Gas and Liquid State*. Van der Waals received a Nobel prize in 1910 "for his work on the equation of state for gases and liquids." He was never satisfied with the accuracy of his equation and yearned for the development of a more general theory of liquids – in his 1910 Nobel lecture he stated that this task "... continually obsesses me, I can never free myself from it, it is with me even in my dreams."

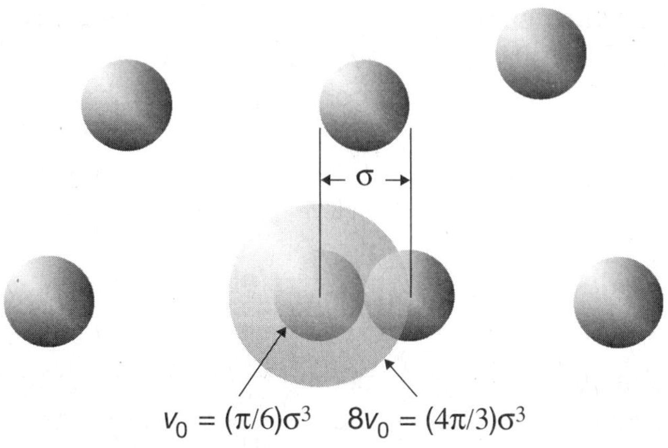

FIGURE 5.1 The volume of a hard sphere of diameter σ is v_0, and volume excluded by a pair of hard spheres is $8v_0$, so the excluded volume per sphere is $8v_0/2 = v_0/4 = \mathbf{b}/\mathcal{N}_A$.

In developing his original molecular theory, van der Waals considered how repulsive (excluded volume) and attractive (cohesive) intermolecular interactions would give rise to deviations from ideal gas behavior. More specifically, van der Waals suggested that the finite sizes of molecules have the effect of reducing the total volume available to a gas from V to $V - \mathbf{b}$, where $\mathbf{b}$ represents the volume from which molecules exclude each other. Thus, if we represent molecules as hard spheres (like ping-pong balls) with a diameter of σ, then the centers of a pair of such spheres cannot approach each other any more closely than σ, as illustrated in Figure 5.1. Such a pair of molecules have an excluded volume of $(4\pi/3)\sigma^3$, and so the excluded volume *per molecule* is half of this volume, $[(4\pi/3)\sigma^3]/2 = (2\pi/3)\sigma^3$. The van der Waals $\mathbf{b}$ parameter is equal to the latter volume expressed in molar units, $\mathbf{b} = \mathcal{N}_A(2\pi/3)\sigma^3$. So, van der Waals predicted that the pressure of a gas composed of hard spheres should be approximately given by the following equation.

$$P = \frac{RT}{\overline{V} - \mathbf{b}} = \frac{[c]RT}{1 - [c]\mathbf{b}} \tag{5.4}$$

Notice that $\overline{V} = (\partial V/\partial n)_{T,P} = V/n = 1/[c]$ is the partial molar volume of the gas.[5] If $\mathbf{b} = 0$, then Eq. 5.4 becomes identical to the ideal gas law $P = RT/\overline{V} = nRT/V = [c]RT = \rho k_B T$, as expected. Thus, the product $[c]\mathbf{b}$ is a unitless quantity that dictates the magnitude of excluded volume contributions to the pressure of a nonideal gas composed of hard-sphere molecules.

[5] In a single-component fluid Eq. 3.52 implies that $V = n\overline{V}$, and so $\overline{V} = (\partial V/\partial n)_{T,P} = V/n$.

Van der Waals further suggested that attractive intermolecular interactions should decrease the pressure of the gas. He incorporated this contribution into his equation by introducing a cohesive parameter $\mathbf{a}$.

$$P = \frac{RT}{\overline{V} - \mathbf{b}} - \frac{\mathbf{a}}{\overline{V}^2}$$
$$= \frac{[c]RT}{1 - [c]\mathbf{b}} - \mathbf{a}[c]^2 \tag{5.5}$$

The reason that van der Waals expected the attractive contribution to the pressure should be proportional to $[c]^2$ will shortly become clear.

Thermodynamic manipulations such as those described in Section 4.1 may be used to determine how the energy of a van der Waals gas depends on $\overline{V}$ or [c]. Recall that the thermodynamic equation of state (Eq. 4.15) may be rearranged to $(\partial U/\partial V)_T = T(\partial P/\partial T)_V - P$. The van der Waals equation implies that

$$\left(\frac{\partial P}{\partial T}\right)_V = \frac{R}{\overline{V} - \mathbf{b}} = \frac{[c]R}{1 - [c]\mathbf{b}} \tag{5.6}$$

and thus

$$\left(\frac{\partial U}{\partial V}\right)_T = \frac{[c]RT}{1 - [c]\mathbf{b}} - \left\{\frac{[c]RT}{1 - [c]\mathbf{b}} - \mathbf{a}[c]^2\right\} = \mathbf{a}[c]^2 \tag{5.7}$$

In other words, the internal pressure of a van der Waals gas is determined entirely by attractive intermolecular interactions. We may now integrate Eq. 5.7 from $[c] = 0$ up to any concentration of interest to obtain the following expression for the dependence of $\overline{U}$ on [c] (at constant T).[6]

$$\int_0^{[c]} d\overline{U} = \overline{U} - \overline{U}_0 = \int_0^{[c]} -\mathbf{a}\, d[c] = -\mathbf{a}[c] \tag{5.8}$$

This result indicates that the effect of intermolecular interactions on the partial molar energy of a van der Waals gas is proportional to the concentration of the gas. That makes sense, since we expect the energy of each molecule in the gas to be proportional to the number of molecules with which it interacts, and that number should scale with the concentration of the gas. Equation 5.8 also makes it clear why van der Waals introduced the term $\mathbf{a}[c]^2$ in Eq. 5.5, as this is required if $\overline{U}$ is to be a linear function of [c].

[6] Since $\overline{V} = 1/[c]$, that implies that $d\overline{V}/d[c] = -1/[c]^2$. Thus, we may reexpress Eq. 5.7 as $(\partial U/\partial V)_T = (\partial \overline{U}/\partial \overline{V})_T = -[c]^2(\partial \overline{U}/\partial [c])_T$ and then multiply both sides by $(-1/[c]^2)\, d[c]$ to obtain $dU|_T = \mathbf{a}[c]^2(-1/[c]^2)\, d[c] = -\mathbf{a}\, d[c]$.

Notice that $\overline{U}_0$ in Eq. 5.8 is the partial molar energy of the system in the low-density (ideal gas) limit.[7] For a monatomic gas we may equate $\overline{U}_0$ with its average translational kinetic energy $\overline{U}_0 = (3/2)RT$, and thus the total energy of a monatomic van der Waals gas may be expressed as follows.[8]

$$\begin{aligned} \overline{U} &= \frac{3}{2}RT - \mathbf{a}[c] \\ &= \frac{3}{2}RT - \frac{\mathbf{a}}{\overline{V}} \end{aligned} \tag{5.9}$$

Figure 5.2 compares experimental (points) and van der Waals equation of state predictions (curves) for two monatomic gases, He and Xe, up to sufficiently high concentrations that they deviate from ideal gas predictions (dashed lines). The van der Waals constants used to generate these predictions were obtained from the experimental values of α_P and κ_T at $T = 300$ K and $[c] = 5$M, combined with Eqs. 5.6 and 5.7, as described in Exercise 5.2 on pages 158–159. The resulting $\mathbf{a}$ and $\mathbf{b}$ values are given in the caption of Figure 5.2.

Although the agreement between the experimental and van der Waals predictions is not perfect, it is remarkably good considering the approximate nature of the van der Waals equation. The global accuracy of the van der Waals equation confirms that the coefficients $\mathbf{a}$ and $\mathbf{b}$ reasonably approximate the influence of attractive (cohesive) and repulsive (excluded volume) intermolecular interactions, respectively. Notice, in particular, that the energy U of Xe is indeed proportional to concentration, as predicted by Eq. 5.9, while the attraction between He atoms is so small that the energy of He is essentially equal to that of an ideal gas (for which $U = \frac{3}{2}nRT$, and is nearly independent of concentration).

Although He behaves nearly ideally, its pressure P and thermal pressure coefficient $(\partial P/\partial T)_V$ are both slightly larger than those of an ideal gas (dashed curves). The van der Waals equation reveals that this positive deviation arises from the finite size of He atoms. The results for Xe are quite different, as both its pressure and energy deviate more significantly from those of an ideal gas. The van der Waals equation indicates that these deviations arise from the stronger attractive interactions between Xe

---

[7] At $[c] = 0$ molecules can no longer interact with each other and so the energy of the gas $\overline{U}_0$ must be identical to that of an ideal gas composed of noninteracting molecules.

[8] This expression for the energy of a monatomic gas does not include the internal electronic energy of each atom. In other words, it is referenced to the energy of stationary gas molecules. For a diatomic or polyatomic gas we could add rotational and vibrational contributions to the energy, treated either classically or quantum mechanically.

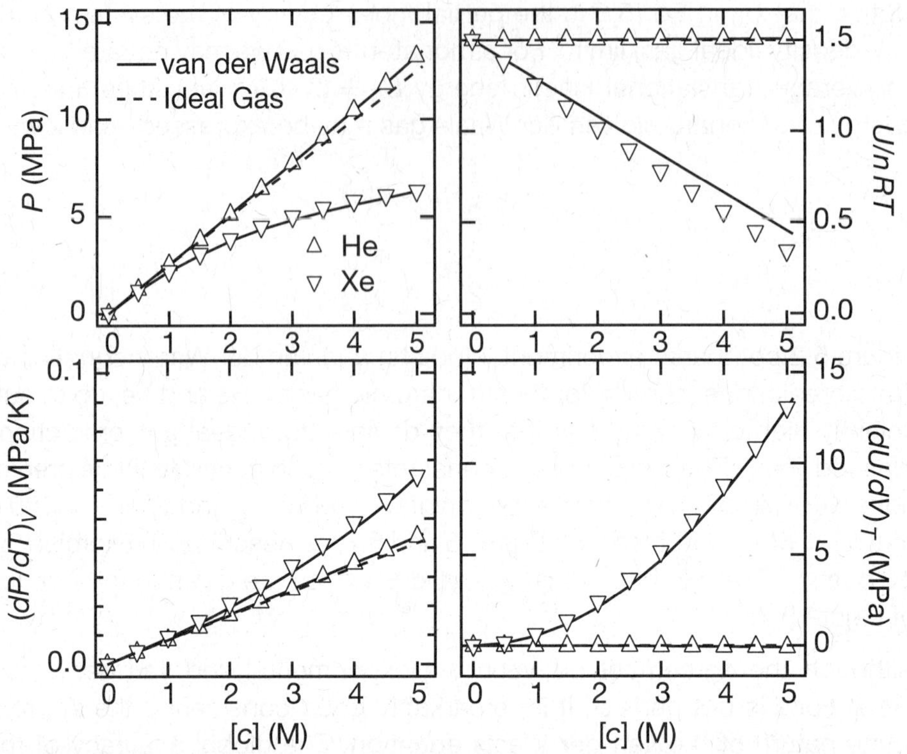

FIGURE 5.2 The thermodynamic properties of He and Xe (triangles) at 300 K are compared with ideal gas (dashed curve) and van der Waals (solid curve) predictions. The van der Waals coefficients used to generate these predictions are $\mathbf{a}_{He} = -0.004$ and $\mathbf{a}_{Xe} = 0.52$ (MPa/M^2), $\mathbf{b}_{He} = 0.010$ and $\mathbf{b}_{Xe} = 0.071$ (1/M); these parameters were obtained from the experimental values of α_P and κ_T at $[c] = 5$ M (using Eqs. 5.6 and 5.7). Data points from E.W. Lemmon, M.O. McLinden and D.G. Friend, "Thermophysical Properties of Fluid Systems" in NIST Chemistry WebBook, NIST Standard Reference Database Number 69, Eds. P.J. Linstrom and W.G. Mallard, National Institute of Standards and Technology, Gaithersburg MD, 20899, http://webbook.nist.gov.

atoms. The larger size and cohesive energy of Xe than He is consistent with the fact that Xe has 54 electrons while He has only 2.

Exercise 5.2

Use the following experimental values of P, α_P, and κ_T for He and Xe at 300 K and $[c] = 5$ M to estimate the van der Waals coefficients of these two gases:

He: $P = 13.23$ MPa, $\alpha_P = 0.003119$ (1/K), and $\kappa_T = 0.071216$ (1/MPa)

Xe: $P = 6.29$ MPa, $\alpha_P = 0.031448$ (1/K), and $\kappa_T = 0.48811$ (1/MPa)

Solution. The experimental values of α_P and κ_T may be used to determine $(\partial P/\partial T)_V = \alpha_P/\kappa_T$, and thus Eq. 5.6 may be solved to obtain:

$$\mathbf{b} = \overline{V} - R\left(\frac{\kappa_T}{\alpha_P}\right) = \frac{1}{[c]} - R\left(\frac{\kappa_T}{\alpha_P}\right)$$

Note that $R(\kappa_T/\alpha_P)$ has units of J/(mol MPa), which may be converted to 1/M units by using the fact that 1 MPa $= 10^6$ J/m^3 $= 10^3$ J/L (since 1 m^3 $= 10^3$ L), and so $R(\kappa_T/\alpha_P)$ [J /(mole MPa)] $\times 10^{-3}$ [MPa/(J/L)] converts $R(\kappa_T/\alpha_P)$ to (1/M) units. Thus, $\mathbf{b} \approx 0.01$ (1/M) for He and $\mathbf{b} \approx 0.071$ (1/M) for Xe.

The van der Waals $\mathbf{a}$ coefficient can be determined using Eq. 5.7, combined with the thermodynamic equation of state (Eq. 4.15 on page 126), to obtain

$$\mathbf{a} = (\partial U/\partial V)_T/[c]^2 = \{T(\alpha_P/\kappa_T) - P\}/[c]^2$$

which may be evaluated using the experimental values α_P, κ_T, P, and [c]. Thus, $\mathbf{a} \approx -0.004$ (MPa/M^2) for He and $\mathbf{a} \approx 0.52$ (MPa/M^2) for Xe.

Generalized van der Waals Equations

Van der Waals himself was well aware of both the strengths and limitations of his nonideal gas equation of state. During his lifetime, as well as over the following 100 years, his original equation was systematically generalized and improved so as to more realistically represent the thermodynamic properties of liquids and solutions, not only under ambient conditions but also over a wide range of temperatures and pressures. This subsection describes one such generalization of the van der Waals equation.

Van der Waals knew that the coefficients $\mathbf{a}$ and $\mathbf{b}$ are only expected to approximate the influence of intermolecular interactions. Most importantly, he understood that the $\overline{V} - \mathbf{b}$ term in Eq. 5.4 is the weakest link in his theory, because $RT/(\overline{V} - \mathbf{b})$ only correctly predicts the influence of molecular excluded volume in moderately low pressure nonideal gases. This can best be seen by performing a Taylor expansion of Eq. 5.4 in powers of [c] – such an expansion is referred to as a *virial expansion* or a *virial equation of state*.

$$P = \frac{[c]RT}{1 - [c]\mathbf{b}} \approx [c]RT\left\{1 + [c]\mathbf{b} + ([c]\mathbf{b})^2 + ([c]\mathbf{b})^3 + \dots\right\} \qquad (5.10)$$

Van der Waals knew that only the first two terms were physically realistic, as he and Ludwig Boltzmann performed quite extensive calculations to demonstrate that the correct density expansion of P has the following first four terms (for a system composed of hard-sphere molecules).

$$P = [c]RT\left\{1 + [c]\mathbf{b} + 0.625([c]\mathbf{b})^2 + 0.28695([c]\mathbf{b})^3 + \dots\right\} \qquad (5.11)$$

In other words, the original van der Waals expression significantly overestimated the higher-order terms on the virial expansion of P.

A remarkably accurate generalization of Eq. 5.4 was developed in 1969 – almost exactly 100 years after van der Waals published his original equation – by two chemical engineers named Norman Carnahan and Kenneth Starling. The resulting Carnahan-Starling (CS) hard-sphere equation of state is most simply expressed in terms of the packing fraction of the hard-sphere fluid

$$\eta \equiv v_0 \rho = \left(\frac{\pi}{6}\right) \rho \sigma^3 \tag{5.12}$$

where $v_0 = (\pi/6)\sigma^3$ is the volume of a single hard sphere (see Fig. 5.1), and $\rho = N/V$ is the fluid's molecular number density. In other words, η represents the fraction of the total volume of the fluid that is occupied by the hard-sphere molecules.[9] Using this variable, the CS equation may be expressed in the following relatively simple form, which predicts how the compressibility factor $Z \equiv P/(\rho k_B T) = P/([c]RT)$ depends on η.

$$\boxed{Z = \frac{1 + \eta + \eta^2 - \eta^3}{(1 - \eta)^3}} \tag{5.13}$$

Computer simulations of systems composed of hard spheres have verified the exceptional accuracy of Eq. 5.13, as shown in Figure 5.3. Notice that the original van der Waals expression $Z = 1/(1 - 4\eta)$ (dashed curve) is only accurate at low packing fractions, while the Carnahan-Starling equation (solid curve) in nearly perfect agreement with the simulation results at all packing fractions, up to the normal freezing point of a hard-sphere fluid (at $\eta \approx 0.494$), and even in the higher-density metastable fluid regime (open points).[10]

If we introduce cohesive interactions to the Carnahan-Starling hard-sphere equation in the same way suggested by van der Waals, we obtain the

---

[9] The packing fraction of liquids are typically close to 0.5, which means that half of the space is occupied by molecules. Since spherical objects cannot completely fill the available space, the highest possible packing fraction is less than 1. More specifically, for a solid in which spheres are arranged in a face-centered, cubic, close-packing structure, the packing fraction is $\pi/(3\sqrt{2}) \approx 0.74$.

[10] The computer simulation results shown in Figure 5.3 are those obtained by J. Kolafa, S. Labik, and A. Malijevsky, published in *Phys. Chem. Chem. Phys.*, volume 6, page 2335 (2004). The metastable fluid results were obtained by compressing the hard-sphere fluid to a density above its normal freezing point sufficiently rapidly that the fluid gets stuck (jammed) in a glassy state that can no longer freeze. In other words, the relaxation time of such metastable fluids is so long that they are kinetically trapped in a noncrystalline state.

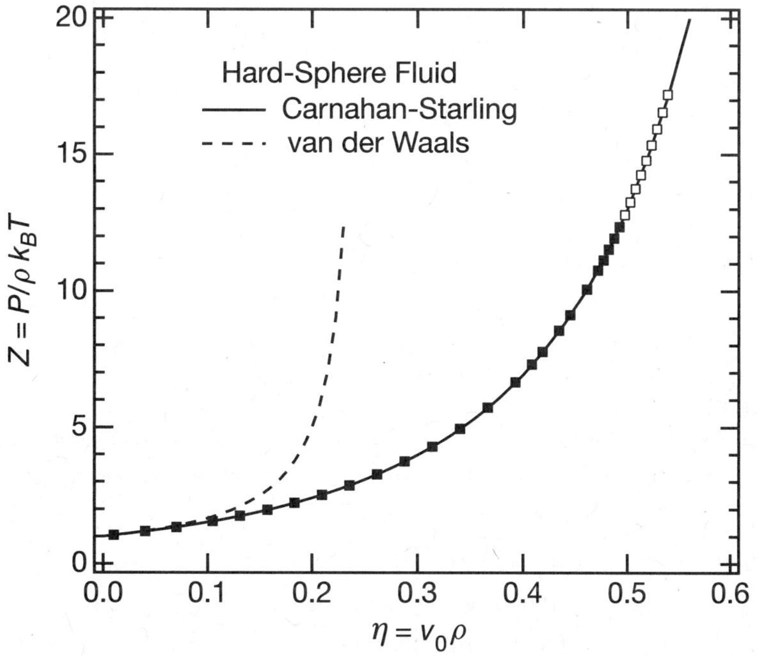

FIGURE 5.3 The compressibility factor $Z = P/\rho k_B T$ of a hard-sphere fluid is plotted as a function of its packing fraction $\eta = v_0\rho$. The solid points are computer simulation results spanning the entire vapor-liquid density range, while the open points extend into the metastable fluid state (see footnote 10 on page 160). The solid curve demonstrates the exceptional accuracy of the Carnahan-Starling equation (while the dashed curve reveals that the original van der Waals excluded volume equation is only accurate at relatively low nonideal gas densities). Data from J. Kolafa, S. Labik, and A. Malijevsky, Accurate equation of state of the hard sphere fluid in stable and metastable regions. *Phys. Chem. Chem. Phys.* **6**, 2335 (2004).

following generalization of the van der Waals equation, which is also referred to as the CSvdW equation of state.

$$\boxed{Z = \frac{PV}{nRT} = \frac{P}{\rho k_B T} = \frac{1+\eta+\eta^2-\eta^3}{(1-\eta)^3} - 4\left(\frac{\tau}{T}\right)\eta} \tag{5.14}$$

Note that in the low-density limit, when $\eta = 0$, the above equation reduces to the ideal gas law $PV/(nRT) = 1$. The two terms on the right-hand side of Eq. 5.14 again describe the nonideal repulsive and attractive contributions to the pressure, respectively. Like the original van der Waals equation, Eq. 5.14 also has only two parameters. One of these is the hard-sphere diameter of each molecule σ (which may be expressed in nm units) and the other is cohesive parameter τ (which has units of absolute temperature K). The definition of the van der Waals $\mathbf{b}$ coefficient implies that

$$\boxed{\sigma = \left(\frac{3\mathbf{b}}{2\pi\mathcal{N}_A}\right)^{1/3}} \tag{5.15}$$

The parameter τ is a measure of the cohesive energy of attraction between molecules.[11] The relationship between τ and $\mathtt{a}$ may be inferred by noting that the attractive contribution to the pressure is $4(\tau/T)\eta[c]RT = R\mathcal{N}_A\tau(2\pi/3)\sigma^3[c]^2 = \mathtt{a}[c]^2$, which, when combined with Eq. 5.15, yields

$$\boxed{\tau = \frac{3\mathtt{a}}{2\pi\sigma^3 R\mathcal{N}_A} = \frac{\mathtt{a}}{R\mathtt{b}}} \tag{5.16}$$

Exercise 5.3

Use Eqs. 5.15 and 5.16, combined with the van der Waals parameters of Xe gas (which are given in Fig. 5.2), to estimate the CSvdW parameters σ (in nm units) and τ (in K units) of Xe.

Solution. Figure 5.2 indicates that the van der Waals parameters of Xe are $\mathtt{a} \approx 0.52$ (MPa/M^2) and $\mathtt{b} \approx 0.071$ (1/M). The van der Waals $\mathtt{b}$ parameter may be used to obtain the CSvdW hard-sphere diameter $\sigma \approx \{3 \times 0.071$ (L/mol) $\times 10^{-3}$ (m^3/L) $\times 10^{27}$ (nm^3/m^3)/[2π 6.02×10^{23} (1/mol)]$\}^{1/3} \approx 0.38$ nm. The van der Waals $\mathtt{a}$ and $\mathtt{b}$ parameters may be combined to obtain the CSvdW cohesive parameter $\tau \approx 0.52$ (MPa/M^2) $\times 10^6$ (Pa/MPa) $\times$ 10^{-3} (m^3/L) / [8.3 (J/K mol) $\times$ 0.071 (1/M)] $\approx$ 880 K. Note that the units in the latter expression are consistent, since 1 Pa = 1 J/m^3 (and 1 M = 1 L/mol).

Since the CSvdW equation more realistically represents hard-sphere-excluded volume contributions to the thermodynamic properties, one might expect Eq. 5.14 to be applicable not only to nonideal gases but also to high-density liquids. That is indeed the case, and in fact Eq. 5.14 may even be applied with surprising success to liquids composed of polyatomic molecules, as shown in Figure 5.4. This figure contains experimental (points) and predicted (curves) pertaining to the equation of state

[11] More specifically, the attractive contribution to the CSvdW equation may be derived by assuming that the potential energy of two atoms has the following form $u(r) = -\epsilon(\sigma/r)^6 = -k_B\tau(\sigma/r)^6$, where r is the separation between the atoms. Thus, when the two atoms are in contact, $r = \sigma$ and $u(\sigma) = -\epsilon = -k_B\tau$, while at larger distances their attractive potential energy drops off as $1/r^6$. The latter distance dependence is the same as that predicted to arise both from molecular electronic polarizability (dispersion interactions) and from interactions between dipolar molecules, when thermally averaged over all orientations. The derivation of the above relation between τ and ϵ further assumes that the molecules are uniformly distributed in space with a concentration equal to [c].

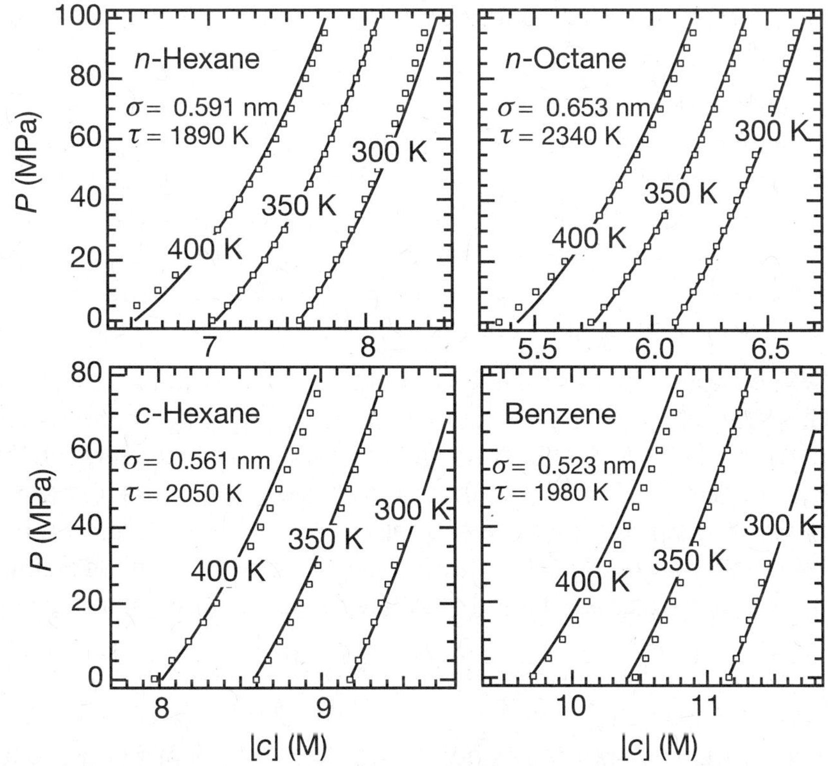

FIGURE 5.4 Experimental (points) equation of state data of liquids ranging from ambient conditions (300 K and 0.1 MPa) to very high pressures and temperatures are compared with predictions obtained using the CSvdW equation (curves). The effective hard-sphere diameter σ and cohesive energy parameters τ were in each case obtained from the experimental data only at ambient conditions (which corresponds to the lowest pressure and temperature point in each panel). Data points from E.W. Lemmon, M.O. McLinden, and D.G. Friend, "Thermophysical Properties of Fluid Systems" in NIST Chemistry WebBook, NIST Standard Reference Database Number 69, Eds. P.J. Linstrom and W.G. Mallard. National Institute of Standards and Technology, Gaithersburg MD, 20899, http://webbook.nist.gov.

of four polyatomic liquids at pressures up to 100 MPa ($\approx$1000 Atm) and temperatures up to 400 K.

The accuracy of the predictions obtained using Eq. 5.14 is quite striking, particularly given that the two coefficients σ and τ were determined using only the experimental values of ρ and $(\partial P/\partial T)_V = \alpha_P/\kappa_T$ of each liquid at ambient temperature and pressure ($T = 300$ K and $P = 0.1$ MPa). The accuracy of the CSvdW predictions shown in Figure 5.4 is remarkable because these molecules clearly are not spherical in shape (while the Carnahan-Starling equation pertains to hard spheres). This suggests that the equation of state of nonspherical molecules is quite similar to that of a fluid composed of spheres. The parameter σ thus represents the *effective hard-sphere diameter* of each of the molecules in Figure 5.4, and $-R\tau[c]$ represents the cohesive energy due to intermolecular attractive interactions.

Notice that the molecules in Figure 5.4 are all nonpolar compounds. In other words, they have no net charge or significantly asymmetric charge distribution (dipole moment), and are also not capable of forming ionic, dipolar, or hydrogen bonds. The isothermal compressibility of polar molecules (and even alcohols) can also be represented using Eq. 5.14. However, for such fluids the values of σ and τ are found to depend on temperature (and so may no longer be treated as constants).[12]

The CSvdW equation may be used to obtain the *excess chemical potential* of a CSvdW fluid, $\mu^{\times} \equiv \mu - \mu^{IG}$, where μ and μ^{IG} are the chemical potentials of the molecule in the fluid and ideal gas states, respectively, where the ideal gas consists of noninteracting molecules *with the same number density as the fluid system*. More specifically, the following expression for $\mu^{\times}$ is obtained by integrating Eq. 5.14 with respect to η to obtain the Helmholtz energy,[13] and then differentiating A with respect to N (at constant T and V).

$$\beta\mu^{\times} = \beta\mu - \beta\mu^{IG} = \frac{\eta(8 - 9\eta + 3\eta^2)}{(1-\eta)^3} - 8\left(\frac{\tau}{T}\right)\eta \qquad (5.17)$$

The two terms on the right-hand side of Eq. 5.17 again represent the repulsive (excluded volume) and attractive (cohesive) contributions to the chemical potential, respectively. Thus, we may use this expression, combined with the CSvdW parameters given in Figure 5.4, to estimate how much the chemical potential of a liquid is influenced by repulsive and attractive intermolecular interactions.[14]

---

[12] Although Eq. 5.14 may be used to approximate the equation of state of liquid water, the strong hydrogen-bonding interactions of water molecules, and other anomalous properties of water (such as its density maximum at $\approx 4°C$), cannot be accurately represented using a simple generalized van der Waals equation.

[13] Note that the thermodynamics identity $P = -(\partial A/\partial V)_T$ implies that $\int dA = \int -P\,dV$, and thus $A - A^{IG} = -\int P\,dV$, where the latter integral is performed starting in an ideal gas state and ending in the actual fluid state of interest. Since Eq. 5.17 pertains to a reference state corresponding to an ideal gas *at the same density* as the fluid, we must also correct the value of A^{IG} so that it pertains to the actual fluid density (using the relations obtained in Problem 10 of Chapter 3).

[14] If we want to obtain the chemical potential of a molecule relative to an ideal gas at the same pressure P as the liquid (rather than the same number density), then we would have to replace $\beta\mu^{IG} = \beta\mu^{IG}(\rho)$ by $\beta\mu^{IG}(P) + \ln[(\rho/(\beta P)]$ on the right-hand side of Eq. 5.17. Note that $k_B T \ln[(\rho/(\beta P)] = k_B T \ln(\rho/\rho_0)$ is the chemical potential change associated with increasing the number density of an ideal gas from its value at the experimental pressure $\rho_0 = \beta P$, to the number density ρ of the liquid at pressure P. The last part of Exercise 5.4 illustrates an application of the above relations.

Exercise 5.4

We may use the CSvdW equation to confirm that Xe behaves like an ideal gas at $P = 0.1$ MPa and $T = 300$ K (where it has an experimental concentration of $[c] \approx 0.0403$ M), as well as to quantify the nonideal behavior of Xe when its pressure is increased to 2 MPa (where its experimental concentration is $[c] \approx 0.902$ M). Recall that the following CSvdW parameters of Xe were obtained in Exercise 5.3 on page 162 (using the van der Waals parameters given in Fig. 5.2).

$$\sigma \approx 0.38 \text{ nm} \quad \tau \approx 880 \text{ K}$$

- Calculate the packing fractions of Xe at the above two concentrations.

  Solution. Recall that the packing fraction is defined as $\eta = (\pi/6)\rho\sigma^3 = v_0\rho = [c]\mathfrak{b}/4$, where $\rho = [c]$ (mol/L) $\times 10^3$ (L/m^3) $\times 10^{-27}$ (m^3/nm^3) $\times$ N_A (1/mol), and thus the packing fractions are $\eta = (\pi/6)(0.0403 \times 0.602)\sigma^3 \approx 0.0007$ at 0.1 MPa and $\eta = (\pi/6)(0.902 \times 0.602)\sigma^3 \approx 0.016$ at 2 MPa.

- Calculate the compressibility factor Z of Xe at the above two packing fractions and compare your results with the experimental values of $Z \approx 0.99$ at 0.1 MPa and $Z \approx 0.90$ at 2 MPa.

  Solution. The compressibility factors may be obtained using Eq. 5.14 (combined with the above values of η and τ), which implies that $Z \approx 0.99$ at 0.1 MPa and $Z \approx 0.88$ at 2 MPa, both of which are in reasonably good agreement with experimental values. The fact that $Z = PV/nRT \approx 1$ is at 0.1 MPa confirms that Xe behaves ideally at this pressure. At the higher pressure of 2 MPa, Xe no longer behaves like an ideal gas, as indicated by the fact that Z is about 10% smaller than its ideal gas value.

- Combine Eqs. 5.1 and 5.17 to predict the fugacity of Xe at 2 MPa, and compare your results with its experimental fugacity of $f \approx 1.8$ MPa at this pressure.

  Solution. Equation 5.1 implies that $f/f_0 \approx e^{\beta(\mu-\mu^0)}$. Since Xe behaves ideally at 0.1 MPa, we may take this as our reference pressure and safely assume that $f_0 = P_0 = 0.1$ MPa. Thus, Eq. 5.1 implies that the fugacity of Xe at 2 MPa is $f \approx 0.1e^{\beta(\mu-\mu^0)}$, where $\mu - \mu^0$ represents the difference between the chemical potential of Xe at 2 MPa and 0.1 MPa. The excess chemical potential $\mu^\times = \mu - \mu^{IG}$ in Eq. 5.17 is the difference between the chemical potential of a molecule in a nonideal gas (or liquid) and that of an isolated (ideal gas) molecule *at the same number density*. Thus, we may obtain $\mu - \mu^0$ from $\mu^\times$ by adding the ideal gas chemical potential change associated with increasing the concentration from $[c]_0$ to $[c]$ (as given by Eq. 3.39 on page 105). More specifically, $\beta(\mu - \mu^0) = \beta\mu^\times + \ln([c]/[c]_0) = -0.24 + 3.11 \approx 2.88$. So, the predicted value of the fugacity of Xe at 2 MPa is $f \approx 0.1e^{2.88} \approx 1.78$ MPa, which is in quite good agreement with the experimental fugacity of 1.80 MPa. Note that the difference between the fugacity and the experimental pressure of 2 MPa is another measure of the noideality of Xe.

More generally, the excess chemical potential $\mu^{\times} = \mu - \mu^{\mathrm{IG}}$ of a solute molecule dissolved in a solvent of packing fraction η may be obtained from the following further generalization of the CSvdW equation to mixtures composed of (spherical) molecules of different sizes and cohesive energies.[15]

$$\beta\mu^{\times} = \frac{d\eta\left[(3+6d-d^2)-(6+9d-6d^2)\eta+(3+3d-3d^2)\eta^2\right]}{(1-\eta)^3} - (1-3d^2+2d^3)\ln(1-\eta) - (1+d)^3\left(\frac{\tau}{T}\right)\eta \tag{5.18}$$

The parameter $d = \sigma_1/\sigma_0$ repesents the ratio of hard-sphere diameters of the solute σ_1 and solvent σ_0 molecules. The parameter τ is a measure of the cohesive interaction energy between the solute and all of the surrounding solvent molecules. The parameter τ may be roughly approximated by the geometric mean of the corresponding pure solvent and pure solute values $\tau \approx \sqrt{\tau_1\tau_0}$. A more accurate measure of τ may be obtained using the experimental solvation free energy (or enthalpy) of a solute at some particular temperature and pressure. Once the parameters d, τ, and η are determined, Eq. 5.18 can be used to predict the temperature and pressure dependence of the solute's chemical potential. Note that since Eq. 5.18 pertains to a dilute solution, the packing fraction is essentially identical to that of the pure solvent $\eta = (\pi/6)\rho\sigma_0^3$ (where ρ is the number density of the pure solvent at the pressure and temperature of interest).

Exercise 5.5

The CSvdW parameters of cyclohexane are $\sigma = 0.561$ nm and $\tau = 2050$ K, and its molarity in the liquid state at 300 K and 0.1 MPa is $[c] \approx 9.18$ M.

- What is the hard-sphere packing fraction η of liquid cyclohexane?

[15] Equation 5.18 is obtained from a generalization of the CS equation to hard-sphere mixtures [see T. Boublik, *J. Chem. Phys.*, vol. 53, p. 471 (1970), and D. Ben-Amotz & D.R. Herschbach, *J. Phys. Chem.*, vol. 97, p. 2295 (1993)] and pertains to a solution in which the solute is sufficiently dilute that one may neglect solute-solute interactions (and μ^{IG} is the chemical potential of an ideal gas *of the same number density* as the solute in the nonideal solution).

Solution. The molecular number density of cyclohexane may be obtained from its molarity $\rho = 9.18$ mol/L $\times 10^3$ L/m^3 $\times$ 6.02 $\times$ 10^{23} (molec/mole) $\times$ 10^{-27} (m^3/nm^3) = 5.5 (molec/nm^3) and so $\eta = (\pi/6)\rho\sigma^3 \approx 0.51$.

- Use the CSvdW equation of state Eq. 5.14 on page 161 to predict the thermal pressure coefficient $(\partial P/\partial T)_V$ of liquid cyclohexane.

Solution. Note that $(\partial P/\partial T)_V = (\partial P/\partial T)_\rho = (\partial P/\partial T)_\eta$ because we assume that the system is closed (and so the number of molecules N is constant), and thus holding V constant is equivalent to holding both $\rho = N/V$ and $\eta = (\pi/6)\rho\sigma^3$ constant. It is not difficult to evaluate the partial derivative of $P = Z\rho k_B T$ with respect to T since the first term on the right-hand side of Eq. 5.14 becomes simply proportional to T when Z is multiplied by $\rho k_B T$, while the second term becomes temperature independent, and so does not contribute at all to the temperature derivative.

$$\left(\frac{\partial P}{\partial T}\right)_V = \rho k_B \left[\frac{1+\eta+\eta^2-\eta^3}{(1-\eta)^3}\right] = [c]R\left[\frac{1+\eta+\eta^2-\eta^3}{(1-\eta)^3}\right]$$

Note that the last equality follows from the fact that $\rho k_B = (N/V)k_B = [(N/\mathcal{N}_A)/V]\mathcal{N}_A k_B = (n/V)R = [c]R$ (although we must express $[c]$ in units of moles/m^3, rather than mol/L, if we wish the answer to be in Pa/K units).

- Use the above results, combined with Eq. 4.15 on page 126, to predict the internal pressure $(\partial U/\partial V)_T$ of liquid cyclohexane.

Solution. Equation 4.15 indicates that $(\partial U/\partial V)_T = T(\partial P/\partial T)_V - P$, and so the above results for η and $(\partial P/\partial T)_V$ can be combined with the experimental pressure of 0.1 MPa = 0.1×10^6 Pa (and 1 Pa = J/m^3) to obtain

$$\left(\frac{\partial U}{\partial V}\right)_T = T\left(\frac{\partial P}{\partial T}\right)_V - P = 318.7\times10^6\,\text{Pa} - 0.1\times10^6\,\text{Pa} \approx 319\,\text{MPa}$$

This prediction is essentially identical to the experimental value of the internal pressure, which is not surprising, since the experimental value of α_P/κ_T was used to obtain the CSvdW parameters. In other words, we have now confirmed that the above values of σ and τ are indeed consistent with experimental values of ρ, $(\partial P/\partial T)_V$, and $(\partial U/\partial V)_T$ of liquid cyclohexane.

- Use either Eq. 5.17 or Eq. 5.18, combined with the above results, to estimate the repulsive and attractive contributions to the solvation energy of cyclohexane in its own pure liquid at 300 K and 0.1 MPa. Note that for such a self-solvation process $d = 1$ and so Eq. 5.18 becomes identical to 5.17.

Solution. Equation 5.17, combined with the previously obtained packing fraction $\eta \approx 0.51$, indicates that $[\eta(8-9\eta+3\eta^2)]/(1-\eta)^3 \approx 18$ is the contribution that intermolecular repulsive interactions make to $\beta\mu^\times = \beta\mu - \beta\mu^{IG}$. The attractive contribution $\beta\mu^\times$ may be obtained from the last term in Eq. 5.17, using the value of $\tau \approx 2050$ K for cyclohexane (see Fig. 5.4). Thus, $-8(\tau/T)\eta \approx -28$ is the attractive contribution to $\beta\mu^\times$.

These results imply that the free energy required to insert a hard sphere the size of cyclohexane into liquid cyclohexane is positive, while attractive interactions between cyclohexane molecules are negative. The larger magnitude of the attractive contributions to μ is what drives cyclohexane to spontaneously condense to a liquid at 300 K.

The CSvdW equation, and its generalization to mixtures, provides a relatively simple and yet remarkably accurate means of predicting the properties of nonideal gases, liquids, and solutions. In order to obtain more accurate results, we might consider further refining the above expressions to more realistically represent molecules with nonspherical shapes and different sorts of cohesive interactions. However, it is often more practical to implement such higher-level calculations by performing computer simulations, using methods such as those described in Section 5.5. In order to understand how this is done we must first set the stage by considering the application of statistical mechanics to fluids (in Section 5.3), as well as revisiting the second law from several different perspectives (in Section 5.4).

5.3 Supermolecule Statistical Mechanics

In Chapter 1 we learned how to calculate the probability that a molecule will occupy a given quantum state, and how to use this probability to calculate the average energy that a molecule will have at a given temperature. Here we will use exactly the same procedure to obtain statistical thermodynamic expressions for all other thermodynamic functions of a macroscopic system.

Recall that in Chapter 1 we focused on the energy levels of an individual molecule and found that the probability of occupying a given quantum state within the molecule is $P(\varepsilon_i) = e^{-\beta\varepsilon_i}/q$ (where $q = \sum_i e^{-\beta\varepsilon_i}$ is the molecule's partition function). We also found that the average energy of a molecule is $\langle\varepsilon\rangle = \sum_i \varepsilon_i P(\varepsilon_i)$, and that the latter sum can be expressed in terms of the partition function $\langle\varepsilon\rangle = -(\partial \ln q/\partial\beta) = -(1/q)(\partial q/\partial\beta)$.[16] Here we will see how these and other relations can be extended to macroscopic systems composed of large numbers of molecules.

In order to apply statistical mechanics to a macroscopic fluid system, it is useful to think of such a system as a single large *supermolecule* – this is the key idea that Gibbs introduced in order to extend Boltzmann's gas theory to liquids (and other nonideal systems). Although such a supermolecule has a much larger number of atoms and energy levels than a

[16] See Eq. 1.23 and footnote 20 on page 28.

small molecule, Gibbs realized that the basic expressions used to describe the probabilities and average energies of molecules can be directly applied to a macroscopic (supermolecule) system.

In order to remember that we are dealing with a supermolecule (macroscopic system), we will use the capital letter $\mathcal{Q}$ (rather than q) to designate its partition function, and we will use the capital letter E_i (rather than ε_i) to designate its individual energy levels. Moreover, we will assume that the container holding our supermolecule has a constant volume and temperature and is closed (so the number of atoms or fundamental particles that make up the supermolecule are constant). Thus, such a supermolecule partition function may more properly be designated as $\mathcal{Q}_{NVT}$ as a reminder that N, V, and T are all held constant. However, to keep our notation simple, we will leave off the subscripts and just call our supermolecule partition function $\mathcal{Q}$. The definition of $\mathcal{Q}$ is exactly the same as that of q (except that the energies are designated by E_i).

$$\boxed{\mathcal{Q} = \sum_i e^{-\beta E_i}} \tag{5.19}$$

Similarly, the probability of finding the system in a given state of energy E_i may be expressed as follows.

$$\boxed{P(E_i) = \frac{e^{-\beta E_i}}{\mathcal{Q}}} \tag{5.20}$$

Since a super molecule is expected to have many very closely spaced energy levels, it is often appropriate to treat them as continuous. Under such conditions it can be convenient to replace the above sum by an integral and thus obtain the following expression for the corresponding probability density function.

$$P(E_i) = \frac{e^{-\beta E_i}}{\int e^{-\beta E_i} d\Gamma} \tag{5.21}$$

The variable Γ represents the (phase space) configuration of the supermolecule. In other words, changing Γ has the effect of varying the positions and momenta of all the particles in the system.[17]

The average energy of a supermolecule is also defined in the same way as that of a single molecule (Eqs. 1.23 or 1.28), except that we may now equate

---

[17] More specifically, Eq. 5.21 represents a highly multidimensional integral over all the possible x, y, and z center of mass coordinates, rotational angles, and momenta of each particle in the system.

the average energy of the supermolecule with the total thermodynamic energy of the system, U.[18]

$$\boxed{U = \langle E \rangle = \frac{\sum_i E_i e^{-\beta E_i}}{\sum_i e^{-\beta E_i}} = \frac{\int E_i e^{-\beta E_i} d\Gamma}{\int e^{-\beta E_i} d\Gamma} = -\frac{1}{\mathcal{Q}}\left(\frac{\partial \mathcal{Q}}{\partial \beta}\right)_{V,N}} \tag{5.22}$$

Note that the latter result may also be expressed in the following alternative forms.

$$U = -\left(\frac{\partial \ln \mathcal{Q}}{\partial \beta}\right)_{V,N} = -k_B\left[\frac{\partial \ln \mathcal{Q}}{\partial(1/T)}\right]_{V,N} = k_B T\left[\frac{T}{\mathcal{Q}}\left(\frac{\partial \mathcal{Q}}{\partial T}\right)_{V,N}\right] \tag{5.23}$$

From thermodynamics we may obtain the following additional relationship between U and A (which is analogous to the Gibbs-Helmholtz relation between H and G, Eq. 4.22).

$$\begin{aligned}
\left[\frac{\partial(A/T)}{\partial(1/T)}\right]_{V,N} &= A\left[\frac{\partial(1/T)}{\partial(1/T)}\right]_{V,N} + \frac{1}{T}\left[\frac{\partial A}{\partial(1/T)}\right]_{V,N} \\
&= A + \frac{1}{T}\left[-T^2\left(\frac{\partial U}{\partial T}\right)_{V,N}\right] \\
&= A - T\left(\frac{\partial U}{\partial T}\right)_{V,N} = A + TS \\
&= U
\end{aligned} \tag{5.24}$$

Comparison of Eqs. 5.23 and 5.24 implies that $-k_B \ln \mathcal{Q} = A/T$, which may be rearranged to obtain the following key statistical thermodynamic relation.[19]

$$\boxed{A = -k_B T \ln \mathcal{Q}} \tag{5.25}$$

We may now combine the above results with $A = U - TS$ to express $S = (U - A)/T$ in terms of $\mathcal{Q}$.

$$\boxed{S = k_B\left[T\left(\frac{\partial \ln \mathcal{Q}}{\partial T}\right)_{V,N} + \ln \mathcal{Q}\right] = k_B\left[\frac{T}{\mathcal{Q}}\left(\frac{\partial \mathcal{Q}}{\partial T}\right)_{V,N} + \ln \mathcal{Q}\right]} \tag{5.26}$$

---

[18] The last equality in Eq. 5.22 is obtained as described in footnote 20 on page 28. Notice that the derivative of $\ln \mathcal{Q}$ with respect to $\beta = 1/k_B T$ is now performed with V held constant, since this is the constraint imposed on our supermolecule.

[19] We might also have included an additional constant of integration A_0 in Eq. 5.25, which represents the value of A at $T = 0$. So, Eq. 5.25 implies that all Helmholtz energies are measured with respect to A_0, or equivalently that $A_0 = 0$.

We can further make use of the thermodynamic identity $P = -(\partial A/\partial V)_{T,N}$ to obtain an expression for P in terms of $\mathcal{Q}$.

$$\boxed{P = k_B T \left(\frac{\partial \ln \mathcal{Q}}{\partial V}\right)_{T,N} = \frac{k_B T}{\mathcal{Q}} \left(\frac{\partial \mathcal{Q}}{\partial V}\right)_{T,N}} \tag{5.27}$$

We can thus also express H and G in terms of $\mathcal{Q}$.

$$\boxed{H = U + PV = k_B T \left[\frac{T}{\mathcal{Q}} \left(\frac{\partial \mathcal{Q}}{\partial T}\right)_{V,N} + \frac{V}{\mathcal{Q}} \left(\frac{\partial \mathcal{Q}}{\partial V}\right)_{T,N}\right]} \tag{5.28}$$

$$\boxed{G = A + PV = k_B T \left[\frac{V}{\mathcal{Q}} \left(\frac{\partial \mathcal{Q}}{\partial V}\right)_{T,N} - \ln \mathcal{Q}\right]} \tag{5.29}$$

The chemical potential of a single molecule within a macroscopic (supermolecule) system may be obtained by utilizing the thermodynamic identity $\mu_i = (\partial A/\partial N_i)_{T,V,N_j \neq N_i}$.

$$\boxed{\mu_i = -k_B T \left(\frac{\partial \ln \mathcal{Q}}{\partial N_i}\right)_{T,V,N_j \neq N_i} = -\frac{k_B T}{\mathcal{Q}} \left(\frac{\partial \mathcal{Q}}{\partial N_i}\right)_{T,V,N_j \neq N_i}} \tag{5.30}$$

Low-Density Gas Systems

The thermodynamics properties of a dilute gas system may be described by treating the system as a collection of isolated (noninteracting) molecules. As a first guess, we might expect that the number of states that are accessible to a system composed of N such molecules would be equal to the product of the number of states accessible to each molecule, and thus that $\mathcal{Q} = q_1 q_2 q_3 \ldots q_N$.[20] This turns out to be exactly right as long as the molecules are each chemically distinguishable (different types of molecules). However, if all the molecules are the same, so $q_1 = q_2 = q_3 = \ldots q$,

[20] Recall that the probabilities of statistically independent events, such as flipping coins, are equal to the product of the individual probabilities. This is also why we expect the partition function of a system composed of independent (uncoupled) molecules to be the product of the corresponding molecular partition functions. You can explicitly prove this result by assuming that the energies E_k of a system composed of two molecules A and B are equal to all the possible combinations of the energies of each molecule. In other words, by writing down all the energies and associated Boltzmann factors, you can show that $\mathcal{Q} = \sum_k g_k e^{-\beta E_k} = (\sum_j g_j^A e^{-\beta \varepsilon_j^A})(\sum_j g_j^B e^{-\beta \varepsilon_j^B}) = q_A q_B$. This procedure could clearly be extended to any number of molecules.

then it turns out that $\mathcal{Q}$ is not equal to q^N, but instead is given by the following expression.

$$\boxed{\mathcal{Q} = \frac{q^N}{N!}} \tag{5.31}$$

The additional factor of $N!$ in the denominator arises because there are $N!$ ways in which N molecules may be permuted (exchanged in position).[21] Since exchanging identical molecules will have no measurable effect on the system, all such states should be treated as identical when calculating $\mathcal{Q}$. In other words, if $N!$ were not included, then the number of states that contribute to $\mathcal{Q}$ would be overcounted by a factor of $N!$. Accounting for the number of indistinguishable permutations in this way is referred to as *Boltzmann statistics*.[22] The appearance of $N!$ in Eq. 5.31 is also closely related to a phenomenon referred to as *Gibbs Paradox*, as explained in Section 5.4.

More generally, for a gas composed of a mixture of N_A molecules of type A, and N_B molecules of type B, etc., the total partition function of the mixture is the product of the partition functions of each component.

$$\boxed{\mathcal{Q} = \mathcal{Q}_A\mathcal{Q}_B\cdots = \left[\frac{(q_A)^{N_A}}{N_A!}\right]\left[\frac{(q_B)^{N_B}}{N_B!}\right]\cdots} \tag{5.32}$$

The following expression for the chemical potential of each molecule in such a system may be obtained by differentiating $\mathcal{Q}$ with respect to N_i.[23]

$$\boxed{\mu_i = -k_BT\ln\left(\frac{q_i}{N_i}\right) = k_BT\ln\rho_i - k_BT\ln q_i'} \tag{5.33}$$

---

[21] Recall that N factorial is defined as $N! = N(N-1)(N-2)(N-3)\ldots 1$.

[22] Equation 5.31 is actually an approximation, but it is so accurate (under typical experimental conditions) that it may be treated as an exact result. More specifically, Boltzmann statistics are expected to hold whenever the number of states that are accessible to each molecule is much larger than 1. Each molecule in a macroscopic gas phase system has more than 10^{30} thermally accessible translational quantum states, as we will see. Since the number of states is far greater than the number of molecules, Boltzmann statistics is clearly applicable.

[23] Equation 5.33 is obtained by combining Eqs. 5.30–5.32 with Stirling's formula $\ln N! \cong N\ln N - N$, which is essentially exact for $N >> 1$ (see the additional note

The last equality makes use of the definition of the internal partition function $q_i^I \equiv q_i/V$, which is a density independent function of temperature only (as we will see in Chapter 10). Thus, the density dependence of the chemical potential of an ideal gas comes entirely from the first term $k_B T \ln \rho_i$, which is consistent with Eq. 3.38.

5.4 Mixed Points of View on Entropy

Entropy of Mixing and Gibbs Paradox

When two different gases are mixed (at constant temperature and pressure), the entropy of the mixture is invariably larger than that of the two separate gases. The resulting entropy change is often referred to as the *entropy of mixing*, and is sometimes described as arising from the higher "disorder" of the final mixture. However, we will see that this entropy increase actually has nothing to do with an increase in disorder, but rather arises simply from the change in the volume accessible to each component.

We will also see that not all gas-mixing processes produce an entropy increase. For example, if two different gases are mixed at constant volume (rather than constant pressure), then the entropy of the mixture will be exactly the same as the total entropy of the two separate gases. Moreover, we will see that Gibbs was interested in explaining why it is that mixing two identical gases should not produce the same entropy change as mixing two different gases. Gibbs's explanation suggests that there is a subtle link between entropy and our experimental ability (or inability) to distinguish configurational microstates.

---

below for more about Stirling's formula).

$$\frac{\mu_i}{k_B T} = -\left(\frac{\partial \ln Q}{\partial N_i}\right) = -\left(\frac{\partial \ln q_i^{N_i}/N_i!}{\partial N_i}\right) = -\left(\frac{\partial N_i \ln q_i}{\partial N_i}\right) + \left(\frac{\partial \ln N_i!}{\partial N_i}\right)$$

$$= -\left(\frac{\partial N_i}{\partial N_i}\right)\ln q_i - N_i\left(\frac{\partial \ln q_i}{\partial N_i}\right) + \left(\frac{\partial N_i}{\partial N_i}\right)\ln N_i + N_i\left(\frac{\partial \ln N_i}{\partial N_i}\right) - \left(\frac{\partial N_i}{\partial N_i}\right)$$

$$= -\ln q_i - 0 + \ln N_i + 1 - 1 = -\ln\left(\frac{q_i}{N_i}\right)$$

The above results were obtained by making use of the fact that q_i is not a function of N_i and so $\left(\frac{\partial \ln q_i}{\partial N_i}\right) = 0$, and also that $N_i\left(\frac{\partial \ln N_i}{\partial N_i}\right) = \frac{N_i}{N_i}\left(\frac{\partial N_i}{\partial N_i}\right) = 1$.

Note that the full Stirling formula is $\ln N! \cong N \ln N - N + \frac{1}{2}\ln(2\pi N)$, but the last term may be neglected when N is very large. When the last term is included, then the Stirling formula is very accurate even when $N = 10$, as $\ln 10! \approx 15.1\ldots$ while $10 \ln 10 - 10 + \frac{1}{2}\ln(20\pi) \approx 15.1$.

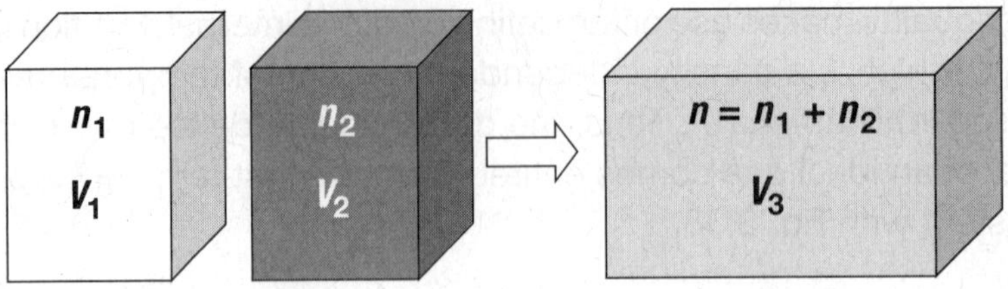

FIGURE 5.5 This diagram illustrates a general mixing process in which two different gases that are initially contained in volumes V_1 and V_2 are mixed in a container of volume V_3. If the mixing process is performed at constant pressure, then the total volume will be the same before and after the mixing process $V_3 = V_1 + V_2$. However, if the pressures of the gases are not all the same, then the volume of the product mixture may not be the same as the sum of the reactant volumes.

Figure 5.5 illustrates a general mixing process in which n_1 moles of one gas and n_2 moles of another gas, which are initially contained in a volumes V_1 and V_2, respectively, are mixed in a container of volume V_3. Since ideal gas molecules do not interact with each other, we expect the total entropy change associated with any such mixing process to be equal to the sum of the entropy changes of the two individual gases. More specifically, the entropy change of the first gas is $\Delta S_1 = n_1 R \ln(V_3/V_1)$, while that for the second gas is $\Delta S_2 = n_2 R \ln(V_3/V_2)$. Thus, we expect the total entropy change in such a general mixing process to be given by the following expression.

$$\Delta S_{\text{mix}} = \Delta S_1 + \Delta S_2 = n_1 R \ln\left(\frac{V_3}{V_1}\right) + n_2 R \ln\left(\frac{V_3}{V_2}\right) \tag{5.34}$$

If the mixing process is carried out at constant pressure $P = P_1 = P_2 = P_3$, then the ideal gas law may be used to determine the volumes, $V_1 = n_1 RT/P$, $V_2 = n_2 RT/P$, and $V_3 = nRT/T = (n_1 + n_2)RT/P = V_1 + V_2$, and so $V_3/V_1 = n/n_1$ and $V_3/V_2 = n/n_2$. We may also express the latter mole ratios in terms of the mole fractions of the two components $\chi_1 = n_1/(n_1 + n_2) = n_1/n$ and $\chi_2 = n_2/(n_1 + n_2) = n_2/n$, or $n_1 = n\chi_1$ and $n_2 = n\chi_2$, and thus obtain the following commonly encountered expression for the entropy of mixing.

$$\boxed{\Delta S_{\text{mix}}|_P = -nR\left(\chi_1 \ln \chi_1 + \chi_2 \ln \chi_2\right)} \tag{5.35}$$

Moreover, since $\chi_1 = n_1/n = N_1/N$ and $\chi_2 = n_2/n = N_2/N$ (where $N = N_A(n_1 + n_2)$ and N_A is Avogadro's number), the above entropy of mixing may also be expressed as $\Delta S_{\text{mix}}|_P = N_1 k_B \ln(N/N_1) + N_2 k_B \ln(N/N_2)$, and then further rearranged to the following form (which is the one that Gibbs preferred to use).

$$\Delta S_{\text{mix}}|_P = k_B\left(N \ln N - N_1 \ln N_1 - N_2 \ln N_2\right) \tag{5.36}$$

More generally, when mixing multiple components at constant pressure, the above formulas may be written more compactly as follows.

$$\Delta S_{mix}|_P = -nR \sum_i \chi_i \ln \chi_i = k_B \left(N \ln N - \sum_i N_i \ln N_i \right) \tag{5.37}$$

Exercise 5.6

Use Eq. 5.34 to calculate the entropy change associated with mixing 1 mole of N_2 with 1 mole of O_2 at a constant pressure of $P = 0.1$ MPa, and verify that the same result is obtained using Eqs. 5.35 and 5.36.

Solution. Since we may safely assume that the N_2 and O_2 behave ideally at this pressure (and are nonreactive), and since $n = n_1 + n_2 = 2$ moles, we expect the volume of the mixture to be twice as large as that of each of the separate pure gases $V_3/V_1 = V_3/V_1 = 2$. Thus, Eq. 5.34 predicts the following entropy of mixing.

$$\Delta S_{mix}|_P = R(\ln 2 + \ln 2) = R(2 \ln 2) \approx 11.5 \text{ J/K}$$

In order to apply Eq. 5.35 we must make use of the fact that each of the components have equal mole fractions of $\chi_1 = \chi_2 = 0.5$, and so $\Delta S_{mix}|_P = -2R(0.5 \ln 0.5 + 0.5 \ln 0.5) = -R(2 \ln 0.5) \approx 11.5$ J/K. Similarly, since $N = 2\mathcal{N}_A$ and $N_1 = N_2 = \mathcal{N}_A$ (where $\mathcal{N}_A$ is Avogadro's number), Eq. 5.36 predicts that $\Delta S_{mix}|_P = k_B[2\mathcal{N}_A \ln(2\mathcal{N}_A) - \mathcal{N}_A \ln \mathcal{N}_A - \mathcal{N}_A \ln \mathcal{N}_A] = \mathcal{N}_A k_B[2 \ln 2 + 2 \ln \mathcal{N}_A - 2 \ln \mathcal{N}_A] = R(2 \ln 2) \approx 11.5$ J/K. Thus, all three formulas predict the same entropy change for this constant pressure mixing process.

It is important to keep in mind that Eqs. 5.35 through 5.37 only hold when the mixing process is carried out at constant pressure. If two gases are mixed at *constant volume*, rather than constant pressure, so that $V_3 = V_2 = V_1$, then Eq. 5.34 implies that $\Delta S|_V = n_1 R \ln 1 + n_2 R \ln 1 = 0 + 0 = 0$. Thus, a constant volume mixing process will not produce any entropy change at all, because the volume accessible to each component is exactly the same before and after the mixing process. This constant volume mixing process also makes it clear that the *entropy of mixing* has nothing to do with the increased disorder associated with allowing two pure gases to become randomly mixed. Rather, the entropy change associated with any such ideal gas-mixing process is entirely due by the entropy change experienced by each component.

In his famous paper *On the Equilibrium of Heterogeneous Substances*, Gibbs included a detailed discussion of the entropy of mixing. Gibbs stressed the fact that Eq. 5.35 should only hold when mixing two gases that are experimentally distinguishable (because they are chemically different types of gases). More specifically, here is how Gibbs discussed this issue in his own words:[24]

> *It is noticed that the value of this expression [Eq. 5.36] does not depend upon the kinds of gases which are concerned . . . except that the gases which are mixed must be of different kinds. If we should bring into contact two masses of the same kind of gas, they would also mix, but there would be no increase of entropy. . . . When we say that two different gases mix we mean that the gases could be separated. . . . But when we say that two gas-masses of the same kind are mixed [we mean that] the separation of the two gases is entirely impossible.*

In other words, if we remove a partition separating two different gases, then they will form a mixture that is experimentally distinguishable from their initial unmixed state. However, if we remove a partition that separates two identical gases, then we will not in any way be able to distinguish the final "mixed" state from the initial "unmixed" state of the gases. Thus, mixing two identical gases at constant pressure should not produce any entropy change.

The following quotation from the subsequent paragraph in Gibbs's paper shows that he was also interested in considering a hypothetical situation in which the difference between two gases becomes arbitrarily small:

> *Now we may without violence to the general laws of gases which are embodied in our equations suppose other gases to exist than such as actually do exist, and there does not appear to be any limit to the resemblance which there might be between two such kinds of gases. But the increase of the entropy due to the mixing of given volumes of the gases at a given temperature and pressure would be independent of the degree of similarity or dissimilarity between them. . . .*

The fact that the entropy of two identical gases will not increase when they are mixed (at constant pressure), while the entropy of two different gases will increase when they are mixed, *even if the difference between the two gases is arbitrarily small*, is often referred to as *Gibbs paradox*. However, Gibbs himself never described this as a paradox, but rather recognized that the entropy of mixing depends on our experimental ability to distinguish the initial and final states of a system.

---

[24] The quoted sentences appear in the two paragraphs immediately following Eq. 297 of Gibbs's paper, *On the Equilibrium of Heterogeneous Substances*. The last two quoted sentences have been condensed for clarity, by removing intervening text. This 321 page paper was published in two parts between 1875 and 1878.

In his later book on *Statistical Mechanics*, Gibbs went beyond the above discussion by suggesting a means of expressing the entropies of gases in a way that produces the correct entropy change, regardless of whether the gases are different or the same.[25] His argument implicitly relied on Boltzmann's expression for entropy $S = k_B \ln \Omega$ (Eq. 2.9) and suggested that Ω (which represents the phase space volume, or the number of configurations, available to a gas) must take into account whether particular configurations of gas molecules are or are not experimentally distinguishable from each other.

Gibbs began by noting that there are $N!$ ways of permuting the locations of the N molecules. He then noted that if all the molecules in a system are identical to each other, then it should not be possible to experimentally distinguish these $N!$ configurations. This implies that the parameter Ω, which appears in Boltzmann's entropy formula, should not represent *all* of the configurations of a gas, but only those configurations that are experimentally distinguishable. In other words, if a gas composed of N *different* (experimentally distinguishable) molecules has $\Omega = \Omega_N$ distinguishable configurations (or microstates), then a gas composed of N *identical* (experimentally indistinguishable) molecules should only have $\Omega = \Omega_N/N!$ distinguishable microstates.

Boltzmann's formula requires that the total entropy change of the system is $\Delta S = k_B \ln(\Omega_f/\Omega_i)$, where Ω_i and Ω_f are the initial and final values of Ω. This implies that when two identical gases are mixed, the factorial factors for the initial and final state will give rise to an entropy change of $k_B \ln(1/N!) - k_B \ln(1/N_1!) - k_B \ln(1/N_2!) = k_B \ln[(N_1!N_2!)/N!] = -k_B \ln[N!/(N_1!N_2!)]$. If we make use of the Stirling approximation $\ln N! \cong N \ln N - N$ (see footnote 23 on pages 172–173), then the latter contribution to $\Delta S_{\text{mix}}|_P$ becomes $-k_B[N \ln N - N_1 \ln N_1 - N_2 \ln N_2]$, which is identical to Eq. 5.36, except for the negative sign. Thus, when we include the appropriate factorial factors in Boltzmann's entropy formula, the entropy change associated with mixing two identical gases vanishes!

In summary, Gibbs found that when the statistical consequences of indistinguishability are correctly taken into account, then mixing *identical gases* produces no entropy change. On the other hand, when two *different gases* are mixed (at constant pressure), then the entropy of the mixture is predicted to be greater than that of the pure gases, as described by Eq. 5.35. The factor of $N!$, which accounts for the indistinguishability of identical

---

[25] Gibbs's more detailed discussion of the effect of molecular distinguishability on entropy is contained in the first few pages of Chapter XV of his book whose full title is *Elementary Principles in Statistical Mechanics Developed with a Special Reference to the Rational Foundation of Thermodynamics*, which was published in 1902, the year before he died at the age of 64.

molecules, is thus required to correctly represent the partition function of a single-component ideal gas, as given by Eq. 5.31.

Boltzmann, Gibbs, and Shannon Entropies

We have now encountered several apparently quite different expressions for entropy, so it is natural to wonder whether they are all actually equivalent to each other. For example, Eq. 5.26 looks very different from Boltzmann's expressions for entropy $S = k_B \ln \Omega$ (Eq. 2.9). However, we will see that all of these expressions are in fact equivalent to each other, as well as to another expression for S that is reminiscent of that obtained from the entropy of mixing formula Eq. 5.37. Moreover, we will also see that all of these expressions for S may be derived from Boltzmann probability formula $P(\varepsilon_i) = e^{-\beta\varepsilon_i} / \sum_i e^{-\beta\varepsilon_i}$ (Eq. 1.14).

The connection between $S = k_B \ln \Omega$ and Eq. 5.26 can easily be demonstrated if we make use of the fact that the energy of a macroscopic isothermal system never deviates significantly from its average value $U = \langle E \rangle$. Such a system will always be found in one of a very large number Ω of configurational states, each of which has essentially the same energy. Thus, Ω represents the effective degeneracy of the system, and so its partition function can be expressed as follows.[26]

$$\mathcal{Q} \approx \Omega e^{-\beta\langle E \rangle} = \Omega e^{-\beta U} \tag{5.38}$$

This expression may be combined with standard thermodynamic identities, and Eq. 5.25, to recover Boltzmann's entropy formula.

$$S = \frac{U - A}{T} = \frac{U + k_B T \ln \mathcal{Q}}{T} = \frac{U - U + k_B T \ln \Omega}{T} = k_B \ln \Omega \tag{5.39}$$

Note that the second of the above equalities is the same as that obtained by combining Eqs. 5.23 and 5.26, thus demonstrating the equivalence of Eq. 5.26 and $S = k_B \ln \Omega$.

Another important expression for S may be obtained by making use of Eq. 5.20, which implies that $-\ln P(E_i) = -\ln[e^{-\beta E_i}/\mathcal{Q}] = \beta E_i + \ln \mathcal{Q}$. If we multiply the latter expression by $k_B P(E_i)$ and then sum over all the quantum states of the system, we obtain the following expression for S.[27]

$$-k_B \sum_i P(E_i) \ln P(E_i) = \frac{1}{T}\frac{\sum_i E_i e^{-\beta E_i}}{\mathcal{Q}} + k_B \ln \mathcal{Q} = \frac{U}{T} - \frac{A}{T} = S \tag{5.40}$$

---

[26] Although Eq. 5.38 is approximate, it is sufficiently accurate that it may be considered to be essentially exact for any macroscopic isothermal system.

[27] The intermediate steps in Eq. 5.40 make use of Eq. 5.20 and the fact that $\sum_i e^{-\beta E_i}/\mathcal{Q} = 1$.

This expression for entropy is attributed to Gibbs, and thus $S = -k_B \sum_i P_i \ln P_i$ is referred to as Gibbs's entropy formula, where P_i is a shorthand notation for $P(E_i)$. Notice how similar this expression is to Gibbs's entropy of mixing formula Eq. 5.37.

The latter expression is also closely related to the dimensionless *Shannon entropy* $-\sum_i P_i \ln P_i$, which was first introduced in a little book entitled *A Mathematical Theory of Communication*, written by Claude Shannon in 1948. The probabilities P_i appearing in Shannon's entropy formula may pertain to a wide variety of phenomena, such as the probability that a given letter will appear in a document or the probability that a given note will appear in a musical score.[28]

To illustrate the connection between the Boltzmann, Gibbs, and Shannon entropy formulas, it is instructive to consider a system that can be in any one of Ω equally probable states, so $P_1 = P_2 = P_3 \cdots = 1/\Omega$. The Gibbs entropy of such a system is $S = -k_B \sum_{i=1}^{\Omega} P_i \ln P_i = -k_B \sum_{i=1}^{\Omega} (1/\Omega) \ln(1/\Omega) = -k_B(\Omega/\Omega) \ln(1/\Omega) = k_B \ln \Omega$. Thus, although Eqs. 2.9, 5.26, and 5.40 look quite different, they are in fact equivalent to each other, as well as to Eqs. 1.14 and 5.20.[29]

5.5 Kirkwood, Widom, and Jarzynski

The results obtained in Sections 5.3 and 5.4 imply that if we knew all the energy levels of a macroscopic (supermolecule) system, we could in principle obtain the partition function $\mathcal{Q}$, and then use $\mathcal{Q}$ to calculate all of its thermodynamic properties. Although these identities are quite general and powerful, they can be difficult to apply, because it is hard to accurately account for all of the quantum states of a macroscopic system. On the other hand, the fundamental connection between $\mathcal{Q}$ and thermodynamics has set the stage for further theoretical advances, as well as the

---

[28] Shannon is considered to be the founder of the field of information theory, and his little book has become so influential that its title has been changed in very a small but significant way to *The Mathematical Theory of Communication*. The key idea behind Shannon's book is that entropy and information are equivalent. For example, if a language had only one letter, it would have a Shannon entropy of zero, and thus could not convey any information. On the other hand, a language with many letters has a large Shannon entropy and thus the capability of communicating much information.

[29] The Gibbs $-k_B \sum_i P_i \ln P_i$ and Clausius $\int \delta Q/T$ expressions for entropy may be applied to the calculation of entropic contributions to protein-ligand binding, as described in a recent paper by Kyle Harpole and Kim Sharp [*J. Phys. Chem. B*, vol. 115, p. 9461 (2011)].

development of modern computer simulation methods. Such simulations make it feasible to predict the thermodynamic properties of complex materials such as molecular solutions, and even biochemical systems.

General Relation Between μ and $\mathcal{Q}$

A key first step leading to the general description of nonideal chemical systems is the recognition that Eq. 5.30 may be manipulated to obtain other expressions for the chemical potential. For example, let's assume that we are interested in describing a liquid composed of N_0 identical solvent molecules and a single solute molecule. Thus, the number of solute molecules is $N_1 = 0$ in the pure solvent and $N_1 = 1$ after the solute has been added. Equation 5.30 indicates that the chemical potential of the solute in this system is $\mu_1 = (\partial A/\partial N_1)_{T,V,N_0}$ (note that the derivative is evaluated while holding the number of solvent molecules, as well as the temperature and volume, constant).

Since we are primarily interested in the influence of intermolecular interactions on μ_1, it is often convenient to subtract the chemical potential of the isolated (ideal gas) solute molecule μ_1^{IG}, and thus obtain the excess chemical potential of the solute $\mu_1^{\times} = \mu_1 - \mu_1^{IG} = (\partial A/\partial N_1)_{T,V,N_0} - (\partial A^{IG}/\partial N_1)_{T,V,N_0} = (\partial A^{\times}/\partial N_1)_{T,V,N_0}$ (where $A^{\times} \equiv A - A^{IG}$). We may express the latter derivative as a difference between the Helmholtz energy of the liquid with and without the added solute molecule.

$$\mu^{\times} = \left(\frac{\partial A^{\times}}{\partial N_1}\right)_{T,V,N_0} = A^{\times}(T,V,N_0,N_1=1) - A^{\times}(T,V,N_0,N_1=0) \quad (5.41)$$

In other words, $A^{\times}(T,V,N_0,N_1=1)$ is the excess Helmholtz energy of a solution containing a single solute molecule (and N_0 solvent molecules), and $A^{\times}(T,V,N_0,N_1=0)$ is the excess Helmholtz energy of the pure solvent (at the same temperature and volume).

The above excess Helmholtz energies may also be expressed in terms of the corresponding partition function $\mathcal{Q}^{\times}$, which represents that part of $\mathcal{Q}$ that results from intermolecular interactions. More specifically, $\mathcal{Q}^{\times} \equiv \mathcal{Q}/\mathcal{Q}^{IG}$, which implies that $-k_BT\ln\mathcal{Q}^{\times} = -k_BT\ln(\mathcal{Q}/\mathcal{Q}^{IG}) = -k_BT(\ln\mathcal{Q} - \ln\mathcal{Q}^{IG}) = A - A^{IG} = A^{\times}$. Moreover, Eq. 5.21 indicates that $\mathcal{Q}^{\times}$ may be expressed as follows (where $E^{\times} = E - E^{IG}$)

$$\mathcal{Q}^{\times} = \int e^{-\beta E^{\times}}\,d\Gamma \quad (5.42)$$

and Eq. 5.25 confirms the relation between $A^{\times}$ and $\mathcal{Q}^{\times}$.

$$A^{\times} = -k_BT\ln\mathcal{Q}^{\times} \quad (5.43)$$

We may now combine Eqs. 5.41 and 5.43 to relate $\mu_1^\times$ to the partition function of the pure solvent $\mathcal{Q}_0^\times$ and that of a solution consisting of the solvent and a single solute molecule $\mathcal{Q}_1^\times$.

$$\boxed{\mu_1^\times = -k_B T\left[\ln \mathcal{Q}_1^\times - \ln \mathcal{Q}_0^\times\right] = -k_B T \ln\left(\frac{\mathcal{Q}_1^\times}{\mathcal{Q}_0^\times}\right)} \tag{5.44}$$

Equation 5.44 is the starting point for the derivation of three remarkable theorems (described in the following three subsections), which form the basis of many modern computer simulation strategies, as well as recent experiments involving biological and synthetic nano-devices.

Kirkwood Reversible Work Theorem

A quite interesting and practically useful expression for the excess chemical potential was obtained in 1935 by Jack Kirkwood – a physical chemistry professor at Cornell University.[30] The theorem Kirkwood derived makes use of the intimate connection between Helmholtz energy and reversible work (Eq. 3.47) to obtain an elegant expression for the chemical potential of a molecule in any system. In essence, this theorem yields a statistical mechanical equation for the amount of work that would be required in order to slowly (reversibly) grow a solute molecule in any system of interest. Although the fundamental importance of Kirkwood's theorem was clear

[30] Kirkwood was born in Oklahoma in 1907 (four years after Willard Gibbs died). As a high school student, Jack demonstrated such a talent for physics and mathematics that he was persuaded by Arthur Amos Noyes, a professor at the California Institute of Technology, to enroll in Caltech before receiving his high school diploma. After receiving his PhD from MIT in 1929, Kirkwood began his academic career as a theoretical physical chemist at Cornell. He later moved to Caltech (where he held the Arthur Amos Noyes chaired professorship), and then to Yale (the same university in which Willard Gibbs spent his entire career). During the years that Kirkwood was at Yale, another brilliant theoretical physical chemist named Lars Onsager was also there. It was Onsager's early development of the adiabatic principle (and the concept of a mean force potential) in 1933 [*Chem. Rev.*, 13, 73 (1933)] that set the stage for Kirkwood's subsequent derivation of the result described in this section [*J. Chem. Phys.*, 3, 300 (1935)]. Onsager and Kirkwood are buried side by side in the Grove Street Cemetery near Yale, where Willard Gibbs is also buried. Kirkwood died of cancer in 1959, and his gravestone contains a long list of his awards and recognitions. Onsager, who died nearly 20 years later, chose to have his gravestone state simply "Nobel Laureate" (although his children later added an asterisk after "Nobel Laureate," and "*etc." in the lower right corner of the gravestone). Gibbs's gravestone identifies him only as a professor of mathematical physics.

from the outset, its applicability to computer simulations was only realized many years later.[31]

The starting point of Kirkwood's analysis is Eq. 5.44, reexpressed in the following suggestive form.[32]

$$\mu_1^\times = -k_B T \ln\left(\frac{\mathcal{Q}_1^\times}{\mathcal{Q}_0^\times}\right) = -k_B T \int_0^1 \frac{\partial \ln \mathcal{Q}_\lambda^\times}{\partial \lambda} d\lambda \tag{5.45}$$

The brilliance of Kirkwood's analysis lies in the special physical significance he assigned to λ, by envisioning it as a coupling parameter that slowly turns on solute-solvent interactions as it varies between 0 and 1.

The excess energy of the system $E_\lambda^\times$ may be expressed as the sum of the excess energy of the pure solvent $E_0^\times$ plus the solute-solvent interaction energy Ψ multiplied times the coupling parameter λ.[33]

$$E_\lambda^\times = E_0^\times + \lambda\Psi \tag{5.46}$$

So, when $\lambda = 0$, then $E_\lambda^\times = E_0^\times$, while when $\lambda = 1$, then $E_\lambda^\times = E_1^\times$. If the parameter λ is varied slowly enough, then the solvent configurations will remain continuously equilibrated around the growing solute. Such a slow, continuously equilibrated process is necessarily reversible. Moreover, Eq. 3.47 implies that the work associated with such a reversible isothermal coupling process must be equal to the corresponding Helmholtz energy change, and Eq. 5.41 identifies that free energy change as $\mu_1^\times$.

---

[31] The first digital computers were developed in the 1940s, and the first application of computers to molecular statistical mechanical calculations was performed in the 1950s, most notably by Bernie Alder and Tom Wainwright who first correctly calculated the equation of state and freezing point of a fluid composed of hard spheres [*J. Chem. Phys.*, 27, 1208 (1957)]. Bernie Alder was a student of Jack Kirkwood, but Kirkwood did not trust Bernie's early computer simulation results and so he held up their publication. An interesting 1997 interview with Bernie Alder about the early history of molecular computer simulations can be found at http://www.computer-history.info/Page1.dir/pages/Alder.html.

[32] The last equality in Eq. 5.45 is obtained using the fact that an integral is an antiderivative, in the sense that $\int (df/dx)dx = f$, and so $\int_0^1 (d \ln \mathcal{Q}_\lambda^\times / d\lambda) d\lambda = \ln \mathcal{Q}_1^\times - \ln \mathcal{Q}_0^\times = \ln(Q_1^\times / \mathcal{Q}_0^\times)$.

[33] The energies $E_\lambda^\times$ and $E_0^\times$ each pertain to a particular configuration of molecules. In order to obtain the corresponding thermodynamic energies U, we must perform an appropriate average $\langle \ldots \rangle$ over all of the equilibrium configurations of the corresponding system, as defined in Eq. 5.47.

Kirkwood's expression for the reversible work associated with slowly introducing a solute is obtained from Eq. 5.45 by first noting that

$$\frac{\partial \ln \mathcal{Q}_\lambda^\times}{\partial \lambda} = \frac{1}{\mathcal{Q}_\lambda^\times}\left(\frac{\partial \mathcal{Q}_\lambda^\times}{\partial \lambda}\right) = \frac{\frac{\partial}{\partial \lambda}\left(\int e^{-\beta E_\lambda^\times} d\Gamma\right)}{\int e^{-\beta E_\lambda^\times} d\Gamma} = \frac{\int \frac{\partial}{\partial \lambda}\left(e^{-\beta E_\lambda^\times}\right) d\Gamma}{\int e^{-\beta E_\lambda^\times} d\Gamma}$$

$$= \frac{\int \frac{\partial}{\partial \lambda}\left[e^{-\beta(E_0^\times+\lambda\Psi)}\right] d\Gamma}{\int e^{-\beta(E_0^\times+\lambda\Psi)} d\Gamma} = \frac{\int (-\beta\Psi) e^{-\beta(E_0^\times+\lambda\Psi)} d\Gamma}{\int e^{-\beta(E_0^\times+\lambda\Psi)} d\Gamma}$$

$$= -\beta \int \Psi P(E_\lambda^\times) d\Gamma \equiv -\beta \langle\Psi\rangle_\lambda \tag{5.47}$$

The quantity $\langle\Psi\rangle_\lambda$ is simply a short-hand notation for the previous integral, which yields the value of Ψ averaged over all configurations of the system equilibrated with a given value of λ. Also, note that Eq. 5.46 implies that $\Psi = \partial E_\lambda^\times/\partial\lambda$ and so $\langle\Psi\rangle_\lambda = \langle \partial E_\lambda^\times/\partial\lambda\rangle_\lambda$ is equivalent to the average force associated with changing λ.

Equation 5.47 may now be combined with Eq. 5.45 to obtain Kirkwood's reversible work theorem, which is also referred to as the Kirkwood coupling-parameter integral expression for $\mu_1^\times$.

$$\boxed{\mu_1^\times = -k_B T \int_0^1 -\beta \langle\Psi\rangle_\lambda \, d\lambda = \int_0^1 \langle\Psi\rangle_\lambda \, d\lambda} \tag{5.48}$$

The latter integral represents the reversible work associated with slowly introducing the solute into the solvent. In other words, $\int_0^1 \langle\Psi\rangle_\lambda \, d\lambda$ is an integral of the force $\langle \partial E_\lambda^\times/\partial\lambda\rangle_\lambda = \langle\Psi\rangle_\lambda$ times the displacement $d\lambda$, and so is equal to the total work associated with the solute coupling process. Since the process is carried out slowly, in such a way that the solute and solvent remain infinitesimally close to equilibrium with each other at every point, the resulting work is equivalent to the reversible work associated with introducing a solute into the system.

$$\int_0^1 \langle\Psi\rangle_\lambda \, d\lambda = W_{\text{rev}} \tag{5.49}$$

Kirkwood's result served as a fruitful starting point for a number of subsequent fundamental theoretical developments. However, it was not until the advent of modern computers that the power of Kirkwood's theorem as a simulation strategy was realized.

Performing Kirkwood coupling parameter simulations requires numerically evaluating $\int_0^1 \langle\Psi\rangle_\lambda \, d\lambda$. More specifically, such simulations begin by equilibrating a box full of solvent molecules (on the computer), and then slowly turning on the interactions between the solute and all the surrounding solvent molecules by gradually increasing the value of λ. After every small change in λ the system is reequilibrated and the associated change in energy $\partial E^{\times}_{\lambda}/\partial\lambda$ is averaged over all configurations to obtain $\langle\Psi\rangle_{\lambda}$. Such simulations require a powerful computer because the average $\langle\Psi\rangle_{\lambda}$ must be accurately determined at each value of λ, and λ must be increased in steps that are small enough to accurately approximate the continuous variation of λ from 0 to 1.

Kirkwood coupling parameter simulations may be used to obtain solvation free energies $\Delta G_S \equiv \overline{G}_1(\text{liq}) - \overline{G}_1(\text{vap}) = \mathcal{N}_A(\mu_1 - \mu_1^{IG}) = \mathcal{N}_A\mu_1^{\times}$. A similar simulation strategy may also be used to determine the free energy associated with bringing two molecules together, such as a protein-ligand binding process, or the free energy associated with changing the configuration of a protein from a folded (native) state to an unfolded (denatured) state, as well as many other sorts of chemical processes.

Widom Insertion Theorem

Another powerful method for obtaining chemical potentials is based on the Widom insertion theorem, whose derivation again begins with Eq. 5.44. This theorem was published in 1963 by Ben Widom, in a paper entitled *Some Topics in the Theory of Fluids*.[34] At the time that Widom derived this theorem it was considered to be an interesting fundamental result, without any suggestion that it would subsequently lead to practical computer simulation strategies.

In deriving his theorem, Widom envisioned a solvation process that is quite different from that suggested by Kirkwood. Rather than slowly turning on the solute-solvent interactions (as Kirkwood imagined doing), Widom considered what would occur if a solute were abruptly inserted into the pure solvent, before the solvent molecules have any time to move. Notice that both Widom's and Kirkwood's solvation processes are hypothetical, in the sense that they cannot be realized experimentally. Nevertheless both processes lead to practical computer simulation strategies, and provide interesting insights into the physical significance of chemical potentials.

A Widom insertion process may be envisioned as beginning with the pure solvent in a given configuration, with a total interaction energy $E_0^{\times}$. We then imagine abruptly attempting to insert the solute (without moving the solvent molecules). The resulting total interaction energy of the solvent plus the solute is $E_1^{\times}$ and the difference between the latter two energies is

[34] Ben Widom obtained his PhD at Cornell in 1953 with professor Simon Bauer. Jack Kirkwood was no longer at Cornell during those years, as he had moved to Caltech in 1947. Widom subsequently joined the faculty at Cornell, where he remains a highly influential theoretical chemist and widely admired lecturer and teacher to this day.

$\Psi \equiv E_1^\times - E_0^\times$. The total energy of the system with a given solvent configuration is thus simply the sum of the pure solvent energy and the change in energy induced by abruptly adding the solute.

$$E_1^\times = E_0^\times + \Psi \tag{5.50}$$

Equation 5.42 implies that the excess partition functions of the solution $\mathcal{Q}_1^\times$ and the pure solvent $\mathcal{Q}_0^\times$ may be expressed as follows.

$$\begin{aligned}\mathcal{Q}_0^\times &= \int e^{-\beta E_0^\times} d\Gamma \\ \mathcal{Q}_1^\times &= \int e^{-\beta E_1^\times} d\Gamma\end{aligned} \tag{5.51}$$

The latter integral may also be expressed as

$$\mathcal{Q}_1^\times = \int e^{-\beta(E_0^\times + \Psi)} d\Gamma = \int e^{-\beta\Psi} e^{-\beta E_0^\times} d\Gamma \tag{5.52}$$

and thus the ratio of the two partition functions becomes

$$\frac{\mathcal{Q}_1^\times}{\mathcal{Q}_0^\times} = \frac{\int e^{-\beta\Psi} e^{-\beta E_0^\times} d\Gamma}{\int e^{-\beta E_0^\times} d\Gamma} \tag{5.53}$$

Recall that Eq. 5.21 implies that $e^{-\beta E_0^\times}/(\int e^{-\beta E_0^\times} d\Gamma) = P(E_0^\times)$ is the probability of finding the pure (equilibrium) solvent in a configuration of energy $E_0^\times$. Thus, evaluating the right-hand side of Eq. 5.53 is equivalent to evaluating the average of $e^{-\beta\Psi}$ over all equilibrium configurations of the *pure solvent*.

$$\frac{\mathcal{Q}_1^\times}{\mathcal{Q}_0^\times} = \int e^{-\beta\Psi} P(E_0^\times) d\Gamma = \langle e^{-\beta\Psi} \rangle_0 \tag{5.54}$$

We may now combine this expression with Eq. 5.44 to obtain the following Widom insertion expression for the solute's excess chemical potential.

$$\boxed{\mu_1^\times = -k_B T \ln\langle e^{-\beta\Psi} \rangle_0} \tag{5.55}$$

This remarkable result implies that one may determine the chemical potential of a solute using the equilibrium configurations of the *pure solvent*, without ever needing to equilibrate the solvent around the solute.

To use the Widom theorem in a computer simulation, one must begin by generating a large number of equilibrium configurations of the pure solvent. This can be done, for example, by allowing the pure solvent system to evolve according to Newton's laws, while maintaining the system at a constant temperature and volume. Once a sufficient number of the pure solvent configurations have been obtained, Eq. 5.55 may be implemented by repeatedly attempting to add the solute at many randomly selected locations within each solvent configuration (without moving the solvent molecules). After each attempted insertion of the solute, the value of $e^{-\beta\Psi}$ is recorded, and then a running average is performed to determine $\langle e^{-\beta\Psi}\rangle_0$. This simulation strategy is found to work beautifully, as long as the solute molecule is not too large (or the density of the solvent is not too high). This is because for very large solutes and/or very high solvent densities all attempted insertions will produce a very large value of Ψ and an essentially zero value of $e^{-\beta\Psi}$, thus making it virtually impossible to accurately evaluate $\langle e^{-\beta\Psi}\rangle_0$.

For small solutes the Widom insertion method is much more efficient than Kirkwood integration. However, the Kirkwood method places no restrictions on the size of the solute and so can still be used when the Widom method fails. As a result, the Widom insertion method is preferred whenever some reasonable fraction of insertion attempts yield a value of Ψ that is not too much larger than $k_B T$ (as is the case for small molecules or in relatively low-density solvents), while Kirkwood's method (and variants thereof) can be used to determine $\mu_1^\times$ for large molecules in high-density systems.

The Widom method is also closely related to another widely used simulation strategy in which a solute is abruptly transformed from one form to another, and an expression very similar to Eq. 5.55 is used to evaluate the difference between the chemical potentials of the two solute species. Such calculations are sometimes referred to as free energy perturbation simulations.[35] For example, one may use this approach to determine the difference between the chemical potentials of hydrocarbons with different head groups. Alternatively, one may use this method to determine the chemical potential increment associated with adding a monomer to the end of a polymer chain. If one starts this process by inserting a single monomer, and then adds additional monomers one at time, then one can obtain the total chemical potential of a long polymer. The latter approach may be viewed as a hybrid of the Kirkwood and Widom methods, as one must equilibrate the solvent around the polymer after each step, before using the Widom method to insert the next monomer.

Jarzynski Irreversible Work Theorem

The Kirkwood and Widom theorems offer two paths to the calculation of the excess chemical potential of a solute molecule $\mu_1^\times$. The Kirkwood

[35] The use of the term *perturbation* to describe such simulations may be traced to the fact that the corresponding analogue of Eq. 5.55 was first obtained by Bob Zwanzig in 1954, as the starting point for his subsequent derivation of an early variant of liquid perturbation theory. However, it is important to stress that while perturbation theory provides approximate results, neither Eq. 5.55 nor Zwanzig's analogous expression are approximate.

expression (Eq. 5.48) requires following a reversible path in which the solute is slowly introduced into the solvent, while the Widom expression (Eq. 5.55) requires performing a highly irreversible abrupt insertion of the solute into the solvent. In 1996 Chris Jarzynski derived a theorem that may be used to evaluate $\mu_1^{\times}$ by following various other paths, which include the Kirkwood and Widom paths as limiting cases.[36]

Jarzynski considered the consequences of turning on a solute coupling parameter in some finite time $\lambda(t)$. As time progresses, the configuration of the solvent molecules need not remain frozen in place (as in the Widom process) but may also not remain perfectly equilibrated with the solute (as in the Kirkwood process). More specifically, at $t = 0$ the intermolecular interaction energy of the pure solvent (in a particular configuration) is $E_0^{\times}$. At some later time t, after the coupling parameter has reached a value of $\lambda(t) = 1$ the intermolecular interaction energy of the system will become $E_1^{\times}$. Moreover, we may further assume that the coupling process is performed adiabatically (with no heat exchange), and thus the first law (Eq. 2.3) implies that the energy change of the system must be equal to the work done on the system.

$$W \equiv \int_0^t \delta W = E_1^{\times} - E_0^{\times}$$

Jarzynski's theorem pertains to the value of $e^{-\beta W}$ averaged over many repeated repetitions of the $\lambda(t)$ coupling process $\langle e^{-\beta W} \rangle_{\lambda(t)}$.[37] Jarzynski's key insight was the realization that the latter average is independent of how fast the coupling process is performed. If the process is performed infinitely

---

[36] Chris Jarzynski was a postdoctoral fellow at Los Alamos National Labs when he derived his now-famous theorem. He subsequently spent several years as a research scientist at the Los Alamos National Laboratory, and is currently a professor of chemistry at the University of Maryland.

[37] The average $\langle e^{-\beta W} \rangle_{\lambda(t)}$ is performed by repeatedly varying the coupling parameter from $\lambda = 0$ to $\lambda = 1$, using the same time-dependent function $\lambda(t)$ to perform the process. Each repeated application of the coupling process will not in general produce exactly the same amount of work because the initial configuration of the solvent (randomly chosen for an equilibrium distribution of solvent configurations) will not be exactly the same. Thus, the coupling process must be repeated a sufficient number of times to get an accurate measure of the average value of the irreversible work associated with a given $\lambda(t)$ coupling process. We may also extend this analysis to isothermal processes, which start and end at the same temperature, by assuming that after the irreversible adiabatic work exchange is performed, the system is allowed to reequilibrate with a bath of the same temperature as the system had initially, $T = \beta/k_B$. Since the latter thermal equilibration process is performed while holding λ constant, it does not involve any additional work exchange and so does not alter the value $\langle e^{-\beta W} \rangle_{\lambda(t)}$.

fast, such that the solvent molecules don't have time to move, then it becomes exactly equivalent to that described by the Widom theorem. More specifically, if the solvent molecules don't move during the coupling process, then the work associated with inserting the solute is exactly equal to the corresponding solute-solvent interaction energy $W = \Psi$ (see Eq. 5.50). Moreover, under these conditions the Jarzynski average $\langle \ldots \rangle_{\lambda(t)}$ is exactly equivalent to the Widom average $\langle \ldots \rangle_0$, since the solvent molecules remain fixed at their initial configuration during the entire coupling process. Thus, $\langle e^{-\beta W} \rangle_{\lambda(t)} = \langle e^{-\beta \Psi} \rangle_0$, and so Eq. 5.55 leads immediately to the Jarzynski theorem (as further described in the next subsection).

$$\boxed{\mu_1^\times = -k_B T \ln \left\langle e^{-\beta W} \right\rangle_{\lambda(t)}} \tag{5.56}$$

Recall that for an infinitely slow (reversible) Kirkwood coupling process, Eqs. 5.48 and 5.49 imply that $\mu_1^\times = W_{\text{rev}}$, or equivalently $e^{-\beta \mu_1^\times} = e^{-\beta W_{\text{rev}}}$. Thus, the Kirkwood (Eq. 5.48), Widom (Eq. 5.55), and Jarzynski (Eq. 5.56) theorems may all be combined to yield the following remarkable sequence of identities.

$$\boxed{e^{-\beta \mu_1^\times} = e^{-\beta W_{\text{rev}}} = \left\langle e^{-\beta \Psi} \right\rangle_0 = \left\langle e^{-\beta W} \right\rangle_{\lambda(t)}} \tag{5.57}$$

Jarzynski's theorem has generated significant interest from both computational and experimental perspectives. Like the Kirkwood and Widom theorems, the Jarzynski irreversible work theorem may be used as the basis for computer simulation strategies.[38] However, while the Widom and Kirkwood chemical potential results involve hypothetical processes that can only be realized using a computer simulation, Jarzynski's theorem also lends itself to direct experimental verification.

In 2002 Carlos Bustamante and coworkers at the University of California in Berkeley performed a remarkable series of experiments involving the unfolding and refolding of a single RNA molecule. When the unfolding was carried out very slowly, the process proved to be nearly perfectly reversible, and so one could obtain the chemical potential change associated with unfolding the RNA molecule. On the other hand, when the ends of the RNA molecule were pulled apart more rapidly, then a larger amount of (irreversible) work was required to unfold the RNA. Repeated unfolding measurements were used to experimentally evaluate the average on

[38] Although the final word is not yet out, it appears likely that the Jarzynski theorem may not provide any significant advantage in terms of computational speed. In other words, it appears to be the case that simulations based on the Jarzynski theorem are not fundamentally more numerically efficient than simulations based on either the Widom or Kirkwood theorem.

the left-hand side of Eq. 5.56 and to experimentally verify the Jarzynski theorem. These elegant experiments confirmed that the same free energy of unfolding may be obtained using either reversible or irreversible work exchange measurements.

More about the Jarzynski Theorem

The following is a more detailed outline of the derivation of Jarzynski's theorem (applied to a solute coupling process).

Let the variable Γ_0 designate a particular (multidimensional) initial configuration of all the molecules in the system, at $t = 0$ just before the start of the solute coupling process. In other words, Γ_0 is equivalent to a point in the configurational *phase space* of a system at the start of the coupling process, and $E_0^\times(\Gamma_0)$ is the solvent interaction energy associated with that configuration. Similarly, the final energy of the solution $E_1^\times(\Gamma_1)$ depends on the configuration Γ_1 of the molecules in the fully coupled system, at the end of the $\lambda(t)$ coupling process.

The first law implies that the work associated with the adiabatic coupling process is exactly equal to the energy difference between the latter two states $W = E_1^\times(\Gamma_1) - E_0^\times(\Gamma_0)$. Note that each initial configuration Γ_0 will lead to a particular final configuration Γ_1, and thus Γ_1 and $E_1^\times$ are both functions of Γ_0, and therefore, so is the solute coupling work $W(\Gamma_0) = E_1^\times(\Gamma_0) - E_0^\times(\Gamma_0)$.

The Jarzynski average $\langle e^{-\beta W}\rangle_{\lambda(t)}$ represents the average of $e^{-\beta W(\Gamma_0)}$ weighted by the probability $P(\Gamma_0)$ of finding the pure solvent in a given initial configuration Γ_0. We may use Eq. 5.21 to express the latter probability as $P(\Gamma_0) = e^{-\beta E_0^\times}/\mathcal{Q}_0^\times$, where $\mathcal{Q}_0^\times = \int e^{-\beta E_0^\times} d\Gamma_0$, and thus obtain the following expressions for the Jarzynski average.

$$\begin{aligned}
\left\langle e^{-\beta W}\right\rangle_{\lambda(t)} &= \int e^{-\beta W} P(\Gamma_0)\, d\Gamma_0 = \frac{\int e^{-\beta W} e^{-\beta E_0^\times} d\Gamma_0}{\int e^{-\beta E_0^\times} d\Gamma_0} \\
&= \frac{\int e^{-\beta(E_1^\times - E_0^\times)} e^{-\beta E_0^\times} d\Gamma_0}{\int e^{-\beta E_0^\times} d\Gamma_0} = \frac{\int e^{-\beta E_1^\times} e^{+\beta E_0^\times} e^{-\beta E_0^\times} d\Gamma_0}{\int e^{-\beta E_0^\times} d\Gamma_0} \\
&= \frac{\int e^{-\beta E_1^\times} d\Gamma_0}{\int e^{-\beta E_0^\times} d\Gamma_0} = \frac{\int e^{-\beta E_1^\times} d\Gamma_1}{\int e^{-\beta E_0^\times} d\Gamma_0} = \frac{\mathcal{Q}_1^\times}{\mathcal{Q}_0^\times}
\end{aligned}$$

The last identity, combined with Eq. 5.44, leads immediately to Eq. 5.56, and thus proves the Jarzynski theorem.

Note that the second to last of the above equalities makes use of a theorem derived by Joseph Liouville in 1835. The Liouville theorem states that the phase space probability density of any mechanical system behaves like an incompressible fluid, in the sense that a given volume element in phase space does not increase or decrease as it evolves in time, and thus $d\Gamma_0 = d\Gamma_1$. It is also important to note that both Γ_0 and Γ_1 must have the same dimensionality, and so each Γ_0 configuration may be envisioned as containing N_0 solvent molecules plus one uncoupled solute *ghost particle*, which does not interact with the solvent molecules, while each Γ_1 configuration contains the same number of molecules, but now with a fully coupled solute.

HOMEWORK PROBLEMS

Problems That Illustrate Core Concepts

1. The following table contains experimental data for CO_2 at $T = 300$ K.

P (MPa)	$[c] = 1/\bar{V}$ (M)	C_V (J/mol-K)	κ_T (1/MPa)	α_P (1/K)
0.1	0.0403	29.016	10.049	0.0033876
1.1	0.467	30.141	0.96484	0.0040214
2.1	0.947	31.438	0.54082	0.0048880

(a) Use the above data to determine the experimental values of $(\partial P/\partial T)_V$ (in units of MPa/K), at each of the three pressures.

(b) Combine the results you obtained in (a) with the thermodynamic equation of state (Eq. 4.15) to evaluate the internal pressure $(\partial U/\partial V)_T$ of CO_2 (in units of MPa), at each of the three experimental pressures.

(c) Verify that the experimental values of $(\partial P/\partial T)_V$ and $(\partial U/\partial V)_T$ at $P = 0.1$ MPa are approximately consisted with those expected for an ideal gas, and determine how much the experimental values of these two partial derivatives differ from ideal gas predictions at the two higher pressures.

2. The Joule-Thompson coefficient $\mu_{JT} = (\partial T/\partial P)_H$ of a nonideal gas determines how much its temperature will change when it expands adiabatically.

(a) Manipulate $(\partial T/\partial P)_H$ to obtain an expression for $(\partial H/\partial P)_T$ in terms of μ_{JT} and C_P.

(b) Now further reduce μ_{JT} to express it in terms of only α_P, κ_T, C_P, T, and V (where both C_P and V are expressed in molar units).

(c) Use the experimental properties of CO_2 gas at $T = 300$ K and $P = 1.1$ MPa (from the table in problem 1) to predict the value of its Joule-Thompson coefficient.

(d) How much do you expect the temperature of CO_2 gas to change when it expands adiabatically from a pressure of 1.1 MPa to a pressure of 0.1 MPa?

3. The following questions pertain to the application of the van der Waals equation of state (Eq. 5.4) to carbon dioxide (CO_2) gas, for which experimental data is provided in problem 1.

(a) Use the van der Waals equation of state (Eq. 5.4) to obtain expressions for $(\partial P/\partial V)_T$ and $(\partial P/\partial T)_V$ in terms of the van der Waals coefficients $\mathbf{a}$ and/or $\mathbf{b}$, and combine those results to obtain an expression for the thermal expansion coefficient α_P of a van der Waals gas.

(b) Combine the thermodynamic equation of state $\left(\frac{\partial U}{\partial V}\right)_T = T\left(\frac{\partial P}{\partial T}\right)_V - P$ with your results in (a) to obtain an equation for the internal pressure, $(\partial U/\partial V)_T$, of a van der Waals gas.

(c) Use the experimental values of $(\partial P/\partial T)_V$ and $(\partial U/\partial V)_T$ at $T = 300$ K and $P = 2.1$ MPa to estimate the van der Waals constants of CO_2, and express $\mathbf{a}$ in units of Pa m^6/mol^2 and $\mathbf{b}$ in units of m^3/mol. Compare your results with the tabulated van der Waals constants of CO_2, $\mathbf{a} = 0.37$ (Pa m^6/mol^2) and $\mathbf{b} = 4.3 \times 10^{-5}$ (m^3/mol), which have been obtained from a fit to the experimental critical point of CO_2, and suggest an explanation for the difference between the two sets of van der Waals constants.

(d) Use both sets of van der Waals constants from (c) to predict the values of P (MPa), $(\partial P/\partial T)_V$ (MPa/K), and $(\partial U/\partial V)_T$ (MPa) for CO_2 at 300 K and $[c] = 0.947$ M, and compare your results with the experimental values obtained in problem 1(c).

4. Given that $\mu_1^\times \equiv \mu_1^{liq} - \mu_1^{IG}$, show that $\mathcal{N}_A \mu_1^\times$ is equivalent to the corresponding Gibbs energy of solvation $\Delta G_S = \overline{G}_1^{liq} - \overline{G}_1^{IG} = RT(\beta\mu_1^\times)$.

5. Use the generalized van der Waals equation of state Eq. 5.14, combined with the information given in Figure 5.4, to predict the internal pressure $(\partial U/\partial V)_T$ of n-hexane at 300 K and 0.1 MPa.

6. The generalized van der Waals chemical potential of a pure solvent and a dilute solute in a solvent are given by Eqs. 5.17 and 5.18.

(a) If the solute and solvent are the same, what is the value of d and how is the value of τ related to that of the pure solvent?

(b) Show that Eq. 5.18 becomes equivalent to Eq. 5.17 when the solute and solvent are the same.

7. The excess chemical potential of acetonitrile (CH_3CN) in its own pure liquid is $\beta\mu^\times \sim -8.4$ at 20°C (293 K) and atmospheric pressure (0.1 MPa). The density of liquid acetonitrile is 0.79 g/ml, and it has a molecular weight of 41 g/mol and a hard-sphere diameter of ~0.424 nm.

(a) What is the packing fraction of liquid acetonitrile?

(b) Use the generalized van der Waals (CSvdW) equation to predict the repulsive contribution to $\beta\mu^\times$ for acetonitrile dissolved in its own pure liquid.

(c) Combine your results in (b) with the total experimental value of $\beta\mu^\times$ to obtain the attractive contribution to $\beta\mu^\times$ and estimate the value of cohesive coefficient τ.

8. High-pressure (supercritical fluid) CO_2 is commonly used as a solvent for extracting organic compounds from geological or biological materials. For example, CO_2 is used to extract caffeine from coffee beans. One might also expect that CO_2 could be used to extract liquid hydrocarbons such as n-octane from crude oil or coal. Your challenge is to use Eqs. 5.17 and 5.18 to estimate the pressure of CO_2 at which such an extraction will become thermodynamically favorable. To simplify the problem, you may assume that the properties of n-octane in a crude oil or coal sample are approximately the same as those of pure liquid n-octane and are essentially pressure independent (over a pressure range of 0–20 MPa). The generalized van der Waals (CSvdW) parameters of n-octane are given in Figure 5.4 and those of CO_2 are $\sigma = 0.350$ nm and $\tau = 834$ K (obtained from a fit to the experimental pressure of CO_2 at 350 K).

(a) Use the information provided in Figure 5.4 to estimate the average packing fraction (η) of liquid n-octane at 350 K over a pressure range of 0–20 MPa (although η varies slightly over this pressure range you may neglect that small variation and treat it as constant). Hint: If the value of η is not similar to 0.5, then you have made a unit conversion mistake.

(b) Use Eq. 5.17, with the above value of η and the τ value given in Figure 5.4, to estimate the solvation Gibbs energy of an n-octane molecule in liquid n-octane $\Delta G_S^{Oct} = RT(\beta\mu^\times)$ at 350 K (see problem 4 above).

(c) Use the CSvdW parameters of CO_2 (along with those of n-octane given in Fig. 5.4) to estimate the values of $d = \sigma_1/\sigma_0$ and $\tau = \sqrt{\tau_1\tau_0}$ for n-octane dissolved in CO_2 (where 1 and 0 pertain to the solute and solvent, respectively).

(d) Use Eq. 5.18 to make a graph of the solvation Gibbs energy $\Delta G_S^{CO_2}$ of n-octane dissolved in in fluid CO_2 at 350 K as a function of the packing fraction η of CO_2 (over the range $0 < \eta < 0.3$) and then use the results you obtained in (b) to determine the packing fraction of CO_2 at which the extraction of n-octane becomes thermodynamically favorable. Note that the Gibbs energy for transferring an n-octane molecule from liquid n-octane to CO_2 is equal to the difference between the corresponding solvation Gibbs energies.

(e) Use Eq. 5.14 to make a plot of the pressure of CO_2 (in MPa units) as a function of its packing fraction, and then use this graph, combined with your results in (d), to estimate the pressure at which the extraction of n-octane will become favorable.

9. For an ideal gas supermolecule system consisting of N identical molecules $\mathcal{Q} = q^N/N!$, where q is the partition function of each molecule (and is independent of N). Recall that Stirling's approximation implies that $\ln(N!) = N \ln N - N$ (when N is large), and $\beta = 1/k_B T$.

(a) Use the relation between U and $\mathcal{Q}$ to verify the following expression for a single-component ideal gas consisting of N molecules: $U = -N(\partial \ln q/\partial\beta)_V = N\langle\epsilon\rangle$.

(b) Use the relation between A and $\mathcal{Q}$ to verify the following expression for the chemical potential of a single-component ideal gas: $\mu^{IG} = (\partial A/\partial N)_{T,V} = -k_B T \ln(q/N)$.

10. Calculate the entropy change when 1 mole of ideal gas A is mixed with 1 mole of ideal gas B at a constant temperature of 300 K, under conditions in which the initial and final total pressures of the gases are the same, $P_A = P_B = P_{A+B} = 0.1$ MPa. How would the result change if the pressure were 1 MPa instead of 0.1 MPa?

11. Given that ideal gases do not interact with each other in any way, calculate the chemical potential change (in J) for each component in the ideal gas process described in problem 10.

12. A computer simulation is performed to generate a large number of statistically independent configurations of 512 water molecules contained in a box of constant volume, which has been equilibrated at a temperature of 300 K and a density equivalent to that of water at 1 Atm. Repeated attempts to insert a methane molecule at random locations in these water configurations leads to various values of the methane-water interaction energy Ψ, from which the following average value is obtained $\langle e^{-\beta\Psi}\rangle_0 = 0.0231$. When a similar simulation is performed using a 1 M NaCl salt solution as the solvent (rather than pure water), then the simulation yields a value of $\langle e^{-\beta\Psi}\rangle_0 = 0.0137$ for the attempted insertion of methane into salt water.

(a) Use the above simulation results for methane in pure water, combined with the Widom theorem (Eq. 5.55), to predict the Gibbs energy of solvation of methane in pure water (as in problem 4 above), and compare your result with methane's experimental hydration Gibbs energy of 8.4 kJ/mol.

(b) Use the above simulation results to calculate the equilibrium concentration ratio $[CH_4]_{liq}/ [CH_4]_{gas}$ (which is also called the Ostwald coefficient), and thus predict the concentration of methane in water when it is in equilibrium with methane gas, which has a concentration of $[CH_4]_{gas} = 0.01$ M.

(c) Now apply the same procedure to the salt water simulation results, in order to predict the concentration of methane in the 1 M NaCl salt solution when it is in equilibrium with methane gas, which has a concentration of $[CH_4]_{gas} = 0.01$ M.

(d) Different salts are known to lead to either *salting out* or *salting in* of organic solutes, depending on whether they decrease or increase the solubility of the solute in water, respectively. Do the above simulation results imply that NaCl will lead to *salting out* or *salting in* of methane?

13. The Jarzynski theorem (Eqs. 5.56 and 5.57) may more generally be expressed as $e^{\Delta A/k_BT} = \langle e^{-W/k_BT}\rangle$, which relates the Helmholtz free energy change of a system ΔA in any isothermal process to the work exchange W associated with that process, where the average $\langle \ldots \rangle$ is performed over repeated applications of the same constraint variation (and the process may or may not be reversible).

(a) Pick any set of five real numbers as input x-values and use these to show that $\langle e^x \rangle \geq e^{\langle x \rangle}$, where $\langle \ldots \rangle$ represents the average value of the indicated quantity. Note that this inequality is a general mathematical result.

(b) Use the above inequality, combined with the Jarzynski theorem, to show that for any isothermal process $\langle W \rangle \geq W_{rev}$. In other words, a reversible process is one in which the maximum amount of work is done *by the system* (recalling that work done by the system is negative).

Problems That Test Your Understanding

14. Which of the following statements are true?

(a) When two *identical* gases are mixed at constant pressure, $\Delta S > 0$.

(b) When two *different* gases are mixed at constant pressure, $\Delta S < 0$.

(c) The Joule-Thompson coefficient (μ_{JT}) of an ideal gas is equal to zero.

(d) The van der Waals **b** coefficient represents the influence of intermolecular cohesive interactions.

(e) The Helmholtz free energy of a single-component ideal gas is $-Nk_BT \ln q + k_BT \ln N!$.

15. The equation of state of a van der Waals gas is given by Eq. 5.5.

(a) Use the van der Waals equation of state to obtain an expression for $(\partial P/\partial T)_V$ (in terms of the van der Waals constants **a** and/or **b**, etc.).

(b) Insert the above result with the thermodynamic equation of state $P = T(\partial P/\partial T)_V - (\partial U/\partial V)_T$ to obtain an expression for $(\partial U/\partial V)_T = (\partial \overline{U}/\partial \overline{V})_T$ (in terms of the van der Waals constants **a** and/or **b**, etc.).

16. The solvation of methane (CH_4) in water (H_2O) at $T = 298$ K (and $P = 0.1$ MPa) has the following experimental solvation thermodynamic properties:

$$\Delta G_s^0 = +8 \text{ kJ/mol} \qquad \Delta H_s^0 = -11 \text{ kJ/mol}$$
$$T\Delta S_s^0 = -20 \text{ kJ/mol}$$

(a) What is the excess chemical potential of methane in water, expressed in unitless form as $\beta\mu^{\times}$?

(b) What is the equilibrium constant for this hydration process, $K_C = [CH_4]_{aq}/[CH_4]_{gas}$?

17. Consider a process in which a solute, A, is dissolved in a solvent, B, at 300 K and 0.1 MPa, where the CSvdW parameters of the pure fluid A are $\sigma_A = 0.36$ nm and $\tau_A = 550$ K, while for pure fluid B, they are $\sigma_B = 0.54$ nm and $\tau_B = 2030$ K.

(a) Estimate the values of d and τ that would be required in order to use the CSvdW equation to evaluate the excess chemical potential of A dissolved in B.

(b) Given that the concentration of the pure solvent is 10.1 M, determine its number density ρ in molecules/nm^3 units.

(c) Use the solvent number density you obtained in (b) to determine its packing fraction η.

(d) Use the following graph of $\beta\mu_A^\times(\eta)$ to estimate the solvation free energy of A in B (express your answer as a number with units).

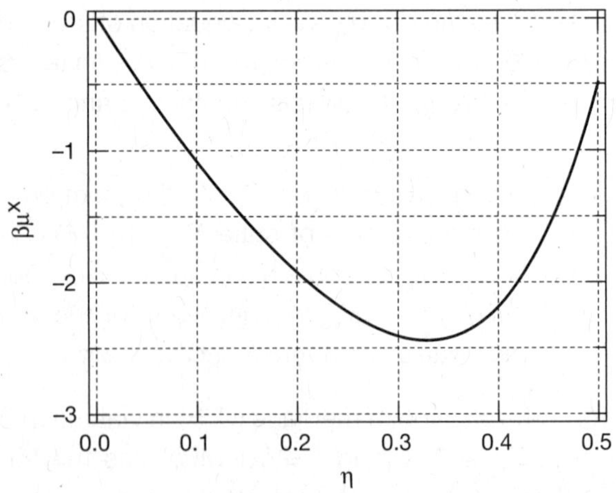

18. Pure liquid cyclohexane in equilibrium with its own vapor phase at 298 K has the following liquid and vapor concentrations: $[\text{c-C}_6\text{H}_{12}]_{\text{liq}} = 9.2$ M and $[\text{c-C}_6\text{H}_{12}]_{\text{vap}} = 0.0052$ M.
   (a) What is the equilibrium constant for the following solvation process?
   $$\text{c-C}_6\text{H}_{12}\ (\text{vapor}) \rightleftarrows \text{c-C}_6\text{H}_{12}\ (\text{liquid})$$
   (b) Use the equilibrium constant obtained in (a) to determine the excess chemical potential $\beta\mu^\times$ of cyclohexane molecule in its own pure liquid.

   (c) Use your result in (b), combined with the Widom theorem, to estimate the value of $\langle e^{-\beta\Psi}\rangle_0$ you would expect to obtain from a computer simulation in which repeated attempts are made to insert a cyclohexane molecule into pure liquid cyclohexane.

   (d) Use your result in (b) to estimate the value of the CSvdW cohesive energy parameter τ for cyclohexane, assuming that the liquid packing fraction is approximately 0.5.

19. Consider a system that initially consists of 1 mole of argon and 1 mole of xenon, each in separate containers, at $T = 300$ K and $P = 10$ Pa (so the gases behave ideally). Recall that the entropy associated with mixing different gases is identical to the total entropy change of the individual gases (use this fact to answer the following questions).
   (a) What is the entropy change associated with mixing the above two gases at constant pressure (and temperature)?

   (b) What is the entropy change associated with mixing the above two gases at constant volume (and temperature)?

20. When xenon (Xe) gas at a concentration of 0.01 M is in equilibrium with liquid water at 298 K, the equilibrium concentration of Xe in water is 0.001 M.
   (a) What is the Gibbs energy of solvation of Xe in water at 298 K (in kJ/mol units)?

   (b) What is the value of $\beta\mu^\times$ for the above process?

   (c) A Widom insertion computer simulation of the above solvation process has obtained a value of $\langle e^{-\beta\Psi}\rangle_0 = 0.135$. What is the predicted Gibbs energy of solvation of Xe in water (in kJ/mol units)?

Introduction to Quantum Mechanics

CHAPTER 6

6.1 The Dawn of Quantum Phenomena

By the late nineteenth century, science had become a remarkably sophisticated and elegant body of knowledge that included classical mechanics, electrodynamics, and thermodynamics. Although there were still many unanswered questions – including whether or not atoms and molecules really existed – it was widely believed that the fundamental laws that govern physics and chemistry were already discovered, and so only the details remained to be worked out. However, as the dawn of the twentieth century approached, it became increasingly evident that something big was still missing from the classical nineteenth-century view of the world.

Evidence that something was amiss came from various sources, including the color of coals glowing in a campfire (blackbody radiation), the line spectra of atoms, the heat capacities of molecules, and the energies of photoelectrons. These and other phenomena led Planck, Einstein, Bohr, de Broglie, and others to realize that the objects of classical physics were not quite what they appeared to be. More specifically, it became clear that particles such as electrons and atoms have wave-like properties, and waves such as light have particle-like properties (as further discussed in Section 1.3). Thus, the classical world-view began to give way to the strange and surprising world of quantum phenomena.

Early attempts to account for quantum phenomena, such as Planck's theory of blackbody radiation, Einstein's quantum theory of light, and Bohr's theory of atomic line spectra, provided important steps in the development of quantum mechanics (see Section 1.3). However, none of these theories provided a self-consistent explanation of all quantum phenomena. In other words, it was obvious that the new theory of quantum mechanics was not yet complete.

A central theme of quantum mechanics is inherent in Planck and Einstein's equation $\varepsilon = h\nu$, which links the wave and particle properties of light. In other words, this equation reflects the fact that the energy of particles such as photons is proportional to the frequency of an associated wave. Expressing this relation as $h = \varepsilon/\nu$ makes it more clear that h is a new fundamental constant that links the particle energy and wave frequency of particle-waves.

Long before the development of quantum mechanics, Boltzmann noted that a classical oscillating system of energy ε and frequency ν will maintain a constant value of ε/ν when it undergoes a reversible adiabatic transformation.[1] Moreover, in Einstein's first paper describing his theory of relativity, he pointed out that electromagnetic radiation (light) has the property that ε/ν is invariant to the reference frame of an observer. In other words, although both the apparent energy and the frequency of light depend on an observer's reference frame, their ratio does not. Thus, clues regarding the importance of ε/ν as an invariant physical quantity are to be found in classical, relativistic, and quantum mechanics.

6.2 The Rise of Wave Mechanics

Bohr's original semiclassical theory of hydrogen (published in 1913) assumed that hydrogen behaved like a classical electron orbiting around a proton, while light was composed of quanta (photons) of energy $h\nu$. The same semiclassical strategy can be used to correctly predict the quantization of other chemically relevant systems (see Section 1.3).

Although Bohr's theory of hydrogen generated tremendous interest in the chemical relevance of quantum phenomena, Bohr himself was not entirely satisfied with his original semiclassical approach. An important clue as to what was missing came from Bohr's comment at the end of his first paper, which pointed out that the quantum states of hydrogen were also ones whose angular momenta differed by multiples of $\hbar = h/2\pi$. This appeared to imply that atomic quantization may arise from a property of electrons themselves, rather than from the quantization of photons that are emitted or absorbed by atoms. This hint, as well as the fact that Bohr's original theory could not explain all of the fine structure in the experimental spectrum

---

[1] Einstein later referred to this theorem of Boltzmann's as the *adiabatic principle*. The Austrian physicist and mathematician Paul Ehrenfest (1880–1933), who was a close friend of both Einstein and Bohr, stressed the importance of this principle in the early development of quantum mechanics.

of hydrogen, made it clear that atomic quantization was not yet entirely understood. It took over a decade of intense effort before our current view of atomic quantization began to take shape.

A key turning point was provided by Louis de Broglie's bold suggestion (in his 1924 PhD thesis) that not only photons but *all* particles of momentum p may be associated with waves of wavelength $\lambda = h/p$ (Eq. 1.11). De Broglie further pointed out that the wave-like properties of electrons could explain the quantization of hydrogen's angular momentum, if it is assumed that the allowed orbits (quantum states) of an electron in hydrogen are those that have an integral number of wavelengths.

The importance of de Broglie's idea was not immediately appreciated. Members of his doctoral thesis examining committee, which included the famous physicists Paul Langevin and Jean Perrin, were skeptical of de Broglie's hypethesis, as were other physicists, including Peter Debye, Max Born, and Werner Heisenberg. A student that attended de Broglie's thesis examination, Robert van der Graaf (who went on to invent the van der Graaf generator) later described the event by saying that "never had so much gone over the heads of so many." However, Langevin must have had some inkling of the importance of de Broglie's work, as he urged de Broglie to send a copy of his thesis to Einstein. As soon as Einstein learned of de Broglie's hypothesis he immediately recognized its significance and famously said that "he has lifted a corner of the great veil."

A few years passed before de Broglie's wave-particle hypothesis was experimentally verified. The first confirmation of his predictions occurred by accident in 1927, when Clinton Davisson and Lester Germer were studying the scattering of electrons by metal powders at the Bell Telephone Laboratories. A technical problem with their apparatus caused the nickel powder to melt and then resolidify to form a large single crystal. When a beam of electrons was directed at the perfect crystalline array of nickel atoms, a diffraction pattern was produced, exactly in the same way that a double-slit (or monochromator grating) diffracts electromagnetic (light) waves. Davisson and Germer did not set out to test de Broglie's hypothesis, and only later came to realize that the strange electron scattering peaks they observed exactly confirmed de Broglie's prediction that $p = h/\lambda$ (Eq. 1.11). In the same year, electron diffraction was independently demonstrated by George Thomson and his student Andrew Reid at the University of Aberdeen. George Thomson was the son of the famous physicist Joseph John (J.J.) Thomson, who received a Nobel prize in 1906 for discovering the electron. After Davisson and Thomson received a Nobel prize (in 1937) for their discovery of electron diffraction, it was noted that J.J. Thomson got a Nobel prize for showing that electrons are particles and his son got a Nobel prize for showing they are waves.

6.3 Wave Equations and Eigenfunctions

Well before the wave-like properties of electrons were experimentally confirmed, Erwin Schrödinger had already succeeded in extending de Broglie's hypothesis to produce a more complete mathematical description of atomic quantization (in 1926). Schrödinger predicted that electronic orbitals (or wavefunctions) should resemble the vibrations of a guitar string or a drum head, rather than looking like planetary particles orbiting around the nucleus. In other words, Schrödinger predicted that the quantum states of electrons in atoms are very similar to the fundamental and overtone vibrations of a musical instrument. In order to understand how Schrödinger arrived at the famous equation that now bears his name, it is useful to first take a detour into the mathematics of waves.

Classical Harmonic Oscillators

A weight hanging from a spring or a swinging pendulum are examples of harmonic oscillators, which undergo simple wave-like motions.[2] In other words, once such a weight is displaced from its equilibrium position, it will tend to vibrate back and forth. If we designate $z(t)$ as the displacement of the weight, then its velocity is $v(t) = dz/dt$ and its acceleration is $a(t) = d^2z/dt^2$. A harmonic oscillator is defined as a system with a quadratic potential energy function, $V(z) = \frac{1}{2}fz^2$ (where f is called the force constant). The kinetic energy of the oscillator is $K = \frac{1}{2}mv^2$ (where m is the mass of the hanging weight). Thus, the total energy of a harmonic oscillator is given by the following Hamiltonian function (see Eq. 1.5).

$$H = K + V = \frac{1}{2}mv^2 + \frac{1}{2}fz^2 \tag{6.1}$$

Since the energy of an isolated (frictionless) harmonic oscillator is conserved, H is time-independent (so $dH/dt = 0$). If we evaluate the time-derivative of the right-hand side of Eq. 6.1, we recover Newton's second law.[3]

$$-fz = m\left(\frac{d^2z}{dt^2}\right)$$
$$F = ma \tag{6.2}$$

---

[2] Note that we have previously encountered classical harmonic oscillators in Chapter 1 (see in particular Exercise 1.2).

[3] The derivation of Eq. 6.2 is very similar to that given on page 9. Recall that Eq. 1.4 implies that the force on the harmonic oscillator is $F(z) = -dV/dz = -fz$.

Equation 6.2 is an example of a second-order differential equation. Solving such an equation requires finding a function $z(t)$ whose second time-derivative is equal to $-(f/m)z(t)$.

Exercise 6.1

Show that $z(t) = A\sin(\omega t)$ is a solution of Eq. 6.2 when $\omega^2 = f/m$ (where A and ω are both constants).

Solution. The first and second time-derivatives of $z(t)$ are $dz/dt = \omega A\cos(\omega t)$ and $d^2z/dt^2 = -\omega^2 A\sin(\omega t) = -\omega^2 z(t)$. Thus, we may plug $z(t)$ into Eq. 6.2 to obtain an identity $-fA\sin(\omega t) = -m\omega^2 A\sin(\omega t)$, or equivalently $(f/m)A\sin(\omega t) = \omega^2 A\sin(\omega t)$, and so $z(t)$ is a solution of Eq. 6.2 when $\omega^2 = f/m$. In other words, we have shown that left and right sides of Eq. 6.2 are equal to each other when $z(t) = A\sin(\omega t) = A\sin(\sqrt{f/m}\,t)$.

More generally, any sinusoidal (harmonic) function such as $z(t) = A\sin(\alpha \pm \omega t)$ (where α is a constant) will be a solution of Eq. 6.2 as long as $\omega^2 = f/m$.[4] This implies that the angular frequency ω of a harmonic oscillator is determined by its force constant f and mass m.

$$\boxed{\omega = \sqrt{\frac{f}{m}}} \tag{6.3}$$

A harmonic oscillator will experience one full vibrational cycle when $\omega t = 2\pi$, so the frequency of the oscillator is $\nu = 1/\tau = \omega/2\pi$, where τ is the time duration of one vibrational cycle (and the frequency ν has units of 1/s or Hz).[5]

$$\boxed{\nu = \frac{\omega}{2\pi} = \frac{1}{2\pi}\sqrt{\frac{f}{m}}} \tag{6.4}$$

Molecular vibrational frequencies are often expressed in *wavenumber* units $\tilde{\nu} = \nu/c = 1/\lambda$, where c is the speed of light and λ is the wavelength of (infrared) light whose frequency is equal to ν.[6]

$$\boxed{\tilde{\nu} = \frac{\nu}{c} = \frac{\omega}{2\pi c} = \frac{1}{2\pi c}\sqrt{\frac{f}{m}}} \tag{6.5}$$

---

[4] Note that $A\cos(\omega t) = A\sin(-\pi/2 + \omega t)$, so functions of the form $A\cos(\alpha \pm \omega t)$ are also solutions of Eq. 6.2.

[5] The parameter $\omega = 2\pi\nu$ has units of radians/sec and is called the angular frequency (also called circular frequency or radial frequency).

[6] See page 33 for a further discussion of wavenumber units and molecular vibrations.

Plane Waves

Waves on a water surface (as well as sound and light waves) are also closely related to harmonic oscillators. If such waves move in only one direction and have a perfect sinusoidal shape, then they are called plane waves. Such waves have a displacement (wavefunction) $D(x, t)$, which is a function of both position x and time t. Plane waves arise as solutions of the following wave equation (where k and ω are both constants).

$$\frac{1}{k^2}\left(\frac{\partial^2 D}{\partial x^2}\right) = \frac{1}{\omega^2}\left(\frac{\partial^2 D}{\partial t^2}\right) \tag{6.6}$$

Exercise 6.2

Show that $D(x, t) = A\sin(kx - \omega t)$ is a solution of Eq. 6.6, and determine the relationship between k, ω, and the phase velocity v of the wave $D(x, t)$.

Solution. Since $\partial^2 D/\partial x^2 = -k^2 A\sin(kx - \omega t)$ and $\partial^2 D/\partial t^2 = -\omega^2 A\sin(kx - \omega t)$, the function $D(x, t)$ is a solution of Eq. 6.6. The phase velocity of a wave is the speed at which its crest moves. The location of the wave crest x_c may be obtained by solving $\partial D/\partial t = A\omega\cos(kx_c - \omega t) = 0$, which will be true if $kx_c - \omega t = \pi/2$ or equivalently $x_c = (\pi/2k) + (\omega/k)t$. So, the velocity of the wave crest is $v = dx_c/dt = \omega/k$.

Various other functions, such as $A\sin(kx \pm \omega t \pm \alpha)$ and $B\cos(kx \pm \omega t \pm \alpha)$ are also solutions to Eq. 6.6 (where k, ω, and α are real positive constants). The parameter $\omega = 2\pi\nu$ again represents the angular (or *circular*) frequency of the wave and λ is its wavelength, while $k = 2\pi\tilde{\nu} = 2\pi/\lambda$ describes the periodicity of the wave in the x-direction (and is referred to as the *wave vector*).[7]

Complex Exponential Waves

The sinusoidal solutions of Eq. 6.6 are closely related to complex exponential functions of the form $D(x, t) = Ce^{i(kx-\omega t)} = Ce^{ikx}e^{-i\omega t}$, where $i = \sqrt{-1}$. It may seem strange to consider such complex-valued exponential functions as possible solutions to the wave equation, since real waves are not imaginary. The reason for considering such solutions is largely a matter of mathematical convenience. Note that it is very easy to take derivatives of

[7] The sign of ωt determines the direction in which the wave is traveling; $A\sin(kx - \omega t)$ represents a wave that is moving in the forward x-direction, while $A\sin(kx + \omega t)$ is moving in the opposite direction.

exponential functions since $de^{ax}/dx = ae^{ax}$. Moreover, the following equation (called Euler's formula) reveals that there is a close connection between trigonometric functions and complex exponential functions (see footnote 18 on page 102).

$$\boxed{e^{i\theta} = \cos\theta + i\sin\theta} \tag{6.7}$$

The above identity may be used to obtain the following useful relations.

$$\cos\theta = \frac{1}{2}\left(e^{i\theta} + e^{-i\theta}\right) \tag{6.8}$$

$$\sin\theta = \frac{1}{2i}\left(e^{i\theta} - e^{-i\theta}\right) \tag{6.9}$$

In other words, complex exponential functions can be viewed as generalized harmonic (sinusoidal) functions. Such functions can be used to represent real waves either by forming linear combinations, such as those in Eqs. 6.8 and 6.9, or by considering the real portion of the complex exponential function to be the physically relevant part of the solution.

Eigenfunctions and Eigenvalues

Another important class of solutions of Eq. 6.6 are ones of the following form.

$$D(x,t) = \Psi(x)e^{-i\omega t} \tag{6.10}$$

These are solutions that have a simple oscillatory (harmonic) time-dependence with an amplitude $\Psi(x)$, which represents the initial shape of the wave (since $e^{-i\omega t} = 1$ when $t = 0$). In other words, Eq. 6.10 describes a class of functions that maintain the same shape as they oscillate with a frequency of $\nu = \omega/2\pi$, just like a vibrating guitar string. Note that if we substitute Eq. 6.10 in Eq. 6.6, the latter wave equation reduces to the following time-independent form.[8]

$$\frac{\partial^2}{\partial x^2}\left[\Psi(x)\right] = -k^2\Psi(x) \tag{6.11}$$

The solution $\Psi(x)$ of such a differential equations is called an *eigenfunction* whose *eigenvalue* is $-k^2$. In other words, a function $\Psi(x)$ is called an eigenfunction of the operator $\frac{\partial^2}{\partial x^2}$ if performing the operation $\frac{\partial^2}{\partial x^2}[\Psi(x)]$ has the effect of regenerating the function $\Psi(x)$ multiplied by a constant that

---

[8] Equation 6.11 is obtained from Eq. 6.6 by equating $\partial^2 D/\partial x^2 = \partial^2/\partial x^2\,[\Psi(x)]\,e^{-i\omega t}$ and $\partial^2 D/\partial t^2 = -\omega^2\Psi(x)e^{-i\omega t}$, and then eliminating $e^{-i\omega t}$ from both sides.

is referred to as the corresponding eigenvalue. Equation 6.11 is an example of a more general class of *eigenvalue equations* all of which have the following general form

$$\boxed{\widehat{A}\Psi = a\Psi} \tag{6.12}$$

where $\widehat{A}$ is an operator and Ψ is an eigenfunction of $\widehat{A}$ whose eigenvalue is equal to the number a. We will soon see that such equations play a central role in quantum mechanics.

Guitar Strings

The solutions to equations such as Eq. 6.11 depend on the boundary conditions imposed on the function $\Psi(x)$. For example, if we wish to describe a vibrating guitar string, then we need to specify that the ends of the string are held fixed. Such boundary conditions restrict the possible solutions of Eq. 6.11 to those for which $\Psi(x) = 0$ at both ends of the string. More specifically, if the string is held fixed at $x = 0$ and $x = L$ (where L is the length of the string), then we require that the solutions of Eq. 6.11 must have the property that $\Psi(0) = 0$ and $\Psi(L) = 0$.

One such solution corresponds to the fundamental vibration of a guitar string whose shape is described by the function $\Psi(x) = A \sin[(\pi/L)x]$ (where A is a constant that represents the maximum amplitude of the string's vibration). If we substitute this function into Eq. 6.11, we find that $k^2 = (\pi/L)^2$ or $k = \pi/L$. Thus, this eigenfunction $\Psi_1 = A\sin(k_1x) = A\sin[(2\pi/\lambda_1)x]$ has a wavelength of $\lambda_1 = 2L$ and a wave vector $k_1 = \pi/L = 2\pi/\lambda_1$.

Other possible solutions of Eq. 6.11 correspond to higher overtones (or harmonics) of the guitar string (as shown in Fig. 6.1). These solutions have shorter wavelengths, corresponding to eigenvalues that are integral multiples of k_1. For example, the second eigenfunction (or first overtone) $\Psi_2(x) = A\sin[(2\pi/L)x] = A\sin(k_2x) = A\sin(2k_1x)$ has a wavelength of $\lambda_2 = \lambda_1/2 = L$, so $k_2 = 2\pi/L$. This eigenfunction again satisfies the boundary conditions at the two ends of the string, since $\Psi_2(0) = A\sin(0) = 0$ and $\Psi_2(L) = A\sin(2\pi) = 0$; it also has a *node* in the middle of the string, since $\Psi_2(L/2) = A\sin(\pi) = 0$.

A similar procedure may be used to generate other overtone vibrational modes of a guitar string (see Fig. 6.1) whose eigenfunctions and eigenvalues may be expressed in the following compact form.

$$\begin{aligned} \Psi_n &= A\sin(k_nx) \\ k_n &= n\left(\frac{\pi}{L}\right) \\ n &= 1, 2, 3\ldots \end{aligned} \tag{6.13}$$

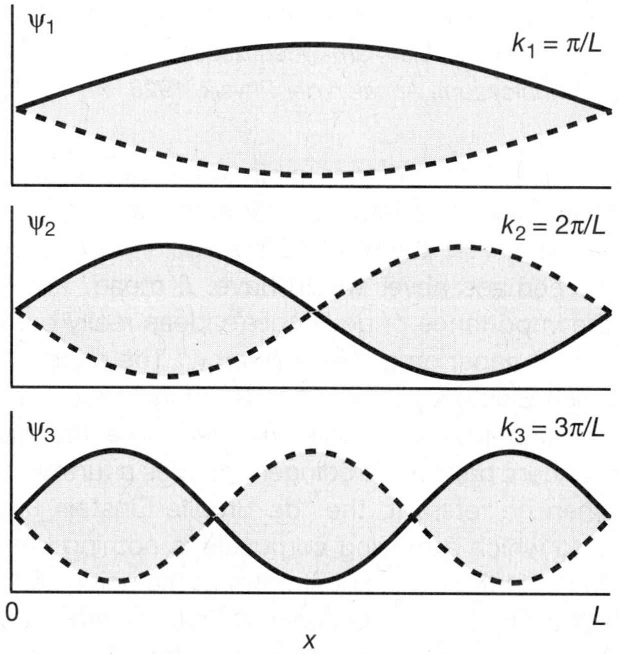

FIGURE 6.1 Normal modes of a guitar string. The solid curves trace the eigenfunctions ψ_n and the shaded areas indicate the region swept out by the string as it vibrates.

Notice that these solutions are quantized, in the sense that the eigenfunction has a wavelength that is determined by the quantum number n. Erwin Schrödinger recognized that the quantization that arises naturally when solving such wave equations could be related to the quantization of electrons in atoms.

6.4 Quantum Operators and Observables

The paper in which Schrödinger first introduced his quantum mechanical wave equation (in 1926), and demonstrated its application to the hydrogen atom, is entitled *Quantization as an Eigenvalue Problem*. His paper begins with the following words, which emphasize the fact that hydrogen's line spectrum (with integer quantum numbers) emerges naturally from the eigenvalues of a wave equation:

> *In this article I wish to show, first of all for the simplest case of the (non-relativistic and unperturbed) hydrogen atom, that the usual rule for quantization can be replaced by another requirement in which there is no longer any mention of "integers." The integral property follows, rather, in the same natural way that, say, the number of nodes of a vibrating string must be an integer. The new interpretation can be generalized and, I believe,*

strikes very deeply into the true nature of the quantization rules. Erwin Schrödinger *(From the German) Quantization as an Eigenvalue Problem (Fourth Communication), Annalen der Physik, 1926, 81(18), 109–139.*

The inspiration that led Schrödinger to his wave equation came from both de Broglie and Einstein. In a letter Schrödinger wrote to Einstein in April of 1926 he states that "... the whole thing would certainly not have originated yet, and perhaps never would have, (I mean, not from me), if I had not had the importance of de Broglie's ideas really brought home to me by your second paper on gas degeneracy." The paper Schrödinger is referring to (which Einstein published in 1925) includes Einstein's cryptic assertion that de Broglie's hypothesis "involved more than merely an analogy." In a subsequent paper Schrödinger provides a further clue regarding his thinking when he refers to the "de Broglie-Einstein undulatory theory, according to which a moving corpuscle is nothing but the foam on a wave of radiation in the basic substratum of the universe."[9] Many years later (in 1961) Paul Dirac, in his obituary to Erwin Schrödinger, recounts a conversation in which Schrödinger explained that his initial breakthrough came when he was trying to generalize de Broglie's waves to the case of bound particles and "finally obtained a neat solution of the problem, leading to the appearance of the energy-levels as eigenvalues of a certain operator."

In order to see how Schrödinger's equation evolved from de Broglie's hypothesis, $p = h/\lambda$, we may begin by noting that the de Broglie relation can be expressed as $p = \hbar k$ (where $\hbar = h/2\pi$ and $k = 2\pi/\lambda$), which also implies that $k = p/\hbar$ or $k^2 = p^2/\hbar^2$. The latter identity may be combined with Eq. 6.11 to obtain $\frac{\partial^2}{\partial x^2}[\Psi(x)] = -(p_x^2/\hbar^2)\Psi(x)$ (where p_x is the particle's momentum in the x-direction), which may be rearranged to the following physically suggestive form.

$$-\hbar^2 \frac{\partial^2}{\partial x^2}[\Psi(x)] = p_x^2 \Psi(x) \tag{6.14}$$

This equation clearly implies that p_x^2 is an eigenvalue associated with the following operator.

$$\boxed{\widehat{p_x^2} \equiv -\hbar^2 \frac{\partial^2}{\partial x^2}} \tag{6.15}$$

This connection between de Broglie's particle-waves and mathematical wave equations is the key insight that led to Schrödinger's wave

[9] This interesting statement appeared in a paper Schrödinger published in early 1926, before the paper in which he first unveiled his wave equation.

mechanics. In other words, Equation 6.15 suggests that both wave and particle properties can be combined in a wave equation in such a way that an observable property is associated with an operator, and the corresponding experimentally measured value appears as an eigenvalue of the operator.

Note that the operator $\widehat{p_x^2}$ may be expressed as a product of two identical operators $-\hbar^2(\partial^2/\partial x^2) = [\pm i\hbar(\partial/\partial x)][\pm i\hbar(\partial/\partial x)]$. This suggests that momentum (in the x-direction) may be equated with the following operator.[10]

$$\boxed{\widehat{p}_x \equiv -i\hbar\frac{\partial}{\partial x} = \frac{\hbar}{i}\frac{\partial}{\partial x}} \tag{6.16}$$

Since the kinetic energy in the x-direction is $K_x = p_x^2/2m$, Eq. 6.15 also implies that the following operator should be associated with kinetic energy.

$$\boxed{\widehat{K}_x \equiv -\frac{\hbar^2}{2m}\frac{\partial^2}{\partial x^2}} \tag{6.17}$$

A free particle – which is defined as a particle that is not influenced by any potential energy function – must have a total energy that is equal to its kinetic energy. Thus, the above results suggest that the total energy or Hamiltonian operator for a free particle-wave should be $\widehat{H} = \widehat{K}_x$. For a particle-wave that is not free, so $V(x) \neq 0$, we may invoke the conservation of energy in order to obtain a more general expression for $\widehat{K}_x$. We know that for any isolated (classical) mechanical system $H = K + V$ is time-independent.[11] This may be rearranged to $K = H - V$, which is consistent with our expectation that the kinetic energy of a particle must decrease as it rolls up a potential energy hill. Thus, Schrödinger boldly suggested equating $\widehat{K}_x = \widehat{H} - V(x)$ in describing a quantum mechanical particle-wave moving in the x-direction on a potential energy surface described by the function $V(x)$. The latter identity may be rearranged to obtain Schrödinger's expression for the Hamiltonian operator pertaining to a particle-wave with a potential energy of $V(x)$.

$$\boxed{\widehat{H} = \widehat{K}_x + \widehat{V}_x = -\frac{\hbar^2}{2m}\frac{\partial^2}{\partial x^2} + V(x)} \tag{6.18}$$

---

[10] Although we might have defined $\widehat{p}_x = +i\hbar\partial/\partial x$, we will see that including the negative sign in Eq. 6.16 is required if the momentum operator is to produce a positive momentum for a particle that is moving in the positive x-direction. The second equality in Eq. 6.16 is a consequence of the fact that $1/i = (1/i)(i/i) = i/i^2 = -i$.

[11] See Eq. 1.6 on page 8, and the subsequent derivation showing that $dH/dt = 0$.

Note that this identity implies that the potential energy operator is an algebraic rather than a differential operator, and so any such function of x, including x itself, should be associated with the following operators.

$$\boxed{\widehat{x} \equiv x\cdot} \tag{6.19}$$

$$\boxed{\widehat{V}_x \equiv V(x)\cdot} \tag{6.20}$$

The kinetic energy operator in three dimensions is obtained simply by adding the kinetic energies in the x-, y-, and z-directions.[12]

$$\widehat{K} \equiv \frac{1}{2m}\left(\widehat{p_x^2} + \widehat{p_y^2} + \widehat{p_z^2}\right) = -\frac{\hbar^2}{2m}\left(\frac{\partial^2}{\partial x^2} + \frac{\partial^2}{\partial y^2} + \frac{\partial^2}{\partial z^2}\right) \equiv -\frac{\hbar^2}{2m}\nabla^2 \tag{6.21}$$

and the potential energy operator is again obtained by simple multiplication.

$$\widehat{V} \equiv V(x, y, z)\cdot \tag{6.22}$$

So, the following Hamiltonian operator pertains to the total energy of a single particle that is influenced by a three-dimensional potential energy function $V(x, y, z)$.

$$\boxed{\widehat{H} \equiv \widehat{K} + \widehat{V} = -\frac{\hbar^2}{2m}\nabla^2 + V(x, y, z)\cdot} \tag{6.23}$$

Exercise 6.3

Show that a $\Psi(x) = A\cos(ax)$ is an eigenfunction of the Hamiltonian of a free-particle confined to the x-axis, but is not an eigenfunction of $\widehat{p_x}$.

Solution. The Hamiltonian operator of a such a particle is $\widehat{H} = \widehat{K} = (-\hbar^2/2m)\partial^2/\partial x^2$ (since $\widehat{V} = V(x) = 0$), and so $\widehat{H}\Psi(x) = (\hbar^2/2m)a^2 A\cos(ax) = (\hbar^2 a^2/2m)\Psi(x)$. Thus, $\Psi(x)$ is an eigenfunction of $\widehat{H}$, with an eigenvalue of $(\hbar a)^2/2m$. However, $\Psi(x)$ is not an eigenfunction of $\widehat{p_x}$ because $\widehat{p_x}\Psi(x) = -i\hbar\partial/\partial x[\Psi(x)] = -i\hbar aA\sin(ax)$ is not equal to $\Psi(x)$ multiplied by a constant.

The wavefunction in the above exercise corresponds to a system that has an energy eigenvalue but not a momentum eigenvalue. Although it is hard to understand how this could be, we will see that such a wavefunction can represent a particle that is bouncing back and forth in a box with a particular kinetic energy but a momentum that can be either positive or negative and so does not have one particular value.

[12] The symbol ∇^2 is a short-hand notation for a three-dimensional second derivative operator, which is also referred to as the *Laplacian* or "del-squared" operator, and is defined by the last identity in Eq. 6.21.

The *time-independent Schrödinger equation* is obtained by incorporating the Hamiltonian operator into a differential equation in which E is the total energy (eigenvalue) and Ψ is the corresponding wavefunction (eigenfunction) of the particle-wave.

$$\boxed{\widehat{H}\Psi = E\Psi} \tag{6.24}$$

By solving the above equation, one may obtain both Ψ and E for a quantum mechanical particle-wave. Obtaining such solutions is the central task of Schrödinger's wave mechanics.

The eigenfunctions of Eq. 6.24 describe the shapes of wavefunctions that have a simple oscillatory time dependence, $e^{-i\omega t}$, as indicated in Eq. 6.10. When we equate the energy difference between two eigenstates with the energy of a photon, we obtain $\Delta E = h\nu = \hbar\omega$, or $\omega = \Delta E/\hbar$, and thus $e^{-i\omega t} = e^{-i(\Delta E/\hbar)t}$. This also suggests that each eigenstate of energy E_i may be associated with an absolute frequency of $\omega_i = E_i/\hbar$ and a time dependence of $e^{-i(E_i/\hbar)t}$.[13] Thus, the full time-dependent wavefunction $\Phi(x, y, z, t)$ of a particle-wave, which is in an eigenstate of energy E, may be expressed as follows.

$$\Phi(x, y, z, t) = \Psi(x, y, z)e^{-i(E/\hbar)t} \tag{6.25}$$

Notice what happens if we evaluate the second time-derivative of Φ.

$$\frac{\partial^2}{\partial t^2}\Phi = -\frac{E^2}{\hbar^2}\Phi = -\frac{\widehat{H}^2}{\hbar^2}\Phi \tag{6.26}$$

The above equation may be rearranged and factored as follows.

$$0 = \left(\widehat{H}^2 + \hbar^2\frac{\partial^2}{\partial t^2}\right)\Phi = \left(\widehat{H} + i\hbar\frac{\partial}{\partial t}\right)\left(\widehat{H} - i\hbar\frac{\partial}{\partial t}\right)\Phi \tag{6.27}$$

This suggests that we may also identify $\widehat{H}$ with the time-derivative operator $\pm i\hbar(\partial/\partial t)$. By selecting the positive sign convention, we obtain the following *time-dependent Schrödinger equation*.[14]

$$\boxed{\widehat{H}\Phi = i\hbar\frac{\partial\Phi}{\partial t}} \tag{6.28}$$

---

[13] However, since the absolute energy of an eigenstate depends on our choice of a reference zero of energy, the same is true for the implied absolute frequency of the eigenstate. Schrödinger himself had speculated that the most natural absolute energy scale is that obtained from the theory of relativity, which implies that a particle whose rest mass is m has a relativistic energy of $E = mc^2$ (Eq. 1.8) and thus an absolute frequency of $\omega = E/\hbar = mc^2/\hbar$.

[14] The positive sign convention implies that a particle described by $\Psi(x) = \Psi(x)e^{-i\omega t}$ will have a positive energy.

This equation may be used to determine the time evolution of any wavefunction. More specifically, we will see that any wavefunction can be represented as a superposition of eigenfunctions, and so the time evolution of such a wavefunction is completely determined by the oscillatory $e^{-i\omega t}$ terms associated with the corresponding eigenfunctions.

What Is Ψ?

Although the elegance and power of Schrödinger's wave mechanics was widely recognized from the moment he first introduced it, the physical significance of the wavefunction Ψ remained the subject of heated speculation. Schrödinger suggested that the charge density of an electron in a hydrogen atom is proportional to the square of the wavefunction $|\Psi|^2$. However, this interpretation was not entirely satisfactory as it did not account for the experimentally observed particle-like properties of electrons.

The currently accepted interpretation of Ψ differs in a subtle but important respect from Schrödinger's view, as $|\Psi|^2$ is now equated with the probability of detecting a particle whose wavefunction is Ψ. However, neither Schrödinger nor Einstein were ever entirely satisfied with this statistical interpretation of Ψ, as reflected by Einstein's famous statement, which is often paraphrased as "God doesn't play dice." The following quotation contains Einstein's actual statement, as it appeared in a letter he wrote to Max Born in December of 1926.

> *Quantum mechanics is certainly imposing. . . . an inner voice tells me that it is not yet the real thing. The theory says a lot, but does not really bring us any closer to the secret of the 'old one'. I, at any rate, am convinced that He does not throw dice.*

Max Born proposed the statistical interpretation of Ψ earlier in 1926; he was also the first person to use the phrase *quantum mechanics* (in 1924) to describe the emerging theoretical description of quantum phenomena. In 1925, Max Born, Werner Heisenberg, and Pascual Jordan developed a matrix formulation of quantum mechanics. Soon thereafter Schrödinger demonstrated the formal equivalence of their matrix mechanics and his wave mechanics.[15]

---

[15] Although Max Born and Albert Einstein firmly disagreed regarding the proper interpretation of Ψ, they were personal friends and enjoyed playing chamber music together, with Einstein on the violin and Born on the piano. From 1921 to 1933, Born was the Director of the Institute of Theoretical Physics at the University of Göttingen. In 1933, Born's professorship was revoked by the Nazi party; although Born was a Lutheran, he was classified as a Jew by the Nazi racial laws due to his ancestry. Born was a determined pacifist, and his international reputation

Although Born was instrumental in the development of matrix mechanics, he adopted Schrödinger's wave mechanics method in order to describe collisions between electrons or α particles and atoms. In explaining why he favored Schrödinger's approach, Born said that "among the various forms of the theory, only Schrödinger's proved itself appropriate for this purpose; for this reason I am inclined to regard it as the most profound formulation of the quantum laws."

However, Born did not agree with Schrödinger's interpretation of Ψ, for reasons that he explained as follows.[16]

> *To us in Göttingen this interpretation seemed unacceptable in the face of well established experimental facts. At that time it was already possible to count particles by means of scintillations or with a Geiger counter, and to photograph their tracks with the aid of a Wilson cloud chamber.*

Born believed that there must be a fundamental connection between Ψ and probability of experimentally detecting a particle. Here is how he explained where he got the idea for his statistical interpretation of Ψ:

> *... an idea of Einstein's gave me the lead. He had tried to make the duality of particles – light quanta or photons – and waves comprehensible by interpreting the square of the optical wave amplitudes as the probability density for the occurrence of photons. This concept could at once be carried over to the Ψ-function: $|\Psi|^2$ ought to represent the probability density for electrons (or other particles). The Statistical Interpretation of Quantum Mechanics, Max Born, Nobel Lecture, 1954. © The Nobel Foundation.*

---

as a physicist was such that he was sought after by numerous universities outside of Germany. In 1933, he accepted a position at the Cambridge University in England. At around the same time his former collaborator Jordan became an active enthusiast of Adolph Hitler, and joined the Nazi storm troupers. Politics had a significant influence on science in Germany during those years, as a growing movement opposed both Jewish and theoretical physics. As a theoretical physicist, Heisenberg was initially subject to suspicion by advocates of this movement, but he remained in Germany throughout World War II, and became the head of the Nazi effort to develop an atomic bomb. In recent years, letters written by Niels Bohr have come to light that describe Heisenberg's efforts to recruit Bohr to join the Nazi side during World War II. However, Bohr remained a steadfast Danish patriot, and helped alert the United States and England to the Nazi atom bomb efforts.

[16] These quotations come from the lecture Born gave when he received a Nobel prize (in 1954) "for his fundamental research in quantum mechanics, especially for his statistical interpretation of the wavefunction."

In other words, Einstein and Born expected particle probabilities to be proportional to $|\Psi|^2$, rather than Ψ, because they knew that the energy of various kinds of waves are proportional to the squares of their amplitudes. For example, the total energy of a classical harmonic oscillator is equal to the value of its potential energy when it reaches its maximum amplitude, $V(x_{max}) = \frac{1}{2} f x_{max}^2$. So, the energy of a harmonic oscillator is proportional to the square of its amplitude. Similarly, the intensity of light is proportional to the square of the amplitude of its electromagnetic field $|E|^2$. More specifically, the intensity of a beam of light is $I = (cn\epsilon_0)/2|E|^2$, which represents the amount of power that shines on a unit area (and has units of $W/m^2 = J/[s\ m^2]$).[17]

More generally, a plane wave described by the function $D(x,t)$ will have an intensity that is proportional to the magnitude of D^2 averaged over a complete oscillation cycle $\tau = 1/\nu$. In other words, $I \propto \langle D^2 \rangle = \frac{1}{\tau}\int_0^\tau |D|^2 dt$. If $D(x,t)$ is a complex-valued function, then $|D|^2 = D^* D$.[18] Thus, it is reasonable to expect the intensity of a particle-wave of the form $\Phi(x,t) = \Psi(x)e^{i\omega t}$ to be proportional to $\Phi^* \Phi = \Psi^* \Psi e^{i\omega t} e^{-i\omega t} = \Psi^* \Psi = |\Psi|^2$.

Born suggested that $|\Psi|^2$ is proportional to the intensity of a beam of particle-waves, or the probability that a particle will be detected in a given region of space. In order to turn this proportionality into an equality, we must ensure that Ψ is properly normalized, so that the total (integrated) probability of finding the particle anywhere is equal to 1.

$$\int_{-\infty}^{\infty} \Psi^* \Psi dx = 1 \tag{6.29}$$

More specifically, if $\psi(x)$ is a wavefunction that is not normalized,[19] then we must determine the value of the normalization constant N, such that $\Psi = N\psi$ is properly normalized. This implies that $\int_{-\infty}^{\infty} (N\psi)^* N\psi dx = N^2 \int_{-\infty}^{\infty} \psi^* \psi dx = 1$, which may readily be rearranged to obtain the following expression for N.

$$N = \frac{1}{\sqrt{\int_{-\infty}^{\infty} \psi^* \psi dx}} \tag{6.30}$$

---

[17] The parameter n is the refractive index, c is the speed of light in a vacuum, and ϵ_0 is the vacuum electric permittivity.

[18] The complex conjugate of any complex number $c = a + ib$ is $c^* = a - ib$. The squared amplitude of a complex number is defined as $|c^2| = c^* \cdot c = (a + ib)(a - ib) = a^2 + b^2$, which is clearly a real number.

[19] In this section the symbol ψ is used to designate wavefunctions that have not yet been normalized, while Ψ is used to designate the corresponding normalized wavefunctions.

More generally, the normalization constant of a three-dimensional particle-wave may be obtained using the same procedure (where $d\tau = dxdydz$ is a spatial volume element, and the integral extends over the entire region that is accessible to the particle-wave).

$$\boxed{N = \frac{1}{\sqrt{\int \psi^* \, \psi d\tau}}} \tag{6.31}$$

Normalization of a wavefunction simply requires multiplying the unnormalized wavefunction by *N*.

$$\boxed{\Psi = N\psi = \frac{\psi}{\sqrt{\int \psi^* \, \psi d\tau}}} \tag{6.32}$$

The product $\Psi^* \, \Psi = |\Psi|^2$ is referred to as a *probability density*, as it determines the probability of observing a particle within a given region of space. The probability density for a particle confined to the *x*-axis is $\rho(x) = \Psi^*(x) \, \Psi(x)$. More generally, a particle in three dimensions described by the wavefunction $\Psi(\tau)$ has the following probability density.[20]

$$\boxed{\rho(\tau) = \Psi^*(\tau) \, \Psi(\tau) = \frac{\psi^*(\tau) \, \psi(\tau)}{\int \psi^*(\tau) \, \psi(\tau) \, d\tau}} \tag{6.33}$$

For example, the probability of observing a particle anywhere between x_1 and x_2 is obtained by integrating $\rho(x)dx$ from x_1 to x_2 .

$$\boxed{P(x_1, x_2) = \int_{x_1}^{x_2} \Psi^* \, \Psi dx = \int_{x_1}^{x_2} \rho(x) dx} \tag{6.34}$$

Similarly, one may determine the probability of finding a three-dimensional particle within a given region of space by integrating $\rho(\tau)d\tau$ over the region of interest.

More generally, one may consider a wavefunction $\Psi(r_1, r_2, r_3 \ldots)$ corresponding to a collection of particles located at positions $r_1, r_2, r_3 \ldots$, where $r_i \equiv (x_i, y_i, z_i)$ represents the three-dimensional coordinate of each particle.

---

[20] The symbol τ is a short-hand notation for the coordinates of the particle in the system, very much in the same way that it is used in Eq. 1.27 on p. 35.

The probability that the collection of particles will have a particular configuration is again determined by the probability density $\rho(\tau)$, as defined in Eq. 6.33.

Exercise 6.4

Consider an unnormalized wavefunction of the form $\psi(x) = e^{-a|x|}$ (plotted below) that is equal to e^{-ax} when $x \geq 0$ and e^{+ax} when $x \leq 0$. Normalize this wavefunction and determine the probability that the corresponding particle will be found within the range $-1/a < x < 1/a$.

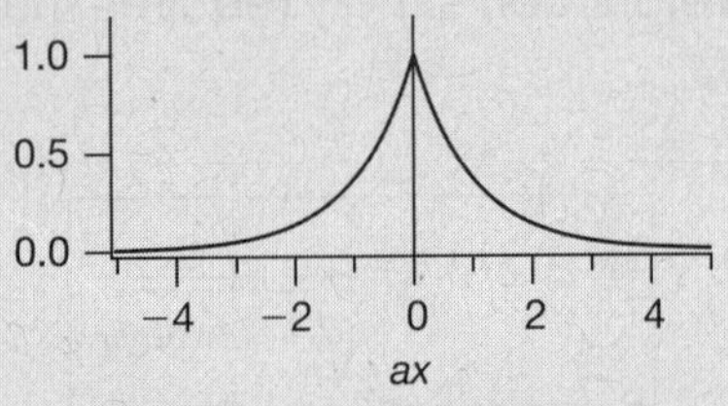

Solution. The square of the above wave function is $\psi^*\psi = \psi^2 = e^{-2a|x|}$, whose total integral is $\int_{-\infty}^{\infty} e^{-2a|x|}dx = \int_0^{\infty} e^{-2ax}dx + \int_{-\infty}^{0} e^{+2ax}dx$. The symmetry of the function implies that the latter two integrals are equal to each other and so the total integral may be evaluated as follows.

$$\int_{-\infty}^{\infty} e^{-2a|x|}dx = 2\int_0^{\infty} e^{-2ax}dx = 2\left(-\frac{1}{2a}\right)e^{-2ax}\Big|_0^{\infty} = \frac{1}{a}$$

We may now normalize the wavefunction to obtain

$$\Psi(x) = \frac{\psi(x)}{\sqrt{\int_{-\infty}^{\infty}\psi(x)^2dx}} = \frac{e^{-a|x|}}{\sqrt{\int_{-\infty}^{\infty}e^{-2a|x|}dx}} = \sqrt{a}e^{-a|x|}$$

and so the probability of finding the particle within the range $-1/a < x < 1/a$ may be determined by integrating the probability density $\rho(x) = \Psi(x)^*\Psi(x) = \Psi(x)^2 = ae^{-2a|x|}$ over the above range, again making use of the symmetry of $\Psi(x)^2$ about zero.

$$\int_{-1/a}^{1/a} \Psi(x)^2dx = 2a\left(-\frac{1}{2a}\right)e^{-2a|x|}\Big|_0^{1/a} = -e^{-2} - (-e^0) = 1 - e^{-2} \approx 0.86$$

Thus, there is an 86% probability of finding the particle between $-1/a < x < 1/a$.

Expectation Values of Experimental Observables

If a particle (or collection of particles) is in an eigenstate of an operator $\widehat{A}$ with an eigenvalue a, then an experimental measurement of the corresponding observable will invariably produce the value a. For example, if a system is described by a wavefunction that is an eigenfunction of $\widehat{H}$ with eigenvalue E, then a measurement of the energy of the system will always produce the value E. Similarly, if a particle's wavefunction is an eigenfunction of $\widehat{p}_x$ with eigenvalue p_x, then a measurement of the momentum of the particle will always produce the value p_x.

But how can we predict the outcome of an experimental measurement that is performed on a particle whose wavefunction is not an eigenfunction of the corresponding operator? We can't avoid dealing with this question since we will see that no particle can ever be described by a wavefunction that is simultaneously an eigenfunction of both $\widehat{x}$ and $\widehat{p}_x$.[21]

Experimental measurements performed on systems that are not in an eigenstate of the corresponding operator are not expected to produce the same measured value every time a measurement is performed. Although we can't predict what value will be measured each time a measurement is performed, we can predict the average of a large number of such measurements.

For example, any delocalized wavefunction $\Psi(x)$ is clearly not an eigenfunction of the position operator $\widehat{x}$, since $\widehat{x}\Psi(x) = x \cdot \Psi(x)$ is a new function of x that is not equal to a constant times Ψ. As a result, a measurement of the position of a particle will produce a distribution of values that are determined by the probability density $\rho(x) = \Psi^*\Psi$. Although we can't predict exactly where the particle will appear next, we can predict the most probable (average) position of the particle, obtained from many repeated measurements of its position. More specifically, the average value, or *expectation value*, of a measurement of the position of a quantum mechanical particle-wave is $\langle x \rangle = \int x\rho(x)dx$. Since $x\rho(x) = x\,\Psi^*\,\Psi = \Psi^* x\,\Psi$, this expectation value may be expressed as follows.

$$\langle x \rangle = \int_{-\infty}^{\infty} \Psi^* \widehat{x}\, \Psi dx = \int_{-\infty}^{\infty} \Psi^* x \Psi dx \tag{6.35}$$

More generally, the expectation value associated with any observable property may be determined by evaluating the following integral (over the entire region accessible to the particle-wave).[22]

$$\boxed{\langle A \rangle = \int \Psi^* \widehat{A}\, \Psi d\tau} \tag{6.36}$$

[21] This conclusion is a consequence of the uncertainty principle, which is in turn a consequence of the fact that x and p_x are conjugate variables whose operators do not commute with each other (as further discussed in the next subsection).

[22] Inserting the operator between the two wavefunctions $\Psi^* \widehat{A}\, \Psi$ rather than somewhere else, such as $\widehat{A}\, \Psi^*\, \Psi$, is required in order to assure that quantum mechanical expectation values are consistent with the corresponding classical values, which is also known as the Ehrenfest theorem. For example, if we require that $\langle p_x \rangle = m(d\,\langle x \rangle\,/dt)$, then it can be shown that we must evaluate $\langle p_x \rangle$ using Eq. 6.36.

It is important to keep in mind that Ψ must be normalized in order to correctly determine an expectation value using Eq. 6.36. However, some quantum mechanical problems, such as those involving free particles or the scattering of photons or electrons by atoms or molecules, give rise to wavefunctions that extend to infinity and so cannot be normalized. The expectation values associated with such wavefunctions can nevertheless be evaluated using the following expressions (in which ψ is not normalized and may even be impossible to normalize).

$$\boxed{\langle A\rangle = \frac{\int \psi^* \widehat{A}\, \psi d\tau}{\int \psi^*\, \psi d\tau}} \tag{6.37}$$

Notice that Eq. 6.37 is essentially equivalent to Eq. 6.36, except that the normalization constant has been explicitly included in Eq. 6.36.

If Ψ happens to be an eigenfunction of $\widehat{A}$, then both Eqs. 6.37 and 6.36 imply that $\langle A\rangle = a$. This can easily be demonstrated by noting that if $\widehat{A}\Psi = a\Psi$, then $\int \Psi^* \widehat{A}\, \Psi d\tau = a \int \Psi^*\, \Psi d\tau$ (since the constant a can be moved outside of the integral), and so the integrals in the numerator and denominator of Eq. 6.37 cancel to obtain $\langle A\rangle = a$.

There are other situations in which expectation values may be determined without needing to explicitly evaluate the integral in Eq. 6.36. For example, if the integrand of Eq. 6.36 is antisymmetric, in the sense that it has positive and negative contributions of identical area, then the integral is necessarily equal to zero, and so $\langle A\rangle = 0$ *by symmetry*.

Exercise 6.5

Show that the normalized wavefunction $\Psi(x) = \sqrt{a}e^{-a|x|}$ (that is further described in Exercise 6.4) is associated with a particle whose position expectation value is equal to zero.

Solution. The position expectation value $\langle x\rangle$ is obtained by integrating $\Psi^* \widehat{x}\, \Psi = axe^{-2a|x|}$ over the entire x-axis. The latter function, plotted in graph (a) below, is anti-symmetric in the sense that it has positive and negative regions of equal area. Thus, the position expectation value $\langle x\rangle$, obtained from

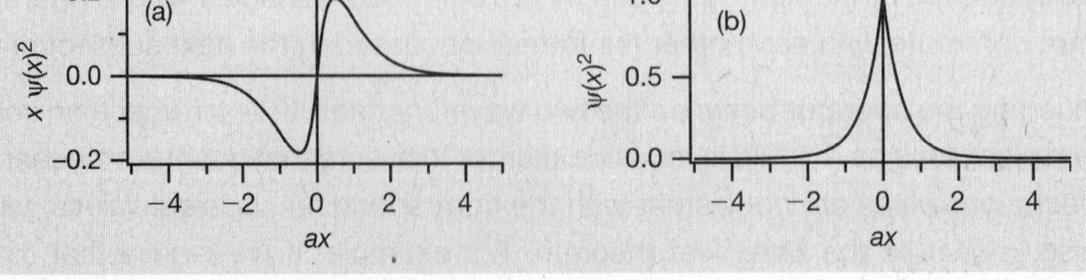

the integral of the function shown in graph (a), is necessarily equal to zero (by symmetry). The fact that $\langle x \rangle = 0$ also makes sense since the particle's probability density $\Psi(x)^2 = ae^{-2a|x|}$, plotted in graph (b), is peaked at $x = 0$ and symmetric in shape.

We will soon see (in Section 6.5) that all experimental observables are represented by a special class of operators, called *Hermitian* operators. We will also see that the eigenvalues of all such operators are necessarily real, and that the corresponding eigenfunctions are necessarily orthogonal, in the sense that $\int \Psi_i^* \Psi_j d\tau$ is equal to zero unless $\Psi_i = \Psi_j$, in which case $\int \Psi_i^* \Psi_i d\tau = 1$. These properties lead to the following very useful result pertaining to expectation values of any wavefunction that is expressed as a linear combination of eigenfunctions.

Consider a wavefunction $\Psi = c_1\Psi_1 + c_2\Psi_2$ where Ψ_1 and Ψ_2 are normalized orthogonal eigenfunctions of $\widehat{A}$ (with different eigenvalues a_1 and a_2), and thus Ψ is not be an eigenfunction of $\widehat{A}$. The following proof demonstrates that the expectation value $\langle A \rangle$ for the wavefunction Ψ is equal to $|c_1|^2 a_1 + |c_2|^2 a_2$.

$$
\begin{aligned}
\langle A \rangle &= \int \Psi^* \widehat{A} \Psi d\tau \\
&= \int (c_1\Psi_1 + c_2\Psi_2)^* \widehat{A} \left(c_1\Psi_1 + c_2\Psi_2\right) d\tau \\
&= \int \left(c_1^*\Psi_1^* + c_2^*\Psi_2^*\right) \widehat{A} \left(c_1\Psi_1 + c_2\Psi_2\right) d\tau \\
&= \int \left(|c_2|^2\, \Psi_1^*\widehat{A}\Psi_1 + c_1^*c_2\Psi_1^*\widehat{A}\Psi_2 + c_2^*c_1\Psi_2^*\widehat{A}\Psi_1 + |c_2|^2\, \Psi_2^*\widehat{A}\Psi_2\right) d\tau \\
&= \int \left(|c_2|^2\, a_1\Psi_1^*\Psi_1 + c_1^*c_2a_2\Psi_1^*\Psi_2 + c_2^*c_1a_1\Psi_2^*\Psi_1 + |c_2|^2\, a_2\Psi_2^*\Psi_2\right) d\tau \\
&= |c_2|^2\, a_1 \cdot 1 + c_1^*c_2a_2 \cdot 0 + c_2^*c_1a_1 \cdot 0 + |c_2|^2\, a_2 \cdot 1 \\
&= |c_2|^2 a_1 + |c_2|^2 a_2
\end{aligned}
$$

The Uncertainty Principle

No aspect of quantum mechanics has attracted more interest and speculation – both within the scientific community and from the general public – than the uncertainty principle. When Werner Heisenberg introduced this restriction on the position and momentum of a quantum mechanical particle, he stated that "the more accurately the position is determined the less accurately the momentum is known and conversely." He based this conclusion on an analysis of a (thought) experiment in which a photon is scattered by an object under a microscope. This analysis suggested that the momentum the photon imparts to the object produces an uncertainty in the object's momentum. Thus, the act of observing the position of an object changes its momentum.[23]

In addition to the theoretical and experimental relevance of the uncertainty principle, it has also generated wide-ranging, and sometimes wild, metaphysical and philosophical speculations. For example, the famous physicist and philosopher of science, Arthur Stanley Eddington, suggested that the uncertainty principle may be the source of human freewill, when he speculated that "surely the human mind will have an equal indeterminacy; for we can scarcely accept a theory which makes out the mind to be more mechanistic than the atom." In a somewhat less far-fetched vein, Niels Bohr argued that the uncertainty principle is rooted in the fundamental impossibility of simultaneously observing both the wave and particle aspects of any object.

The validity of the uncertainty principle is not contingent on any such philosophical speculations. We will see that the uncertainty principle may be obtained from a mathematical theorem pertaining to the spectral analysis of any function. We will also see that the uncertainty principle reflects the fact that quantum mechanical operators do not in general commute with each other.

A mathematical procedure called Fourier transformation[24] may be used to describe functions as a superposition of sinusoidal waves. The relationship between a function and its Fourier transform is illustrated in Figure 6.2, which shows several wavefunctions (on the left) and their frequency spectra (on the right). In other words, the peaks in the spectra on the right indicate the frequency components that contribute to the corresponding wavefunction. The spectra are obtained by Fourier transformation of the wavefunctions, and conversely, the wavefunctions may be obtained by inverse Fourier transformation of the spectra.

---

[23] Locating an object more precisely requires a photon of shorter wavelength and larger momentum $p = h/\lambda$. Thus, the more accurately one determines an object's position, the more uncertain its momentum becomes.

[24] The French mathematician and physicist Jean Baptiste Joseph Fourier (1768–1830) is credited with originating the concept of Fourier transformation, which he used to analyze problems associated with heat flow. Fourier is also credited with the discovery of the greenhouse effect. Fourier went with Napoleon Bonaparte on his Egyptian expedition in 1798, and was made governor of Lower Egypt and secretary of the Institut d'Égypte, which was formed by Napoleon to carry out research during his Egyptian campaign.

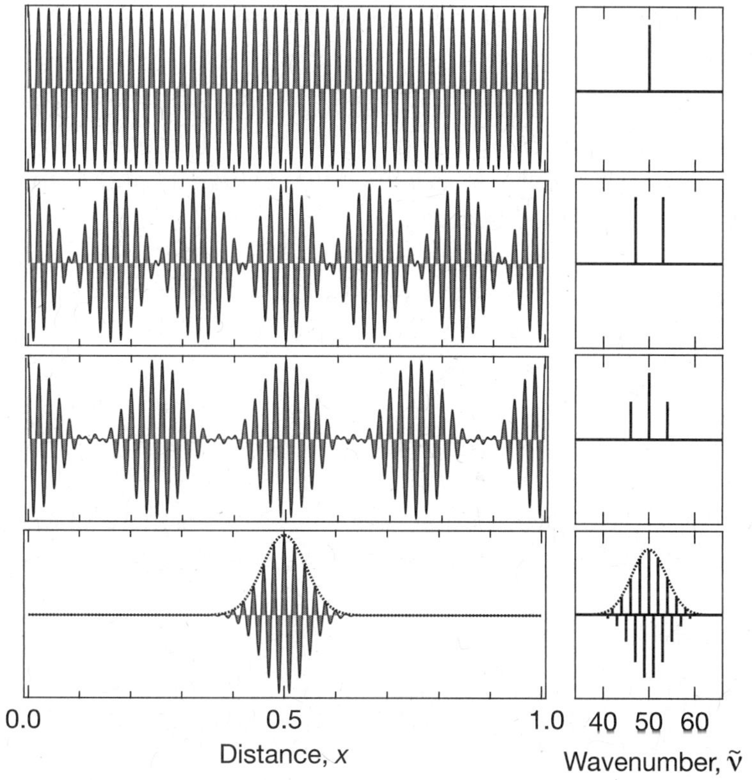

FIGURE 6.2 Fourier transformation of the wavefunctions on the left produces the frequency spectra shown on the right. The wavefunction on the top is a simple plane wave with a single frequency component. As more frequency components are added, the wavefunctions develop a beating pattern (as the result of constructive and destructive interference between the component waves).

More specifically, the following integral may be used to calculate the Fourier (cosine) transform of any function $f(x)$.

$$g(\tilde{\nu}) = \int f(x)\cos(2\pi x\tilde{\nu})dx \tag{6.38}$$

Similarly, $f(x)$ may be regenerated by performing an inverse Fourier transform of $g(\tilde{\nu})$.[25]

$$f(x) = \int g(\tilde{\nu})\cos(2\pi x\tilde{\nu})d\tilde{\nu} \tag{6.39}$$

Notice that the integral in Eq. 6.39 may be viewed as the limit of a sum $\sum g(\tilde{\nu}_i)\cos(2\pi x\tilde{\nu}_i)$ of waves, $\cos(2\pi x\tilde{\nu}_i)$, with different frequencies $\tilde{\nu}_i$ and

---

[25] The variable $\tilde{\nu}$ is equivalent to the wavenumber $1/\lambda$, where λ is the wavelength of the corresponding plane wave.

amplitudes $g(\tilde{\nu}_i)$. In other words, $g(\tilde{\nu})$ is the amplitude of the frequency spectrum of $f(x)$.[26]

If we assume that $f(x)$ and its Fourier transform $g(\tilde{\nu})$ both have normalized intensities, $\int |f(x)|^2\,dx = 1$ and $\int |g(\tilde{\nu})|^2\,d\tilde{\nu} = 1$, then it can be shown that the following inequality invariably holds (where x_0 and $\tilde{\nu}_0$ may be any particular constant values of x and $\tilde{\nu}$).

$$\left(\int (x - x_0)^2 |f(x)|^2\,dx\right)\left(\int (\tilde{\nu} - \tilde{\nu}_0)^2 |g(\tilde{\nu})|^2\,d\tilde{\nu}\right) \geq \frac{1}{16\pi^2} \tag{6.40}$$

If we choose $x_0 = \langle x \rangle$ and define

$$\langle x \rangle \equiv \int x |f(x)|^2\,dx$$

then the first of the above integrals becomes equivalent to the *variance* of x (that is equal to the square of the *standard deviation* of x).

$$\int (x - \langle x \rangle)^2 |f(x)|^2\,dx = \left\langle x^2 \right\rangle - 2\langle x \rangle \langle x \rangle + \langle x \rangle^2 = \left\langle x^2 \right\rangle - \langle x \rangle^2 \equiv \sigma_x^2$$

Similarly, if we choose $\tilde{\nu}_0 \equiv \langle \tilde{\nu} \rangle$ and define

$$\langle \tilde{\nu} \rangle \equiv \int \tilde{\nu} |g(\tilde{\nu})|^2\,d(\tilde{\nu})$$

the second integral in Eq. 6.40 yields the following variance of $\tilde{\nu}$.

$$\int (\tilde{\nu} - \langle \tilde{\nu} \rangle)^2 |g(\tilde{\nu})|^2\,d\tilde{\nu} = \left\langle \tilde{\nu}^2 \right\rangle - \langle \tilde{\nu} \rangle^2 \equiv \sigma_{\tilde{\nu}}^2$$

Thus, Eq. 6.40 implies that $\sigma_x^2 \sigma_{\tilde{\nu}}^2 \geq 1/(16\pi^2)$, which, after taking the square root of both sides, yields the following *uncertainty relation* that indicates that the product of the standard deviations of x and $\tilde{\nu}$ cannot be smaller than $1/(4\pi)$.

$$\sigma_x \sigma_{\tilde{\nu}} \geq \frac{1}{4\pi} \tag{6.41}$$

The results shown at the bottom of Figure 6.2 illustrate how the above uncertainty relation applies to a wavefunction with a Gaussian amplitude and a Gaussian frequency spectrum.[27] Gaussian functions are unique in that they are the only functions that have an uncertainty product that is exactly equal to its minimum possible value $\sigma_x \sigma_{\tilde{\nu}} = 1/(4\pi)$. This implies that a Gaussian wavefunction that is more sharply localized (has a smaller

---

[26] More generally, Fourier transforms may be represented as integrals involving complex exponential waves $g(\tilde{\nu}) = \int f(x) e^{-2\pi i x \tilde{\nu}}\,dx$ and $f(x) = \int g(\tilde{\nu}) e^{2\pi x \tilde{\nu}}\,d\tilde{\nu}$.

[27] Recall that a Gaussian function is defined as $e^{-(x-x_0)^2/(2\sigma^2)}$, where $x_0 = \langle x \rangle$ is the mean value and $\sigma = \sqrt{\left\langle x^2 \right\rangle - \langle x \rangle^2}$ is the standard deviation of the function.

σ_x) must have a broader frequency spectrum (a larger $\sigma_{\tilde{\nu}}$). Conversely, a highly delocalized sinusoidal wavefunction, such as that shown at the top of Figure 6.2, must have a very sharp frequency spectrum.

The quantum mechanical implications of Eq. 6.41 become evident when we make use of the de Broglie relation $p = h/\lambda = h\tilde{\nu}$, which suggests that we may obtain the following uncertainty relation pertaining to quantum mechanical particle-waves simply by multiplying both sides of Eq. 6.41 by Planck's constant (and equating $\hbar = h/2\pi$).

$$\boxed{\sigma_x \sigma_p \geq \tfrac{1}{2}\hbar} \qquad (6.42)$$

This famous uncertainty relation may also be shown to be a consequence of the fact that the quantum mechanical operators x and p_x do not commute with each other, so $\widehat{x}\widehat{p}_x \neq \widehat{p}_x\widehat{x}$.

Exercise 6.6

Demonstrate that the operators $\widehat{x}$ and $\widehat{p}_x$ do not commute with each other by showing that $\widehat{x}\widehat{p}_x - \widehat{p}_x\widehat{x}$ is not equal to zero.

Solution. Note that the operators $\widehat{x}\widehat{p}_x$ and $\widehat{p}_x\widehat{x}$ do not have the same effect when applied to an arbitrary wavefunction $\Psi(x)$.

$$\widehat{x}\widehat{p}_x[\Psi] = x(-i\hbar)\frac{\partial}{\partial x}[\Psi(x)]$$

$$\widehat{p}_x\widehat{x}[\Psi] = (-i\hbar)\frac{\partial}{\partial x}[x\Psi(x)] = (-i\hbar)[\Psi(x)] + x(-i\hbar)\frac{\partial}{\partial x}[\Psi(x)]$$

This implies that $(\widehat{x}\widehat{p}_x - \widehat{p}_x\widehat{x})\,\Psi = i\hbar\Psi$ and so $\widehat{x}\widehat{p}_x - \widehat{p}_x\widehat{x} = i\hbar \neq 0$, which proves that the operators $\widehat{x}$ and $\widehat{p}_x$ do not commute.

The above exercise implies that the *commutator* of $\widehat{x}$ and $\widehat{p}_x$ may be expressed as follows.

$$\boxed{[\widehat{x}, \widehat{p}_x] \equiv \widehat{x}\widehat{p}_x - \widehat{p}_x\widehat{x} = i\hbar} \qquad (6.43)$$

More generally, the commutator of any two operators is defined as

$$\boxed{[\widehat{A}, \widehat{B}] \equiv \widehat{A}\widehat{B} - \widehat{B}\widehat{A}} \qquad (6.44)$$

For any pair of operators corresponding to quantum mechanical observables A and B, it can be shown that the following inequality relates the expectation value of the operator $[\widehat{A}, \widehat{B}]$ to the uncertainties of A and B.[28]

$$\boxed{\sigma_A \sigma_B \geq \frac{1}{2}\left|\langle[\widehat{A}, \widehat{B}]\rangle\right|} \tag{6.45}$$

Equation 6.45 may be used to derive uncertainty relations for any pair of quantum mechanical observables whose operators do not commute with each other.[29] If the operators corresponding to two observables do commute with each other, then $[\widehat{A}, \widehat{B}] = 0$, in which case $\sigma_A \sigma_B$ could have any arbitrarily small (nonnegative) value.

Exercise 6.7

Show that Eq. 6.45 may be used to obtain the position-momentum uncertainty relation..

Solution. If we identify the operators in Eq. 6.45 as $\widehat{A} = \widehat{x} = x$ and $\widehat{B} = \widehat{p_x} = -i\hbar\frac{\partial}{\partial x}$, then Eq. 6.43 indicates that $[\widehat{A}, \widehat{B}] = i\hbar$. Thus, $\frac{1}{2}|\langle[\widehat{A}, \widehat{B}]\rangle| = \frac{1}{2}\sqrt{(i\hbar)^* i\hbar} = \frac{1}{2}\hbar$ and so Eq. 6.45 requires that $\sigma_x \sigma_{p_x} \geq \frac{1}{2}\hbar$, which is equivalent to Eq. 6.42.

The uncertainty principle is sometimes incorrectly assumed to apply to individual particle detection events. In other words, Eq. 6.42 need not imply that a single particle cannot simultaneously have both a precise position and momentum. The derivations of Eqs. 6.42 and 6.45, combined with Born's statistical interpretation of Ψ, make it clear that σ_x and σ_p pertain to the standard deviations of many repeated measurements performed on identically prepared particle-waves. When individual particles (such as photons or electrons) are detected, their positions appear to be highly localized, and their momenta and energies appear to be precisely conserved, exactly like classical particles. However, the experimentally observed *statistical distributions of the positions and momenta* of many identically prepared particles are well described by probability densities that are proportional to the square of the corresponding wavefunctions.

---

[28] The uncertainties of A and B are again defined as the standard deviations of the corresponding experimentally measured values, $\sigma_A \equiv \sqrt{\langle A^2\rangle - \langle A\rangle^2}$ and $\sigma_B \equiv \sqrt{\langle B^2\rangle - \langle B\rangle^2}$.

[29] The right-hand side of Eq. 6.45 is invariably a nonnegative real number. So, if $\widehat{A}$ and $\widehat{B}$ do not commute, then Eq. 6.45 imposes a positive lower bound on $\sigma_A \sigma_B$.

Exercise 6.8

Determine the uncertainty in the position of a particle whose normalized wavefunction is $\Psi(x) = \sqrt{a}e^{-a|x|}$ (as further described in Exercises 6.4 and 6.5).

Solution. In Exercise 6.5, we found that $\langle x \rangle = 0$. Evaluating $\langle x^2 \rangle$ requires integrating the function $\Psi^* x^2 \Psi = ax^2 e^{-a|x|}$, plotted below. Since this function is symmetric about $x = 0$, we may evaluate it by integrating over positive x-values and then doubling the result.

$$\langle x^2 \rangle = 2a \int_0^\infty x^2 e^{-ax} dx = \frac{4}{a^2}$$

The last equality is obtained by consulting the integrals given in Appendix B, one of which implies that $\int_0^\infty x^2 e^{-ax} dx = \Gamma(2+1)/a^{2+1} = 2!/a^{2+1} = 2/a^3$. Thus, the uncertainty in the particle's position is $\sigma_x = \sqrt{\langle x^2 \rangle - \langle x \rangle^2} = 2/a$.

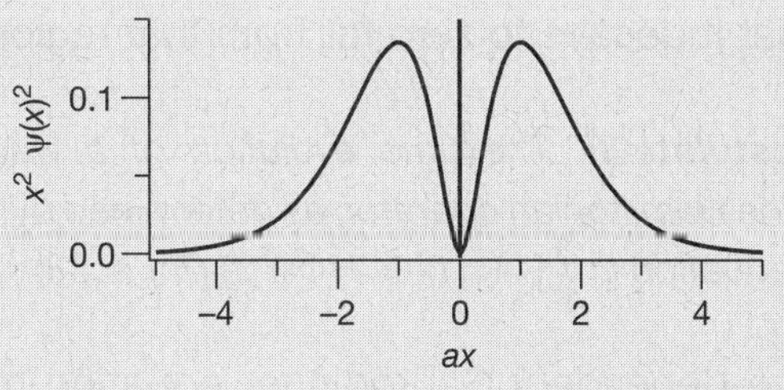

6.5 Formal Postulates of Quantum Mechanics

Both classical and quantum mechanics (as well as thermodynamics and statistical mechanics) can be derived from a set of foundational postulates. For example, the following two postulates of classical mechanics may be used to describe the instantaneous state and time-evolution of any classical system.

> ***Classical Postulate I.*** *All physical objects are composed of particles whose state at any time t is completely determined by specifying the position $r(t)$ and momentum $p(t) = mv = m(dr/dt)$ of each particle.*[30]
>
> ***Classical Postulate II.*** *The time evolution of the position and momentum of any particle is determined by the force acting on it, as determined by Newton's second law $F(t) = ma = m(d^2r/dt^2)$ (Eq. 1.3).*

[30] More precisely the position and momentum may be represented as vectors in three-dimensional space, $r = \vec{r}$ and $p = \vec{p}$.

The following two postulates are the quantum analogues of the above classical postulates, while the third quantum postulate does not have a classical counterpart.

> ***Quantum Postulate I.*** *The state of any system consisting of particles whose positions at time t are described by the configuration $\tau = r_1, r_2, r_3 \ldots$ is determined by a well-behaved wavefunction $\Phi(\tau, t)$, such that $\rho(\tau)d\tau = \Phi^* \Phi\, d\tau$ is equal to the probability that the particles will be found within $d\tau$ of a particular configuration τ.*[31]

A function is "well-behaved" if it is single-valued, continuous, continuously differentiable, and normalizable. A single-valued function is one for which $\Phi(\tau)$ has a single value at each particle configuration τ. Normalizable functions have the property that they are square integrable, in the sense that $\int \Phi^* \Phi d\tau$ is equal to a real finite number, and so can be normalized. The latter requirement also implies that $\Phi^* \Phi$ must be localized to some region of space, such that it decays to zero far from that region.[32]

> ***Quantum Postulate II.*** *The time evolution of a wavefunction $\Phi(\tau, t)$ is governed by the Hamiltonian operator, as determined by the time-dependent Schrödinger equation $i\hbar \frac{\partial}{\partial t}[\Phi(\tau, t)] = \widehat{H}\Phi(\tau, t)$ (Eq. 6.28).*

Note that the time-dependent Schrödinger equation may be rearranged to $d\Phi/\Phi = (\widehat{H}/i\hbar)dt = (-i\widehat{H}/\hbar)dt$, and then integrated to obtain $\ln \Phi = -i\widehat{H}t/\hbar$. Thus, when the latter integral is evaluated from 0 to t, one obtains $\ln[\Phi(\tau, t)/\Phi(\tau, 0)] = -i\widehat{H}t/\hbar$ and so $\Phi(\tau, t)/\Phi(\tau, 0) = e^{-i\widehat{H}t/\hbar}$ or equivalently $\Phi(\tau, t) = e^{-i\widehat{H}t/\hbar}[\Phi(\tau, 0)]$. Thus, $e^{-i\widehat{H}t/\hbar}$ is referred to as the *time-evolution operator*, as it has the effect of evolving a wavefunction from its initial state to its state at time t. Note that if the system is initially in an eigenstate of energy E then the time-evolution operator becomes $e^{-iEt/\hbar}$, which simply has the effect of harmonically modulating the amplitude of the wavefunction with a frequency of $\omega = E/\hbar$.[33]

---

[31] The symbol τ here stands for a multidimensional variable describing the positions of every particle in the system, very much in the same way as it does in Eq. 1.27 on p. 35.

[32] Unless stated otherwise, all integrals are assumed to extend over the entire configuration space available to the particles.

[33] When the system is in a nonstationary state (in other words, if it is not in an eigenstate), then its wavefunction may be represented as a linear combination of eigenstate wavefunctions. The time evolution of such a nonstationary state is again dictated by the fact that each eigenstate is harmonically modulated with a frequency that is determined by its energy. Differences between the energies of each pair of eigenstates or eigenstate wavefunctions? give rise to frequency

The following measurement postulate has no classical counterpart, because classical mechanics implicitly assumes that all dynamical variables have values that can be measured precisely (and that such measurements can in principle be performed without changing the state of the system).

Quantum Postulate III. *Every dynamical variable is represented by a linear Hermitian operator $\widehat{A}$ (with normalized eigenfunctions Ψ_i and eigenvalues a_i) such that an arbitrary state described by a wavefunction Φ has the following properties:*

1. *The outcome of a measurement of the dynamical variable is always one of the eigenvalues of $\widehat{A}$.*

2. *The probability of measuring the eigenvalue a_i is $\left|\int \Psi_i^* \Phi \, d\tau\right|^2$.*

3. *The state of the system after a measurement that produced the value a_i is described by the eigenfunction Ψ_i.*

A dynamical variable is a quantity such as position, momentum, or some combination thereof. A linear operator is one with the property that $\widehat{A}[\Psi_a + \Psi_b] = \widehat{A}\Psi_a + \widehat{A}\Psi_b$. A Hermitian operator is one with the property that $\int \Psi_a^*(\widehat{A}\Psi_b)d\tau = \int (\widehat{A}\Psi_a)^*\Psi_b d\tau$. The following is a proof of the important, and practically useful, facts that the eigenvalues of a Hermitian operator are all real and the eigenfunctions associated with different eigenvalues are orthogonal to each other.[34]

If Ψ_i and Ψ_j are two eigenfunctions of a Hermitian operator, $\widehat{A}$, whose eigenvalues are a_i and a_j, respectively, then

$$\begin{aligned} \widehat{A}\Psi_i &= a_i\Psi_i \\ \text{so, } \Psi_j^*(\widehat{A}\Psi_i) &= \Psi_j^*(a_i\Psi_i) \end{aligned} \tag{6.46}$$

and similarly,

$$\begin{aligned} \widehat{A}\Psi_j &= a_j\Psi_j \\ \text{so, } (\widehat{A}\Psi_j)^* &= (a_j\Psi_j)^* \\ \text{and thus, } (\widehat{A}\Psi_j)^*\Psi_i &= (a_j\Psi_j)^*\Psi_i\,. \end{aligned} \tag{6.47}$$

---

differences of $\Delta\omega = \Delta E/\hbar$. It is these energy differences, rather than the absolute values of the energies, which dictate the time evolution of nonstationary states. Thus, the time evolution of a nonstationary state does not depend on the value of the reference energy with respect to which eigenstate energies are measured.

[34] Two wavefunctions are orthogonal if $\int \Psi_a^*\Psi_b d\tau = 0$ unless $\Psi_a = \Psi_b$ (in which case $\int \Psi_a^*\Psi_a d\tau = 1$, assuming that Ψ_a is normalized).

We may now subtract Eq. 6.47 from Eq. 6.46 and integrate to obtain the following identity.

$$\int \left[\Psi_j^*(\widehat{A}\Psi_i) - (\widehat{A}\Psi_j)^*\Psi_i\right] d\tau = \int \left[a_i\Psi_j^*\Psi_i - a_j^*\Psi_j^*\Psi_i\right] d\tau$$

$$0 = (a_i - a_j^*) \int \Psi_j^*\Psi_i d\tau \qquad (6.48)$$

Note that the left-hand side is equal to zero, because $\widehat{A}$ is Hermitian, and so $\int \Psi_j^*(\widehat{A}\Psi_i)d\tau = \int (\widehat{A}\Psi_j)^*\Psi_i d\tau$. Thus, the following two important conclusions are direct consequences of Eq. 6.48.

- If $\Psi_i = \Psi_j$, then $a_i = a_j$ (and $a_i^* = a_j^*$), and so Eq. 6.48 implies that $a_i = a_i^*$ (and $a_j = a_j^*$), thus proving that *the eigenvalues of $\widehat{A}$ are real.*
- If $\Psi_i \neq \Psi_j$ and $a_i \neq a_j$, then Eq. 6.48 requires that $\int \Psi_j^*\Psi_i\, d\tau = 0$, so *the eigenfunctions of $\widehat{A}$ are orthogonal.*

Note that degenerate eigenfunctions (which have the same eigenvalue) need not be orthogonal, but it is always possible to represent degenerate eigenfunctions as a linear combination of orthogonal eigenfunctions.

HOMEWORK PROBLEMS

Problems That Illustrate Core Concepts

1. A molecular C-H stretch vibration may be approximated as a harmonic oscillator with a mass equal to that of a hydrogen atom $m \approx 1.7 \times 10^{-27}$ kg. The experimentally observed frequency of such vibrations is $\tilde{\nu} \approx 3000$ cm^{-1}, where $\tilde{\nu} = \nu/c = \omega/(2\pi c)$ and $c \approx 3 \times 10^{10}$ cm/s is the speed of light.

(a) Use the above information to estimate the force constant of a C-H bond (in N/m units).

(b) What wavelength λ of light (in nm units) would be required in order to induce a C-H vibrational transition, such that $h\nu = \hbar\omega = hc/\lambda = \varepsilon_1 - \varepsilon_0$?

2. Consider a plane wave $\Phi(x,t) = C\cos(ax - bt)$ (where C, a, and b are constants).

(a) Show that $\Phi(x,t)$ is an eigenfunction of the operator $\frac{\partial^2}{\partial x^2}$ and determine the corresponding eigenvalue.

(b) Show that $\Phi(x,t)$ is an eigenfunction of the operator $\frac{\partial^2}{\partial t^2}$ and determine the corresponding eigenvalue.

(c) Use the above results to show that $\Phi(x,t)$ is a solution of Eq. 6.6, and determine the relationship between the parameters a and b, and the wave's phase velocity v.

(d) Determine the values of the parameters a and b for an ocean wave that is approximately described by $\Phi(x,t)$ and has an experimentally measured wavelength of 350 m and phase velocity of 23 m/s.

3. Consider the following wave functions (where A, B, k, and a are constants):

(i) $\Psi(x) = A\sin(kx)$ (iv) $\Psi(x) = Ae^{-ikx}$
(ii) $\Psi(x) = B\cos(kx)$ (v) $\Psi(x) = Be^{+ikx}$
(iii) $\Psi(x) = a/x$ (vi) $\Psi(x) = Be^{-ax^2}$

(a) Determine which of the above functions are eigenfunctions of the momentum operator, $\hat{p}_x$, and identify the corresponding eigenvalue.

(b) Determine which of the above functions are eigenfunctions of the kinetic energy operator, $\hat{K}_x$, and determine the corresponding eigenvalue.

4. Evaluate the momentum expectation values of the free particle wavefunctions e^{-ikx} and e^{+ikx} (where k is a positive constant). What do your results tell you about the direction in which free particles described by each of these wavefunctions are moving?

5. The following problems pertain to a wavefunction of the form $\psi(x) = e^{-x^2}$ (which is not yet normalized).

(a) Sketch $\psi(x)$ (over a range of $-2 < x < 2$).

(b) Normalize $\psi(x)$.

(c) What is the probability that the particle will be found at some positive value of x?

(d) Evaluate the expectation values $\langle x \rangle$ and $\langle x^2 \rangle$, and use these to determine the uncertainty in the particle's position $\sigma_x = \sqrt{\langle x^2 \rangle - \langle x \rangle^2}$.

(e) Evaluate the uncertainty in the particle's momentum $\sigma_{p_x} = \sqrt{\langle p_x^2 \rangle - \langle p_x \rangle^2}$, and show that this is exactly equal to the minimum value allowed by the position-momentum uncertainty relation (Eq. 6.42).

6. Determine the conditions under which a wavefunction of the form $\psi(x) = e^{-ax^2}$ will be an eigenfunction of the operator $\widehat{A} = \frac{\partial^2}{\partial x^2} - bx^2$, where a and b are positive real numbers, and express the resulting eigenvalue in terms of b.

7. The results of problem **6.** above indicate that the normalized Gaussian wavefunction $\Psi(x) = (2a/\pi)^{1/4} e^{-ax^2}$ is an eigenfunction of an operator that closely resembles the Hamiltonian of a harmonic oscillator $\widehat{H} = \widehat{K} + \widehat{V} = -\frac{\hbar^2}{2m}\frac{\partial^2}{\partial x^2} + \frac{1}{2}fx^2$.

(a) Divide $\widehat{H}$ by $-\hbar^2/2m$ to convert it to the same form as the operator in problem **6.**, and then use the result you obtained in problem **6.** to show that $a = \sqrt{fm}/2\hbar$ and $E = \langle H \rangle = \frac{1}{2}\hbar\omega$ (where $\omega = \sqrt{f/m}$).

(b) Use the virial theorem (Eq. 1.46 on page 41) to show that $\langle K \rangle = \langle V \rangle$ for a harmonic oscillator and then use this result, combined with the fact that $\langle H \rangle = \langle K \rangle + \langle V \rangle$, to infer the values of $\langle K \rangle$ and $\langle V \rangle$.

(c) Determine $\langle V \rangle$ by directly evaluating the integral corresponding to this expectation value, and confirm that your result is equivalent to that which you obtained in (b).

8. A Hermitian operator $\widehat{A}$ is defined as one for which $\int_{-\infty}^{\infty} \Psi_a^*(\widehat{A}\Psi_b)dx = \int_{-\infty}^{\infty} (\widehat{A}\Psi_a)^*\Psi_b dx$, where Ψ_a and Ψ_b are any two wavefunctions, and $[\ldots]^*$ stands for a complex conjugate of $[\ldots]$ in which $i = \sqrt{-1}$ is replaced by $-i$.

(a) Show that the position operator $\widehat{x} = x\cdot$ is Hermitian. Hint: Since x is a real number $x^* = x$, and multiplication by x is commutative, so $\Psi \cdot x = x \cdot \Psi$.

(b) Show that the momentum operator $\widehat{p_x}$ is Hermitian. Hint: Use the formula for integration by parts $\int udv = uv - \int vdu$, with $u = \Psi_a^*$ and $v = -i\hbar\Psi_b$ so $dv = (dv/dx)dx$, as well as the fact that any wavefunction corresponding to a single particle must have a normalizable wavefunction, which requires that $\Psi(-\infty) = \Psi(\infty) = 0$.

Problems That Test Your Understanding

9. Consider the wavefunction $\Psi(x) = Ce^{-ax}$ (where both C and a are positive real constants).

(a) Is $\Psi(x)$ an eigenfunction of $\widehat{A} = \partial/\partial x$ and if so, what is its eigenvalue?

(b) Is $\Psi(x)$ an eigenfunction of $\widehat{B} = \partial^2/\partial x^2$ and if so, what is its eigenvalue?

10. Consider the wavefunction Be^{-ibx} for a free electron, of mass 9.1×10^{-31} kg, whose de Broglie wavelength is 1 nm (where B and b are positive real constants).

(a) What is the momentum of the electron?

(b) What is the value of b?

(c) What is the value of $\langle K_x \rangle$?

11. The diatomic molecule CO has an experimentally observed vibrational frequency of $\tilde{\nu} \sim 2140$ cm^{-1}, and may be approximated as a harmonic oscillator with a mass of 7×10^{-27} kg.

(a) Determine the wavelength of light that would be required to induce a vibrational transition of CO (expressed in nm units).

(b) Determine the vibrational force constant of CO (in N/m units).

12. Consider the normalized wavefunction $\Psi(x) = \sqrt{\frac{15}{16}}(1 - x^2)$, which is only nonzero over the range $-1 \leq x \leq 1$ (so Ψ is equal to zero when $x < -1$ and $x > 1$).

(a) Confirm that $\Psi(x)$ is normalized by showing that the total probability of finding the particle within the range $-1 \leq x \leq 1$ is equal to 1.

(b) What is $\langle x \rangle$?

(c) What is σ_x?

13. Determine the minimum value of the uncertainty product pertaining to the operators $\widehat{A} = x\cdot$ and $\widehat{B} = \partial/\partial x$.

14. Consider a green photon with a wavelength of 500 nm.

(a) What is the momentum of the photon (in kg m/s units)?

(b) What is the energy of the photon (in J units)?

(c) What is the frequency of the photon (in s^{-1} units)?

(d) What is the wavenumber of the photon (in cm^{-1} units)?

15. A given particle has the following expectation values: $\langle x \rangle = 0$, $\langle x^2 \rangle = a^2$, $\langle p \rangle = 0$. What is the smallest possible *momentum squared expectation value* that this particle could have?

16. Consider a wavefunction that is proportional to $\sin[(2\pi/a)x]$ and extends over the range $-\frac{a}{2} < x < \frac{a}{2}$ (and is equal to zero everywhere else).

(a) Normalize the above wavefunction.

(b) Sketch the probability density of the above wavefunction.

(c) What is the probability of finding the above particle at some positive x value (anywhere in the range $0 \leq x \leq \frac{a}{2}$) ?

17. The ground state of a particle confined to a box that extends over the range $0 < x < a$ has a normalized wavefunction of $\Psi_1(x) = \sqrt{\frac{2}{a}} \sin[(\pi/a)x]$.

(a) Evaluate the expectation value of the *kinetic energy* of the particle.

(b) What is the expectation value of the *momentum squared* of the particle?

(c) What is the expectation value of the *position* of the particle?

18. A given particle-wave has a (normalized) Gaussian probability density $\frac{1}{a\sqrt{2\pi}}e^{-x^2/(2a^2)}$, where a = 1 Å. What are the standard deviations of the *position* and *momentum* of this particle?

19. Consider an operator of the form $\widehat{A} = a\frac{\partial^2}{\partial x^2} + \frac{b}{x}$ and a function of the form $\Psi(x) = xe^{-cx}$.

(a) Determine the conditions under which $\Psi(x)$ will be an eigenfunction of $\widehat{A}$.

(b) Express the resulting eigenvalue in terms of the constant a, b, and/or c.

Simple Systems and Chemical Applications

CHAPTER 7

Now that we have identified the basic rules that govern the behavior of particle-waves, we are ready to apply our understanding to problems of chemical interest. These include the scattering of electrons and atoms, the quantization of atomic and molecular energy levels, and the shapes of the electronic and vibrational wavefunctions. In this section we will focus on systems whose Schrödinger equation can be solved analytically (as opposed to numerically). Although there are relatively few such systems, we will see that they may be used to gain a deep understanding of the role of quantum phenomena in a wide variety of chemical processes.

7.1 Free, Confined, and Obstructed Particles

We will first focus on systems involving either no potential energy, $V(x) = 0$, or very simple (piecewise-constant) potential energy functions. Such systems can be used to understand the quantum mechanical properties of gas phase (free) electrons and other particles, as well as electrons that are bound within the (negative) potentials of atoms or that encounter (positive) potential energy barriers. For example, we will see how the confinement of a particle-wave to a small region of space leads to energy quantization. Thus, as the space available to an electron decreases, the gap between its electronic quantum states increases. We will also see how particle-waves are able to tunnel through walls. This surprising behavior explains processes ranging from proton transport in water to nuclear radioactivity, and plays a key role in devices such as scanning tunneling microscopes and quantum well lasers.

Free Particles

In both classical and quantum mechanics a free particle is defined as one whose total energy is equal to its kinetic energy. In other words, the Hamiltonian of a free particle is $H = K$. To keep things simple, lets initially consider a single particle that is moving along the x-direction, in which case its classical Hamiltonian is $H = \frac{1}{2}mv_x^2 = p_x^2/2m$, and its quantum Hamiltonian is $\widehat{H} = \widehat{K}_x = -(\hbar^2/2m)(\partial^2/\partial x^2)$. Thus, such a particle is described by the following Schrödinger equation.

$$\widehat{H}\,[\Psi(x)] = -\frac{\hbar^2}{2m}\frac{\partial^2}{\partial x^2}\,[\Psi(x)] \tag{7.1}$$

It is easy to show that any one of the following wavefunctions is a viable solution of the above differential equation (where A and k are real constants).

$$\begin{aligned}\Psi(x) &= A\sin(kx)\\ \Psi(x) &= A\cos(kx)\\ \Psi(x) &= Ae^{ikx}\\ \Psi(x) &= Ae^{-ikx}\end{aligned} \tag{7.2}$$

Notice that the energy eigenvalues of all of the above functions are $E = (\hbar k)^2/2m$.[1]

The last two wavefunctions in Eq. 7.2 are eigenfunctions of $\widehat{p}_x$, with eigenvalues of $+\hbar k$ and $-\hbar k$, and thus correspond to particles moving to the right or left, respectively. The first two eigenfunctions in Eq. 7.2 are not eigenfunctions of $\widehat{p}_x$, but have a momentum expectation value of $\langle p_x\rangle = 0$. Since these wavefunctions have positive (nonzero) root-mean-squared momenta $\sqrt{\langle p_x^2\rangle} = \sqrt{2m\,\langle K_x\rangle} = \hbar k$, they apparently correspond to particles that are going both left and right with equal probability (as further described in exercise 6.3 on page 206).[2] Although it is difficult to picture such a counter-propagating particle, we will shortly see that such wavefunctions also arise

---

[1] These wavefunctions are not normalizable, because they extend over the entire x-axis. However, the energy eigenvalue associated with these waves does not depend on the value of A. We also know (see p. 214) that the expectation value of any eigenfunction of $\widehat{H}$ is equal to its energy eigenvalue so $\langle H\rangle = E = (\hbar k)^2/2m$ for all these free particle-waves.

[2] The fact that $A\sin(kx)$ or $A\cos(kx)$ represent particles moving both left and right may also be inferred by using Euler's formula (Eq. 6.7) to represent these two wavefunctions as a linear combination of the last two wavefunctions in Eq. 7.2.

in describing a particle that is trapped in a box, and so has either positive or negative momentum (with equal probability) as it bounces back and forth.

Since there is nothing in Eq. 7.1 that restricts the values of k, the energies of perfectly free particle-waves are *not* quantized. In other words, the kinetic energy of a free particle-wave can take on any real positive value, just like a classical particle. However, a free particle-wave is associated with an eigenfunction of wavelength $\lambda = 2\pi/k = h/p_x$, which gives rise to highly nonclassical behavior, such as the diffraction of electrons and atoms by solid crystals (as previously described in Section 1.3).

Particle in a Box

The binding of electrons to atoms or molecules is induced by the attractive (coulombic) potential energy between negative electrons and positive nuclei. So, we expect the Hamiltonian associated with a bound electron to have the form $H = K + V$. The simplest example of such a Hamiltonian is one in which V is a one-dimensional box extending from $x = 0$ to $x = L$, such that $V(x)$ is equal to zero inside the box and infinity outside the box.

$$V(x) = \begin{cases} 0 & \text{if } 0 < x < L \\ \infty & \text{if } x \leq 0 \text{ or } x \geq L \end{cases} \tag{7.3}$$

The Schrödinger equation pertaining to such a particle-in-a-box system may be expressed as follows.

$$\begin{aligned} \widehat{H}\Psi &= \left[\widehat{K}_x + V(x)\right]\Psi \\ &= -\frac{\hbar^2}{2m}\frac{\partial\Psi^2}{\partial x^2} = E\Psi \quad \text{inside the box} \end{aligned} \tag{7.4}$$

In other words, within the box the Hamiltonian looks exactly like that of a free particle (Eq. 7.1). Since the potential is infinite outside the box, we expect the particle's probability density (and wavefunction) to be entirely confined to the region within the box. Moreover, since Ψ cannot be discontinuous at the edge of the box, it must be that $\Psi(0) = \Psi(L) = 0$. This constraint on Ψ is an example of the role of *boundary conditions* in the solution of differential equations. These play exactly the same role in this problem as they did in the mathematical description of the normal modes of a vibrating guitar string (see Fig. 6.1 and the associated discussion). In fact, Eqs. 7.4 and 6.11 are essentially identical (except for a multiplicative constant). Thus, we expect the eigenfunctions of the Schrödinger equation for a particle in a box to look identical to those of the guitar string.

Confinement of a particle-wave within a box of length L thus has the effect of restricting the allowed values of k to $k_n = n\pi/L$ (where n is a positive whole number). More specifically, the normalized eigenfunctions of a particle confined to a box that extends from $x = 0$ to $x = L$ may be expressed as follows.

$$\Psi_n = \sqrt{\frac{2}{L}}\sin(k_n x) = \sqrt{\frac{2}{L}}\sin\left[\left(\frac{n\pi}{L}\right)x\right] \quad \text{where } n = 1, 2, 3\ldots \tag{7.5}$$

Notice that this implies the energies of a particle in a box are quantized, as the eigenvalues associated with these eigenfunctions are $E_n = (\hbar k_n)^2/2m$, which may be expressed as follows (using the fact that $\hbar = h/2\pi$ and $k_n = n\pi/L$).

$$E_n = \frac{n^2 h^2}{8mL^2} \tag{7.6}$$

Note that Eq. 7.6 is identical to Eq. 1.12, which was obtained using Bohr's semiclassical treatment of a particle of mass m bouncing back and forth in a box of length L (whose energy spectrum was restricted by requiring that photons behave like particles of energy $\varepsilon = h\nu$).

The mathematical form of the wavefunctions of a particle in a box depends on exactly where the ends of the box are located. If the box begins at $x = 0$ and ends at $x = L$, then the eigenfuctions are described by Eq. 7.5. However, if we assume that the center of the box is located at $x = 0$, so its ends are located at $x = -L/2$ and $x = L/2$, then we would expect its ground state wavefunction Ψ_1 to be proportional to $\cos(k_1 x) = \cos(\pi x/L)$. Although $\cos(k_1 x)$ and $\sin(k_1 x)$ are different functions, the way that Ψ_1 looks does not depend on the location of the center of the box, and the same is true for all of the higher-order eigenfunctions Ψ_n, as illustrated in Figure 7.1.[3] Moreover, the energies E_n and wavelengths $\lambda_n = 2\pi/k_n$ of the particle-wave eigenfunctions do not depend on where the center of the box is located.

Recall that neither $\sin(kx)$ nor $\cos(kx)$ are eigenfunctions of $\widehat{p}_x$, although they are both eigenfunctions of $\widehat{p_x^2}$. Thus, although a particle in a box has nonzero values of $\langle p_x^2\rangle = 2mE_n = (\hbar k_n)^2$ and $\sqrt{\langle p_x^2\rangle} = \sqrt{2mE_n} = \hbar k_n$, a measurement of its momentum is equally likely to yield either a positive or a negative result, $\pm\hbar k_n$, and so $\langle p_x\rangle = 0$.

Although a particle in a box is a simple (idealized) quantum mechanical system, its solution illustrates several important principles, and also has practical applications. For example, the above results may be used to predict the influence of a particle's mass, as well as the size of the region

---

[3] The eigenfunctions of a box that is centered at $x = 0$ alternate between sine and cosine functions as n increases, but their shapes look identical to the functions described by Eq. 7.5, and shown in Figure 7.1.

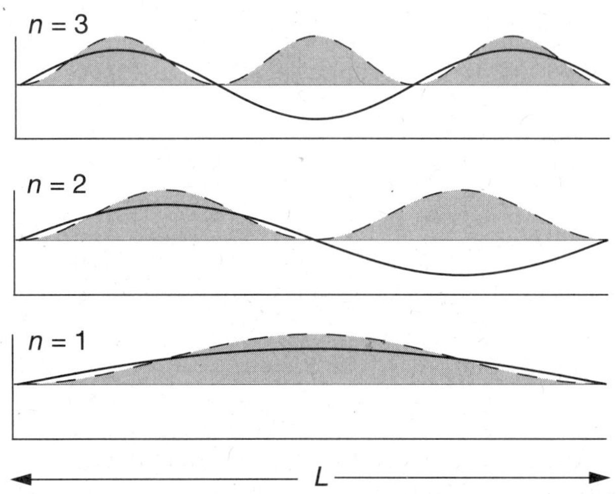

FIGURE 7.1 The first three eigenfunctions Ψ_n of a particle in a box (solid curves), and the corresponding probability densities Ψ_n^2 (dashed curves).

within which it is confined, on the energy of the resulting quantum states. More specifically, Eq. 7.6 may be used to explain the orders of magnitude of all the quantum state spacings listed in Figure 1.2 on page 20, as well as to explain and predict molecular electronic absorption spectra, as we will see.

Particle in a Ring

We might expect the properties of a particle confined to a box of length L to be closely related to those of a particle that is confined to a circular ring of circumference L (such as an object that is orbiting around a central point). However, the boundary condition pertaining to a circular ring is not the same as that associated with a linear box. More specifically, while the particle-in-a-box boundary condition requires that Ψ must be equal to zero at the ends of the box, the *cyclic boundary condition* of a particle in a ring does not require that the particle must be equal to zero anywhere in the ring, but rather that Ψ is a continuous and single-valued function of the particle's angular position, and thus satisfies the condition that $\Psi(\phi) = \Psi(\phi + 2\pi)$ (as further explained below).

The circumference of a ring of radius r is $2\pi r$, and the distance a particle travels when it sweeps through an angle ϕ is ϕr (see Fig. 7.2). We may use these relations to equate $L = 2\pi r$ and $x = \phi r$, and thus reexpress the particle-in-a-box wavefunctions as $\Psi_n \propto \sin[(n\pi/L)x] = \sin[(n/2)\phi]$. However, these wavefunctions are not entirely satisfactory for several reasons. On one hand, they correspond to particles that are rotating both clockwise and countclockwise at the same time, and on the other hand, they are arbitrarily pinned to zero at $\Psi_n(0) = \Psi_n(L) = 0$.

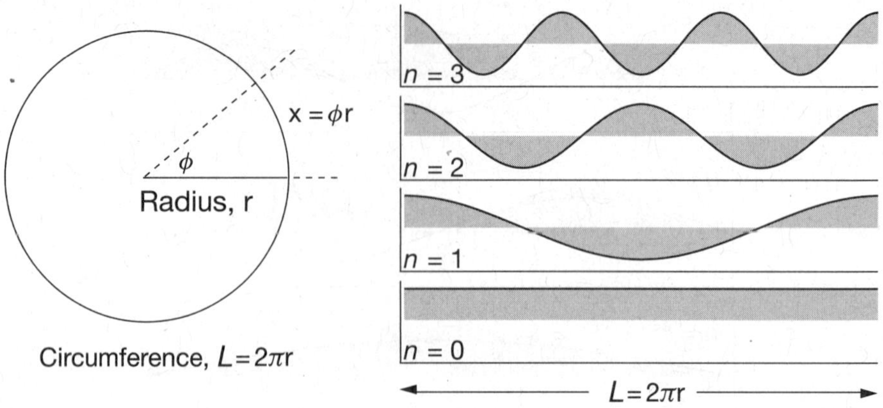

FIGURE 7.2 A particle-wave confined to a ring of circumference L is similar to one that is confined to a box of length L, except that the cyclic boundary conditions now require that Ψ must be continuous (and single valued). The curves on the left illustrate the real portion of the first four eigenfunctions of a particle in a ring, $\Psi_n = \sqrt{1/2\pi}\, e^{in\phi} = \sqrt{1/2\pi}\,[\cos(n\phi) + i\sin(n\phi)]$.

In order to obtain a more acceptable expression for the particle-in-a-ring eigenfunctions, we may begin by assuming that they have a general oscillating (complex exponential) form $\Psi_\phi \propto e^{ib\phi}$ (where the value of b is yet to be determined). The cyclic boundary condition constraint requires that $e^{ib\phi} = e^{ib(\phi+2\pi)} = e^{ib\phi}e^{ib2\pi}$, and so Eq. 6.7 implies that $1 = (e^{i\pi})^{2b} = (-1)^{2b}$. Thus, the particle-in-a-ring boundary conditions will be satisfied whenever b is an integer, as that will assure that $(-1)^{2b} = 1$. This requires that the normalized eigenfunctions of a particle in a ring must have the following form.

$$\boxed{\Psi_n(\phi) = \sqrt{\frac{1}{2\pi}}\, e^{in\phi} \quad \text{where } n = 0, \pm 1, \pm 2, \pm 3 \ldots} \tag{7.7}$$

Since these wavefunctions are complex exponentials, they are difficult to picture. However, it is easy enough to plot the real portion of the wavefunctions $Re[\Psi_n] = \sqrt{1/2\pi}\cos(n\phi)$, as shown in Figure 7.2. Note that the probability densities of all the particle-in-a-ring eigenstates are constant (independent of ϕ), since $\Psi_n^*\Psi_n = 1/2\pi$. Thus, the probability of detecting the particle is the same at all angles ϕ. This makes sense since there is no reason to expect the particle to have a preferred angular position.

The Hamiltonian operator for a particle in a ring is $\widehat{H} = \widehat{K}$, just as it was for a particle in a box. By identifying $x = \phi r$, we may transform the variable x to ϕ, and thus obtain the following expression for the kinetic energy operator expressed in angular units.

$$\boxed{\widehat{K}_\phi = -\frac{\hbar^2}{2mr^2}\frac{\partial^2}{\partial\phi^2} = -\frac{\hbar^2}{2I}\frac{\partial^2}{\partial\phi^2}} \tag{7.8}$$

The parameter $I = mr^2$ is the *moment of inertia* of a mass m rotating around a radius r. Note that the kinetic energy of a classical rotating particle is $K = p^2/2m = L^2/2I$, where L is the particle's *angular momentum*.[4] This suggests that we may express the quantum mechanical kinetic energy operator associated with circular motion as $\widehat{K} = \widehat{L^2_\phi}/2I$, by defining the angular momentum squared operator as follows.

$$\boxed{\widehat{L^2_\phi} = -\hbar^2 \frac{\partial^2}{\partial \phi^2}} \tag{7.9}$$

We may also factor the angular momentum square operator $\widehat{L^2_\phi} = \widehat{L}_\phi \widehat{L}_\phi$, to obtain the following angular momentum operator.[5]

$$\boxed{\widehat{L}_\phi = -i\hbar \frac{\partial}{\partial \phi}} \tag{7.10}$$

The Schrödiger equation for a particle in a ring is obtained by equating $\widehat{H} = \widehat{K}_\phi$.

$$\widehat{H}\Psi = -\frac{\hbar^2}{2I}\frac{\partial^2 \Psi}{\partial \phi^2} = E\Psi \tag{7.11}$$

---

[4] Angular momentum is defined classically as $\vec{L} = \vec{r} \times \vec{p} = |r||p| \sin\phi \, \vec{n}$, where $\vec{r}$ and $\vec{p}$ are the position and momentum vectors and $\times$ is the vector cross-product (and $\vec{n}$ is a unit vector perpendicular to both $\vec{r}$ and $\vec{p}$, as given by the *right-hand rule*, as further described in footnote 5). The kinetic energy of a rotating particle may be expressed as $K = \frac{1}{2}mv^2 = \frac{1}{2}mr^2\omega^2 = \frac{1}{2}I\omega^2$, where $\omega = d\phi/dt$ is the particle's angular velocity (or angular frequency). In other words, L, I, and ω are analogous to p, m, and v, respectively, in that the dynamical variables associated with circular motion look the same as those for linear motion if we replace p by L, m by I, and v by ω. Thus, for example, the angular momentum may be expressed as $L = I\omega$, which is analogous to the linear momentum $p = mv$.

[5] The negative sign in the expression for $\widehat{L}_\phi$ indicates a particle that is rotating counterclockwise is defined as having a positive angular momentum (while one moving clockwise has a negative angular momentum). This is also consistent with the *right-hand rule*, stating that a particle rotating in the same direction as the fingers of the right hand has an angular momentum vector that points along the thumb.

The wavefuctions described by Eq. 7.7 are eigenfunctions of this Hamiltonian, and produce the following energy eigenvalues.[6]

$$\boxed{E_n = \frac{n^2\hbar^2}{2I} = \frac{n^2\hbar^2}{2mr^2}} \tag{7.12}$$

Exercise 7.1

Confirm that the energies given in Eq. 7.12 are the energy eigenvalues of the corresponding wavefunctions.

Solution. In order to determine the energy eigenvalues of the wavefunctions in Eq. 7.7, we must operate on these wavefunctions using the Hamiltonian in Eq. 7.11. Since $\frac{\partial}{\partial\phi}[e^{in\phi}] = ine^{in\phi}$, the first and second derivatives of Ψ_n are $\frac{\partial}{\partial\phi}[\Psi_n] = in\Psi_n$ and $\frac{\partial^2}{\partial\phi^2}[\Psi_n] = (in)^2\Psi_n = -n^2\Psi_n$, respectively. The latter second derivative may now be used to evaluate Eq. 7.11, and thus confirm that the energies in Eq. 7.12 are the same as the corresponding energy eigenvalues.

$$\widehat{H}\Psi_n = \left(-\frac{\hbar^2}{2I}\right)(-n^2)\Psi_n = \frac{n^2\hbar^2}{2I}\Psi_n$$

The fact that n may take on both positive and negative values has an interesting physical significance. Notice that the eigenfuctions of the particle-in-a-ring Hamiltonian are also eigenfunctions of the angular momentum operator (Eq. 7.10) with eigenvalues of $n\hbar$. Thus, the angular momentum expectation values of a rotating particle-wave are simply integral multiples of $\hbar$.

$$\langle L_\phi \rangle = n\hbar \quad \text{where } n = 0, \pm 1, \pm 2, \pm 3 \ldots \tag{7.13}$$

[6] Notice that the energy levels in Eq. 7.12 are identical to those obtained using the particle-in-a-box formula Eq. 7.6, when replacing $L \rightarrow 2\pi r$ and $n \rightarrow 2n$ (as required in order to conform with the cyclic boundary conditions). It is also interesting to note that Eq. 7.12 may be obtained using Bohr's old quantization procedure (described in Section 1.3, pages 18–22). More specifically, this assumes that a photon that could be absorbed by the rotating particle must have a frequency equal to the average of the initial and final rotational frequencies of the particle. The correct particle-in-a-ring energy levels are be obtained by assuming that the particle is initially stationary (has no angular momentum). A sequence of photon absorption steps lead exactly to the energies given by Eq. 7.12.

In other words, positive values of n correspond to a particle that has a positive angular momentum (and so is rotating counterclockwise), and negative values of n correspond to a particle rotating in the other direction (while $n = 0$ corresponds to a particle with no angular momentum). This suggests that angular momentum is quantized, in the sense that it comes in packets of size $\hbar$, which is reminiscent of the quantization of photons in packets of energy $h\nu$.

The above particle-in-a-ring results may be used to approximate the quantization of electrons within atoms, as well as the quantization molecular rotational motions. The primary difference between the latter two types of quantization is that the first involves the rotation of electrons of mass $m = m_e$, while the latter involves the rotation of molecules with a much larger mass $m >> 1000\,m_e$. Thus, Eq. 7.12 implies that electronic rotational quantization will have much larger energy spacings than molecular rotational quantization. More specifically, the rotational states of electrons within atoms typically have energy spacings equivalent to visible or ultraviolet photons, while molecular rotational quantum transitions occur in the microwave spectral region (see Fig. 1.2 on page 20). Although accurately describing atomic orbitals and molecular rotational quantization requires extending the above results to three dimensions (as described in Section 7.5 and Chapter 8), we may immediately use what we already know to obtain reasonable estimates of the molecular electronic spectra, as described below.

Applications to Electronic Spectroscopy

The particle-in-a-box model may be used to estimate the optical absorption and emission wavelengths of conjugated polyene chains, such as 1,3-butadiene and 1,3,5-hexatriene. In order to apply a one electron model to such multi-electron systems, we must take into account the fact that no more than two electrons may occupy a given molecular orbital (eigenstate).[7] Thus, if a conjugated polyene chain has a total of N carbons, each of which contributes one electron to the molecule's π orbital, then the lowest energy $N/2$ eigenstates will each be filled with a pair of electrons when the molecule is in its ground electronic state. The molecule's lowest energy optical absorption will thus promote one electron from the highest occupied molecular orbital (HOMO) to the lowest unoccupied molecular

[7] The reason for this restriction is related to the Pauli exclusion principle, which states that no two Fermi particles (such as electrons) can occupy the same quantum state, as further explained in Section 8.2. Since electrons have two possible spin states (as we will later see), each electronic orbital may be occupied by up to two (spin-paired) electrons.

orbital (LUMO). The energy difference between the LUMO and HOMO state is the corresponding photon energy, as illustrated in the following exercise.

Exercise 7.2

Consider a conjugated polyene chain that has a total length L and contains an even number N of carbon atoms, each of which contributes one electron to its conjugated π electronic structure.

- Use the particle-in-a-box model to determine the quantum number $n = n_H$ of the highest occupied molecular orbital (HOMO) in the ground state of such a molecule.

  Solution. Since no more than two electrons can occupy each electronic quantum state, the HOMO orbital must have a quantum number of $n_H = \frac{N}{2}$.

- What is the quantum number $n = n_L$ of lowest unoccupied molecular orbital (LUMO) of this system?

  Solution. The LUMO orbital is the next quantum state above the LUMO orbital, so $n_L = n_H + 1 = \frac{N}{2} + 1$.

- Obtain expressions for the energy ε, frequency ν, and wavelength λ of a photon whose energy is matched to the difference between the LUMO and HOMO orbital energies.

  Solution. The energy of such a photon may be obtained using Eq. 7.6, by subtracting the LUMO and HOMO energies.

$$\varepsilon = E_{n_L} - E_{n_H} = \frac{h^2(n_L^2 - n_H^2)}{8m_eL^2}$$

  Notice that the mass m_e is that of the electron, since that is the particle whose energy is quantized as a result of its confinement within the polyene chain. Since the energy of a photon is $h\nu$, a photon that could induce an optical absorption from the HOMO to the LUMO orbital is predicted to have a frequency of $\nu = \varepsilon/h$ and a wavelength of $\lambda = c/\nu = hc/\varepsilon$, where c is the speed of light.

The same strategy may be applied to the particle-in-a-ring model in order to predict the optical absorption energy and wavelength of an aromatic molecule such as benzene, whose six carbon atoms contribute six electrons to its aromatic π orbitals.

Exercise 7.3

Use the particle-in-a-ring model to predict the HOMO to LUMO optical absorption energy of benzene, which has six π electrons confined to a ring with a radius approximately equal to one C-C bond length $r \approx 1.4$ Å.

Solution. The HOMO state will be one in which the first three quantum states are each filled by pairs of electrons. The lowest energy $n = 0$ state will have two electrons, and the remaining four electrons will go into the two (degenerate) states with $n = n_H = \pm 1$. The next higher-energy LUMO orbitals have quantum numbers $n = n_L = \pm 2$. Thus, Eq. 7.12 implies that benzene will have the optical absorption energy.

$$E_{n_L} - E_{n_H} = (n_L^2 - n_H^2)\frac{\hbar^2}{2I} = \frac{3\hbar^2}{2m_e r^2}$$

When the experimental bond length of benzene $r \approx 1.4 \times 10^{-10}$ m and the mass of one electron $m_e \approx 9.1 \times 10^{-31}$ kg (and $\hbar \approx 1.05 \times 10^{-34}$ J/s) are inserted into the above expression, the optical transition energy is predicted to be about 9.3×10^{-19} J or 5.8 eV, which is remarkably close to benzene's experimental optical transition energy of 6.2 eV.

A more detailed analysis of molecular optical absorption and emission spectra is required in order to understand why some states absorb more strongly than others, and why molecular fluorescence emission maxima typically have longer wavelengths than the corresponding absorption maximum, as well as why fluorescence and absorption spectra often look like mirror images of each other. These and other related topics are further discussed in Chapter 9 (particularly Section 9.4).

Tunneling Through Barriers and Its Applications

Another interesting and important difference between classical and quantum particles is that particle-waves can tunnel through walls. In other words, although a classical particle can only overcome a potential energy barrier (wall) if its kinetic energy exceeds the potential energy at the top of the wall, a quantum particle-wave may pass through a wall even when its total energy is less than the height of the wall.

Consider a one-dimensional potential energy barrier of height V and width b (where both V and b are real positive constants, as described by Eq. 7.14 and illustrated in Fig. 7.3).

$$V(x) = \begin{cases} 0 & \text{if } x \leq 0 \\ V & \text{if } 0 < x < b \\ 0 & \text{if } x \geq b \end{cases} \tag{7.14}$$

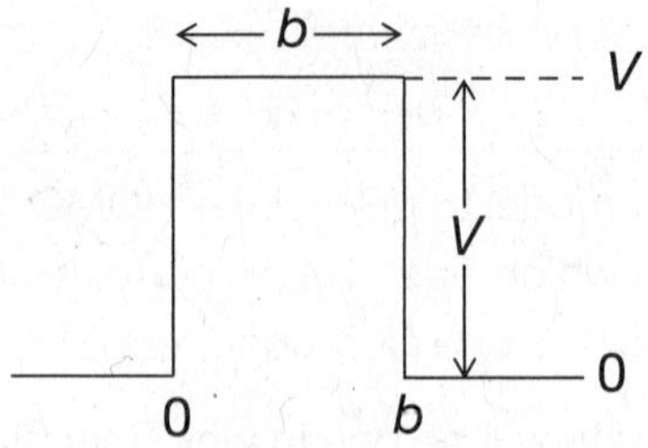

FIGURE 7.3 A one-dimensional potential energy barrier of height V and width b.

If a particle approaches the barrier from the left ($x < 0$) with an energy of $E = (\hbar k)^2/2m$ that is less than the barrier height, then $\epsilon \equiv E/V < 1$. Under these conditions a classical particle would simply bounce off of the barrier, and would never be found either within the barrier (at $0 < x < b$) or on the far side of the barrier (at $x > b$). However, we will see that a quantum mechanical particle-wave can tunnel through a barrier, as it has a finite probability of emerging on the other side.

On the left and right sides of the barrier $V(x) = 0$, so in those regions the Hamiltonian is the same as that of a free particle $\widehat{H} = \widehat{K}_x$. However, inside the barrier the potential energy is equal to V and so $\widehat{H} = \widehat{K}_x + V$. Thus, the Schrödinger equation inside the barrier is $\widehat{H}\Psi = \widehat{K}_x\Psi + V\Psi = -(\hbar^2/2m)(\partial^2\Psi/\partial x^2) + V\Psi = E\Psi$, which may be rearranged to the following form.

$$\frac{\partial^2\Psi}{\partial x^2} = \frac{2m}{\hbar^2}(V - E)\Psi = \kappa^2\Psi \tag{7.15}$$

The coefficient $\kappa^2 = (2m/\hbar^2)(V - E) = (2m/\hbar^2)V(1 - \epsilon)$ is a positive constant. If it were negative, then the solution to the Schrödinger equation would be an oscillating function such as $e^{\pm i\kappa x}$. However, since the constant is positive, the solution must have a non-oscillating exponential form $Ce^{\pm\kappa x}$, where $\kappa = \sqrt{2m(V - E)}/\hbar$ is a real number (and C is a constant). As the barrier becomes very tall or very wide, this wavefunction must have a negative exponent $Ce^{-\kappa x}$, because the probability that the particle will penetrate through the wall must decrease as the height or width of the wall increases.[8] Such a decaying exponential wavefunction is also referred to as a *tunneling tail*.

The complete solution of the barrier penetration problem includes the wavefunction of the incoming particles Ae^{ikx} and the particles that emerge on the other side of the wall Be^{ikx} (as well as particles that are back-reflected by the barrier $\propto e^{-ikx}$). The relationships between the amplitudes of these components may be determined by requiring that both the value

[8] Recall that the particle-in-a-box results indicate that when the wall is infinitely high, the particle would not penetrate into the wall at all.

and slope of the complete wavefunction remain continuous at $x = 0$ and $x = b$. Doing so leads to the following expression for the barrier transmission coefficient τ, which represents the probability that the incoming particle will penetrate through the barrier.

$$\tau = \frac{|B^2|}{|A^2|} = \left\{1 + \frac{\sinh^2(\kappa b)}{4\epsilon(1-\epsilon)}\right\}^{-1} \tag{7.16}$$

The function sinh(x) is defined as $\frac{1}{2}(e^x - e^{-x})$ (see Appendix B for further details). Notice that if the barrier is sufficiently high and wide, so that $\kappa b >> 1$, then the above expression reduces to the following simpler form.

$$\tau \approx 16\epsilon(1-\epsilon)e^{-2\kappa b} \tag{7.17}$$

In other words, the transmission coefficient decreases exponentially as the width of the barrier increases.

A scanning tunneling microscope (STM) is an instrument that may be used to visualize individual atoms. The key element of an STM is an atomically sharp metal tip. As this tip approaches a conductive solid surface, the electrons in the tip may tunnel into the surface. Since the resulting tunneling current depends very sensitively on the distance between the tip and the surface, an STM is capable of resolving atomic features. More specifically, an STM is often operated in *constant-current mode*, in which a piezoelectric crystal is used to move the STM tip up and down (with sub-nm resolution) so as to maintain a fixed tunneling current as the tip is scanned over a surface, thus producing an atomic contour map.

Tunneling also plays an interesting role in the description of particle-waves that are confined to a box whose walls have a finite (rather than infinite) height V. The quantized eigenfunctions of particles whose energies are less than V look similar to those described by Eq. 7.5, except that they have exponential tunneling tails that penetrate into the walls of the box. The quantum states of such particles are similar to those of electrons trapped in semiconductor quantum well devices, which are used to produce solid-state diode lasers and other devices.[9]

The radioactive decay of atoms can also be described as a quantum mechanical tunneling process. Marie Curie coined the term *radioactivity*

---

[9] Semiconductor quantum wells may be produced by coating a surface with alternating layers of semiconductors with different bandgaps (such as a GaAs quantum well embedded in AlGaAs). The thickness of such quantum wells are typically of the order of 2 nm to 20 nm, and so have visible or near-infrared optical transitions. The energy levels of electrons in such systems are similar to those given by Eq. 7.6, but not exactly the same, because the latter equation pertains to a single particle confined in a box with walls of infinite height.

and performed brilliantly designed experimental studies of this phenomena, which led her to discover numerous new elements, including polonium (Po), named after her native country of Poland. Marie Curie was the first person ever to earn two Nobel prizes, one in Physics (in 1903) and one in Chemistry (in 1911).[10] The finite probability that an element such as polonium ($^{210}_{84}Po$) will emit an alpha particle ($^{4}_{2}He^{2+}$, a helium nucleus with two protons and two neutrons) may be viewed as a tunneling process.[11] More specifically, an alpha particle that is initially bound within a polonium nucleus behaves like a particle confined within a very small box, and thus has a ground state with tremendously high kinetic energy. Although this kinetic energy is not high enough to overcome the much higher barrier that binds the alpha particle to the nucleus, there is a finite probability that the alpha particle will penetrate through the barrier and emerge with its dangerously high kinetic energy ($\sim$5.3 MeV).

7.2 Quantum Harmonic Oscillators

A wide variety of phenomena, including light and molecular vibrations, may be represented as harmonic oscillators. The quantization of electromagnetic waves gives rise to photons, while that of molecular vibrations gives rise to the nonclassical heat capacities of molecules and solids (as discussed in Section 1.4).

Recall that a harmonic oscillator is defined as having a quadratic potential energy function, $V(x) = \frac{1}{2}fx^2 = \frac{1}{2}m\omega^2x^2$ (where $\omega = 2\pi\nu = \sqrt{f/m}$).[12] The Hamiltonian of such a system $\widehat{H} = \widehat{K} + \widehat{V}$ is obtained by equating $\widehat{V}$ with the

---

[10] Marie Curie was born in Warsaw but studied at the Sorbonne in Paris, where she remained for most of her illustrious research career. Her early studies of the radioactivity of uranium (U) and thorium (Th) led her, along with her husband Pierre Curie, to discover polonium (Po) and radium (Ra) in 1898. Her first Nobel prize was awarded jointly with Pierre and Henri Becquerel (who, in 1896, first observed what Marie later came to understand as the radioactive decay of uranium). Pierre Curie was killed in 1906, in a fluke accident, when he was run over by a horse-drawn carriage as he crossed Rue Dauphine in Paris, at night in the pouring rain. Marie's second Nobel prize was awarded to her alone "in recognition of her services to the advancement of chemistry by the discovery of the elements radium and polonium, by the isolation of radium and the study of the nature and compounds of this remarkable element."

[11] The explanation of radioactive decay as a quantum mechanical tunneling process was first proposed in 1928 by George Gamow, a Ukrainian physicist.

[12] The relation between f, m, ω, and ν follows from Eqs. 6.3 and 6.4. Note that we may also express the potential energy as $V(x) = \frac{1}{2}I\omega^2$ where $I = mx^2$ is the moment of inertia of the oscillator.

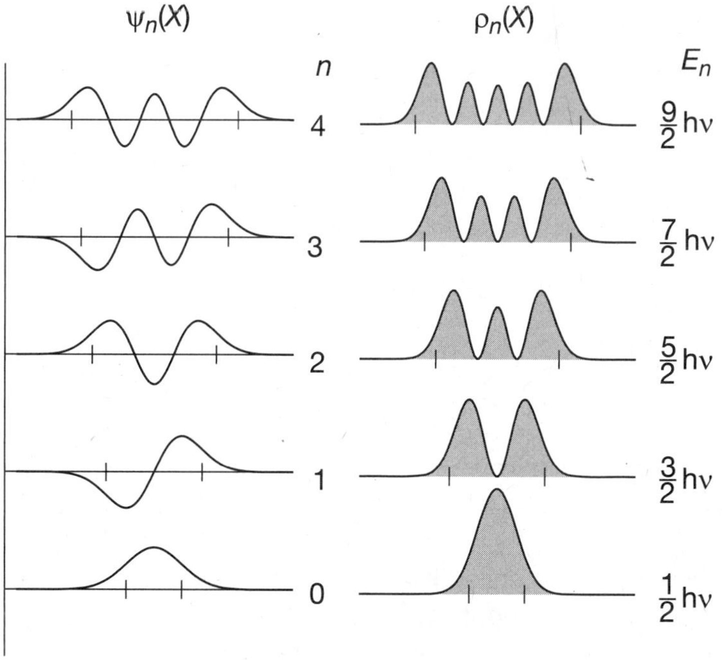

FIGURE 7.4 Wavefunctions, probability densities, and energies of the first five eigenstates of the harmonic oscillator Hamiltonian. The tick marks on the x-axis indicate the corresponding classical turning points (at which $E_n = \frac{1}{2}fx_{max}^2$).

above harmonic potential $V(x)$, and thus lead to the following Schrödinger equation.

$$\widehat{H}\Psi = \left[\frac{1}{2m}\widehat{p_x^2} + \frac{f}{2}\widehat{x^2}\right]\Psi = \frac{1}{2m}\left[(mw)^2 x^2 - \hbar^2\frac{\partial^2}{\partial x^2}\right]\Psi = E\Psi \qquad (7.18)$$

The solutions of the harmonic oscillator Schrödinger equation look quite similar to those of the particle in a box (see Fig. 7.4). We may view a harmonic oscillator as a particle in a "soft" box with quadratic (rather than infinitely steep) walls. This analogy between a harmonic oscillator and a particle in a box also implies that a harmonic oscillator should have an evenly spaced ladder of energies, as may be inferred from the following simple argument.

The quadratic shape of a harmonic potential is such that it behaves like a box whose width increases with increasing energy. More specifically, the total energy of a classical harmonic oscillator is equal to the value of its potential energy when it has reached its maximum extension, $E = \frac{1}{2}fx_{max}^2$. Thus, the box length associated with a harmonic oscillator extends approximately from $-x_{max}$ to $+x_{max}$, and so $L \approx 2x_{max} \propto \sqrt{E}$. Replacing L by $\sqrt{E}$ in Eq. 7.6 implies that for a harmonic oscillator $E \propto n^2/L^2 = n^2/E$, or $E^2 \propto n^2$, which implies that $E \propto n$. Thus, we expect E to be proportional to n for a harmonic oscillator (rather than to n^2 as it is for a particle in a box with infinitely steep walls).

The eigenfunctions of a harmonic oscillator are described by the following mathematical expressions (and are plotted in Fig. 7.4).

$$\boxed{\Psi_n = N_n\, H_n(y)\, e^{-y^2/2} \quad \text{where } n = 0, 1, 2\ldots} \tag{7.19}$$

$$y = x/\sigma \qquad N_n = \sqrt{\frac{1}{2^n\, n!\, \sqrt{\pi}\,\sigma}} \qquad H_0(y) = 1$$

$$\sigma = \sqrt{\frac{\hbar}{m\omega}} \qquad H_n(y) = (-1)^n e^{y^2} \frac{d^n}{dy^n}\left[e^{-y^2}\right] \qquad H_1(y) = 2y \qquad H_2(y) = 4y^2 - 2$$

The functions $H_n(y)$ are called Hermite polynomials, the first few of which are shown above on the right, while others can be derived using the above general expression for $H_n(y)$, or by making use of the recursion relation $H_{n+1}(y) = 2yH_n(y) - dH_n(y)/dy$.

Exercise 7.4

Use Eq. 7.19 to determine the functional form of the first four eigenfunctions of a harmonic oscillator.

Solution. By inserting $n = 0$ into the expression for N_n we obtain

$$N_0 = \sqrt{\frac{1}{\sqrt{\pi}\sigma}} = \left(\frac{m\omega}{\pi\hbar}\right)^{\frac{1}{4}}$$

The subsequent values of N_n may be obtained by noting that Eq. 7.19 implies that $N_n = N_{n-1}/\sqrt{2n} = N_0/\sqrt{2^n n!}$. The first three Hermite polynomials $H_n(y)$ are given in Eq. 7.19. The fourth one may be obtained either by directly evaluating $(-1)^4 e^{y^2} \frac{d^3}{dy^3}\left[e^{-y^2}\right] = 8y^3 - 12y$ or by making use of the recursion relation, which indicates that $H_3(y) = 2yH_2(y) - \frac{d}{dy}H_2(y) = (8y^3 - 4y) - 8y$. Thus, the first four eigenfunctions of the harmonic oscillator may be expressed as follows.

$$\Psi_0(x) = N_0 e^{-y^2/2} = N_0 e^{-x^2/2\sigma^2}$$

$$\Psi_1(x) = N_1(2y)e^{-y^2/2} = N_0\sqrt{2}\left(\tfrac{x}{\sigma}\right)e^{-x^2/2\sigma^2}$$

$$\Psi_2(x) = N_2(4y^2 - 2)e^{-y^2/2} = N_0\frac{1}{\sqrt{2}}\left[2\left(\tfrac{x}{\sigma}\right)^2 - 1\right]e^{-x^2/2\sigma^2}$$

$$\Psi_3(x) = N_3(8y^3 - 12y)e^{-y^2/2} = N_0\frac{1}{\sqrt{3}}\left[2\left(\tfrac{x}{\sigma}\right)^3 - 3\left(\tfrac{x}{\sigma}\right)\right]e^{-x^2/2\sigma^2}$$

These four eigenfunctions (as well as Ψ_4) are plotted in Figure 7.4.

The eigenvalues of a harmonic oscillator may be obtained by inserting the eigenfunctions Ψ_n into Eq. 7.18.

$$\boxed{E_n = h\nu\left(n+\frac{1}{2}\right) = \hbar\omega\left(n+\frac{1}{2}\right) \quad \text{where } n = 0, 1, 2\ldots} \tag{7.20}$$

Notice that Eq. 7.20 confirms our expectation that a harmonic oscillator has an evenly spaced ladder of energies, and we now see that the lowest energy state is located exactly $\frac{1}{2}h\nu$ above the minimum of the quadratic potential energy function.[13] This implies that the zero-point energy of a harmonic oscillator has a finite vibrational amplitude (and thus nonzero kinetic and potential energy expectation values, as further discussed below). Moreover, since Ψ_0 is a Gaussian function, the uncertainty principle implies that $\sigma_x\sigma_{p_x} = \hbar/2$.[14]

Equations 7.19 and 7.20 may be used to evaluate the expectation values of various observables. Since the functions Ψ_n are eigenfunctions of the Hamiltonian operator, we know that $\langle H\rangle = E_n = \hbar\omega(n+\frac{1}{2})$. The expectation value of the potential energy is $\langle V\rangle = \frac{1}{2}m\omega^2\langle x^2\rangle = \frac{1}{2}\hbar\omega(n+\frac{1}{2})$. This clearly implies that $\langle K\rangle = \langle H\rangle - \langle V\rangle = \frac{1}{2}\hbar\omega(n+\frac{1}{2})$, and so $\langle V\rangle = \langle K\rangle = \frac{1}{2}\langle H\rangle$, which is consistent with the predictions of the virial theorem when applied to a quadratic potential (see Eq. 1.46 on page 41). It is also interesting to note that the lowest energy state of a quantized harmonic oscillator has a zero-point energy of $E_0 = \frac{1}{2}h\nu$, and so has a nonzero average kinetic and potential energy of $\langle K\rangle = \langle V\rangle = \frac{1}{4}h\nu$. Thus, even at zero temperature, a quantized harmonic oscillator cannot remain entirely motionless. The zero-point energies of electromagnetic waves also influence atomic and molecular emission rates, as well as other interesting experimentally measurable phenomena (to which we will return in Section 10.4).

Application to Molecular Vibrational Spectroscopy

One of the most important chemical applications of the harmonic oscillator is the prediction of molecular vibrational quantization and the interpretation of vibrational spectra. For example, if we assume that the bond

---

[13] It is also interesting to note that Eq. 7.20 is exactly the same as that obtained when we applied Bohr's original semiclassical method to a vibrational harmonic oscillator (see Section 1.3, and particularly page 22).

[14] Note that the standard deviation of the position of the harmonic oscillator is not exactly equal to σ in Eq. 7.19 because the probability density of the harmonic oscillator ground state is $[e^{-x^2/2\sigma^2}]^2 = e^{-x^2/\sigma^2} = e^{-x^2/2\sigma_x^2}$, which implies that $\sigma_x = \sigma/\sqrt{2}$.

connecting the two atoms of a diatomic molecule behaves like a harmonic spring, then the associated force constant f may be determined from the measured vibrational frequency of the diatomic using $f = \omega^2 \mu$ (since $\omega = 2\pi\nu = \sqrt{f/\mu}$),[15] where μ is the reduced mass of the diatomic whose atomic masses are m_1 and m_2.

$$\mu = \frac{m_1 m_2}{m_1 + m_2} \tag{7.21}$$

In other words, a vibrating diatomic is mechanically equivalent to a single particle of mass μ hanging from a spring of force constant f, whose displacement x is equivalent to $R - R_e$ where R is the diatomic's internuclear distance and R_e is its equilibrium (minimum energy) bond length.

Thus, a homonuclear diatomic such as H_2 behaves like a harmonic oscillator whose mass is half that of a single hydrogen atom $\mu_{H_2} = m_H^2/2m_H = m_H/2$. On the other hand, a diatomic such as HI has a reduced mass that is very similar to that of a single hydrogen atom, because iodine is so much heavier than hydrogen that $\mu_{HI} = m_H m_I/(m_H + m_I) \approx m_H(m_I/m_I) = m_H$. This is consistent with our expectation that the heavy iodine atom would remain essentially stationary as the light hydrogen atom vibrates.

Although vibrational quantization may often be accurately approximated using the harmonic oscillator model, molecular vibrations are never perfectly harmonic, as illustrated in Figure 7.5. An anharmonic potential energy function is required to more accurately describe high energy vibrational quantum states, particularly as the vibrational energy approaches the bond dissociation energy D_e (defined as the energy difference between the minimum of the potential energy curve and the energy of the two dissociated atoms).

A Morse potential of the following mathematical form may be used to more realistically predict the anharmonic vibrational states of a diatomic molecule (see Fig. 7.5).

$$V(R) = D_e[1 - e^{-a(R-R_e)}]^2 \tag{7.22}$$

The three parameters of the Morse potential are R_e, D_e, and $a = \sqrt{f/2D_e}$, which is related to the corresponding harmonic vibrational frequency $\omega_0 = 2\pi\nu_0 = \sqrt{f/\mu} = a\sqrt{2D_e/\mu}$. The vibrational energy eigenvalues of a Morse oscillator may be expressed as follows.

$$E_v = h\nu_0\left(v + \frac{1}{2}\right) - \frac{\left[h\nu_0\left(v + \frac{1}{2}\right)\right]^2}{4D_e} \tag{7.23}$$

[15] Recall that molecular vibrational frequencies are often reported in wavenumber units $\tilde{\nu} = \nu/c$ (where c is the speed of light). In other words, $\tilde{\nu} = 1/\lambda$, where λ is the wavelength of a photon whose frequency is resonant with the molecular vibration.

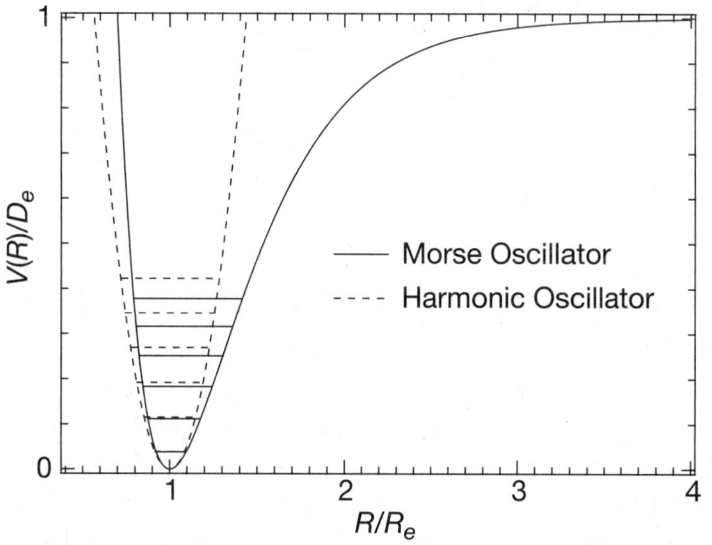

FIGURE 7.5 Comparison of the potential energies and first six vibrational quantum states of harmonic and anharmonic (Morse) oscillators representing the vibration of HCl. The Morse potential for this system is characterized by $a \approx 2.3/R_e$ and $h\nu_0 \approx D_e/13$ (obtained using $R_e \approx 1.27$Å, $D_e \approx 450$ kJ/mol, $f \approx 480$ N/m, and $\tilde{\nu}_0 \approx 2900$ cm^{-1}).

Note that the vibrational quantum number is designated by the symbol v (rather than n) as this is the usual convention in vibrational spectroscopy. Thus, the first term on the right-hand side of Eq. 7.23 is identical to the harmonic oscillator energy spectrum (Eq. 7.20). The additional negative term on the right-hand side implies that the energies of an anharmonic Morse oscillator are no longer evenly spaced, but become more closely spaced as v increases, $E_{v+1} - E_v = h\nu_0 - (v+1)(h\nu_0)^2/2D_e$. Equation 7.23 may be used to predict the quantum state energies of a Morse oscillator up to a value of v equal to $v_{\max} = \frac{2D_e}{h\nu_0} - 1$, as that is the last bound vibrational quantum state whose energy is less than D_e.

The vibrational quantum state energies shown in Figure 7.5 depend not only on the shape of the potential energy function $V(R)$, but also on the reduced mass μ of the vibrating atoms. So, the zero-point energy and quantum state spacings of DCl are predicted to be about $1/\sqrt{2}$ smaller than those of HCl, although the two diatomics have the same $V(R)$. As the vibrational quantum state spacing decreases, so does the influence of anharmonicity. This is why hydrogen vibrations tend to be more strongly influenced by anharmonicity than vibrations involving atoms with a larger reduced mass.

At ambient temperatures, most of the vibrational population will be in the ground (*zero-point energy*) vibration quantum state. The peaks appearing in a vibrational spectrum are determined by selection rules that depend on the molecule's symmetry and the nature of its interaction with light. For a harmonic oscillator, the only allowed vibrational transitions are those with $\Delta v = \pm 1$, while for an anharmonic oscillator this selection rule is somewhat relaxed, leading to the appearance of weak vibrational overtone transitions (with $\Delta v > \pm 1$). The theoretical basis for such selection rules and other factors that influence the intensities of vibrational spectra are further discussed in Chapter 9 (particularly Section 9.4).

7.3 Raising and Lowering Operators

Greater insight into the quantum harmonic oscillator may be obtained by introducing two new operators, $\widehat{a}^+$ and $\widehat{a}$. We will see why these new operators are called raising and lowering operators (or creation and annihilation operators), and how they can be used to obtain remarkably simple derivations of both the energies and eigenstates of a harmonic oscillator. The two new operators are defined in the following way, in terms of the position $\widehat{x}$ and momentum $\widehat{p}_x$ operators.

$$\widehat{a} = \frac{m\omega\widehat{x} + i\widehat{p}_x}{\sqrt{2m\hbar\omega}} = (2m\hbar\omega)^{-\frac{1}{2}}\left(m\omega x + \hbar\frac{\partial}{\partial x}\right) \tag{7.24}$$

$$\widehat{a}^+ = \frac{m\omega\widehat{x} - i\widehat{p}_x}{\sqrt{2m\hbar\omega}} = (2m\hbar\omega)^{-\frac{1}{2}}\left(m\omega x - \hbar\frac{\partial}{\partial x}\right) \tag{7.25}$$

Notice that $\widehat{a}$ and $\widehat{a}^+$ are both real (rather than complex) operators, because $i\widehat{p}_x = \hbar\partial/\partial x$.

The harmonic oscillator Schrödinger equation (Eq. 7.18) may be expressed quite nicely in terms of $\widehat{a}^+\widehat{a}$.[16]

$$\widehat{H}\Psi = \hbar\omega\left(\widehat{a}^+\widehat{a} + \frac{1}{2}\right)\Psi = E\Psi \tag{7.26}$$

This expression, combined with Eq. 7.20, clearly implies that $\widehat{a}^+\widehat{a}\Psi_n = n\Psi_n$. Thus, the operator $\widehat{a}^+\widehat{a} \equiv \widehat{n}$ is referred to as the *number operator*, because it returns an eigenvalue of n when it operates on Ψ_n. In other words, $\widehat{n}$ is a dimensionless (scaled and shifted) representation of $\widehat{H}$, and all the eigenfunctions of $\widehat{H}$ are also eigenfunctions of $\widehat{n}$.

$$\widehat{n}\Psi_n = \left(\frac{\widehat{H}}{\hbar\omega} - \frac{1}{2}\right)\Psi_n = \left(\frac{E_n}{\hbar\omega} - \frac{1}{2}\right)\Psi_n = n\Psi_n \tag{7.27}$$

---

[16] The derivation of Eq. 7.26 requires making use of the commutation relation $[\widehat{x}, \widehat{p}_x] = i\hbar$ (Eq. 6.43). More specifically, $\widehat{a}^+\widehat{a} = [(m\omega x)^2 + \widehat{p_x^2}]/(2m\hbar\omega) + \frac{i}{2\hbar}[\widehat{x}, \widehat{p}_x] = \widehat{H}/\hbar\omega - \frac{1}{2}$. The operator $\widehat{a}^+$ is also called the adjoint of $\widehat{a}$, where the adjoint of any operator is defined by the relation $\widehat{A}\Psi = (\widehat{A}^+\Psi)^*$ (and thus a Hermitian operator is equal to its own adjoint).

The operator $\hat{a}^+$ is called the *raising operator* because it has the effect of promoting an eigenfunction up to the next quantum state, as demonstrated by the following sequence of identities (which make use of the fact that $[\hat{a}, \hat{a}^+] = 1$).[17]

$$\hat{n}(\hat{a}^+\Psi_n) = \hat{a}^+\hat{a}\hat{a}^+\Psi_n = \hat{a}^+(\hat{a}^+\hat{a} + 1)\Psi_n = \hat{a}^+(n + 1)\Psi_n = (n + 1)(\hat{a}^+\Psi_n)$$

In other words, the wavefunction $\hat{a}^+\Psi_n$ is an eigenfunction of $\hat{n}$ whose eigenvalue is $n + 1$, which implies that $\hat{a}^+\Psi_n \propto \Psi_{n+1}$. A similar argument may be used to show that $\hat{a}$ is a *lowering operator* such that $\hat{a}\Psi_n \propto \Psi_{n-1}$. The required constants of proportionality may be inferred from the fact that $\hat{n}\Psi_n = \hat{a}^+\hat{a}\Psi_n = n\Psi_n$.[18]

$$\hat{a}\Psi_n = \sqrt{n}\Psi_{n-1} \tag{7.28}$$

and

$$\hat{a}^+\Psi_n = \sqrt{n+1}\Psi_{n+1}\,. \tag{7.29}$$

Exercise 7.5

Confirm that Eqs. 7.28 and 7.29 are consistent with the fact that $\hat{n}\Psi_n = n\Psi_n$.

Solution. We may first replace $\hat{n}$ by $\hat{a}^+\hat{a}$ and then operate on Ψ_n sequentially using Eqs. 7.28 and 7.29, as follows.

$$\hat{n}\Psi_n = \hat{a}^+\hat{a}\Psi_n = \hat{a}^+\sqrt{n}\Psi_{n-1} = \sqrt{n}\hat{a}^+\Psi_{n-1} = \sqrt{n}\sqrt{n}\Psi_n = n\Psi_n$$

The ground-state eigenfunction of a harmonic oscillator may be obtained by making use of the fact that $\hat{n}\Psi_0 = \hat{a}^+\hat{a}\Psi_0 = 0\Psi_0 = 0$, which implies that $\hat{a}\Psi_0 = 0$ and so $[m\omega x + \hbar(\partial/\partial x)]\Psi_0 = 0$. The latter identity may be rearranged to the following form.

$$\frac{d\Psi_0}{\Psi_0} = -\left(\frac{m\omega}{\hbar}\right) x dx$$

This equation can readily be integrated to obtain $\ln \Psi_0 = -(m\omega/\hbar)\frac{1}{2}x^2 = -x^2/2\sigma^2$ (where $\sigma^2 = \hbar/m\omega$), thus demonstrating that the ground-state

---

[17] The commutation relation $[\hat{a}, \hat{a}^+] = 1$ follows immediately from Eqs. 7.24 and 7.25, combined with Eq. 6.43.

[18] A detailed derivation of Eqs. 7.28 and 7.29 may by carried out by identifying the constants of proportionality as a_n and a_n^+, such that $\hat{a}\Psi_n = a_n\Psi_{n-1}$ and $\hat{a}^+\Psi_{n-1} = a_n^+\Psi_n$. Thus, $\hat{a}^+\hat{a}\Psi_n = \hat{a}^+a_n\Psi_{n-1} = a_n\hat{a}^+\Psi_{n-1} = a_n a_n^+\Psi_n = n\Psi_n$ and so $a_n a_n^+ = n$. The latter identity, combined with the fact that $\hat{a}_n^+$ is the adjoint of $\hat{a}$, implies that $a_n = a_n^+ = \sqrt{n}$, as will become more clear in the following subsection (see footnote 23 on page 251).

wavefunction of a harmonic oscillator must have a Gaussian form (where N_0 is the appropriate normalization constant).[19]

$$\Psi_0 = N_0 e^{-x^2/2\sigma^2} \tag{7.30}$$

Exercise 7.6

Combine Eqs. 7.29 and 7.30 to obtain the first excited state eigenfunction Ψ_1 of a harmonic oscillator.

Solution. Equation 7.29 indicates that $\Psi_1 = \widehat{a}^+\Psi_0/\sqrt{1} = \widehat{a}^+\Psi_0$ and so the Ψ_0 given in Eq. 7.30 (combined with Eq. 7.25) can be used to generate Ψ_1.

$$\Psi_1 = \widehat{a}^+\Psi_0 = (2m\hbar\omega)^{-\frac{1}{2}}\left(m\omega x - \hbar\frac{\partial}{\partial x}\right)\Psi_0 = (2m\hbar\omega)^{-\frac{1}{2}}\left(m\omega x + \frac{\hbar x}{\sigma^2}\right)\Psi_0$$

The above expression may be further simplified by using the fact that $\sigma^2 = \hbar/m\omega$

$$\Psi_1 = (2m\hbar\omega)^{-\frac{1}{2}}(2m\omega x)\Psi_0 = N_0\sqrt{2}\left(\tfrac{x}{\sigma}\right)e^{-x^2/2\sigma^2}$$

which is equivalent to the Ψ_1 in Eq. 7.19 (and Exercise 7.4).

The remaining eigenfunctions may be obtained by successively applying the raising operator.

$$\Psi_n = \frac{1}{\sqrt{n}}\widehat{a}^+\Psi_{n-1} = \frac{1}{\sqrt{n!}}(\widehat{a}^+)^n\Psi_0 \tag{7.31}$$

The resulting harmonic oscillator eigenfunctions are all identical to those described by Eq. 7.19.

7.4 Eigenvectors, Brackets, and Matrices

It is often convenient to represent quantum mechanical integrals using the following Dirac bracket notation.

$$\langle i|j\rangle \equiv \int \Psi_i^*\Psi_j d\tau \tag{7.32}$$

$$\langle i|\widehat{A}|j\rangle \equiv \int \Psi_i^*\widehat{A}\Psi_j d\tau \tag{7.33}$$

[19] Note that integration of the above equation includes a constant of integration C, whose value is determined by normalizing the wavefunction $N_0 = e^C = 1/\sigma^{\frac{1}{2}}\pi^{\frac{1}{4}}$, which is consistent with Eq. 7.19.

The quantities $\langle i|$ and $|j\rangle$ are also called *bra* and *ket* vectors, respectively.[20] For example, if Ψ_i and Ψ_j are normalized, orthogonal eigenvectors of $\widehat{A}$ (with eigenvalues a_i and a_j), then $\langle i|j\rangle = 0$ unless $\Psi_i = \Psi_j$, in which case $\langle i|i\rangle = 1$ and $\langle i|\widehat{A}|j\rangle = a_i$.

The integral $\int \Psi_i^* \widehat{A} \Psi_j d\tau$ is also referred to as a *matrix element* A_{ij}.

$$A_{ij} \equiv \langle i|\widehat{A}|j\rangle \tag{7.34}$$

In other words, A_{ij} is equivalent to the ij'th element of a matrix corresponding to the operator $\widehat{A}$. Notice that each diagonal element of the matrix $A_{ii} = \langle i|\widehat{A}|i\rangle$ is equivalent to the expectation value of the operator $\widehat{A}$ associated with the wavefunction Ψ_i. Moreover, if the wavefunctions Ψ_i are eigenfunctions of $\widehat{A}$, then the matrix is necessarily diagonal (so all of its off-diagonal elements are equal to zero). Conversely, finding the eigenfunctions and eigenvalues of an operator is equivalent to the process of diagonalizing the corresponding matrix. Stated in yet another way, every operator can be equated with a matrix, and the wavefunctions $\langle i|$ and $|j\rangle$ are equivalent to horizontal and vertical vectors (one-dimensional matrices), such that the product of the three matrices $\langle i|\widehat{A}|j\rangle$ is equal to the matrix element A_{ij}.

It is easier to see the above connections if we consider a specific example, such as a harmonic oscillator. The diagonalized form of the harmonic oscillator Hamiltonian looks like this

$$\widehat{H} = \begin{pmatrix} \frac{1}{2}h\nu & 0 & 0 & \cdots \\ 0 & (1+\frac{1}{2})h\nu & 0 & \cdots \\ 0 & 0 & (2+\frac{1}{2})h\nu & \cdots \\ \vdots & \vdots & \vdots & \ddots \end{pmatrix} \tag{7.35}$$

while the corresponding $\widehat{n}$ matrix is even simpler.

$$\widehat{n} = \begin{pmatrix} 0 & 0 & 0 & \cdots \\ 0 & 1 & 0 & \cdots \\ 0 & 0 & 2 & \cdots \\ \vdots & \vdots & \vdots & \ddots \end{pmatrix} \tag{7.36}$$

[20] This nomenclature relies on the mathematical isomorphism (one-to-one correspondence) between functions and vectors. For example, a function $f(x)$ with only three values f_1, f_2, and f_3 may be represented as a three-dimensional vector extending from the origin to the point (f_1, f_2, f_3). Similarly, a function with a larger number of values may be represented as a vector in a higher dimensional space.

Notice that the above matrix representations of $\widehat{H}$ and $\widehat{n}$ are consistent with Eq. 7.26.[21]

The bra vectors associated with the harmonic oscillator eigenstates may be represented as follows.

$$\langle 0| = (1 \quad 0 \quad 0 \quad \dots)$$
$$\langle 1| = (0 \quad 1 \quad 0 \quad \dots)$$
$$\langle 2| = (0 \quad 0 \quad 1 \quad \dots) \tag{7.37}$$

The corresponding ket vectors look the same, except that they are vertical rather than horizontal one-dimensional matrices. So, $\widehat{H}\,|0\rangle = E_0\,|0\rangle$ is equivalent to the following matrix expression.[22]

$$\begin{pmatrix} \frac{1}{2}h\nu & 0 & 0 & \dots \\ 0 & (1+\frac{1}{2})h\nu & 0 & \dots \\ 0 & 0 & (2+\frac{1}{2})h\nu & \dots \\ \vdots & \vdots & \vdots & \ddots \end{pmatrix} \begin{pmatrix} 1 \\ 0 \\ 0 \\ \vdots \end{pmatrix} = \begin{pmatrix} \frac{1}{2}h\nu \\ 0 \\ 0 \\ \vdots \end{pmatrix} = \frac{1}{2}h\nu \begin{pmatrix} 1 \\ 0 \\ 0 \\ \vdots \end{pmatrix}$$

We may also multiply both sides of the above equation by $\langle 0|$ on the left to obtain the following matrix expression for $\langle 0|\widehat{H}|0\rangle = H_{00} \equiv E_0$.

$$(1 \quad 0 \quad 0 \quad \dots) \begin{pmatrix} \frac{1}{2}h\nu & 0 & 0 & \dots \\ 0 & (1+\frac{1}{2})h\nu & 0 & \dots \\ 0 & 0 & (2+\frac{1}{2})h\nu & \dots \\ \vdots & \vdots & \vdots & \ddots \end{pmatrix} \begin{pmatrix} 1 \\ 0 \\ 0 \\ \vdots \end{pmatrix} = \frac{1}{2}h\nu$$

The operators $\widehat{a}$ and $\widehat{a}^+$ do not commute with $\widehat{H}$ and $\widehat{n}$, so the raising and lowering operators cannot be diagonal with respect to the above eigenvectors. Since Eqs. 7.28 and 7.29 require that $\widehat{a}\,|n\rangle = \sqrt{n}\,|n-1\rangle$ and $\widehat{a}^+\,|n\rangle = \sqrt{n+1}\,|n+1\rangle$, the operators $\widehat{a}$ and $\widehat{a}^+$ must have the following form.

$$\widehat{a} = \begin{pmatrix} 0 & \sqrt{1} & 0 & 0 & \dots \\ 0 & 0 & \sqrt{2} & 0 & \dots \\ 0 & 0 & 0 & \sqrt{3} & \dots \\ 0 & 0 & 0 & 0 & \dots \\ \vdots & \vdots & \vdots & \vdots & \ddots \end{pmatrix} \tag{7.38}$$

---

[21] More specifically, Eqs. 7.35 and 7.36 are consistent with the fact that $\widehat{H} = \hbar\omega(\widehat{n} + \frac{1}{2}\widehat{1})$, where $\widehat{1}$ is a diagonal unit matrix, all of whose diagonal elements are equal to 1.

[22] See Appendix B regarding the multiplication of matrices.

$$\widehat{a}^+ = \begin{pmatrix} 0 & 0 & 0 & 0 & \dots \\ \sqrt{1} & 0 & 0 & 0 & \dots \\ 0 & \sqrt{2} & 0 & 0 & \dots \\ 0 & 0 & \sqrt{3} & 0 & \dots \\ \vdots & \vdots & \vdots & \vdots & \ddots \end{pmatrix} \tag{7.39}$$

Notice that multiplication of the above matrices confirms that $\widehat{a}^+\widehat{a} = \widehat{n}$, where the $\widehat{n}$ matrix is given by Eq. 7.36. By changing the order of multiplication, we may also confirm that $\widehat{a}\widehat{a}^+ = \widehat{a}^+\widehat{a} + 1$ or $[\widehat{a}, \widehat{a}^+] = 1$.[23]

Raising and lowering operators may be used to describe transitions between states in other types of quantum mechanical systems. For example, when describing light, the operators $\widehat{a}^+$ and $\widehat{a}$ are sometimes referred to as photon creation and annihilation operators, respectively. In Chapter 8 we will see that similar operators may also be used to describe angular momentum or electron spin-flip transitions.

7.5 Three-Dimensional Systems

All of the quantum mechanical systems we have discussed so far have been one-dimensional, in the sense that they only involved a single degree of freedom (either x or ϕ). Extending these results to systems with two or more degrees of freedom requires using the appropriate kinetic and potential energy operators to describe the corresponding Hamiltonian. We will see that many of the results obtained for one-dimensional systems can readily be extended to higher dimensions. On the other hand, we will also see that some new phenomena emerge, such as the connection between the symmetry of a system's Hamiltonian and the degeneracy of the corresponding eigenvalues, as well as the appearance of new quantum numbers and relationships between these quantum numbers.

Particle in an *n*-Dimensional Box

One of the simplest examples of a quantum system with more than one spatial dimension is a particle confined to a two-dimensional box. More specifically, we may characterize such a box by its lengths in the x and y directions, L_x and L_y, respectively. The potential energy is assumed to be equal to zero inside the box and infinity elsewhere. So, inside the box,

---

[23] The symmetry of the matrices corresponding to the $\widehat{a}$ and $\widehat{a}^+$ operators also reflects the fact that $\widehat{a}^+$ is the adjoint of $\widehat{a}$, as the adjoint of a matrix is obtained by transposing the rows and columns and then taking the complex conjugate of each matrix element, $a^+_{ij} = a^*_{ji}$. Since $\widehat{a}$ and $\widehat{a}^+$ are both real, the $\widehat{a}^+$ matrix must be equivalent to the transpose of the $\widehat{a}$ matrix. It is this symmetry that requires $a_n = a^+_n = \sqrt{n}$, as discussed in footnote 18 on page 247.

the particle only has kinetic energy, $\widehat{H} = \widehat{K} = \widehat{K}_x + \widehat{K}_y = -(\hbar^2/2m)[\partial^2/\partial x^2 + \partial^2/\partial y^2]$. Thus, the Schrödinger equation of this system may be expressed as follows.

$$\widehat{H}\Psi = \widehat{K}_x\Psi + \widehat{K}_y\Psi = -\frac{\hbar^2}{2m}\frac{\partial^2\Psi}{\partial x^2} - \frac{\hbar^2}{2m}\frac{\partial^2\Psi}{\partial y^2} = E\Psi \tag{7.40}$$

Since the Hamiltonian is expressible as the sum of separate x- and y-dependent terms, we might expect Ψ to be expressible as a product of separate x- and y-dependent wavefunctions, $\Psi(x, y) = X(x)Y(y)$. The reason that this makes sense becomes evident when we insert a solution of this form into Eq. 7.40.

$$\begin{aligned}
\widehat{H}\Psi &= \widehat{K}_x X(x)Y(y) + \widehat{K}_y X(x)Y(y) \\
&= \left[\widehat{K}_x X(x)\right] Y(y) + \left[\widehat{K}_y Y(y)\right] X(x) \\
&= [E_x X(x)]\, Y(y) + [E_y Y(y)]\, X(x) \\
&= (E_x + E_y)\, X(x)Y(y) \\
&= (E_x + E_y)\, \Psi
\end{aligned} \tag{7.41}$$

In other words, since $\widehat{K}_x$ only operates on $X(x)$ and $\widehat{K}_y$ only operates on $Y(y)$, the two-dimensional Hamiltonian is equivalent to the sum of two one-dimensional Hamiltonians. Moreover, since we have already solved the Schrödinger equation for a one-dimensional particle in a box, we know what its eigenfunctions and eigenvalues look like. More specifically, if we assume that the box extends from $0 < x < L_x$ and $0 < y < L_y$, then the solutions of Eq. 7.44 must have the following form.

$$\begin{aligned}
X(x) &= \sqrt{\frac{2}{L_x}}\sin(k_{n_x}x) = \sqrt{\frac{2}{L_x}}\sin\left[\left(\frac{n_x\pi}{L_x}\right)x\right] \\
Y(y) &= \sqrt{\frac{2}{L_y}}\sin(k_{n_y}y) = \sqrt{\frac{2}{L_y}}\sin\left[\left(\frac{n_y\pi}{L_y}\right)y\right]
\end{aligned} \tag{7.42}$$

Inserting $\Psi(x, y) = X(x)Y(y)$ into Eq. 7.40 yields the following eigenvalues.

$$E_{n_x,n_y} = \frac{n_x^2h^2}{8mL_x^2} + \frac{n_y^2h^2}{8mL_y^2} \quad \text{where } n_x = 1, 2, 3\ldots \text{ and } n_y = 1, 2, 3\ldots \tag{7.43}$$

The wavefunctions and probability densities corresponding to the first four eigenstates of such a two dimensional particle in a box are shown in Fig. 7.6. Notice that the degeneracy of the energy levels depends on the

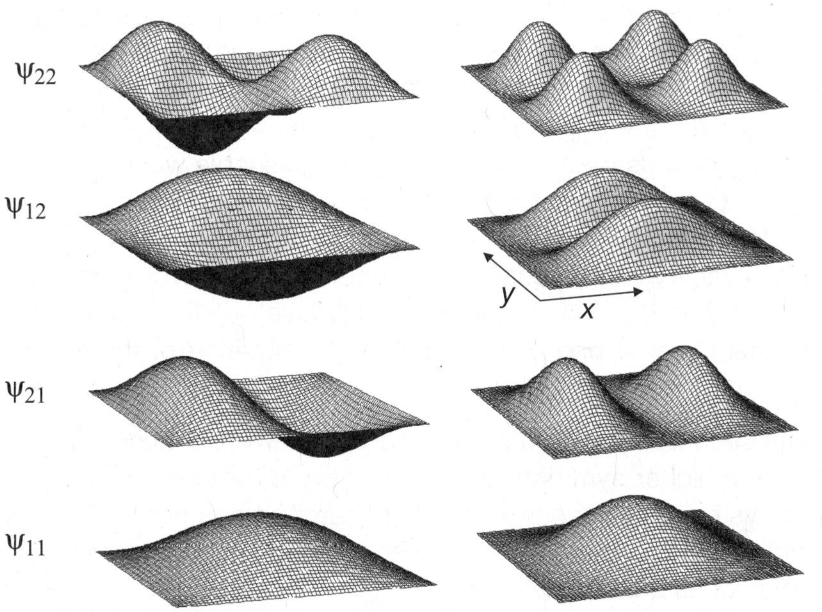

FIGURE 7.6 Eigenfunctions $\Psi_{n_x n_y}$ and probability densities $\rho(x, y) = \Psi^*_{n_x n_y}\Psi_{n_x n_y}$ of a particle in a two-dimensional box, with quantum numbers $n_x = 1, 2$ and $n_y = 1, 2$.

relative lengths of the sides of the box. If the box has a square shape with $L_x = L_y = L$, then the eigenfunctions Ψ_{12} and Ψ_{21} will have exactly the same energy $E_{12} = E_{21} = (1^2 + 2^2)(h^2/8mL^2) = 5h^2/(8mL^2)$. So, the first excited state of a particle in a square box is twofold degenerate ($g = 2$). But, if $L_x \neq L_y$, then these two eigenstates will no longer be degenerate.

The above particle-in-a-box results can clearly be extended to three dimensions simply by specifying the length of the three sides of the box and introducing the kinetic energy operator in the z-direction.

$$
\begin{aligned}
\widehat{H}\Psi &= \widehat{K}\Psi = -\frac{\hbar^2}{2m}\nabla^2\Psi \\
&= -\frac{\hbar^2}{2m}\left[\frac{\partial^2}{\partial x^2} + \frac{\partial^2}{\partial y^2} + \frac{\partial^2}{\partial z^2}\right]X(x)Y(y)Z(z) \\
&= \left[E_x + E_y + E_z\right]X(x)Y(y)Z(z) = E_{n_x,n_y,n_z}\Psi
\end{aligned}
\tag{7.44}
$$

The energy eigenvalues are again simply the sum of three one-dimensional particle-in-a-box energies.

$$
E_{n_x,n_y,n_z} = \frac{n_x^2 h^2}{8mL_x^2} + \frac{n_y^2 h^2}{8mL_y^2} + \frac{n_z^2 h^2}{8mL_z^2} \quad \text{where } n_i = 1, 2, 3 \ldots \tag{7.45}
$$

If the box is a cube $L_x = L_y = L_z = L$, then permuting the quantum numbers n_x, n_y, and n_z will leave the energy unchanged, and thus lead to degeneracy, while a less symmetrical box will have reduced degeneracy. More specifically, the degeneracy of each energy level of a particle in a cubical box depends on the number of distinguishable ways in which the three quantum numbers may be permuted. If the three quantum numbers are all different (such as 1, 2, 3), then they may be permuted in 3! = 6 ways. However, if all three quantum numbers are the same (such as 1, 1, 1 or 2, 2, 2), then the corresponding energy level will be nondegenerate, because there is only one distinguishable permutation of three identical numbers.

The connection between symmetry and degeneracy is quite general, as systems with higher symmetry invariably have states of higher degeneracy than systems with lower symmetry. Stated in another way, reducing the symmetry of a system has the effect of reducing (or "breaking") the degeneracy of its quantum states.[24]

Particle on a Sphere

A particle that is confined to the surface of a sphere may be viewed as a generalization of a particle-in-a-ring system. The angular momentum operator corresponding to the rotational angle ϕ is given by Eq. 7.10. The angle ϕ is conventionally associated with rotation about the z-axis, and so the associated angular momentum is referred to as L_z.[25]

$$\boxed{\widehat{L_z} = -i\hbar\frac{\partial}{\partial\phi} = -i\hbar\left[x\frac{\partial}{\partial y} - y\frac{\partial}{\partial x}\right] = \widehat{x}\,\widehat{p_y} - \widehat{y}\,\widehat{p_x}} \tag{7.46}$$

The last two equalities indicate that the classical mechanical identity $L_z = |\vec{r} \times \vec{p}| = xp_y - yp_x$ also holds in quantum mechanics, as long as each of the latter quantities is replaced by the corresponding operator.

Since the axes x, y, and z are equivalent, similar expressions pertain to the angular momentum about the x- and y-axes.

$$\widehat{L_x} = -i\hbar\left[y\frac{\partial}{\partial z} - z\frac{\partial}{\partial y}\right] = \widehat{y}\widehat{p_z} - \widehat{z}\widehat{p_y} \tag{7.47}$$

$$\widehat{L_y} = -i\hbar\left[z\frac{\partial}{\partial x} - x\frac{\partial}{\partial z}\right] = \widehat{z}\widehat{p_x} - \widehat{x}\widehat{p_z} \tag{7.48}$$

---

[24] On rare occasions, a system that is not highly symmetrical may have an *accidental degeneracy* if two states with different quantum numbers happen to have the same energy.

[25] See footnote 4 on page 233 for more about angular momentum.

Squaring the above operators produces the corresponding $\widehat{L}_i^2$ operators, whose sum is equal to the total angular momentum squared operator $\widehat{L}^2$.

$$\widehat{L}^2 = \widehat{L}_x^2 + \widehat{L}_y^2 + \widehat{L}_z^2 \tag{7.49}$$

The location of a particle anywhere on the surface of a sphere may be determined by specifying the two angles ϕ and θ, corresponding to the rotation around the z-axis and the tilt angle with respect to the z-axis, respectively. Transforming from Cartesian x, y, and z coordinates to polar r, θ, and ϕ coordinates can be performed using the following trigonometric relations.

$$\begin{aligned} x &= r \sin\theta \cos\phi \\ y &= r \sin\theta \sin\phi \\ z &= r \cos\theta \end{aligned} \tag{7.50}$$

These transformations may be used to express $\widehat{L}^2$, in terms of the angles ϕ and θ.[26]

$$\boxed{\widehat{L}^2 = \frac{-\hbar^2}{\sin^2\theta}\left[\frac{\partial^2}{\partial\phi^2} + \sin\theta\frac{\partial}{\partial\theta}\left(\sin\theta\frac{\partial}{\partial\theta}\right)\right]} \tag{7.51}$$

Using these identities we may express the kinetic energy operator of a particle confined to the surface of a sphere (of constant radius) as follows.

$$\begin{aligned} \widehat{K} &= -\frac{\hbar^2}{2m}\nabla^2 = -\frac{\hbar^2}{2m}\left(\frac{\partial^2}{\partial x^2} + \frac{\partial^2}{\partial y^2} + \frac{\partial^2}{\partial z^2}\right) \\ &= \frac{\widehat{L}^2}{2mr^2} = \frac{1}{2I}\left(\widehat{L}_x^2 + \widehat{L}_y^2 + \widehat{L}_z^2\right) \\ &= -\frac{\hbar^2}{2I\sin^2\theta}\left[\frac{\partial^2}{\partial\phi^2} + \sin\theta\frac{\partial}{\partial\theta}\left(\sin\theta\frac{\partial}{\partial\theta}\right)\right] \end{aligned} \tag{7.52}$$

---

[26] The expression for $\widehat{L}_z$ in terms of ϕ is given by Eq. 7.46, while the corresponding expressions for $\widehat{L}_x$ and $\widehat{L}_y$ may be expressed as follows.

$$\widehat{L}_x = i\hbar\left(\sin\phi\frac{\partial}{\partial\theta} + \frac{\cos\phi}{\tan\theta}\frac{\partial}{\partial\phi}\right)$$

$$\widehat{L}_y = -i\hbar\sin\phi\left(\cos\phi\frac{\partial}{\partial\theta} - \frac{\sin\phi}{\tan\theta}\frac{\partial}{\partial\phi}\right)$$

A *particle on a sphere* is defined as a free particle (with no potential energy) confined to the surface of a sphere. Thus, the Hamiltonian of such a particle is again $\widehat{H} = \widehat{K}$, and so its Schrödinger equation may be expressed as follows.

$$\widehat{H}\Psi(\phi, \theta) = \widehat{K}\Psi(\phi, \theta) = E\Psi(\phi, \theta) \tag{7.53}$$

Since the two degrees of freedom ϕ and θ are independent of each other (just like the variables x, y of a particle in a two-dimensional box) we expect the eigenfunctions to be expressible as a product of functions of the two angles $\Psi(\phi, \theta) = \Phi(\phi)\Theta(\theta)$. Moreover, since $\widehat{H}$ depends on two angles, we expect the corresponding eigenfunctions to depend on two quantum numbers. These quantum numbers are conventionally identified as the *orbital* angular momentum quantum number ℓ, and the *magnetic* quantum number m_ℓ.[27] The solutions of Eq. 7.53 may be expressed as follows.

$$\Psi_{\ell,m_\ell} = \Phi_{m_\ell}(\phi)\Theta_{\ell m_\ell}(\theta)$$

$$\Phi_{m_\ell}(\phi) = \frac{1}{\sqrt{2\pi}} e^{im_\ell\phi} \tag{7.54}$$

$$\Theta_{\ell m_\ell}(\theta) = (-1)^{(m_\ell+|m_\ell|)/2}\sqrt{\frac{(2\ell+1)(\ell-|m_\ell|)!}{2(l+|m_\ell|)!}}P_\ell^{|m_\ell|}(\cos\theta)$$

$$\ell = 0, 1, 2, 3, \ldots \quad \text{and} \quad m_\ell = 0, \pm 1, \pm 2, \cdots \pm \ell$$

Notice that the magnitude of quantum number m_ℓ is bounded by the value of ℓ. Thus, the possible values of m_ℓ are limited to the range $-\ell \leq m_\ell \leq \ell$. The functions $P_\ell^{|m_\ell|}(\cos\theta)$ are called associated Legendre functions, the first three of which have the following simple form.[28]

$$\begin{aligned} P_0^0(\cos\theta) &= 1 \\ P_1^0(\cos\theta) &= \cos\theta \\ P_1^1(\cos\theta) &= -\sin\theta \end{aligned} \tag{7.55}$$

---

[27] As these names imply, we will see that ℓ determines the total orbital angular momentum of an electron, while m_ℓ determines the projection of the angular momentum along the z-axis, and thus dictates the influence a magnetic field along the z-axis will have in splitting the degeneracy of the *magnetic sublevels* associated with each value of the total angular momentum.

[28] The higher-order functions may be determined using the following expression.

$$P_\ell^{|m_\ell|}(x) = \frac{(-1)^{|m_\ell|}}{2^\ell \ell!}(1-x^2)^{|m_\ell|/2}\frac{d^{\ell+|m_\ell|}}{dx^{\ell+|m_\ell|}}(x^2-1)^\ell$$

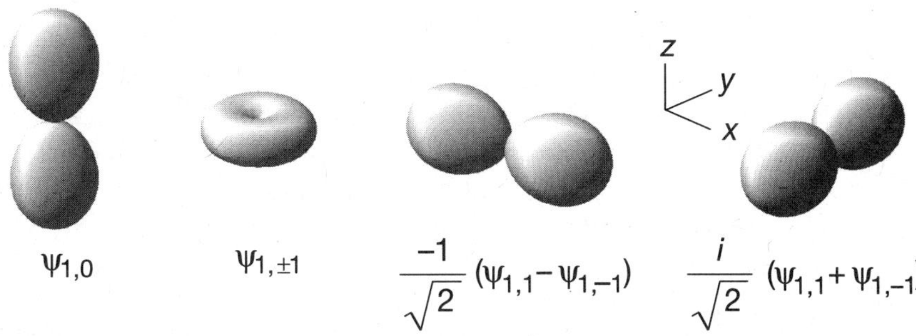

FIGURE 7.7 Spherical harmonic probability densities corresponding to $\ell = 1$ and $m_\ell = 0, \pm1$, and their linear combinations.

These functions, combined with Eq. 7.54, yield the first four eigenfunctions of a particle-on-a-sphere Ψ_{ℓ,m_ℓ}, which are also referred to as *spherical harmonics* $Y_\ell^{m_\ell}$.[29]

$$\Psi_{0,0} = Y_0^0 = \frac{1}{\sqrt{4\pi}}$$

$$\Psi_{1,0} = Y_1^0 = \frac{1}{2}\sqrt{\frac{3}{\pi}}\cos\theta$$

$$\Psi_{1,+1} = Y_1^{+1} = -\frac{1}{2}\sqrt{\frac{3}{2\pi}}\sin\theta\, e^{+i\phi}$$

$$\Psi_{1,-1} = Y_1^{-1} = \frac{1}{2}\sqrt{\frac{3}{2\pi}}\sin\theta\, e^{-i\phi} \tag{7.56}$$

Figure 7.7 shows plots of the probability densities for all the spherical harmonics with $\ell = 1$. More specifically, the first two objects on the left are the probability densities of the $\ell = 1$ eigenfunctions in Eq. 7.56, while the two objects on the right are probability densities derived from linear combinations of the latter eigenfuctions. It is pretty easy to show that these linear combinations are also eigenfunctions of Eq. 7.52. More generally, it is also easy to prove that any linear combination of *degenerate* eigenfunctions of any Hamiltonian must be eigenfunctions of that Hamiltonian.[30]

---

[29] The following is a list of all the spherical harmonic functions with $\ell = 2$.

$$\Psi_{2,0} = Y_2^0 = \frac{1}{4}\sqrt{\frac{5}{\pi}}(3\cos^2\theta - 1)$$

$$\Psi_{2,\pm1} = Y_2^{\pm1} = \mp\frac{1}{2}\sqrt{\frac{15}{2\pi}}\,(\sin\theta\,\cos\theta)\, e^{\pm i\phi}$$

$$\Psi_{2,\pm2} = Y_2^{\pm2} = \frac{1}{4}\sqrt{\frac{15}{2\pi}}\left(\sin^2\theta\right) e^{\pm 2i\phi}$$

[30] In order to prove that this is true, we may consider any function $\Psi = C_a\Psi_a + C_b\Psi_b$ (where Ψ_a and Ψ_b are eigenfunctions of $\widehat{H}$ with eigenvalue E). Thus,

The eigenfunctions Y_1^0, $-(1/\sqrt{2})(Y_1^1 + Y_1^{-1})$, and $(i/\sqrt{2})(Y_1^1 - Y_1^{-1})$ are clearly identical in shape, but are oriented along different axes (see Fig. 7.7). These eigenfunctions are also all real (as opposed to complex), as can be shown by replacing $e^{\pm i\phi}$ by $\cos\phi \pm i\sin\phi$ (using Euler's formula, Eq. 6.7). Moreover, it is no accident that the above three functions closely resemble atomic p_x, p_y, and p_z orbitals, as we will see that the shapes of the latter orbitals are determined by the above spherical harmonic functions.

The energy eigenvalues of a particle on a sphere may be obtained by inserting the eigenfunctions in Eq. 7.54 into Eq. 7.53.

$$E_\ell = \left(\frac{\hbar^2}{2I}\right)\ell\,(\ell+1) \tag{7.57}$$

Notice that these eigenvalues only depend on ℓ. In other words, all the eigenfunctions with the same ℓ value but different m_ℓ values have the same energy. Since there are $2\ell+1$ allowed m_ℓ values corresponding to each value of ℓ, the *degeneracy* of the corresponding rotational states is $2\ell+1$.

Further insight into the physical significance of rotational degeneracies can be gained by noting that Eq. 7.57, combined with Eqs. 7.52 and 7.53, implies that the (root-mean-square) magnitude of the angular momentum is

$$\sqrt{\langle L^2\rangle} = \sqrt{2IE_{\ell,m_\ell}} = \hbar\sqrt{\ell\,(\ell+1)} \tag{7.58}$$

while applying the operator $\widehat{L_z}$ (Eq. 7.46) to Ψ_{ℓ,m_ℓ} yields the following expression for the expectation value of the z-component of the angular momentum.[31]

$$\langle L_z\rangle = \hbar m_\ell \tag{7.59}$$

Thus, states with different m_ℓ values, but the same ℓ value, have angular momentum vectors of the same magnitude, but are pointing in different directions with respect to the z-axis. In other words, the quantum number ℓ determines the magnitude of the total angular momentum, while m_ℓ determines the projection of the angular momentum along the z-axis. This also makes it clear why the allowed values of m_ℓ are bounded by the value of ℓ, as the projection of the angular momentum along the z-axis cannot be larger than the total magnitude of the angular momentum.

$\widehat{H}\Psi = (C_a\Psi_a + C_b\Psi_b)\Psi = C_aE\Psi_a + C_bE\Psi_b = E(C_a\Psi_a + C_b\Psi_b) = E\Psi$, which proves that Ψ is also an eigenfunction of $\widehat{H}$ and that it has the same eigenvalue as Ψ_a and Ψ_b. It is important to keep in mind that this theorem only applies to degenerate eigenfunctions.

[31] Notice that the functions Ψ_{ℓ,m_ℓ} are eigenfunctions of both $\widehat{L^2}$ and $\widehat{L_z}$, so the associated expectation values are equal to the corresponding eigenvalues.

Application to Molecular Rotational Spectroscopy

The above particle-on-a-sphere results may be used to predict the rotational quantum states of molecules. For example, the rotation of a diatomic molecule of bond length R_0 and reduced mass $\mu = m_1m_2/(m_1 + m_2)$ has a moment of inertia of $I = \mu R_0^2$.[32] Thus, a diatomic is predicted to have rotational quantum state energies given by Eq. 7.57. In describing molecular rotational transition, the quantum number ℓ is conventionally called J. Moreover, Eq. 7.57 is often expressed as $E_J = BJ(J + 1)$, where $B = \hbar^2/2I$ is called the *rotational constant*. Furthermore, Eq. 7.57 is often expressed as

$$E_J/hc = \widetilde{B}J(J + 1) \tag{7.60}$$

where $\widetilde{B} = B/hc = \hbar/(4\pi Ic)$ is the rotational constant in wavenumber units.[33]

Rotational states have a degeneracy of $2J + 1$, which again reflects the number of m_J rotational sublevels with the same energy. The selection rules for rotational (microwave) absorption dictate that transitions can take place between any rotational states that differ by $\Delta J = \pm 1$ (as explained in Section 9.4 of Chapter 9). Thus, a transition from the ground to the first excited rotational state of a diatomic molecule has an energy of $\Delta E = E_1 - E_0 = (\hbar^2/2I)[1(1 + 1) - 0(0 + 1)] = \hbar^2/I = 2B$, which corresponds to a photon of frequency of $\tilde{\nu} = 2\widetilde{B}$ (in wavenumber units).

Since rotational energy spacings are typically small compared to k_BT, many vibrational states are populated at ambient temperatures. Thus, the $\Delta J = \pm 1$ selection rule implies that rotational spectra will contain a series of approximately evenly spaced lines corresponding to transitions from states with different values of J, since $E_{J+1} - E_J = (\hbar^2/I)(J + 1) = 2B(J + 1)$ (as further described in Section 9.4 of Chapter 9). Careful measurements also reveal that the spacings between rotational lines are not precisely evenly spaced, but typically decrease slightly with increasing J. This is because the centrifugal force experienced by molecules increases as they rotate more rapidly, thus slightly increasing their bond lengths. Such effects can be accounted for by including the following centrifugal correction to the rotational quantum state energies of a diatomic molecule, $E_J = BJ(J + 1) - D[J(J + 1)]^2$. This effect is quite small because the centrifugal distortion constant D is typically ten thousand times smaller than the rotational constant B.

---

[32] The symbol R_0 is used to denote the most probable bond length of the diatomic in its ground vibrational (zero-point energy) state. This bond length is typically very similar to the classical equilibrium (potential energy minimum) bond length R_e.

[33] As in the case of vibrational transitions, rotational transition frequencies are often expressed as $\tilde{\nu} = 1/\lambda$ (in cm^{-1} units), where λ is the wavelength of a photon that is resonant with the rotational transition $\Delta E = hc\tilde{\nu} = h\nu = \hbar\omega$.

HOMEWORK PROBLEMS

Problems That Illustrate Core Concepts

1. Consider a particle confined in a box that extends from $x = 0$ to $x = b$.
   (a) Sketch the probability densities for the first five quantum states of the above particle.
   (b) Based on your sketches, can you guess the value of n for which there is exactly a 25% probability of finding the particle somewhere in the range $0 \leq x \leq b/4$?
   (c) Confirm your guess by calculating the probability of finding the particle within the range $0 \leq x \leq b/4$ for that value of n.

2. Consider a particle that is confined to a box of length a, which extends from $x = -a/2$ to $x = +a/2$, so that it has a potential energy of $V(x) = 0$ inside the box (when $-a/2 < x < +a/2$) and infinite outside the box (when $x \leq -a/2$ and $x \geq a/2$). Recall that the Hamiltonian for a particle inside such a box is $\widehat{H} = \widehat{K_x}$, and the infinite potential energy at the boundaries of the box require that $\Psi(x) = 0$ at $x = -a/2$ and $x = +a/2$. In other words, the particle can only exist inside the box, and so you may assume that all integrals associated with this problem extend from $x = -a/2$ to $x = +a/2$.
   (a) Sketch the ground-state wavefunction $\Psi_1(x)$ of this system, determine its correct functional form, and then normalize it.
   (b) Determine the energy eigenvalue of $\Psi_1(x)$.
   (c) What is $\langle K \rangle$?
   (d) What is $\langle V \rangle$?
   (e) What is $\langle H \rangle$?
   (f) What is $\langle x \rangle$?
   (g) What is $\langle x^2 \rangle$?
   (h) What is σ_x?
   (i) What is $\langle p_x \rangle$?
   (j) What is $\langle p_x^2 \rangle$?
   (k) What is σ_{p_x}?
   (l) Show that $\sigma_x \sigma_{p_x}$ is consistent with the uncertainty relation Eq. 6.42.

3. Use the following procedure to demonstrate that the eigenfunctions of a particle in a box may be expressed as a linear combination of free-particle wavefunctions, and to illustrate a physical correspondence between the classical and quantum particle-in-a-box systems.
   (a) Use Eulers formula, $e^{\pm ikx} = \cos kx \pm i \sin kx$, to show that $\Psi_1(x)$ in problem 2 can be expressed as $B(e^{+ikx} + e^{-ikx})$, and relate the values of k and B to the box length a.
   (b) Show that the functions e^{ikx} and e^{-ikx} are orthogonal over the range of the box in problem 2 (using the value of k identified above).
   (c) Use the above results to show that momentum expectation value of a particle confined to a box such as that in problem 2 can be expressed as the sum of four integrals, two of which are equal to zero because e^{+ikx} and e^{-ikx} are orthogonal, and the other two of which are each proportional to $\pm\hbar k$. Hint: Use Euler's formula (Eq. 6.7) to expand Ψ before applying the momentum operator (and then make use of the fact that $e^{\pm ikx}$ are momentum eigenfunctions).
   (d) What do the above results imply about the direction(s) in which the particle that is confined within the box in problem 2 is moving?
   (e) How do the above results explain the fact that a particle in a box can have $\langle p_x \rangle = 0$ but $\langle K_x \rangle \neq 0$?

4. The particle-in-a-box energy formula Eq. 7.6 may be used to predict the wavelengths of photons required to optically excite such as quantum system.
   (a) Obtain an expression for the wavelength λ of a photon required to induce a particle-in-a-box to undergo a transition from an initial state with quantum number n_i to a final state with quantum number n_f, expressed

as a function of the particle mass m and the length of the box L (as well as other constants).

(b) Use the above result to predict the wavelength (in nm units) of a photon that would be required to promote an electron from a state with $n_i = 5$ to $n_f = 6$, if it is confined within a one-dimensional box of 1 nm length.

5. Use the particle-in-a-box model to estimate the average carbon-carbon bond length of hexatriene, given that its experimental optical absorption wavelength is $\lambda \approx 250$ nm.

6. Use the particle-in-a-ring model to estimate the carbon-carbon bond length of benzene, given that its experimental optical absorption wavelength is $\lambda \approx 200$ nm.

7. Verify that the particle-in-a-ring wavefunctions in Eq. 7.7 are normalized and show that the corresponding probability densities are independent of ϕ (for all the quantum states).

8. Verify that particle-in-a-ring energy eigenfunctions are also eigenfunctions of the angular momentum operator, and show that states with positive and negative values of n correspond to particles that are rotating in opposite directions.

9. The wavefunctions of a particle that is confined within a box whose walls have a finite height will look different from those corresponding to a box with infinitely high walls.

(a) Make a sketch illustrating the difference between the ground-state wavefunctions of a particle in these two kinds of box potentials.

(b) Which of the two systems do you expect to have a lower ground-state energy (assuming the two boxes have the same length)? Hint: Recall the relationship between the kinetic energy and wavelength of a particle-wave, and think about which of the two ground-state wavefunctions will have a longer wavelength.

10. Equation 7.16 represents the quantum mechanical prediction for the tunneling probability, τ, when a particle of mass m and energy E encounters a barrier of height $V > E$ and width b:

(a) Verify that Eq. 7.16 reduces to Eq. 7.17 when the barrier is very high and wide.

(b) Use Eq. 7.17 to predict the tunneling probability produced by an STM tip that is 1 Å (10^{-10} m) away from a metal surface. Note that the tunneling particle is an electron ($m = m_e$) and a typical barrier height is $V \approx 5$ eV (and 1 eV $\approx 1.6 \times 10^{-19}$ J). You may assume that the electrons have a typical thermal energy of $E \approx k_B T = k_B$ 300 K, which is equivalent to $E \approx \frac{1}{40}$ eV, and thus $V - E \approx V$ (because $V >> E$).

(c) Use Eq. 7.17 to estimate the tunneling probability of a proton with thermal energy $E \approx k_B T = k_B$ 300 K, when it encounters a barrier of height $V \approx 2E$ and width of 1 Å (as is typically the case for proton tunneling in a hydrogen bonded system).

11. Verify that the ground-state wavefunction of a harmonic oscillator is an eigenfunction of the harmonic oscillator Hamiltonian, and that its eigenvalue is $\frac{1}{2}h\nu$.

12. Evaluate the expectation values $\langle x \rangle$ and $\langle x^2 \rangle$ to obtain an expression for the standard deviation (uncertainty) of the position σ_x for the ground state of a harmonic oscillator (to verify that $\sigma_x = \sigma/\sqrt{2} = \sqrt{\hbar/2m\omega}$). Why is σ_x smaller than σ by a factor of $1/\sqrt{2}$? Hint: Think about how the ground-state wavefunction is related to the corresponding probability density.

13. Use the general definition of the Hermite polynomials $H_n(y)$ in Eq. 7.19 to obtain $H_4(y)$, and then verify that it may also be obtained from $H_3(y)$ (which is given in exercise 7.4 on page 242) using the recursion relation $H_{n+1}(y) = 2yH_n(y) - dH_n(y)/dy$. Plot $H_4(y)\, e^{-y^2/2}$ over the range $-6 < y < 6$ to confirm that it looks like Ψ_4 in Figure 7.4.

14. Use Eqs. 7.24 and 7.25 to show that $[\hat{a}, \hat{a}^+] = 1$ (by making use of the fact that $[\hat{x}, \hat{p}_x] = i\hbar$).

15. Use the matrix representations of $\hat{n}$, $\hat{a}$, and $\hat{a}^+$ (Eqs. 7.36, 7.38, and 7.39) to confirm that $\hat{a}^+\hat{a} = \hat{n}$ and $[\hat{a}, \hat{a}^+] = 1$. Note that the number 1 may be equated with a diagonal unit matrix (whose diagonal elements are all equal to 1, and off-diagonal elements are all equal to zero).

16. Use the matrix representations of the harmonic oscillator Hamiltonian (Eq. 7.37) and eingenvectors (Eq. 7.35) to verify that $E_2 = \langle 2|\hat{H}|2\rangle = \hbar\omega(2 + \frac{1}{2}) = h\nu(2 + \frac{1}{2})$.

17. The diatomic molecule HF has a vibrational transition frequency of $\tilde{\nu} = \nu/c = 4138\ \text{cm}^{-1}$ (where c is the speed of light). Note that the full width at half maximum (FWHM) of any peaked function $f(x)$ is defined as the difference between the x values of the two points (on the left and right sides of the maximum) at which the function is equal to one-half of its maximum value.

(a) What is the reduced mass, μ (kg), of HF?

(b) What is the force constant, f (N/m), of HF?

(c) What is the FWHM (in Å units) of the vibrational ground-state probability density of HF?

(d) Predict frequency $\tilde{\nu}$ (cm^{-1}) of DF (where D is a deuterium atom, which has twice the mass of a hydrogen atom), assuming it has the same force constant as HF.

18. The diatomic molecules F_2, O_2, and N_2 have vibrational frequencies of 2359 cm^{-1}, 917 cm^{-1}, and 1580 cm^{-1} (but not in that order).

(a) Could the above frequency differences be attributed only to differences in μ (if you assume that all three molecules have the same force constant)?

(b) Match the molecules with the vibrational frequencies (using only what you know about the bond-orders of the three molecules).

19. Consider a particle in a cubic three-dimensional box (with $L_x = L_y = L_z = L$).

(a) What is the degeneracy of the energy level for which the three quantum numbers (n_x, n_y, and n_z) are 1, 2, and 3, and are combined in any order?

(b) Compare the above degeneracy with the fact that the number of permutations of n different (distinguishable) objects is $n!$.

(c) What would the energies and degeneracies of the states in part (a) become if one side of the box was twice as long as the other two sides, so that $L_x = L_y = L$ and $L_z = 2L$?

20. Consider the angular momentum expectation values for a particle on a sphere.

(a) Calculate the expectation values of the total (root-mean-squared) angular momentum, $\sqrt{\langle L^2\rangle}$, and the projection of the angular momentum onto the z-axis, $\langle L_z\rangle$, for all the quantum states of a particle on a sphere with $\ell \leq 3$.

(b) Why is it impossible for the quantum number m_ℓ to be larger than 3 when the orbital angular momentum quantum number is $\ell = 3$? In other words, what is the largest value of m_ℓ for which the projection of the angular momentum along the z-axis does not exceed the total (root-mean-squared) angular momentum?

(c) Count the number of allowed m_ℓ values associated with $\ell = 3$ to verify that the degeneracy of the corresponding energy level is $2\ell + 1$.

21. The rotation of a diatomic molecule (of fixed bond length R_0) is mechanically equivalent to a single particle of mass μ (where μ is the reduced mass of the diatomic) rotating on a sphere of radius R_0. Notice that the reduced masses of the diatomics HF and DF are the same as those used to calculate the vibrational quantum states of these molecules (as in problem 17 above).

(a) Given that the bond length of HF is 0.92 Å, calculate its moment of inertia (in units of kg m^2).

(b) What is the rotational constant of HF, expressed in both energy B (J) and wavenumber $\widetilde{B}$ (cm^{-1}) units?

(c) Predict the first rotational transition energy, $\Delta E = E_1 - E_0$, of HF in both J and cm^{-1} units, and compare your result with the experimental value of 41.9 cm^{-1}.

(d) How much do you expect the first rotational transition energy of DF to differ from that of HF, assuming that both molecules have the same bond length?

Problems That Test Your Understanding

22. Consider an electron in a one-dimensional box that extends from x=0 to x=1 nm.

(a) Draw a diagram illustrating the shape of the ground-state wavefunction for this electron confined to a one-dimensional box.

(b) What is the de Broglie wavelength of the electron in the above ground-state eigenfunction (in nm units)?

(c) What is the energy eigenvalue of the electron in the ground state (in J units)?

(d) What is the expectation value of the kinetic energy of the electron in the ground state (in J units)?

23. The following questions pertain to a one-dimensional box of 1 nm length that is used to predict the π electron energies in an eight-carbon conjugated polyene chain (CH_2=CH–CH=CH–CH=CH–CH=CH_2).

(a) What are the values of the particle-in-a-box quantum number for the highest occupied molecular orbital (HOMO) and the lowest unoccupied molecular orbital (LUMO) of the above molecule?

(b) Predict the wavelength of light that would be required in order to induce a HOMO to LUMO transition (expressed in nm units)?

24. A hydrogen atom may be roughly approximated as an electron that is confined within a cubical box of dimensions $L_x = L_y = L_z = 0.3$ nm.

(a) What is the energy difference between the ground and first excited state of this system (expressed as a number with units)?

(b) What is the degeneracy of the first excited state of this system?

25. The following questions pertain to a harmonic oscillator.

(a) Use the matrix representation of the Hamiltonian operator to evaluate the expectation value $\langle 2|\widehat{H}|2\rangle$.

(b) Use the matrix representation of the operator $\widehat{a}$ to complete the following expression: $\widehat{a}|3\rangle = |\ \ \rangle$.

26. Consider the diatomic molecule HD (composed of a hydrogen atom whose mass is approximately equal to m_p and a deuterium whose mass is twice as large) that has a bond length of approximately 0.074 nm.

(a) What is the moment of inertia of HD (expressed as a number with units)?

(b) What is the value of the quantum number J, and the degeneracy, of the rotational quantum state of HD whose energy is $3\hbar^2/I$?

(c) What is the root-mean-squared magnitude of the angular momentum of HD in the same rotational state as in (b)?

(d) What are all the possible expectation values of the projection of the angular momentum of HD along the z-axis when it is in its second excited rotational state (two levels above the ground state)?

(e) What frequency of light (expressed in cm^{-1} units) would be required to excite HD from its *ground* to its *first excited* rotation state?

27. The vibrational frequency of the diatomic molecules I_2 is 214 cm^{-1}. What is the expectation value of the *potential energy* of I_2 when it is in its *third* vibrational excited state (above the ground state)?

28. Consider a particle of mass m confined to a three-dimensional box whose dimensions are $L_x = 1$ nm, $L_y = 2^{1/2}$ nm, and $L_z = 2$ nm. How many different quantum states of this system have an energy of $7h^2/(8mL_x^2)$, and what are the quantum numbers n_x, n_y, and n_z for each of these state?

29. Sketch the shapes of the ground-state wavefunctions for a particle confined to a box with walls of either infinite or finite height, so as to illustrate the differences between the shapes and ground-state energies of these two systems.

30. Use the matrix (and vector) representations of the Hamiltonian operator and eigenvectors of the harmonic oscillator to obtain the following results.

(a) Write down the vector (one-dimensional matrix) corresponding to $|2\rangle$, and show that $\langle 2|2\rangle = 1$.

(b) Evaluate the matrix expression corresponding to $\widehat{H}|2\rangle = E_2|2\rangle$ in order to determine the eigenvalue E_2.

31. Given that H_2 has vibrational frequency of 4400 cm^{-1}, predict the vibrational frequency of HD (assuming that both molecules have the same force constant, and given that $M_H = 1$ g/mol and $M_D = 2$ g/mol).

32. Carbon monoxide CO has an experimentally measured rotational constant of $\widetilde{B} = 1.93$ cm^{-1} (and atomic molar masses of $M_C = 12$ g/mol and $M_O = 16$ g/mol).

(a) What is the reduced mass of CO expressed in both g/mol and kg per molecule units?

(b) What is the bond length of CO (in Å units)?

(c) What are the expectation values of the total (root-mean-squared) angular momentum and all of the allowed projections of the angular momentum along the z-axis, when CO is in its first excited rotational state (immediately above the ground state)?

33. Consider a three-dimensional particle-in-a-box model for the two electrons of ethylene $CH_2{=}CH_2$, with the following parameters: $L = L_x = 3$ Å and $L_y = L_z = 1$ Å $= L/3$.

(a) What are the values of quantum numbers, n_x, n_y, and n_z, for the ground and first excited states of this system?

(b) What is the energy difference between the ground and first excited state of this system (expressed in J units)?

34. Consider a four-dimensional hyper-cube whose sides all have the same length.

(a) How many quantum numbers are required to specify each quantum state of a particle-wave contained in such a hyper-cube?

(b) What is the degeneracy of the first excited state (immediately above the ground state) of such a particle-wave?

35. Verify that the spherical harmonic function with $\ell = 1$ and $m_\ell = 0$ is an eigenfunction of the following particle-on-a-sphere kinetic energy operator (Eq. 7.52) $\widehat{K} = -\frac{\hbar^2}{2I\sin^2\theta}\left[\frac{\partial^2}{\partial\phi^2} + \sin\theta\frac{\partial}{\partial\theta}\left(\sin\theta\frac{\partial}{\partial\theta}\right)\right]$, and determine the corresponding eigenvalue.

Atoms and Spinning Particle-Waves

CHAPTER 8

8.1 The Hydrogen Atom

A hydrogen atom is a three-dimensional quantum mechanical system composed of a single electron bound to a single proton. We might expect this system to resemble a particle on a sphere, except for the fact that the distance between the electron and proton is no longer fixed, so the system now has three independent degrees of freedom r, θ, and ϕ. Thus, we expect that the eigenfunctions and eigenvalues of the hydrogen atom will depend on three quantum numbers, two of which are the same as those of a particle on a sphere ℓ and m_ℓ, and one of which is a new *principal* quantum number n.

The conventional notation used to designate atomic orbitals associates the quantum numbers $\ell = 0, 1, 2, 3, \ldots$ with the letters $s, p, d, f, \ldots$, respectively. Thus, a hydrogen 1s orbital corresponds to the eigenfunction with $n = 1$ and $\ell = 0$, while a 2s orbital has $n = 2$ and $\ell = 0$. The three 2p orbitals arise from the three allowed values of $m_\ell = 0, \pm 1$ (with $n = 2$ and $\ell = 1$). More specifically, the $2p_z$ orbital corresponds to $n = 2$, $\ell = 1$, and $m_\ell = 0$, while the $2p_x$ and $2p_y$ orbitals are real linear combinations of the complex $m_\ell = \pm 1$ wavefunctions, as shown in Fig. 7.7.[1]

[1] A similar linear combination procedure may be used to generate real representations of the five 3d orbitals. The orbital conventionally designated as $3d_z^2$ has $n = 3$, $\ell = 3$, and $m_\ell = 0$, while the other four three-dimensional orbitals derive from real linear combinations of $m_\ell = \pm 1$ and ± 2 eigenfunctions.

The attractive interaction between the electron and proton is described by a coulombic potential energy function, $V(r) = -(e^2/4\pi\epsilon_0)(1/r)$, where $e = 1.6 \times 10^{-19}$ C is the absolute value of the charge of an electron and $\epsilon_0 = 8.85 \times 10^{-12}$ C^2/(J m) is the permittivity of free space. Thus, if r is expressed in meters, then $V(r)$ will have units of joules.

The Hamiltonian for the hydrogen atom includes both kinetic and potential energy contributions.[2]

$$\boxed{\widehat{H} = \widehat{K} + \widehat{V} = -\frac{\hbar^2}{2\mu}\nabla^2 - \frac{e^2}{4\pi\epsilon_0}\frac{1}{r}} \tag{8.1}$$

The kinetic energy operator may be expressed either in Cartesian coordinates x, y, and z, or in polar coordinates r, θ, and ϕ (see Eq. 7.52). However, since r is no longer constant, the expression for $\widehat{K}$ in polar coordinates is somewhat more complicated than it was for a particle on a sphere.[3]

$$\begin{aligned}\widehat{K} &= -\frac{\hbar^2}{2\mu}\nabla^2 = -\frac{\hbar^2}{2\mu}\left(\frac{1}{r}\frac{\partial^2}{\partial r^2}r\right) + \frac{\widehat{L^2}}{2\mu r^2} \\ &= -\frac{\hbar^2}{2I}\left\{2r\frac{\partial}{\partial r} + r^2\frac{\partial^2}{\partial r^2} + \frac{1}{\sin^2\theta}\left[\frac{\partial^2}{\partial\phi^2} + \sin\theta\frac{\partial}{\partial\theta}\left(\sin\theta\frac{\partial}{\partial\theta}\right)\right]\right\}\end{aligned} \tag{8.2}$$

The eigenfunctions of the Schrödinger equation for hydrogen are separable into products of radial $R_{n,\ell}(r)$ and angular (spherical harmonic) $Y_\ell^{m_\ell}(\theta,\phi)$ functions.[4]

$$\boxed{\Psi_{n,\ell,m_\ell}(r,\theta,\phi) = R_{n,\ell}(r)\,Y_\ell^{m_\ell}(\theta,\phi)} \tag{8.3}$$

The energies (eigenvalues) of the hydrogen atom depend only on the principal quantum number n.

$$\boxed{E_n = -\frac{hcR_H}{n^2} \quad \text{where } n = 1, 2, 3, \ldots} \tag{8.4}$$

---

[2] The mass μ that appears in Eq. 8.1 is the reduced mass of the electron-proton pair, $\mu = m_e m_p/(m_e + m_p)$. Since the mass of the electron is nearly 2000 times smaller than the proton, μ is virtually identical to m_e, and thus a hydrogen atom is essentially equivalent to an electron bound to a stationary positive charge.

[3] The parameter $I = \mu r^2$ is the moment of inertia associated with hydrogen's electron-proton system. Note that the operator associated with the radial kinetic energy may be expressed in the following equivalent ways $-\frac{\hbar^2}{2\mu}\left(\frac{1}{r}\frac{\partial^2}{\partial r^2}r\right) = -\frac{\hbar^2}{2\mu r^2}\left(r\frac{\partial^2}{\partial r^2}r\right) = -\frac{\hbar^2}{2I}\frac{\partial}{\partial r}\left[r^2\frac{\partial}{\partial r}\right] = -\frac{\hbar^2}{2I}\left(2r\frac{\partial}{\partial r} + r^2\frac{\partial^2}{\partial r^2}\right)$.

[4] The functions $Y_\ell^{m_\ell}(\theta,\phi)$ are identical to the particle-on-a-sphere eigenfunctions Eq. 7.56 (and footnote 29).

Notice that these energies are all negative, as they represent the binding energy of the electron (relative to the energy of an unbound free electron at $r = \infty$). The *Rydberg constant* $R_H = m_e e^4/8h^3 c\epsilon_0^2 = 109,737\ \text{cm}^{-1}$ was originally determined empirically, from experimental measurements of hydrogen's line spectrum.[5] Niels Bohr was the first to accurately predict the value of R_H using his original semiclassical theory of hydrogen (see pages 18–22). The formula that Bohr derived may be expressed as follows.

$$E_n = -\left(\frac{\hbar^2}{2m_e a_0^2}\right)\frac{1}{n^2} \tag{8.5}$$

The constant a_0 is known as the *Bohr radius*, and is equivalent to the radius at which hydrogen's ground-state electron is most likely to be found, as we will see.

$$a_0 = \frac{4\pi\varepsilon_0\hbar^2}{m_e e^2} = 0.529\ \text{Å} \tag{8.6}$$

Quite remarkably, Bohr's theoretical expression for R_H is exactly the same as that which was later obtained by solving the Schrödinger equation.

The allowed values of the quantum numbers ℓ and m_ℓ are determined by the value of the principal quantum number: $\ell = 0, 1, 2 \ldots (n-1)$ and $-\ell \leq m_\ell \leq \ell$.[6] As in the case of a particle on a sphere, the bounds on m_ℓ reflect the fact that the projection of the angular momentum cannot exceed the magnitude of the total angular momentum. Similarly, the bound on ℓ is consistent with the fact that hydrogen's rotational kinetic energy cannot exceed its total kinetic energy.[7]

---

[5] The Rydberg constant (times hc) is equivalent to the ionization energy of an electron in hydrogen's ground state, $hcR_H \approx 13.6$ eV or 2.18×10^{-18} J or 1,313 kJ/mol. The Rydberg constant is also closely related to the *Hartree* energy unit $E_h = 2hcR_H$ (see Appendix B). The exact value of the constant appearing in Eq. 8.4 differs (very slight) from R_H because μ is not exactly equal to m_e. In other words, the values of E_n are more accurately expressed by replacing m_e by μ in each of the above expressions.

[6] Thus, for each value of n there are a total of $\sum_{\ell=0}^{n-1} 2\ell + 1 = n^2$ eigenfunctions of the same energy. This would correspond to the electronic degeneracy of a single electron with a given n value were it not for the two possible spin states of an electron. Thus, the total degeneracy (including spin) of a single electron with a given value of n is $2n^2$.

[7] However, since hydrogen's total energy includes both (negative) potential energy and (positive) kinetic energy contributions, it is not so obvious why the total energy should impose a bound on the total angular momentum. The key to understanding this connection becomes clearer when we recall that the virial theorem (1.46 on page 41) relates the average values of the potential and kinetic energies.

The following expressions describe hydrogen's radial wavefunctions $R_{n,\ell}(r)$ with principal quantum numbers up to $n = 3$.

$$R_{1,0} = 2\left(\frac{1}{a_0}\right)^{3/2} e^{-r/a_0}$$

$$R_{2,0} = \left(\frac{1}{2a_0}\right)^{3/2}\left(2 - \frac{r}{a_0}\right)e^{-r/2a_0}$$

$$R_{2,1} = \left(\frac{1}{2a_0}\right)^{3/2}\left(\frac{r}{\sqrt{3}\,a_0}\right)e^{-r/2a_0}$$

$$R_{3,0} = \left(\frac{1}{3a_0}\right)^{3/2}\left[1 - \frac{2}{3}\frac{r}{a_0} + \frac{2}{27}\left(\frac{r}{a_0}\right)^2\right]e^{-r/3a_0}$$

$$R_{3,1} = \left(\frac{1}{3a_0}\right)^{3/2}\frac{4\sqrt{2}}{3}\left(1 - \frac{1}{6}\frac{r}{a_0}\right)e^{-r/3a_0}$$

$$R_{3,2} = \left(\frac{1}{3a_0}\right)^{3/2}\frac{2\sqrt{2}}{27\sqrt{5}}\left(\frac{r}{a_0}\right)^2 e^{-r/3a_0} \qquad (8.7)$$

The above functions may be used to calculate the probability of finding an electron at a given distance r. Recall that $\Psi^*\Psi = |\Psi|^2$ represents the probability of finding hydrogen's electron at some particular location in space. If we are interested in knowing the probability of finding the electron at a distance r from the nucleus, then we must integrate $|\Psi|^2$ over the angles θ and ϕ. Doing so yields the following expression for the *radial probability density* $P_{n,\ell}(r)$ of hydrogen.[8]

$$\boxed{P_{n,\ell}(r) = R^2_{n,\ell}(r)\,r^2} \qquad (8.8)$$

The probability of finding the electron within a shell of thickness dr located at a distance r from the nucleus is $P_{n,\ell}(r)dr = R^2_{n,\ell}(r)\,r^2dr$. Thus, the total probability of finding the electron within a sphere of radius r may be obtained by evaluating the integral $\int_0^r R^2_{n,\ell}(r)\,r^2dr$. Similarly, the probability

---

Although this theorem was originally derived for classical mechanical systems, it also holds for quantum systems, where $\langle \ldots \rangle$ represents a quantum mechanical expectation value (rather than a classical time, or Boltzmann, average). For the coulomb potential of hydrogen $V(r) \propto r^{-1}$, so the virial theorem implies that $\langle V \rangle / \langle K \rangle = -2$, and thus, $E_n = \langle H \rangle = \langle K \rangle + \langle V \rangle = \langle K \rangle - 2\langle K \rangle = -\langle K \rangle > 0$. It is this link between E_n and $\langle K \rangle$ that imposes an upper bound on hydrogen's rotational kinetic energy, and thus on the quantum number ℓ.

[8] The derivation of Eq. 8.8 is very similar to that used to obtain Eq. 1.32 on page 37. The factor of 4π does not appear in Eq. 1.32 because the spherical harmonic functions are already normalized, and so yield a value of 1 rather than 4π when integrated over all angles θ and ϕ.

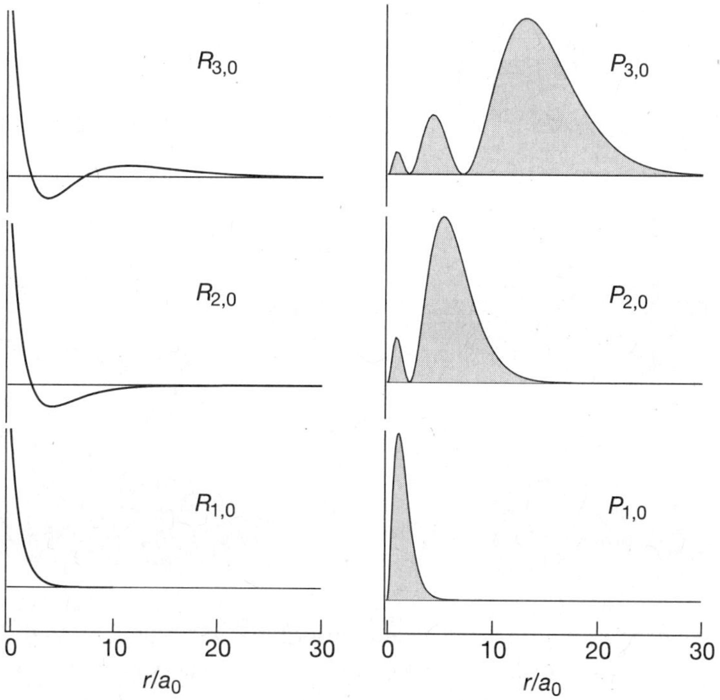

FIGURE 8.1 Radial wavefunctions $R_{n,\ell}$ and radial probability distribution functions $P_{n,\ell}$ for the first three *s* orbitals of hydrogen (with $n = 1, 2,$ or 3 and $\ell = 0$).

of finding the electron *outside* of the sphere of radius r is $1 - \int_0^r R^2_{n,\ell}(r)\, r^2 dr$ (since the total probability of finding the electron anywhere is 1).

The curves in Figure 8.1 show the radial wavefunctions and radial probability densities associated with the 1*s*, 2*s*, and 3*s* orbitals of hydrogen. Notice that increasing the value of *n* increases the number of nodes (zero-crossing points) of the radial wavefunction and increases the most probable distance of the electron from the nucleus (as indicated by the location of the maximum value of the radial probability density).

The full radial and angular wavefunctions of hydrogen $\Psi_{n,\ell,m_\ell}(r,\theta,\phi)$ are obtained by multiplying $R_{n,\ell}(r)$ (Eq. 8.7) by the spherical harmonic functions $Y_\ell^{m_\ell}(\theta,\phi)$ (Eq. 7.56). For example, the following are the full eigenfunctions of hydrogen corresponding to $n = 1$ and 2.

$$
\begin{aligned}
\Psi_{1,0,0} &= \frac{1}{\sqrt{\pi}}\left(\frac{1}{a_0}\right)^{3/2} e^{-r/a_0} \\
\Psi_{2,0,0} &= \frac{1}{4\sqrt{2\pi}}\left(\frac{1}{a_0}\right)^{3/2}\left(2-\frac{r}{a_0}\right) e^{-r/2a_0} \\
\Psi_{2,1,0} &= \frac{1}{4\sqrt{2\pi}}\left(\frac{1}{a_0}\right)^{3/2}\left(\frac{r}{a_0}\right) e^{-r/2a_0}\cos\theta \\
\Psi_{2,1,\pm 1} &= \mp\frac{1}{8\sqrt{\pi}}\left(\frac{1}{a_0}\right)^{3/2}\left(\frac{r}{a_0}\right) e^{-r/2a_0}\sin\theta\, e^{\pm i\phi}
\end{aligned}
\tag{8.9}
$$

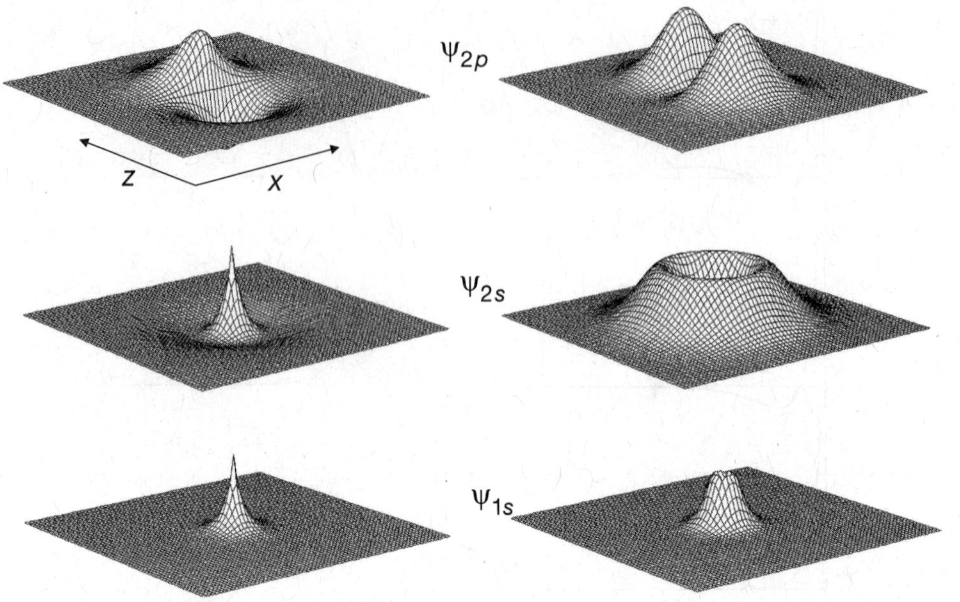

FIGURE 8.2 Hydrogen atom orbital slices through the *x-z* plane. The left-hand plots show $\Psi_{1,0} = \Psi_{1s}$, $\Psi_{2,0} = \Psi_{2s}$, and $\Psi_{2,1} = \Psi_{2p_z}$. The right-hand plots show the corresponding probability densities (weighted by r^2), $\Psi_{1s}^2 r^2$, $\Psi_{2s}^2 r^2$, and $\Psi_{2p_z}^2 r^2$.

Figure 8.2 shows plots of the 1*s*, 2*s*, and 3p_z orbitals (and probability densities) in the *x-z* plane. The plots on the left show Ψ, while those on the right show $\Psi^2 r^2$.

Since any linear combination of degenerate eigenfunctions of $\widehat{H}$ is also an eigenfunction of $\widehat{H}$, we can construct various sorts of *hybrid orbitals* from eigenfunctions with the same *n* value. For example, we can add and subtract a 2*s* and 2p_z orbital to obtain two new *sp* orbitals, $\frac{1}{\sqrt{2}}(\Psi_{1s} + \Psi_{2p_z})$ and $\frac{1}{\sqrt{2}}(\Psi_{1s} - \Psi_{2p_z})$. These two orbitals look identical, except that they are rotated by 180° with respect to each other. We can also combine a 2*s* orbital with a 2p_x and a 2p_y orbital to produce three sp^2 orbitals; these are again identical looking (and co-planar), except that they are rotated by 120° with respect to each other. Similarly, we can combined all four $n = 2$ orbitals to produce the following four equivalent sp^3 orbitals, which are oriented tetrahedrally with respect to each other, as shown in Fig. 8.3.

$$\tfrac{1}{2}\left(\Psi_{2s} + \Psi_{2p_x} + \Psi_{2p_y} + \Psi_{2p_z}\right)$$
$$\tfrac{1}{2}\left(\Psi_{2s} - \Psi_{2p_x} - \Psi_{2p_y} + \Psi_{2p_z}\right)$$
$$\tfrac{1}{2}\left(\Psi_{2s} - \Psi_{2p_x} + \Psi_{2p_y} - \Psi_{2p_z}\right)$$
$$\tfrac{1}{2}\left(\Psi_{2s} + \Psi_{2p_x} - \Psi_{2p_y} + \Psi_{2p_z}\right)$$

These four identical looking hybrid orbitals are all necessarily eigenfunctions of $\widehat{H}$. Thus, the degeneracy of the $n = 2$ orbitals is not accidental, but

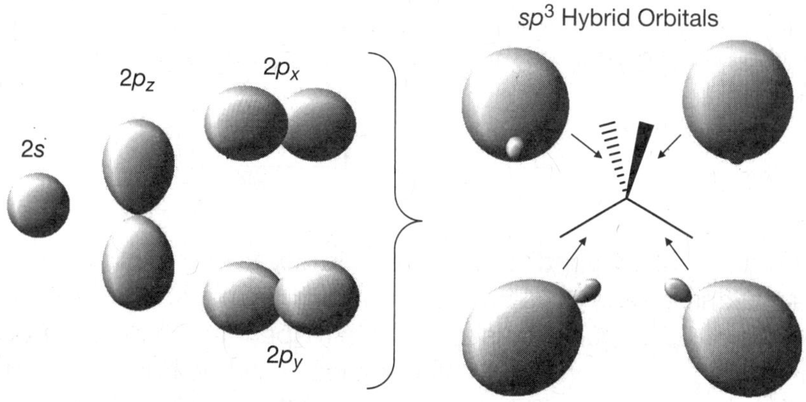

FIGURE 8.3 Tetrahedral sp^3 hybrid orbitals are produced from linear combinations of $2s$, $2p_x$, $2p_y$, and $2p_z$ wavefunctions. The plotted functions correspond to the probability densities produced by forming linear combinations of the spherical harmonic $Y_\ell^{m_\ell}$ contributions to each of the $n = 2$ hydrogen orbitals.

reflects a deeper symmetry relationship between the four Ψ_{2,ℓ,m_ℓ} wavefunctions.[9] Such hybrid orbitals also play an important role in facilitating the formation of covalent bonds and determining the geometries of polyatomic molecules.

The solution of the Schrödinger equation for hydrogen may also be used to predict the orbitals and energies of one-electron ions such as He^+ and Li^{2+}. The only difference between these ions and hydrogen is that they have a different nuclear charge Z (and mass). The orbitals of hydrogen-like atoms can be obtained simply by replacing a_0 by a_0/Z in the expressions for the hydrogen wavefunctions Ψ_{n,ℓ,m_ℓ}. Making this substitution in Eq. 8.5 also yields the following correct expression for the energies of hydrogen-like atoms.

$$E_n(Z) = -\left(\frac{\hbar^2}{2m_e a_0^2}\right)\frac{Z^2}{n^2} = -hcR_H\left(\frac{Z^2}{n^2}\right) \tag{8.10}$$

Bohr obtained this same formula using his original semiclassical quantum theory. He then used his predictions to demonstrate that some previously detected astronomical spectral lines had been incorrectly assigned to hydrogen. More specifically, he showed that although those lines did not agree with Eq. 8.4, they agreed perfectly with Eq. 8.10 when $Z = 2$, and so he correctly assigned the lines to He^+ ions.

The eigenfunctions of hydrogen-like atoms (Eq. 8.9) may also be used to determine the expectation values of experimentally relevant quantities.

[9] This deeper symmetry relation was first recognized by Pauli in 1926, and was later described in detail by Bander and Itzykson in a 1966 article in the *Reviews of Modern Physics*.

For example, by evaluating the integral $\langle r\rangle = \langle n\,\ell\,m_\ell|\,r\,|n\,\ell\,m_\ell\rangle$, one may obtain the following expression for the mean radius of various hydrogen-like orbitals.

$$\langle r\rangle = \left(\frac{a_0}{Z}\right)n^2\left\{1+\frac{1}{2}\left[1-\frac{\ell(\ell+1)}{n^2}\right]\right\} \tag{8.11}$$

Notice that $\langle r\rangle$ only depends on n and ℓ. This is because $\langle r\rangle$ is averaged over all orientations and so does not depend on m_ℓ (because m_ℓ describes the orientation of the angular momentum with respect to the z-axis). Equation 8.11 indicates that orbitals with higher angular momentum (larger ℓ) have a smaller average radius. This is consistent with the behavior of a classical electron rotating around a nucleus that rotates more rapidly when it is closer to the nucleus. The same is also true of the orbital period of planets rotating around the Sun, as the inner planets have shorter periods than the outer planets. This similarity in behavior results from the fact that both coulomb and gravitational potentials are proportional to $1/r$.

The expectation value of the potential energy may be obtained by evaluating the following integral $\langle V\rangle = -\frac{e^2}{4\pi\varepsilon_0}\left\langle\frac{1}{r}\right\rangle = -\frac{e^2}{4\pi\varepsilon_0}\langle n\,\ell\,m_\ell|\,\frac{1}{r}\,|n\,\ell\,m_\ell\rangle$. The result is surprisingly simple, as it indicates $\langle V\rangle = 2\langle H\rangle$. In other words, the average potential energy of a hydrogen-like orbital is exactly twice its total energy $\langle V\rangle = 2E_n$. Moreover, since $\langle H\rangle = \langle K\rangle + \langle V\rangle$, the above result implies that $\langle K\rangle = \langle H\rangle - \langle V\rangle = E_n - 2E_n = -E_n$. This is again consistent with the behavior of a classical system with $V(r) \propto 1/r$, for which the virial theorem (Eq. 1.46 on page 41) predicts that $\langle V\rangle/\langle K\rangle = -2$.

8.2 Spin Angular Momentum

The orbital angular momentum of an electron is analogous to that which the Earth has when it rotates around the Sun. It turns out that electrons also have *spin* angular momentum, which is analogous to the rotation of the Earth about its own axis. However, electron spin is inherently quantum mechanical and relativistic in origin and so behaves very differently from a classical spinning object. For example, the measurement of an electron's spin angular momentum along any axis can only produce one of two values, referred to as *spin up* and *spin down*, which differ by exactly $\hbar$.[10]

The idea that electrons could possess an intrinsic degree of freedom with two possible values was first proposed by Wolfgang Pauli in 1924. Soon thereafter, George Uhlenbeck and Samuel Goudsmit (while they were students of Paul Ehrenfest) suggested that the fine-structure of hydrogen's

[10] Since angular momentum states differ by $\hbar$, they would vanish as $\hbar \to 0$. This is why electron spin is said to be a fundamentally quantum mechanical kind of angular momentum that vanishes in the classical limit.

spectrum could be explained if one assumed that electrons possess two spin states. More specifically, the existence of electron spin was suggested by the splitting of atomic lines in a magnetic field, which implied that electrons have an intrinsic magnetic moment, similar to that produced by a rotating charge (or electromagnet).[11]

The first experiment that clearly revealed the strange properties of spinning electrons was performed by Otto Stern and Walther Gerlach in 1922 – before anyone suspected that electrons have spin states! Stern and Gerlach were intending to test earlier predictions pertaining to orbital angular momentum. More specifically, in 1920 the same Wolfgang Pauli (while he was still a graduate student) had suggested that the experimental magnetization of metals like iron could be explained by introducing *space quantization*. This strange concept seemed to imply that not only is angular momentum quantized but also its projection along any (laboratory) axis. Since this suggestion did not make any physical sense, it was assumed to be only a mathematical relation whose physical significance remained to be discovered.

Early one cold winter morning in 1920, while lying in bed daydreaming under his warm blankets, Otto Stern wondered how he might perform an experiment to determine whether space quantization actually exists. The scheme he proposed for doing this was very simple. He thought that if atoms possess a magnetic moment, then he could use a magnetic field to measure the projections of their magnetic moments onto the external magnetic field axis. Stern knew that if a magnet were placed in an inhomogeneous (spatially varying) magnetic field, the magnet would experience a force very similar to that of a ball rolling down a gravitational field gradient.

---

[11] Ralph Kronig had earlier proposed a similar idea, but Pauli convinced him not to publish it, because he did did not believe it. However, some years later, in 1927, Pauli finally accepted the idea and worked out a detailed mathematical description of spin. About one year after that, Paul Dirac showed that electron spin arises naturally as a prediction of relativistic quantum theory. The experimental value of the spin magnetic moment of an electron is twice as large as would be expected if it were nonrelativistic. Before Dirac developed his relativistic theory of electron spin, Niels Bohr had suspected that this factor of two was a relativistic effect. This was subsequently confirmed by Llewellyn Hilleth Thomas in 1926, while he was a graduate student at Cambridge. Thomas went on to play a key role in the development of quantum statistics (i.e., the Thomas-Fermi model), which is a predecessor of modern density functional theory. Thomas also contributed significantly to the development of computers – he invented the first core memory in 1946 while he was a professor of Physics at the Watson Labs in Columbia University (where he also taught the first course ever offered on the numerical solution of differential equations using computers).

If the magnet was aligned along the field, then it would experience a force driving it in the higher field direction; whereas if it was aligned against the field, it would be driven in the lower field direction. If space quantization was real, then a beam of atoms with quantized angular momentum projections would behave like magnets pointing in particular (discrete) directions in space. Thus, such a beam of atoms should split into spatially separated beams as it passed through an inhomogeneous magnetic field.

When Stern told his former boss Max Born about his plans to perform such an experiment, here is how Born described his reaction:

> *It took me quite some time before I took the idea seriously. I thought always that [space] quantization was a kind of symbolic expression for something which you don't understand. But to take this literally like Stern did, this was his own idea...I tried to persuade Stern that there was no sense [in it], but then he told me that it was worth a try. Reprinted with permission from Bretislav Friedrich, Dudley Herschbach, Stern and Gerlach: How a Bad Cigar Helped Reorient Atomic Physics, Physics Today, Dec. 1, 2003, American Institute of Physics.*

In designing his experiment, Stern chose to pass a beam of silver (Ag) atoms through an inhomogeneous magnetic field. A beam of hydrogen atoms might have been theoretically preferable, but such a beam would be more difficult to generate, as silver atoms could be produced simply by evaporation of molten silver. Moreover, it was known that an isolated Ag atom behaved in many respects like it had a single hydrogen-like electron. For example, many of the optical transitions of both hydrogen and silver had a doublet (two line) fine structure.[12] Before the development of the Schrödinger equation, such doublets were thought to arise from the orbital angular momentum of the electron. Thus, before performing his experiment, Stern predicted that space quantization should split his beam of silver atoms into two beams.[13] However, we now know that Ag atoms have no net orbital angular momentum, but they do have one unpaired electron with spin angular momentum. So, Stern's choice of Ag for his experimental test of space quantization was really a lucky accident.

---

[12] When a highly excited H atom undergoes a transition from a 3*s* state to a 2*p* state, the spectrum appears as a doublet centered around 656.3 nm. We now know that this fine structure arises from the interaction between the orbital angular momentum of the 2*p* states and the spin angular momentum of the electron. In other words, the electron experiences a magnetic field arising from its own orbital angular momentum, and is thus split into two states arising from the two (spin up and spin down) projections of the electron spin onto the orbital magnetic field.

[13] Stern published a paper in 1921 in which he described the experiment that he and Gerlach intended to perform. This paper included his calculations based on the old Bohr-Sommerfeld atomic theory, which implied that a beam of neutral silver atoms should be split into two beams by an inhomogeneous magnetic field.

It took Stern and Gerlach over a year to set up their experiment. Their initial results were inconclusive, as they could not see evidence of any silver atoms deposited on the detection plate. Here is how Stern described the events that led to the ultimate success of their experiment.

> *After venting to release the vacuum, Gerlach removed the deflector flange. But he could see no trace of the silver atom beam and handed the flange to me. With Gerlach looking over my shoulder as I peered closely at the plate, we were surprised to see gradually emerge the trace of the beam.... Finally we realized what [had happened]. I was then the equivalent of an assistant professor. My salary was too low to afford good cigars, so I smoked bad cigars. These had a lot of sulfur in them, so my breath on the plate turned the silver into silver sulfide, which is jet black, so easily visible. It was like developing a photographic film. Reprinted with permission from Bretislav Friedrich, Dudley Herschbach, Stern and Gerlach: How a Bad Cigar Helped Reorient Atomic Physics, Physics Today, Dec. 1, 2003, American Institute of Physics.*

What they saw on the developed plates clearly revealed that the silver atoms were split into two beams by the inhomogeneous magnetic field, just as Stern had predicted. Thus, the angular momentum of each silver atom did indeed appear to be quantized along the axis defined by the external magnetic field. This result was quite astonishing to other prominent physicists, as it implied that the projection of an electron's angular momentum along *any axis* really was spatially quantized.[14] It took several years before it was realized that the two spots Stern and Gerlach observed arose from the two spin states of silver's one unpaired electron.

Recall that the degeneracy of orbital angular momentum quantum states is $2\ell + 1$. In an external magnetic field, this degeneracy is broken to reveal $2\ell + 1$ magnetic sublevels. Thus, the observation of two spots in the Stern-Gerlach experiment suggests that $2\ell + 1 = 2$, which is only possible if $\ell = 1/2$. This strange half-integral quantum number is now recognized as a new spin angular momentum quantum number s, which only has one possible value.

$$s = \frac{1}{2}$$

[14] The splitting of the beam of silver atoms seemed to imply that the atoms were in some way able to align themselves to the imposed magnetic field. However, Einstein and Ehrenfest, among others, were not satisfied by this explanation because there was no mechanism by which the randomly oriented silver atoms could align themselves to the magnetic field under the collision-free experimental conditions. For these and other reasons, Stern and Gerlanch's results made it clear that something very strange was going on.

The projections of an electron's spin angular momentum along the z-axis are again bounded by the values of s, such that $-s \le m_s \le s$. Since m_s can only increase or decrease by whole numbers, it can only have two possible values.

$$m_s = \pm\frac{1}{2}$$

In other words, the total root-mean-squared magnitude of an electron's spin angular momentum is $\hbar\sqrt{s(s+1)} = \hbar\sqrt{3}/2$, while the projection of an electron's spin angular momentum can only have two possible values $\pm m_s\hbar = \pm\frac{1}{2}\hbar$. The latter two eigenvalues are referred to as the *spin-up* and *spin-down* states of an electron. It is these spin states that give rise to the pairing of electron spins in atoms and molecules, as we will see.

Pauli Spin Matrices

The orbital angular momentum operators $\widehat{L^2}, \widehat{L}_x, \widehat{L}_y, \widehat{L}_z$ (see Eqs. 7.46–7.49 on page 254–255) may be represented as matrices. This is particularly convenient when applied to spin angular momentum, in which case the operators $\widehat{L}_x$, $\widehat{L}_y$, and $\widehat{L}_z$ may be represented by the following Pauli spin matrices.

$$\widehat{S}_x = \frac{\hbar}{2}\begin{pmatrix} 0 & 1 \\ 1 & 0 \end{pmatrix}$$

$$\widehat{S}_y = \frac{\hbar}{2}\begin{pmatrix} 0 & -i \\ i & 0 \end{pmatrix}$$

$$\widehat{S}_z = \frac{\hbar}{2}\begin{pmatrix} 1 & 0 \\ 0 & -1 \end{pmatrix} \tag{8.12}$$

The two spin states $|s, m_s\rangle = \left|\frac{1}{2}, \pm\frac{1}{2}\right\rangle$ are often represented using the shorthand notation $|+\rangle = \left|\frac{1}{2}, \frac{1}{2}\right\rangle$ (spin up), and $|-\rangle = \left|\frac{1}{2}, -\frac{1}{2}\right\rangle$ (spin down).[15] These spin eigenfunctions may also be represented as column vectors that, when operated on by $\widehat{S}_z$, yield the corresponding eigenvalues $m_s\hbar = \pm\frac{\hbar}{2}$.

$$\widehat{S}_z\,|+\rangle = \frac{\hbar}{2}\begin{pmatrix} 1 & 0 \\ 0 & -1 \end{pmatrix}\begin{pmatrix} 1 \\ 0 \end{pmatrix} = +\frac{\hbar}{2}\begin{pmatrix} 1 \\ 0 \end{pmatrix} = \frac{\hbar}{2}\,|+\rangle$$

$$\widehat{S}_z\,|-\rangle = \frac{\hbar}{2}\begin{pmatrix} 1 & 0 \\ 0 & -1 \end{pmatrix}\begin{pmatrix} 0 \\ 1 \end{pmatrix} = -\frac{\hbar}{2}\begin{pmatrix} 0 \\ 1 \end{pmatrix} = -\frac{\hbar}{2}\,|-\rangle$$

The operator associated with the square of the spin angular momentum $\widehat{S}^2$ may be obtained from the above spin matrices (using Eq. 7.49).

$$\widehat{S}^2 = \widehat{S}_x^2 + \widehat{S}_y^2 + \widehat{S}_z^2 = \tfrac{3}{4}\hbar^2\begin{pmatrix} 1 & 0 \\ 0 & 1 \end{pmatrix} \tag{8.13}$$

Note that $\frac{3}{4}\hbar^2 = s(s+1)\hbar^2 = \frac{1}{2}(\frac{1}{2}+1)\hbar^2$ is the eigenvalue of $\widehat{S}^2$. Thus, when $\widehat{S}^2$ is applied to either one of the two spin states, the result is $\widehat{S}^2\,|\pm\rangle =$

[15] The two spin states are sometimes referred to as $\alpha = |+\rangle$ and $\beta = |-\rangle$.

$\frac{3}{4}\hbar^2 |\pm\rangle$, which is consistent with the fact that both spin vectors have a root-mean-squared angular momentum expectation value of $\sqrt{\langle S^2\rangle} = \frac{\sqrt{3}}{2}\hbar$.

Angular Momentum Ladder Operators

The harmonic oscillator raising and lowering operators (see Eqs. 7.24 and 7.25 on page 246) may be generalized to obtain the following raising and lowering operators for rotation about the z-axis.

$$\begin{aligned}\widehat{l^+} &= \widehat{L_x} + i\widehat{L_y} \\ \widehat{l^-} &= \widehat{L_x} - i\widehat{L_y}\end{aligned} \tag{8.14}$$

When the $\widehat{l^+}$ and $\widehat{l^-}$ operators are applied to the spherical harmonic functions $\Psi_{\ell,m_\ell} \equiv |\ell, m_\ell\rangle$ (Eqs. 7.54 and 7.56), they have the effect of raising or lowering the m_ℓ quantum number, $\widehat{l^+}|\ell, m_\ell\rangle \propto |\ell, m_\ell + 1\rangle$ and $\widehat{l^-}|\ell, m_\ell\rangle \propto |\ell, m_\ell - 1\rangle$.[16]

The corresponding spin raising and lowering operators may be obtained in the same way from the $\widehat{S}_x$ and $\widehat{S}_y$ matrices.

$$\begin{aligned}\widehat{S}^+ &= \widehat{S}_x + i\widehat{S}_y = \frac{\hbar}{2}\left[\begin{pmatrix}0 & 1\\ 1 & 0\end{pmatrix} + \begin{pmatrix}0 & 1\\ -1 & 0\end{pmatrix}\right] = \hbar\begin{pmatrix}0 & 1\\ 0 & 0\end{pmatrix} \\ \widehat{S}^- &= \widehat{S}_x - i\widehat{S}_y = \frac{\hbar}{2}\left[\begin{pmatrix}0 & 1\\ 1 & 0\end{pmatrix} - \begin{pmatrix}0 & 1\\ -1 & 0\end{pmatrix}\right] = \hbar\begin{pmatrix}0 & 0\\ 1 & 0\end{pmatrix}\end{aligned} \tag{8.15}$$

It is easy to verify that these raising and lowering operators have the expected effect on the up and down spin states.

Exercise 8.1

Use the matrix representations of the spin raising and lowering operators to show that $\widehat{S}^+|-\rangle = \hbar|+\rangle$, $\widehat{S}^+|+\rangle = 0$, $\widehat{S}^-|+\rangle = \hbar|-\rangle$, and $\widehat{S}^-|-\rangle = 0$.

Solution.

$$\widehat{S}^+|-\rangle = \hbar\begin{pmatrix}0 & 1\\ 0 & 0\end{pmatrix}\begin{pmatrix}0\\ 1\end{pmatrix} = \hbar\begin{pmatrix}1\\ 0\end{pmatrix} = \hbar|+\rangle$$

$$\widehat{S}^+|+\rangle = \hbar\begin{pmatrix}0 & 1\\ 0 & 0\end{pmatrix}\begin{pmatrix}1\\ 0\end{pmatrix} = \hbar\begin{pmatrix}0\\ 0\end{pmatrix} = 0$$

$$\widehat{S}^-|+\rangle = \hbar\begin{pmatrix}0 & 0\\ 1 & 0\end{pmatrix}\begin{pmatrix}1\\ 0\end{pmatrix} = \hbar\begin{pmatrix}0\\ 1\end{pmatrix} = \hbar|-\rangle$$

[16] The constant of proportionality corresponding to $\widehat{l^+}$ may be shown to be $\hbar\sqrt{\ell(\ell+1) - m_\ell(m_\ell+1)}$, while that corresponding to $\widehat{l^-}$ is $\hbar\sqrt{\ell(\ell+1) - m_\ell(m_\ell-1)}$.

$$\widehat{S}^- \, |-\rangle = \hbar \begin{pmatrix} 0 & 0 \\ 1 & 0 \end{pmatrix} \begin{pmatrix} 0 \\ 1 \end{pmatrix} = \hbar \begin{pmatrix} 0 \\ 0 \end{pmatrix} = 0$$

8.3 Fermi, Bose, and Pauli Exclusion

So far, we have focused on systems whose quantum mechanical wavefunctions describe individual particles (such as a single electron). However, in order to treat multielectron atoms and molecules, we must establish a connection between the wavefunctions of individual electrons and those describing multiple electrons. In particular, we will see that the symmetry of such multiparticle wavefunctions determines whether two particles can or cannot occupy the same state. This is why two electrons may only occupy the same atomic or molecular wavefunction if they have different m_s values (and so are in different spin states).

The simplest multielectron atom is helium (He), which consists of two electrons bound to a positive nucleus composed of two protons and two neutrons (an alpha particle). If we assume for the moment that the two electrons do not interact with each other, but only with the nucleus, then the Hamiltonian associated with each electron would look identical to that of a hydrogen-like atom. Thus, if we label the position of one of the electrons as r_1 and the other as r_2, then the Hamiltonian of two noninteracting electrons would be $\widehat{H} = \widehat{H}_1 + \widehat{H}_2$, where $\widehat{H}_1 = \widehat{K}_1 + \widehat{V}_1 = -(\hbar^2/2m)\nabla_1^2 + V(r_1)$ and $\widehat{H}_2 = -(\hbar^2/2m)\nabla_2^2 + V(r_2)$. Under these conditions it is easy to demonstrate that the combined wavefunction Ψ of the two electrons may be expressed as a product of the wavefunctions ψ_i of the two individual electrons. For example, if $\psi_a(r_1)$ and $\psi_b(r_2)$ are two single-electron eigenfunctions with eigenvalues E_a and E_b, respectively, then the eigenfunction of the combined two-electron system would be $\Psi(r_1, r_2) = \psi_a(r_1)\psi_b(r_2)$ and would have an eigenvalue of $E_a + E_b$.

Exercise 8.2

Show that $\Psi(r_1, r_2) = \psi_a(r_1)\psi_b(r_2)$ is an eigenfunction of the Hamiltonian for the above system, and determine the energy eigenvalue corresponding to the ground state of the system. Compare your results with the true ground-state energy of helium ($\approx -7{,}620$ kJ/mol) and suggest a possible reason for the discrepancy between the two values.

Solution. The Hamiltonian for such a system is

$$\widehat{H} = \widehat{H}_1 + \widehat{H}_2 = [-(\hbar^2/2m)\nabla_1^2 + V(r_1)] + [-(\hbar^2/2m)\nabla_2^2 + V(r_2)]$$

So, when $\widehat{H}$ operates on $\Psi(r_1, r_2) \propto \psi_a(r_1)\psi_b(r_2)$, then $\widehat{H}_1$ will operate only on $\psi_a(r_1)$ and $\widehat{H}_2$ will operate only on $\psi_b(r_2)$ to yield $\widehat{H}\Psi = E_a\Psi + E_b\Psi = (E_a + E_b)\Psi$.

The energies E_a and E_b for such noninteracting electrons may be obtained using Eq. 8.10. Thus, if both electrons are in the lowest energy ($n = 1$) orbital (with opposite spins), then the total energy of this two-electron system would be

$$E = E_a + E_b = -hcR_H\left(\frac{Z^2}{n^2}\right) - hcR_H\left(\frac{Z^2}{n^2}\right) = -2hcR_H\left(\frac{2^2}{1^2}\right) = -8hcR_H$$

$$\sim -8(1,313\ \text{kJ/mol}) \sim -10,500\ \text{kJ/mol}$$

The true ground-state energy of helium is nearly 30% smaller in magnitude than the above prediction, because we have neglected electron-electron interactions, which add an additional repulsive (positive) contribution to the energy.

Let's now consider what would happen if we exchanged one electron with the other to obtain $\Psi(r_1, r_2) = \psi_a(r_2)\psi_b(r_1)$. Since the two electrons are the same, the result should not depend on which one we label as electron 1 or 2, and so the latter wavefunction is also an acceptable eigenfuction of $\widehat{H}$, and again has an eigenvalue of $E_a + E_b$. More generally we could represent Ψ as any linear combination of the above two degenerate eigenfunctions to produce two other acceptable eigenfunctions $\Psi(r_1, r_2) = \frac{1}{\sqrt{2}}[\psi_a(r_1)\psi_b(r_2) \pm \psi_a(r_2)\psi_b(r_1)]$.[17]

Whether we choose to define Ψ as the sum or difference of $\psi_a(r_1)\psi_b(r_2)$ and $\psi_a(r_2)\psi_b(r_1)$ turns out to have surprisingly important chemical consequences. Notice that if we assume that $\Psi(r_1, r_2) = \frac{1}{\sqrt{2}}[\psi_a(r_1)\psi_b(r_2) + \psi_a(r_2)\psi_b(r_1)]$, then exchanging the electron subscripts yields $\Psi(r_2, r_1) = \frac{1}{\sqrt{2}}[\psi_a(r_2)\psi_b(r_1) + \psi_a(r_1)\psi_b(r_2)] = \Psi(r_1, r_2)$. However, if we assume that $\Psi(r_1, r_2) = \frac{1}{\sqrt{2}}[\psi_a(r_1)\psi_b(r_2) - \psi_a(r_2)\psi_b(r_1)]$, then $\Psi(r_2, r_1) = \frac{1}{\sqrt{2}}[\psi_a(r_2)\psi_b(r_1) - \psi_a(r_1)\psi_b(r_2)] = -\Psi(r_1, r_2)$. In other words, the sign of Ψ either inverts or remains unchanged when we exchange the electrons. We might expect that the observable properties of the system would not be affected by such a sign change, since the probability density $\Psi^*\Psi$ is independent of the sign of Ψ and so is the same for both the symmetric (+) and antisymmetric (−) linear combinations of electron wavefunctions. However, it turns out that the symmetry of such wavefunctions plays a critical role in dictating the structures of atoms and molecules.

[17] If we operate on $\Psi(r_1, r_2)$ with $\widehat{H} = \widehat{H}_1 + \widehat{H}_2$, it is again easy to show that $\widehat{H}\Psi(r_1, r_2) = (E_a + E_b)\Psi(r_1, r_2)$. Notice that the wavefunction $\psi_a(r_1)\psi_b(r_2)$ should be normalized to two, since this combined wavefunction describes two electrons. When constructing other linear combinations such as $\frac{1}{\sqrt{2}}[\psi_a(r_1)\psi_b(r_2) \pm \psi_a(r_2)\psi_b(r_1)]$, we must introduce a factor of $\frac{1}{\sqrt{2}}$ so that the resulting wavefunction remains properly normalized (to two).

Experimental evidence indicates that some kinds of particles can only occupy antisymmetric multiparticle wavefunctions, while others can only occupy symmetric wavefunctions. Particles whose wavefunctions are antisymmetric, and so change sign when swapping identities, are called *Fermi* (or Fermi-Dirac) particles, while those that do not change sign are called *Bose* (or Bose-Einstein) particles. Most of the particles from which atoms and molecules are composed, including electrons, protons, and neutrons, are Fermi particles, while photons and molecular vibrations behave like Bose particles.[18]

The chemical importance of the distinction between Fermi and Bose particles becomes evident if we consider what would happen if we assumed that the two electrons in a helium atom have identical wavefunctions (with identical quantum numbers, including spin) so $\psi_a = \psi_b = \psi$. Since electrons are Fermi particles, this would imply that $\Psi = \frac{1}{\sqrt{2}}[\psi\psi - \psi\psi] = 0$, which indicates that there cannot be any electrons in such a wavefunction. In other words, *no two Fermi particles can occupy a state with an identical set of quantum numbers*.

At the start of this section, we assumed that the two electrons in a helium atom do not interact with each other, and so the eigenfunctions of helium could be described as the products of one-electron wavefunctions, and the associated symmetric or antisymmetric linear combinations. However, the symmetry properties of the wavefunctions of Fermi and Bose particles are not restricted to systems with separable solutions. In other words, it is always the case that *no two Fermi particles can occupy the same quantum state*. Thus, Fermi particles behave as if they exclude each other, while Bose particles do not. This property of Fermi particles is known as the *Pauli exclusion principle*. It plays a tremendously important role in chemistry, as it requires that no more than two electrons (of opposite spin) can fill each atomic orbital. Thus, the structure of the periodic table is dictated in large part by the fact that electrons are Fermi particles (as further described in Section 8.4).

Nuclear Spin and Molecular Rotational Quantization

The Pauli principle also plays a surprisingly important role in molecular rotational quantization. This is because atomic nuclei may have either half-integral or integral values of nuclear spin (depending on the number and

---

[18] More generally, any particles that have a half-integral spin ($S = \frac{1}{2}, \frac{3}{2}, \frac{5}{2} \ldots$) are found to behave like Fermi particles (or Fermions), while particles with integral spin ($S = 1, 2, 3 \ldots$) are found to behave like Bose particles (or Bosons). Atomic nuclei with an even number of protons plus neutrons must have an integral total spin and so behave like Bosons, while those with an odd number of protons plus neutrons behave like Fermions.

alignment of the proton and neutron spins in the nucleus). For example, a hydrogen atom $H = {}^1H$ (whose nucleus contains a single proton) has a nuclear spin of $\frac{1}{2}$, while a deuterium atom $D = {}^2H$ (whose nucleus contains one proton and one neutron) has a nuclear spin of 1; so, ^1H is a Fermion, while D is a Boson.[19]

Since homonuclear diatomics are composed of two identical atoms, their rotational quantum states are influenced by nuclear spin statistics (while the rotational states of heteronuclear diatomics are not). Note that the rotational eigenfunctions of homonuclear diatomics may be either symmetric (for $J = 0, 2, 4\ldots$) or antisymmetric (for $J = 1, 3, 5\ldots$) with respect to exchange of the two nuclei. Thus, the Pauli exclusion principle requires that a particular homonuclear diatomic, with a given nuclear spin configuration, can only have rotation states of either even or odd J.

For example, H_2 may exist in two different nuclear spin states; if its two nuclear spins are aligned in opposite directions, it is called *para*-H_2, while if its two nuclear spins are aligned in the same direction, it is called *ortho*-H_2. Since the nuclear spins of *para*-H_2 are antisymmetric with respect to nuclear exchange, *para*-H_2 can only have symmetric rotational states with even J values. On the other hand, the aligned nuclear spins of *ortho*-H_2 are symmetric with respect to nuclear exchange, and so *ortho*-H_2 can only have antisymmetric rotational states with odd J values. Thus, the ground rotational state of *para*-H_2 is the $J = 0$ state, while the ground state of *ortho*-H_2 is the $J = 1$ state.[20]

The consequences of nuclear spin statistics can be detected spectroscopically, since nuclear statistics restricts the allowed values of J for symmetric molecules (see for example Figure 9.4 in Chapter 9). Nuclear spin statistics also influences molecular partition functions since, for example, homonuclear diatomics have only half as many rotational states as heteronuclear

---

[19] All other atomic nuclei also have a fixed nuclear spin quantum number. For example, ^{12}C and ^{13}C have nuclear spins of 0 and $\frac{1}{2}$, respectively, while ^{14}N and ^{16}O have nuclear spins of 1 and 0, respectively.

[20] At room temperature the difference between the ground-state energies of *para*-H_2 and *ortho*-H_2 is too small to make a significant difference in their relative populations. However, *para*-H_2 has nuclear spin degeneracy of 1 (because it has a total nuclear spin of $S = \frac{1}{2} - \frac{1}{2} = 0$ and so a spin degeneracy of $2S + 1 = 1$), while *ortho*-H_2 has a nuclear spin degeneracy of 3 (because it has a total nuclear spin of $S = \frac{1}{2} + \frac{1}{2} = 1$ and so a spin degeneracy of $2S + 1 = 3$). This factor of 3 difference between the degeneracies of the two forms of H_2 means that the equilibri population of *ortho*-H_2 is three times greater than that of *para*-H_2. More gen ally, a homonuclear diatomic composed of atoms, each of which has a nucle spin of S, will have $(S + 1)(2S + 1)$ symmetric nuclear spin states and $S(2S + 1$ antisymmetric nuclear spin states.

diatomics (as further discussed in Chapter 10, on pages 345–347). However, both homonuclear and heteronuclear diatomics are predicted (using Eq. 1.23) to have an average rotational energy of RT and a rotational heat capacity of R, in the high temperature limit.

8.4 Multielectron Atoms and the Periodic Table

The structure of the periodic table is dictated by the electronic properties of atoms. The electronic structure of each atom is influenced by the Pauli exclusion principle, since no more than two electrons (of opposite) spin can occupy a given atomic orbital. Atomic structures are also influenced by space quantization, as this restricts the relative orientations of electron angular momentum vectors to discrete (quantized) values.

The Filling Up Principle

The order in which atomic orbitals are filled with electrons – known as the *Aufbau* ("filling up") principle – was originally formulated by Bohr and Pauli. The *Aufbau* principle makes use of the fact that the ground state of an atom is that state with the lowest total energy, while conforming to the Pauli exclusion principle.

In order to predict the ground-state structure of a multielectron atom, we must consider the role of electron-electron interactions in breaking the degeneracy of orbitals with the same principal quantum number n. Recall that the mean radius of hydrogen-like orbitals decreases as ℓ increases (see Eq. 8.11). This implies that electrons in states with a larger ℓ value (but the same n value) are closer together, and so have greater (more positive) electron-electron repulsion. This is why the $2p$ electrons of carbon (with $n = 2$ and $\ell = 1$) are higher in energy than the $2s$ electrons (with $n = 2$ and $\ell = 0$).

The fact that atomic orbital energies increase with increasing ℓ implies that at some point they will become greater than the next higher s orbital. The ℓ value at which this occurs may be predicted using the so-called "$n + \ell$ rule" (proposed by Erwin Madelung), which states that orbitals with a smaller value of $n + \ell$ have lower energies. In cases where two orbitals have the same $n + \ell$ value (but different n and ℓ values), the orbital with the lowest value of n is expected to have a lower energy. This rule is consistent with the filling sequence indicated by the diagram at the top of Fig. 8.4. The physical basis for the $n + \ell$ rule may be traced to the fact that the number of (radial and angular) nodes in an atomic orbital increases as $n + \ell$ increases, and wavefunctions with a larger number of nodes typically have higher energies (as is the case for a particle in a box and a harmonic oscillator, etc.).

The sequence in which atomic orbitals are filled determines the structure of the periodic table, as indicated in Fig. 8.4. For example, the $3d$

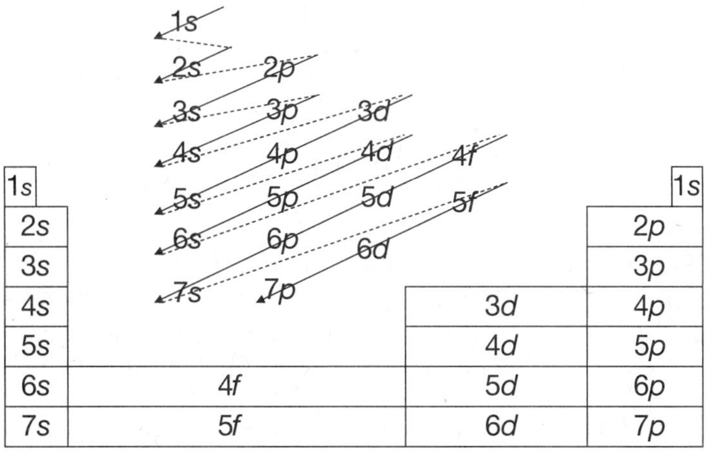

FIGURE 8.4 Atomic ground-state energies, and the structure of the periodic table, are dictated by the Pauli exclusion principle. The filling sequence indicated by the diagram (above) and the periodic table (below) is that predicted by the $n + \ell$ rule. However, the energies of the orbitals connected by dashed lines are so close to each other that some atoms do not fill in the indicated order (as is the case, for example, for the coinage metals Cu, Ag, and Au).

orbital of potassium (K) has a higher energy than its 4*s* orbital. So, the ground state of K is one in which its 19 electrons have a configuration of $1s^2 2s^2 2p^6 3s^2 3p^6 4s^1$, with one lone (unpaired) electron in a 4*s* (rather than a 3*d*) orbital. The electron configuration of K can also be designated as $[\text{Ar}]4s^1$, since the core electrons of potassium have the same closed-shell configuration as argon (Ar).

Some atoms have electron configurations that do not conform to this simple rule. One such exception is Ag, whose 47 electrons have a ground-state configuration of $[\text{Kr}]5s^1 4d^{10}$. The fact that the unpaired electron is in a 5*s* rather than a 4*d* state is inconsistent with the $n + \ell$ rule (and the filling sequence indicated in Fig. 8.4, which implies that the 5*s* orbital should fill before the 4*d* orbital).[21]

The reason for such exceptions to the $n + \ell$ rule may be traced to the fact that the orbitals that are connected by dashed lines in Fig. 8.4 have very similar energies. The actual order in which electrons fill these orbitals is determined by a delicate balance of competing electron-electron repulsion and spin-orbit coupling interactions, as well as relativistic effects. These interactions also dictate how electrons will distribute themselves within a partially filled subshell. Predicting this distribution requires considering the influence of both orbital and spin angular momentum on the relative energies of various valence electron configurations.

---

[21] The other *coinage metals* (Cu and Au) and some other atoms (such as Cr) also have ground-state electron configurations that don't conform to the $n + \ell$ rule.

Atomic Term Symbols

The orbital angular momentum of an atom arises from the vector sum of the orbital angular momentum vectors of each of its electrons. The resulting total orbital angular momentum may be described by the quantum number number L. For example, for two electrons with orbital angular momentum quantum numbers ℓ_1 and ℓ_2, the possible values of L are $\ell_1 + \ell_2 \geq L \geq |\ell_1 - \ell_2|$. The actual value of L is determined by the relative alignments of the orbital angular momentum vectors of the electrons. In other words, each value of L corresponds to a different set of electron m_ℓ values ($-\ell \leq m_\ell \leq \ell$). The space quantization of m_ℓ implies that L can also only take on integer values between $\ell_1 + \ell_2$ and $|\ell_1 - \ell_2|$.

Similarly, the total spin quantum number S of an atom is obtained from the corresponding electron spin quantum numbers s_i. Thus, two electrons (with spin quantum numbers $s_1 = \frac{1}{2}$ and $s_2 = \frac{1}{2}$) can combine to produce S values of either $s_1 + s_2 = 1$ or $|s_1 - s_2| = 0$.

If two electrons are spin-paired, then they have no net spin $S = 0$. The number of magnetic sublevels associated with such an electron pair is $2S + 1 = 1$, and so they are said to have a spin multiplicity of one, or to be in a *singlet* state.

If the two electron spins are aligned in the same direction, then they have a total spin quantum number of $S = 1$ and $2S + 1 = 3$, and so are described as being in a *triplet* state.[22]

A single (unpaired) electron must have a value of $S = \frac{1}{2}$, since in this case $S = s = \frac{1}{2}$. Such an unpaired electron has a spin multiplicity of $2S + 1 = 2$, and so is said to be in a *doublet* state. For example, the one unpaired electron of Ag is in such a doublet state, and that is why it produced two spots in the Stern-Gerlach experiment.

The total angular momentum of an atom – including both orbital and spin contributions – is dictated by its total angular momentum quantum number $J = L + S$. The allowed values of J are determined by the relative alignments of the orbital and spin angular momenta, since space quantization again requires that J may only take on values of $L + S \geq J \geq |L - S|$ that differ by integer increments. As a result, each value of J gives rise to $2J + 1$ magnetic sublevels. In other words, if a Stern-Gerlach experiment were performed on an atom with a total angular momentum quantum number of J, then one would expect to observe $2J + 1$ separate spots on the detection plate. Since the ground state of Ag has $S = \frac{1}{2}$ and $L = 0$, its J value must be $J = L + S = \frac{1}{2}$ and so $2(\frac{1}{2}) + 1 = 2$, which is again consistent with the fact that two spots were observed.

---

22 Such a triplet state can only occur when the two electrons have different values of n, ℓ, and/or m_ℓ, as the Pauli exclusion principle requires that no two electrons can have the same values of n, ℓ, m_ℓ, and m_s.

The projections of the orbital and spin angular momenta onto the z-axis are determined by the quantum numbers M_S and M_L. Just as there are $2S+1$ values of M_S associated with every value of S, there are also $2L+1$ values of M_L associated with a given value of L. The specific values of M_S and M_L are determined by the sum of the corresponding m_s and m_ℓ values of all the electrons, $M_S = m_{s,1} + m_{s,2} + m_{s,3} \ldots$ and $M_L = m_{\ell,1} + m_{\ell,2} + m_{\ell,3} \ldots$, and so the projection of the total angular momentum along the z-axis is determined by the quantum number $M_J = M_L + M_S$.

The total angular momentum operators $\widehat{J}^2$ and $\widehat{J}_z$ commute with $\widehat{H}$. This implies that eigenfunctions of $\widehat{H}$ are also eigenfunctions of $\widehat{J}^2$ and $\widehat{J}_z$. Thus, the quantum numbers J and M_J are referred to as "good" atomic quantum numbers. The quantum numbers L, S, M_L, and M_S do not have the same status because the corresponding operators do not in general commute with $\widehat{H}$. In other words, orbital and spin angular momenta are coupled to each other and so the corresponding multielectron wavefunctions are not separable into a product of orbital and spin wavefunctions. However, since spin-orbit coupling is often relatively weak, the latter quantum numbers may nevertheless be useful in describing the states of multielectron atoms. Thus, the following *Russell-Saunders term symbols* have been developed as a short-hand notation for specifying the values of S, L, and J associated with a particular multielectron configuration.

$$\boxed{{}^{(2S+1)}L_J} \tag{8.16}$$

The left-superscript $2S+1$ indicates the spin-multiplicity. The orbital angular momentum quantum number $L = 0, 1, 2, 3, \ldots$ is designated by the letters S, P, D, F,... (following the same convention used to designate the ℓ values of hydrogen-like atoms). The right-hand subscript indicates the total angular momentum quantum number J. Electrons in filled subshells (with the same n and ℓ value) have no net spin or orbital angular momentum. Thus, rare gases, which have completely filled shells with $S = L = J = 0$, necessarily have term symbols of ${}^1S_0$. For all other atoms, we only need to consider electrons in partially filled subshells when deriving atomic term symbols.

Exercise 8.3

Obtain the term symbols for the ground states of helium and lithium.

Solution. The ground state of a He atom has two spin-paired electrons in a $1s^2$ configuration, and so $S = |s_1 - s_2| = 0$ (and $2S+1 = 1$), $L = \ell_1 + \ell_2 = 0 + 0 = 0$, and $J = L + S = 0$. Thus, the ground state of He has a ${}^1S_0$ ("singlet S zero") term symbol. The ground state of a Li atom has a configuration of $1s^2 2s^1$ with a single unpaired electron in a $2s$ orbital, and so $S = s = \frac{1}{2}$, $L = 0$, and $J = L + S = \frac{1}{2}$. Thus, Li has a ${}^2S_{1/2}$ ("doublet S one-half") ground-state term symbol.

A subshell that is missing one electron has the same set of possible angular momentum values as a subshell containing a single electron. For example, a p ($\ell = 1$) orbital has three magnetic sublevels ($2\ell + 1 = 3$) and so can hold up to six electrons. Thus, a p^5 configuration has the same possible values of S, L, and J as a p^1 configuration (and the same is true of p^4 and p^2 configurations). For a half-filled subshell, such as p^3, the only value of L that is consistent with the Pauli exclusion principle is $L = 0$ (and so $S = J$).

Exercise 8.4

What are the possible values of S, L, and J for fluorine (F) that are consistent with the $n + \ell$ rule?

Solution. The $n + \ell$ rule (and the diagrams in Fig. 8.4) indicates that the ground state of fluorine (F) has five $2p^5$ valence electrons. Since this configuration is missing one electron, the resulting quantum numbers are the same as those of a $2p^1$ configuration. Thus, the possible values of the spin and orbital angular momentum quantum numbers are $S = s = \frac{1}{2}$ and $L = \ell = 1$. The values of L and S place the following bounds on the total angular momentum $L + S \geq J \geq |L - S|$, and thus there are two possible values of $J = \frac{3}{2}$ or $\frac{1}{2}$.

Which of the above total angular momentum quantum numbers correspond to the actual ground-state fluorine is determined by a balance of competing interactions, whose relative magnitudes may be predicted using Hund's rules, as described below.

Atomic Ground-State Electron Structures

Multielectron atoms with partially filled subshells can have various possible term symbols. Each such term symbol corresponds to a particular arrangement of valence electrons. The term symbol corresponding to the lowest energy (ground) state of an atom is that which minimizes electron-electron repulsion and spin-orbit interactions.

One of the important contributions to electron-electron repulsion comes from *spin-correlation* (or *exchange stabilization*) energy. This arises from the fact that the Pauli exclusion principle prevents electrons with parallel spins from occupying states with the same n, ℓ, and m_ℓ values. Thus, unpaired electrons tend to stay farther apart from each other than electrons with paired spins. As a result, unpaired electrons have lower electron-electron repulsion, and so the ground states of atoms tend to favor configurations with higher spin multiplicity.

For electrons with a given set of ℓ values, the states of larger L tend to have lower overlap between electron wavefunctions, and so have lower energy than states with smaller L. However, in some cases the state with the largest value of L may be forbidden by the Pauli exclusion principle (as is the case for carbon, as we will see).

Spin-orbit interactions reduce the energy of atomic states in which the orbital and spin angular momentum vectors are aligned in opposite directions. Thus, spin-orbit coupling tends to stabilize states with lower J values. However, when orbitals are more than half full, the situation is inverted so states with higher J have lower energy.

The following set of *Hund's rules* (proposed by Friedrich Hund) have been developed to take the above effects into account and provide a simple procedure for predicting the ground-state term symbols of multielectron atoms:

1. For a given electron configuration, the term symbol with the largest S (highest spin multiplicity $2S + 1$) has the lowest energy.
2. For a given value of S, the term symbol with the largest L has the lowest energy.
3. The ground-state value of J is determined as follows:
   (a) If a valence shell is less than half full, then the state with lowest J has the lowest energy.
   (b) If a valence shell is more than half full, then the state with the highest J has the lowest energy.
   (c) If a valence shell is half filled, then the ground-state (with largest S) will always have $L = 0$ as the only state that is consistent with the Pauli exclusion principle, and so $J = S$.

Exercise 8.5

Use Hund's rules to predict the term symbol for the ground-state of a carbon atom.

Solution. The $n + \ell$ rule (and the diagrams in Fig. 8.4) indicates that an isolated carbon atom should have a ground-state electron configuration of $1s^2 2s^2 2p^2$. The two valence electrons each have a spin of $s = \frac{1}{2}$. These may either be paired and thus have a total spin of $S = 0$ or unpaired to produce a total spin of $S = 1$. The first of the above rules indicates that the ground-state should be a triplet state with $S = 1$ and $2S + 1 = 3$. The orbital angular momentum quantum numbers of the two valence electrons are both $\ell = 1$, and so the possible values of L are 0, 1, and 2. The second Hund rule indicates that the ground-state should be that with the largest possible value of L. However, the $L = 2$ configuration is one in which both valence electrons have $n = 2$, $\ell = 1$, $m_\ell = 1$, and $m_s = \frac{1}{2}$, which is forbidden by the Pauli exclusion principle. Thus, the ground-state configuration of carbon must have $L = 1$. Finally, the possible values of J range from $L + S = 2$ to $|L - S| = 0$. Since the valence shell is less than half full, the third Hund rule indicates that the ground-state will be that with $J = 0$. Thus, the ground-state of carbon is correctly predicted to be a $^3P_0$ ("triplet P zero") state.

Ground-state term symbols may also be derived using the following procedure, which makes it easier to recognize (and avoid) configurations that are forbidden by the Pauli exclusion principle. This begins by assigning values of $m_s = \frac{1}{2}$ and $m_\ell = \ell$ (the maximum possible value of m_ℓ) to the first electron in the partially filled (valence) subshell. Then the other subshell electrons are sequentially added to the remaining m_ℓ sublevels in decreasing order, $m_\ell = (\ell - 1) \cdots - \ell$. If the subshell is more than half filled, then the additional electrons are assigned $m_s = -\frac{1}{2}$ and sequentially paired with the electrons in the above m_ℓ sublevels in the same order (starting with $m_\ell = \ell$). The ground-state values of S and L may then be determined simply by adding up the m_ℓ and m_s values of all the electrons, $S = \sum m_s$ and $L = \sum m_\ell$. Notice that this procedure necessarily produces the maximum values of S and L, which are consistent with the Pauli exclusion principle. The ground-state value of J depends on whether the subshell is more or less than half full (as specified by Hund's rules).

Exercise 8.6

Use the above procedure to predict the ground-state configuration of the $2p^2$ valence electrons of a C atom.

Solution. Since $\ell = 1$, there are three possible values of $m_\ell = 1$, 0, and -1, which you could represent by three boxes, as shown below. Now put the two valence electrons into the first two boxes (with spin up).

$$m_\ell = \quad 1 \quad\quad 0 \quad\quad -1 \quad\quad L = 1 + 0 = 1$$

$$\boxed{\uparrow}\ \boxed{\uparrow}\ \boxed{}$$

$$m_s = \quad \tfrac{1}{2} \quad\quad \tfrac{1}{2} \quad\quad\quad S = \tfrac{1}{2} + \tfrac{1}{2} = 1$$

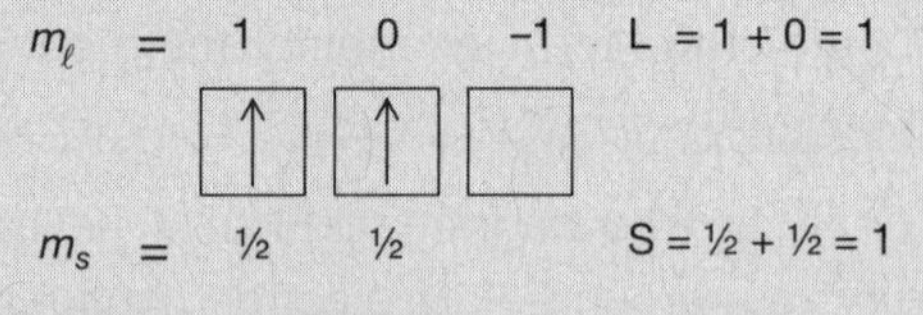

This diagram makes it easy to see that $S = \sum m_s = 1$ and $L = \sum m_\ell = 1$. Since the subshell is less than half full, the third Hund's rule requires that $J = |L - S| = 0$, and so the ground-state term symbol of C is correctly predicted to be $^3P_0$. This graphic procedure makes it very easy to correctly determine the ground-state term symbol for any atom, given its ground-state electron configuration.

The ground-state electron configuration of Ag is $[\text{Kr}]5s^1 4d^{10}$, which is exceptional in that it cannot be correctly predicted using the $n + \ell$ filling rule (as discussed on page 283). However, given the latter electron configuration, we may easily determine the ground-state term symbol of Ag.

Exercise 8.7

Use Hund's rules to determine the ground-state term symbol of the (Kr)$5s^1 4d^{10}$ ground-state configuration of a silver atom.

Solution. Since the filled $4d$ subshell of Ag has no net spin or orbital angular momentum, the lone $1s$ electron requires that $L = \ell = 0$, $S = s = \frac{1}{2}$, and so $J = L + S = \frac{1}{2}$ and therefore, the ground-state term symbol of Ag is ${}^2S_{1/2}$.

It is important to keep in mind that Hund's rules only apply to atomic ground states, and so cannot be used to correctly predict the term symbols of atomic excited states. Moreover, the above *LS coupling scheme* breaks down for very heavy atoms (such as those with $6p$ or higher valence electrons), for which relativistic spin-orbit interactions are sufficiently large that L and S are no longer useful quantum numbers. For such atoms it is more appropriate to use a *jj coupling scheme* in which each electron is first assigned a total one-electron angular momentum quantum number $\ell + s \geq j \geq |\ell - s|$, and then these quantum numbers are combined to obtain the total J and M_J values corresponding to different relative alignments of the electron angular momentum vectors.

HOMEWORK PROBLEMS

Problems That Illustrate Core Concepts

1. Note that all of the eigenfunctions of the hydrogen atom Hamiltonian, $\widehat{H}$, are also eigenfunctions of $\widehat{L}^2$.

(a) What are the values of $\langle H \rangle = \langle E \rangle$ (in J units) and $\sqrt{\langle L^2 \rangle}$ (in J s units) for the following hydrogen orbitals?

(i) $1s$ (ii) $2s$ (iii) $2p$ (iv) $3p$ (v) $3d$

(b) Apply the $\widehat{L}_z$ operator to the $\Psi_{2,1,0}$, $\Psi_{2,1,1}$, and $\Psi_{2,1,-1}$ orbitals of hydrogen, to demonstrate that they are each eigenfunctions of $\widehat{L}_z$ with eigenvalues equal to $m_\ell \hbar$.

(c) Without performing any calculations, explain why only one of the three real $2p$ orbitals of hydrogen, $2p_x$, $2p_y$, and $2p_z$, is an eigenfunction of $\widehat{L}_z$, and predict the expectation value $\langle L_z \rangle$ for each of the three $2p$ orbitals (again without doing any explicit calculations). Hint: Recall that two of the $2p$ orbitals are real linear combinations of the complex $\Psi_{2,1,1}$, and $\Psi_{2,1,-1}$ orbitals, and make use of the theorem that is proven on p. 215.

2. Show that the expectation value of the potential energy $\langle V \rangle$ of hydrogen in a $1s$ orbital is equal to twice its total energy E_1, and thus that the expectation value of the kinetic energy is $\langle K \rangle = -E_1$. Confirm that this result is consistent with the predictions of the virial theorem (Eq. 1.46 on page 41).

3. Show that the most probable value of r for an electron in the $1s$ orbital of hydrogen is a_0.

4. Show that the expectation value of r (or mean radius) of hydrogen in the 1s orbital is $\frac{3}{2}a_0$.

5. What is the probability of finding a hydrogen 1s electron within the so-called van der Waals radius of hydrogen $r_{vdW} \approx 1.2$ Å?

6. How would the results obtained in problems 3 and 4 differ for the one electron of He^+?

7. The Balmer series of spectral lines for hydrogen-like atoms correspond to transitions between $n = 2$ and states with larger values of n. For hydrogen the first few lines in this series have wavelengths of 656 nm (red), 486 nm (blue-green), and 434 nm (blue). Predict the wavelengths of these Balmer series lines for He^+.

8. The following questions pertain to the ground (1s) and first two excited (2s and 2p) states of a lithium cation (Li^{2+}).

(a) What is the expectation value of the radius in the above three states?

(b) What is the expectation value of the potential energy in each state?

(c) What is the expectation value of the kinetic energy in each state?

(d) Which of the above states has the largest angular (rotational) kinetic energy?

(e) What is the term symbol of a Li^{2+} ion in a 1s state?

(f) If the Stern-Gerlach experiment were performed with Li^{2+} ion in a 1s state how many spots would be produced?

(g) What are the two possible term symbols of a Li^{2+} ion in a 2p state?

(h) How could a Stern-Gerlach experiment be used to determine which of the above two term symbols is the correct one for the 2p state of a Li^{2+} ion?

9. Use the Pauli spin matrices (Eq. 8.12) to obtain a matrix expression for the operator $\widehat{S}^+ = \widehat{S}_x + i\widehat{S}_y$ and demonstrate what effect this operator has when it is applied to an electron in a *spin-down* state.

10. Consider a system consisting of a pair of *noninteracting* electrons confined to a one-dimensional box of length L (extending from $x = 0$ to $x = L$).

(a) What is the total ground-state energy of this system?

(b) What is the only possible value of the total spin quantum number S in the ground-state of this system, which is consistent with the Pauli exclusion principle?

(c) Express the ground-state wavefunction for this system as a linear combination of the wavefunctions of a single electron confined within this same box, in a way that is consistent with the fact that electrons are Fermi particles.

(d) If you assumed that both electrons are in exactly the same state, so all of their quantum numbers (including spin) were identical, show that the corresponding two-electron wavefunction would necessarily be equal zero (and thus no such state could exist).

11. The following questions pertain to an isolated boron (B) atom.

(a) Use the $n + \ell$ rule to predict the ground-state electron configuration of B.

(b) What are all the possible values S, L, and J for B.

(c) Use Hund's rules to predict the term symbol corresponding to the lowest energy (ground) state of B.

12. Complete Exercise 8.4 (on page 286) by using Hund's rules to predict the term symbol for the ground state of a fluorine (F) atom.

13. The following questions pertain to an isolated oxygen (O) atom.

(a) Use the $n + \ell$ rule to predict the ground-state electron configuration of an O atom.

(b) What are all the possible values S, L, and J for an O atom?

(c) Use Hund's rules to predict the term symbol corresponding to the lowest energy (ground) state of an O atom.

Problems That Test Your Understanding

14. Consider a hydrogen atom that is in a $2p$ excited state.

(a) What are all the possible values of the quantum numbers n, ℓ, m_ℓ, s, and m_s for such an excited H atom?

(b) What are all the possible term symbols for such an excited H atom?

(c) Imagine that you performed a Stern-Gerlach experiment on such excited H atoms, all of which are in a $2p$ state, and your results revealed two spots on the detector plate. Could this information be used to determine the precise term symbol that describes the electronic configuration of these excited H atoms, and if so, what is that term symbol?

15. A hydrogen anion H^- (which is also known as the hydride ion) has two electrons bound to one proton. In answering the following questions you may assume that H^- may be approximated by a system consisting of two noninteracting electrons whose wavefunctions are identical, except for their spin. In other words, each electron can either be in a spin up $\alpha = |+\rangle$ or spin down $\beta = |-\rangle$ state, so α_1 and β_1 represent the two possible spin states of one electron, while α_2 and β_2 represent the two possible spin states of the other electron.

(a) Express the ground-state wavefunction of H^- in terms of the one-electron states α_1, β_1, α_2, and β_2 in a way that is consistent with the fact that electrons are Fermi particles.

(b) Use an expression similar to what you obtained in (a) to show that the two ground-state electrons cannot have the same spin.

16. Consider a C^{5+} ion (which has one electron and a nucleus containing six protons).

(a) What is the expectation value of the Hamiltonian of C^{5+} in its ground state?

(b) What are the expectation values of both the *kinetic* and *potential* energies of the C^{5+} ion in its ground state?

(c) What wavelength of light would be required in order to excite C^{5+} from a $1s$ to a $2p$ state?

17. Predict the ground-state term symbol of Si ($Z = 14$).

18. Determine the total probability that the ground-state electron of a hydrogen atom will be found within a sphere of radius $r = a_0$ centered on the nucleus.

19. The following questions pertain to the $3d$ excited state of a hydrogen atom.

(a) What are the values of the quantum numbers n, ℓ, m_ℓ, and s?

(b) What is the expectation value of the Hamiltonian (in J units)?

(c) What are the expectation values of the potential and kinetic energies (in J units)?

(d) What is the expectation value of the radius (in Å units)?

20. The following questions pertain to the ground state of a neutral fluorine F ($Z = 9$) atom.

(a) What is the ground-state electron configuration of F?

(b) What is the ground-state term symbol of F?

21. The experimental ground-state electronic configuration of a silver (Ag) atom is exceptional in that it does not agree with the configuration predicted by the $n + \ell$ rule. The following questions pertain to the ground state of a hypothetical Ag atom, whose electron configuration *does* conform to the $n + \ell$ rule. Recall that Ag has 47 electrons.

(a) What ground-state electron configuration would Ag have if it conformed to the $n + \ell$ rule?

(b) What would the ground-state term symbol of Ag be if it conformed to both the $n + \ell$ and Hund's rules?

(c) How many spots would have been observed in the Stern-Gerlach experiment if the ground-state term symbol of Ag were the same as what you obtained in (b)?

22. The following questions pertain to the ground state of a beryllium Be^{3+} ion (which has a single electron and a nuclear charge of 4).

(a) Show that the most probable value of the radius, r_{max}, is equal to $a_0/4$.

(b) Determine the probability of finding the electron outside of a sphere of radius $r_{max} = a_0/4$.

23. Use the three Pauli spin matrices $\widehat{S}_x$, $\widehat{S}_y$, and $\widehat{S}_z$ (given below) to obtain a matrix expression for the spin angular momentum squared operator $\widehat{S}^2$, given that $\widehat{S}^2 = \widehat{S}_x^2 + \widehat{S}_y^2 + \widehat{S}_z^2$, and show that the eigenvalues of $\widehat{S}^2$ are equal to $s(s+1)\hbar^2$.

$$\widehat{S}_x = \frac{\hbar}{2}\begin{pmatrix} 0 & 1 \\ 1 & 0 \end{pmatrix} \qquad \widehat{S}_y = \frac{\hbar}{2}\begin{pmatrix} 0 & -i \\ i & 0 \end{pmatrix}$$

$$\widehat{S}_z = \frac{\hbar}{2}\begin{pmatrix} 1 & 0 \\ 0 & -1 \end{pmatrix}$$

Covalent Bonding and Optical Spectroscopy

CHAPTER 9

9.1 Covalent Bond Formation

The electronic structures of molecules have many similarities to those of atoms, as in both cases the eigenfunctions and eigenvalues represent solutions of the corresponding Schrödinger equation. The key difference between atomic and molecular systems is that the potential energy functions of molecules have a more complicated shape than those of atoms. In other words, while the nuclear charge of an atom is located at a single point, that of a molecule is located at different points in space. The wavefunctions of electrons around each atomic nucleus within a molecule often look quite similar to the corresponding atomic wavefunctions. However, they are not exactly the same because each electron feels the effects of its interaction with all of the other nuclei and electrons in the molecule. These interactions, and the resulting electron delocalization, lead to the decrease in energy that is associated with the formation of chemical bonds.

In order to understand chemical bonding it is useful to begin by considering the simplest of all possible molecules, H_2^+, which consists of a single electron shared by two protons. The Schrödinger equation for this molecule is sufficiently simple that it can be solved exactly. For larger diatomic and polyatomic molecules (with many electrons), solving the Schrödinger equation requires making use of various approximation strategies. We will illustrate how this can be done using H_2 as an example, and then see how the same approach may be extended to predict molecular orbital and bond formation energies for other diatomic and polyatomic molecules.

The electronic Hamiltonian $\widehat{H}_e$ associated with the interaction between the one electron of H_2^+ and its two protons may be expressed as follows.

$$\widehat{H}_e = \widehat{K}_e + \widehat{V}_e = -\frac{\hbar^2}{2m_e}\nabla_e^2 - \frac{e^2}{4\pi\epsilon_0}\left(\frac{1}{r_1} + \frac{1}{r_2}\right) \quad (9.1)$$

Note that the potential energy is the sum of two terms, each of which is identical to that of a single hydrogen atom (see Eq. 8.1 on page 266), but now depends on the distances r_1 and r_2 between the electron and each of the two nuclei.

We must also consider the nuclear Hamiltonian $\widehat{H}_n$, including the kinetic energies of the two nuclei and the repulsive potential energy due to their interaction with each other, that depends on the internuclear distance $R = |R_2 - R_1|$ (where R_1 and R_2 are the positions of the two nuclei).

$$\widehat{H}_n = \widehat{K}_n + \widehat{V}_n = -\frac{\hbar^2}{2m_p}\left(\nabla_{n,1}^2 + \nabla_{n,2}^2\right) + \frac{e^2}{4\pi\epsilon_0}\left(\frac{1}{R}\right) \quad (9.2)$$

Notice that the nuclear kinetic energy operator is the sum of the kinetic energies of the two nuclei. The potential energy operator contains a single term representing the repulsive (coulombic) interaction between the two positive protons. Thus, the complete Hamiltonian for H_2^+ is the sum of the nuclear and electronic contributions $\widehat{H} = \widehat{H}_n + \widehat{H}_e$.

$$\widehat{H} = -\frac{\hbar^2}{2}\left[\frac{1}{m_p}\left(\nabla_{n,1}^2 + \nabla_{n,2}^2\right) + \frac{1}{m_e}\nabla_e^2\right] + \frac{e^2}{4\pi\epsilon_0}\left(\frac{1}{R} - \frac{1}{r_1} - \frac{1}{r_2}\right) \quad (9.3)$$

Born-Oppenheimer Approximation

In general, we would expect the eigenfunctions of Hamiltonians such as Eq. 9.3 to depend on all the electronic and nuclear coordinates of the system $\Psi(r_1 \ldots, R_1 \ldots)$. A significant simplification of molecular quantum calculations may be obtained using a physically reasonable, and often highly accurate, approximation developed by Max Born and Robert Oppenheimer in 1927. This *Born-Oppenheimer approximation* implies that molecular eigenfunctions can be expressed as a product of electronic and nuclear eigenfunctions.

$$\Psi = \psi_e(r_1, r_2)\psi_n(R_1, R_2) \quad (9.4)$$

The accuracy of this approximation is linked to the fact that nuclei are much heavier than electrons, and so move much more slowly. Thus, electrons can respond to changes in nuclear position essentially instantaneously.

This implies that we may treat the nuclei as if they are stationary when calculating electronic wavefunctions.[1]

The Born-Oppenheimer approximation implies that the electronic contribution to molecular orbital energies may be obtained by first solving the electronic Schrödinger equation. For an H_2^+ molecule, for example, we would use Eq. 9.1 and equate $\widehat{H} = \widehat{H}_e$ (at a fixed internuclear separation R), and then solve the following Schrödinger equation.

$$\widehat{H}\psi_e = \widehat{K}_e\psi_e + \widehat{V}_e\psi_e = E_e\psi_e \tag{9.5}$$

The total energy of H_2^+, with a given internuclear separation R, is obtained by adding the electronic energy E_e and internuclear repulsive potential energy V_n. By repeating this calculation at various values of R, we obtain the molecular *potential-energy function* $V(R)$ of H_2^+.

$$V(R) = E_e + V_n = E_e + \frac{e^2}{4\pi\epsilon_0}\left(\frac{1}{R}\right) \tag{9.6}$$

The resulting potential-energy function is shown in Fig. 9.1 (solid curve). Note that this potential energy curve looks quite similar to an anharmonic Morse potential, such as that shown in Fig. 7.5 (on page 245), which is often used to approximately describe the vibrational potential energy of diatomic molecules. The H_2^+ potential shown in Fig. 9.1 has a minimum at $R_e = 2.00a_0 = 1.06$ Å and a dissociation energy of $D_e = V(R_e) - E_{1s} = 0.1025\,E_h = 266.5$ kJ/mol (where $E_h = -2E_{1s}$ is a Hartree energy unit, and E_{1s} is the ground-state energy of an isolated hydrogen atom).[2]

---

[1] Note that when the nuclear kinetic energy operator is applied to $\Psi = \psi_e\psi_n$, one would expect to obtain the following three terms: $\nabla_n^2\Psi = \psi_e\nabla_n^2\psi_n + \psi_n\nabla_n^2\psi_e + 2\nabla_n\psi_e\nabla_n\psi_n$. The last two terms include responses of the electronic wavefunctions to the nuclear motion ($\nabla_n^2\psi_e$ and $\nabla_n\psi_e$). It is these terms that are neglected in the Born-Oppenheimer approximation. If they were not neglected, then the electronic and nuclear degrees of freedom would be coupled to each other and so Ψ would not be separable into a product of nuclear and electronic wavefunctions.

[2] These R_e and D_e values are those obtained from experimental measurements of the electronic, vibrational, and rotational spectrum of H_2^+, and are in excellent agreement with results obtained using the Born-Oppenheimer approximation. Note that D_e is not exactly the same as the experimental bond dissociation energy because D_e measures the dissociation energy from the electronic potential energy minimum, while the experimental dissociation energy is smaller than D_e by an amount that is exactly equal to the molecule's vibrational ground-state (zero-point) energy.

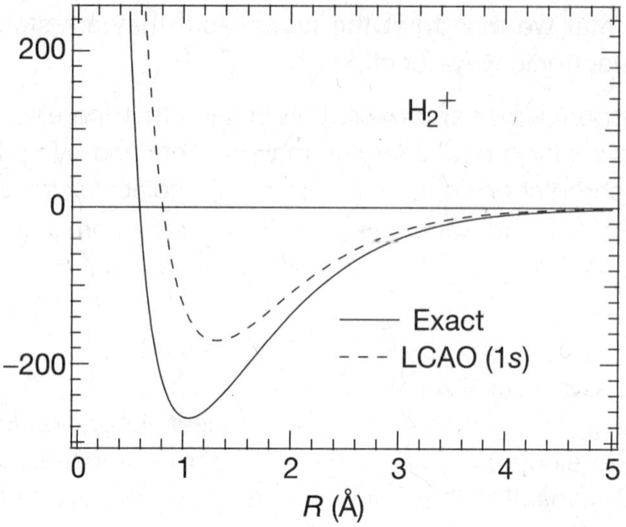

FIGURE 9.1 The potential-energy function $V(R)$ for H_2^+ obtained by solving the Schrödinger equation (solid curve) is essentially identical to the experimental potential function. The dashed curve is obtained using the simplest possible LCAO approximation with a hydrogen 1*s* atomic orbital basis set.

Variational Principle

The variational principle provides a powerful means for obtaining increasingly accurate estimates of the true ground-state eigenfunctions and energies of atomic and molecular systems.[3] This is based on the *variational theorem*, which states that the exact ground-state energy of any quantum system E_0 is invariably smaller than the approximate

[3] This principle also plays an important role in many other areas of science. Examples include Fermat's principle in optics, and the principle of least action in mechanics. Fermat's principle states that "the path taken between two points by a ray of light is the path that can be traversed in the least time." In other words, it implies that light will travel along a straight line in free space, and it may also be used to derive the formula for optical reflection from a mirror, as well as Snell's Law for optical refraction. The closely related mechanical principle of least action applies to the paths followed by classical particles (and is the basis of the Hamilton and Lagrange formulations of classical mechanics). These and other applications of the variational principle make use of the calculus of variations in order to obtain optimal solutions based on the variational principle. There is a school of thought that holds that the variational principle expresses a deep unifying truth underlying all physical laws. Euler (see footnote 18 on page 102) expressed this idea particularly eloquently as follows: "Since the fabric of the Universe is most perfect and is the work of a most wise Creator, nothing whatsoever takes place in the Universe in which some relation of maximum and minimum does not appear."

ground-state energy $E_{0'}$ obtained using any approximate trial ground-state wavefunction $\Psi_{0'}$.

$$\boxed{E_0 \leq E_{0'}} \tag{9.7}$$

The proof of this important theorem begins by noting that any trial wavefunction $\Psi_{0'}$ can in general be expressed as a linear combination of the true eigenfunctions Ψ_n of the system.[4]

$$\Psi_{0'} = \sum_{n=0}^{\infty} c_n \Psi_n \tag{9.8}$$

Since the functions Ψ_n are the exact eigenfunctions of the system, we know that $\widehat{H}\Psi_n = \widehat{H}\,|n\rangle = E_n\,|n\rangle$, and the exact eigenvalues are equal to the corresponding energy expectation values $E_n = \langle n\,|H|\,n\rangle$. Moreover, these energies are necessarily all greater than or equal to the ground-state energy (as the ground-state eigenvalue E_0 is, by definition, that with the lowest energy).

$$E_n = \langle n\,|H|\,n\rangle \geq E_0 \tag{9.9}$$

It is now easy to prove that $E_{0'} \geq E_0$, by showing that $E_{0'} - E_0 \geq 0$.

$$\begin{aligned} E_{0'} - E_0 &= \left\langle 0' \left| \left(\widehat{H} - E_0\right) \right| 0' \right\rangle \\ &= \int \Psi_{0'}^{*} \left(\widehat{H} - E_0\right) \Psi_{0'} d\tau \\ &= \int \left(\sum_{n=0}^{\infty} c_n^{*} \Psi_n^{*}\right) \left(\widehat{H} - E_0\right) \left(\sum_{n=0}^{\infty} c_n \Psi_n\right) d\tau \\ &= \sum_{n=0}^{\infty} c_n^{*} c_n \left(E_n - E_0\right) \geq 0 \end{aligned} \tag{9.10}$$

The last identity in Eq. 9.10 is obtained by making use of the fact that the eigenfunctions Ψ_n are orthogonal and normalized (and so $\int \Psi_i^{*}\Psi_j d\tau = 0$ if $i \neq j$ and 1 when $i = j$). The final inequality follows from the fact that $c_n^{*}c_n \geq 0$ and $E_n - E_0 \geq 0$.

The power of the variational theorem derives from the fact that it may be used to systematically improve the calculation of ground-state wavefunctions and energies. In other words, since the exact ground-state wavefunction has the lowest energy, any adjustment of a trial wavefunction $\Psi_{0'}$ that produces a decrease in $E_{0'}$ is necessarily an improvement. Thus, we can start with any functional form for $\Psi_{0'}$ that depends on some parameters, and then adjust the parameters so as to minimize $E_{0'}$. This variational

---

[4] Although we may not know what the Ψ_n functions are, we know that they form a complete basis set whose linear combinations can describe any other wavefunction of the system.

approximation strategy is clearly quite general, as the above proof placed no restrictions on the form of $\widehat{H}$ or on the types of basis functions used to approximate its eigenstates.

Linear Combination of Atomic Orbitals

Since molecules are composed of atoms, it is not surprising that molecular orbitals resemble a superposition of the corresponding atomic orbitals. Thus, one might expect that it would be useful to represent molecular orbitals as a linear combination of atomic orbitals – which is known as the *LCAO approximation*.

It is easiest to see how this works by applying it to H_2^+, whose electronic Hamiltonian is given in Eq. 9.1. We are interested in finding eigenfunctions of the corresponding Schrödinger equation $\widehat{H}_e\Psi = E\Psi$. The LCAO approximation implies that the ground-state eigenfunction $\Psi_{0'} = |0'\rangle$ can be represented as a linear combination of hydrogen 1*s* orbitals $\Psi_{1s} = |1s\rangle$. If we refer to the 1*s* orbitals centered on each of the two nuclei as $|a\rangle$ and $|b\rangle$, then $|0'\rangle$ may be expressed as follows.

$$|0'\rangle = c_a\,|a\rangle + c_b\,|b\rangle \tag{9.11}$$

We next operate on this trial wavefunction with the exact electronic Hamiltonian of H_2^+ to obtain the corresponding energy $\widehat{H}_e\,|0'\rangle = E\,|0'\rangle = E(c_a\,|a\rangle + c_b\,|b\rangle\,)$, and then multiply by $\langle a|$ to obtain the following expression for $\langle a|\,\widehat{H}_e\,|0'\rangle$.

$$c_a\,\langle a|\,\widehat{H}_e\,|a\rangle + c_b\,\langle a|\,\widehat{H}_e\,|b\rangle = E\left(c_a\,\langle a|a\rangle + c_b\,\langle a|b\rangle\right)$$

The above result may be expressed in the following more compact form.

$$c_aH_{aa} + c_bH_{ab} = E\left(c_a + c_bS_{ab}\right)$$

The parameter $S_{ab} = \langle a|b\rangle$ is referred to as the *overlap integral* between orbitals centered on the two different protons (and $S_{aa} = \langle a|a\rangle = 1$ since the hydrogen orbitals are normalized).

The following additional identity is obtained in exactly the same way by multiplying by $\langle b|$ to evaluate $\langle b|\,\widehat{H}_e\,|0'\rangle$.

$$c_aH_{ba} + c_bH_{bb} = E\left(c_aS_{ba} + c_b\right)$$

The above two equations are equivalent to the following pair of identities.

$$\begin{aligned}\left(H_{aa} - E\right)c_a + \left(H_{ab} - ES_{ab}\right)c_b &= 0\\ \left(H_{ba} - ES_{ba}\right)c_a + \left(H_{bb} - E\right)c_b &= 0\end{aligned} \tag{9.12}$$

This set of equations may also be expressed in matrix form.

$$\begin{pmatrix} H_{aa}-E & H_{ab}-ES_{ab} \\ H_{ba}-ES_{ba} & H_{bb}-E \end{pmatrix}\begin{pmatrix} c_a \\ c_b \end{pmatrix} = \begin{pmatrix} 0 \\ 0 \end{pmatrix} = 0 \tag{9.13}$$

The solution of such a system of linear (homogeneous) equations only exists if the following determinant is equal to zero (see Appendix B for more about determinants).

$$\begin{vmatrix} H_{aa}-E & H_{ab}-ES_{ab} \\ H_{ba}-ES_{ba} & H_{bb}-E \end{vmatrix} = 0 \tag{9.14}$$

Moreover, because the two atoms in H_2^+ are identical, we may further simplify the above notation by equating $H_{aa} = H_{bb} = \alpha$ and $H_{ab} = H_{ba} = \beta$ and $S_{ab} = S_{ba} = S$, where α is the energy associated with a single 1*s* atomic orbital and β is called the *resonance energy*. Using this simplified notation, the above *secular equation* may be expressed as follows.

$$\begin{vmatrix} \alpha-E & \beta-ES \\ \beta-ES & \alpha-E \end{vmatrix} = (\alpha-E)^2 - (\beta-ES)^2 = 0 \tag{9.15}$$

The only variable in this equation is E (since the other parameters represent integrals whose values may be independently determined). Thus, the two solutions of this quadratic equation may be expressed as follows.

$$E_{\pm} = \frac{\alpha \pm \beta}{1 \pm S} \tag{9.16}$$

The full ground-state potential-energy function obtained using the above approximation is equal to $E_+ + e^2/(4\pi\epsilon_0 R)$, where the second term corresponds to the repulsive proton-proton coulombic potential energy. The required integrals $\alpha = H_{aa}$, $\beta = H_{ab}$, and $S = S_{ab}$ may be evaluated explicitly to obtain the following expression for the full ground-state potential-energy function of H_2^+, where $\mathcal{R} = R/a_0$ (and E_h is a Hartree energy unit equivalent to 2,625.5 kJ/mol).[5]

$$V(\mathcal{R}) = E_h \left\{ \frac{1}{\mathcal{R}} - \frac{1}{2} + \frac{(\mathcal{R}+1)e^{-2\mathcal{R}} - \mathcal{R}(\mathcal{R}+1)e^{-\mathcal{R}} - 1}{\mathcal{R}\left[\left(1+\mathcal{R}+\frac{1}{3}\mathcal{R}^2\right)e^{-\mathcal{R}}+1\right]} \right\} \tag{9.17}$$

Figure 9.1 compares the true potential-energy function of H_2^+ (solid curve) with that obtained using Eq. 9.17 (dashed curve). Notice that this is the simplest possible LCAO approximation, as it uses an atomic orbital basis set consisting only of hydrogen 1*s* atomic orbitals, and yet the resulting $V(R)$ has all of the right qualitative features, including a reasonable estimate of the equilibrium bond length and energy.

---

[5] The ground-state potential corresponds to the energy $E_+ = (\alpha+\beta)/(1+S)$, where $\alpha = E_h\left[-\frac{1}{2} - \frac{1}{\mathcal{R}} + \left(1+\frac{1}{\mathcal{R}}\right)e^{-2\mathcal{R}}\right]$, $\beta = E_h\left[-\frac{1}{2}S_{ab} - (1+\mathcal{R})\,e^{-\mathcal{R}}\right]$, and $S = \left(1+\mathcal{R}+\frac{1}{3}\mathcal{R}^2\right)e^{-\mathcal{R}}$. Note that the repulsive coulombic potential energy can be expressed as $E_h(1/\mathcal{R})$.

Far more accurate LCAO predictions may be obtained if the atomic orbital basis set is expanded to include more hydrogen eigenfunctions. By using a larger number of basis functions, one may represent molecular orbitals of different shape, and the variational theorem may be used to determine which shape best approximates the true ground-state eigenfunction. In other words, we may extend Eq. 9.11 to represent $|0'\rangle$ by the following sum of a greater number of atomic orbitals.[6]

$$|0'\rangle = \sum_i c_{a,i} |a\rangle_i + \sum_j c_{b,j} |b\rangle_j \tag{9.18}$$

The coefficients $c_{a,i}$ and $c_{b,j}$ may be treated as adjustable parameters whose optimal values are determined using the variational theorem (by minimizing the ground-state energy).

Other Basis Functions

Since the variational theorem is quite general, it places no restriction on the types of basis functions that may be used to represent molecular orbitals. In practice, molecular quantum calculations are often performed using basis functions that resemble atomic orbitals, but have simpler, or more mathematically convenient, functional forms.

One such set of basis functions are called *Slater orbitals* (first introduced by John Slater in 1930), which have the following functional form.

$$Nr^{\eta} e^{-\zeta r} Y_{\ell}^{m_\ell}(\theta, \phi) \tag{9.19}$$

These orbitals look very similar to hydrogen orbitals; they have exactly the same angular (spherical harmonic) wavefunctions $Y_{\ell}^{m_\ell}$ (see Eq. 8.9), but somewhat simpler radial wavefunctions (see Eq. 8.7). The exponents η and ζ may be treated as adjustable parameters in variational calculations (and any number of such functions with different η and ζ values may be combined to obtain increasingly accurate representations of molecular orbitals).

An even more widely used set of basis functions are called *Gaussian orbitals*, which have the following general form.

$$Nx^a y^b z^c e^{-\alpha r^2} Y_{\ell}^{m_\ell}(\theta, \phi) \tag{9.20}$$

Notice that they contain Gaussian $e^{-\alpha r^2}$, rather than simple exponential $e^{-\zeta r}$, radial functions. Again, the exponents a, b, c, and α may be treated as adjustable parameters in variational calculations, and the basis set may

[6] The ket vectors $|a\rangle_i$ and $|b\rangle_j$ again represent atomic orbitals centered on the a and b atoms, respectively.

be expanded to include linear combinations of many such functions. The reason that Gaussian functions are useful is that they make it easier to calculate various integrals (such as overlap integrals) and thus lead to a significant reduction in the time required to perform molecular energy and structure calculations.[7]

Antibonding and Multiple-Bonding

All of the electronic states of H_2^+ resemble linear combinations of atomic hydrogen orbitals. Bonding molecular orbitals are produced when atomic orbitals are combined with the same phase (to produce a molecular orbital with fewer nodes). Antibonding orbitals are produced when atomic orbitals are combined with the opposite phase (to produce a molecular orbital with more nodes).

For example, the ground-state bonding molecular orbital of H_2^+ may be expressed as the following linear combination of 1*s* orbitals centered on each of the two atoms.[8]

$$\sigma = \sigma_g 1s = N\left(1s_a + 1s_b\right) \tag{9.21}$$

The subscript *g* indicates that the ground-state orbital is symmetrical with respect to inversion of the orbital about a point located at the center of the molecule.[9] In other words, the above orbital looks identical when the *a* and *b* atoms are interchanged.

If the 1*s* orbitals are subtracted (combined with the opposite phase), they produce the following antibonding molecular orbital.

$$\sigma^* = \sigma_u 1s = N\left(1s_a - 1s_b\right) \tag{9.22}$$

In this case exchanging the *a* and *b* atoms changes the sign of the wavefunction and so it is antisymmetric with respect to inversion, as indicated by the subscript *u*.[10] This orbital has a node and so resembles the first excited state of a particle in a box. It is referred to as an antibonding orbital, because it has a higher energy than two separate hydrogen atoms (in their ground 1*s* states). Thus, if an H_2^+ molecule were excited to this state (by

---

[7] The application of Gaussian basis sets in quantum calculations was pioneered by the theoretical chemist John Pople, who obtained a Chemistry Nobel prize in 1998 "for his development of computational methods in quantum chemistry."

[8] The notation used in Eq. 9.21 is equivalent to that used in Eq. 9.11. In other words, Eq. 9.21 could have been written as $\sigma_g 1s = N\left(|a\rangle + |b\rangle\right)$.

[9] The letter *g* stands for the German word *gerade*, which means symmetric.

[10] The letter *u* stands for the German word *ungerade*, which means antisymmetric.

absorption of a photon, for example), then it would fly apart (dissociate) to form a free hydrogen atom and a proton, $H_2^+ \rightarrow H + H^+$.

We may construct additional excited states by forming the following linear combinations of higher-energy hydrogen orbitals.

$$\begin{aligned}
\sigma = \sigma_g 2s &= N\left(2s_a + 2s_b\right) \\
\sigma^* = \sigma_u 2s &= N\left(2s_a - 2s_b\right) \\
\sigma_p = \sigma_g 2p_z &= N\left(2p_{za} - 2s_{zb}\right) \\
\sigma_p^* = \sigma_u 2p_z &= N\left(2p_{za} + 2s_{zb}\right) \\
\pi = \pi_u 2p_x &= N\left(2p_{xa} + 2s_{zb}\right) \\
\pi = \pi_u 2p_y &= N\left(2p_{ya} + 2s_{yb}\right) \\
\pi^* = \pi_g 2p_x &= N\left(2p_{xa} - 2s_{xb}\right) \\
\pi^* = \pi_g 2p_y &= N\left(2p_{za} - 2s_{zb}\right)
\end{aligned}$$

Notice that the subscripts g and u again indicate whether the corresponding orbital is symmetric or antisymmetric with respect to inversion. In combining $2p$ orbitals, it is assumed that the z-axis is parallel to the bond-axis, while the y- and x-axes are perpendicular to the bond-axis. The $2p_z$ orbitals combine to form a bonding $\sigma_p = \sigma_g 2p_z$ orbital (with two nodes) and an antibonding $\sigma_p^* = \sigma_u 2p_z$ orbital (with three nodes). The functions designated as π orbitals correspond to linear combinations of atomic $2p_x$ or $2p_y$ orbitals, which are perpendicular to the bond-axis. Note that the two bonding $\pi = \pi_u$ orbitals must have the same energy, as must the two antibonding $\pi^* = \pi_g$ orbitals.

One may represent the electronic states of other homonuclear diatomics using the above molecular orbitals. As in the case of multielectron atoms, the electrons may occupy each state in pairs (with opposite spin). The ground-state electron configuration is again dictated by the Aufbau principle (which specifies that orbitals are filled in order of increasing energy).

Figure 9.2 shows how the relative energies of the valence electrons change for homonuclear diatomics formed from atoms in the second row of the periodic table. The large variation in the energy of the bonding $\sigma = \sigma_g 2p_z$ orbital arises because this energy is very sensitive to small changes in the degree of overlap between the p_z orbitals on one atom and the s orbitals on the other atom.

The twofold degeneracy of the π_u and π_g orbitals plays an important role in dictating diatomic ground-state electronic structure, as a twofold degenerate state can hold up to four (spin-paired) electrons. In keeping with the first Hund's rule (see page 287), when a twofold degenerate state is partially filled, the two electrons will have a lower energy if they remain unpaired.

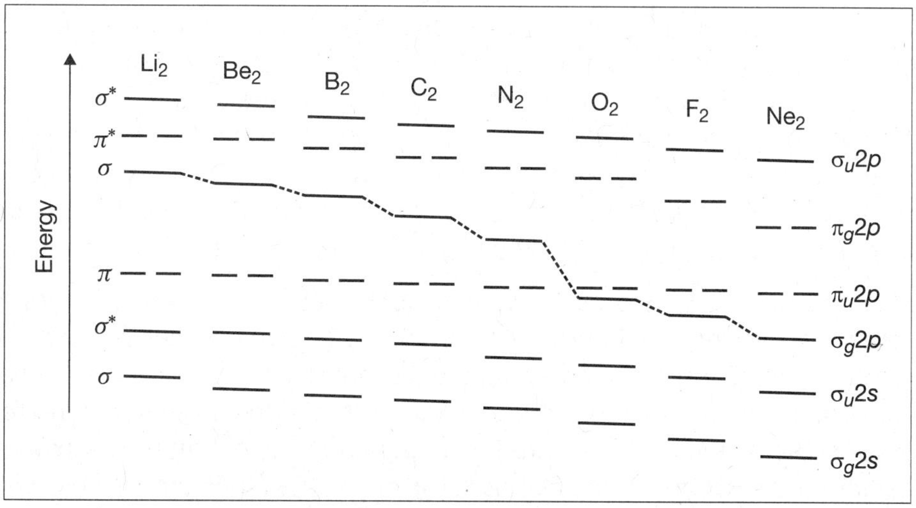

FIGURE 9.2 The relative energies of the valence electrons for the second row homonuclear diatomics change when the σ ($\sigma_g 2p$) orbital crosses below the two π ($\pi_u 2p$) orbitals. The two lowest energy core σ ($\sigma_g 1s$) and σ^* ($\sigma_u 1s$) orbitals are not shown in this diagram.

The *bond order* of a diatomic is defined as one half of the difference between the number of bonding n_B and antibonding n_A electrons. However, molecules with identical bond orders need not have identical bond dissociation energies. This is due in part to the fact that the stabilization energy of bonding electrons is often slightly smaller in magnitude than the destabilizing influence of antibonding electrons. Thus, for example, although both C_2 and O_2 have the same bond order (see Exercise 9.1), O_2, which has a total of six antibonding electrons, has a smaller bond dissociation energy ($\sim$ 498 kJ/mol) than C_2 ($\sim$618 kJ/mol), which has a total of four antibonding electrons.

Exercise 9.1

Predict the bond order of diatomic C_2, N_2, O_2, and F_2.

Solution. Since C_2 has a total of 12 electrons, these will fill all of the above molecular orbitals up to (and including) the two bonding π ($\pi_u 2p$) orbitals. Thus, C_2 is predicted to have a bond order of $(n_B - n_A)/2 = (8 - 4)/2 = 2$. N_2, O_2, and F_2 have 14, 16, and 18 electrons, respectively. Thus, the ground state of N_2 will be filled up to the bonding σ ($\sigma_g 2p$) orbital, and so is predicted to have a bond order of 3. The additional two electrons of O_2 will occupy the two antibonding π^* ($\pi_g 2p$) orbitals, and so O_2 is predicted to have have a bond order of 2. The two additional antibonding electrons of F_2 produce a bond order of 1.

Note that the two highest energy electrons in the ground state of O_2 occupy the degenerate pair of antibonding π^* orbitals, and so are expected to remain unpaired. Thus, the ground state of O_2 is predicted to be a triplet state, as confirmed experimentally. This diradical character of O_2 is responsible for its high reactivity, as well as the fact that it is paramagnetic; in other words, O_2 is attracted towards regions of high magnetic field as the energy of its unpaired spins is lowered by alignment with the field.

Molecular bond orders are also roughly correlated with the corresponding vibrational frequencies $\tilde{\nu}$ and force constants. This makes sense because one would expect molecules with a greater bond order to have a potential function $V(R)$ with a deeper well, and thus also a higher curvature (greater force constant) and a higher vibrational frequency. For example, the experimental vibrational frequencies of N_2 (2358 cm^{-1}), C_2 (1855 cm^{-1}), O_2 (1580 cm^{-1}), and F_2 (892 cm^{-1}) are roughly correlated with their relative bond orders. The same is true of the corresponding vibrational force constants 2295 N/m, 1216 N/m, 1177 N/m, 470 N/m, respectively.[11]

9.2 Molecular Bonding Made Easy

Various semiempirical methods have been devised for predicting molecular energies and wavefunctions. These strategies incorporate empirical (experimental) information into quantum calculations. Such methods can be very computationally efficient and predictively accurate. However, semiempirical methods are only expected to be reliable when applied to molecules that are sufficiently similar to those used to determine the empirical input parameters.

One particularly useful (and simple) semiempirical method is called the *Hückel approximation*, as it was developed by the German physical chemist Erich Hückel in 1937. This method is often used to describe the *resonance stabilization* associated with conjugated double bonds and aromatic π molecules. We will see that the same approach can also be used to accurately predict the relative energies of other sorts of molecules and ions.

The starting point of the Hückel approximation is the LCAO method described in Section 9.1 (pages 298–300). However, the Hückel approximation neglects electron-electron repulsion and thus assumes that $S = 0$.

---

[11] There are many reasons why one should not expect bond orders, vibrational frequencies, and force constants to be perfectly correlated with each other. The precise balance of attractive and repulsive interactions that determine the shape of $V(R)$ depends on the total number (and relative energetic contributions) of the bonding and antibonding electrons. In addition, molecular vibrational frequencies depend on the reduced mass of the corresponding atoms since $\omega = 2\pi hc\tilde{\nu} = \sqrt{f/\mu}$.

Moreover, only nearest neighbor resonance energy integrals are considered, so the Hückel approximation assumes that $\beta = 0$ for all non-nearest neighbor atoms. The value of β for nearest neighbor atoms is treated as an empirically adjustable parameter. Thus, β may be determined from the experimental enthalpies of formation of one or more molecules, and then used to predict the ground-state energies of a wide variety of other similar types of molecules.

It is easiest to see how the Hückel approximation works by first applying it to simple diatomics such as H_2^+ or H_2. Such molecules give rise to the following 2×2 *secular determinant*.

$$\begin{vmatrix} \alpha - E & \beta \\ \beta & \alpha - E \end{vmatrix} = (\alpha - E)^2 - \beta^2 = 0 \qquad (9.23)$$

Note that Eq. 9.23 is the same as Eq. 9.15, except that we have set $S = 0$, in accordance with the Hückel approximation. Since the two atoms in a diatomic are nearest neighbors, the parameter β remains nonzero.

The two roots of Eq. 9.23 are equivalent to those in Eq. 9.16 (when $S = 0$).

$$E_+ = \alpha + \beta \quad \text{and} \quad E_- = \alpha - \beta \qquad (9.24)$$

The value of the empirical parameter β may be determined, for example, from the bond formation energy of H_2^+. Once β is determined, then the Hückel approximation may be used to predict the bond formation energy of H_2, as well as various other molecules and ions composed entirely of hydrogen atoms, as we will see.

Exercise 9.2

Use the Hückel approximation to express the bond formation energy, E_B, of H_2^+ in terms of the Hückel parameters α and/or β. Note that E_B is equivalent to the energy of the reaction $H + H^+ \rightarrow H_2^+$.

Solution. The bond formation energy E_B is equal to the difference between the energy of the one electron bound to the product H_2^+, and that of the one electron bound to the reactant H atom. Equation 9.24 indicates that the energy of one electron in the ground state of H_2^+ is $E_+ = \alpha + \beta$, while the energy of an isolated H atom is α. Thus, the Hückel approximation predicts that the bond formation energy of H_2^+ may be expressed as $E_B = E_+ - \alpha = (\alpha + \beta) - \alpha = \beta$.

The experimental bond formation energy of H_2^+ is $E_B = -210$ kJ/mol. Thus, the results of the above exercise imply that we may use this experimental energy to fix the empirical value of the Hückel resonance integral $\beta \approx E_B \approx -210$ kJ/mol.

We may now use the Hückel approximation to predict the bond formation energy of H_2, which is equivalent to the energy of the reaction $H + H \rightarrow H_2$.

Since there are now two (paired) electrons in the product molecule, its ground-state energy is predicted to be $2(\alpha + \beta)$. The two reactant atoms each have an energy of α, and so the bond formation energy of H_2 is predicted to be $2(\alpha + \beta) - 2\alpha = 2\beta \approx -420$ kJ/mol, which differs by less than 4% from the experimental value of $E_B = -436$ kJ/mol. This agreement is quite remarkable given that electron-electron repulsion has been neglected, and that the bond length of H_2 is about 25% shorter than that of H_2^+.[12]

Given the accuracy of the above prediction, one may hope to be able to make use of the Hückel approximation in predicting the relative stabilities of more unusual molecules and ions composed of hydrogen atoms. For example, we might wonder whether a molecule such as H_3 or an ion such as H_3^+ is predicted to be stable or unstable with respect to dissociation. Notice that such triatomic molecules and ions could perhaps exist either as linear chains or as cyclic triatomic rings. We will see that the Hückel approximation may be used to correctly predict which of these structures has the lowest energy.

The Hückel determinants for linear and cyclic triatomic structures are not the same because the two structures do not have the same number of nearest neighbor atoms. More specifically, since the first and third atoms in a linear triatomic are not nearest neighbors, the 13 and 31 matrix elements in the Hückel secular determinant should be set to zero. Thus, the Hückel secular equation for a linear triatomic may be expressed as follows (and the evaluation of 3 × 3 determinants is explained in Appendix B).

$$\begin{vmatrix} \alpha - E & \beta & 0 \\ \beta & \alpha - E & \beta \\ 0 & \beta & \alpha - E \end{vmatrix} = 0 \tag{9.25}$$

$$(\alpha - E)^3 - 2\beta^2(\alpha - E) = 0$$

$$(\alpha - E)[(\alpha - E) - \sqrt{2}\beta][(\alpha - E) + \sqrt{2}\beta] = 0$$

[12] One might also wonder if the same Hückel model could be used to predict the bond formation energy of H_2^-, which would involve adding a third electron to the excited state to yield a net bond formation energy of $E_B = 2(\alpha + \beta) + (\alpha - \beta) - 3\alpha = \beta \approx -210$ kJ/mol. This suggests that H_2^- should have the same bond formation energy as H_2^+. However, higher-level calculations indicate that H_2^- is unstable and thus the Hückel approximation evidently cannot be relied on to accurately predict the energies of antibonding states. This makes sense because the Hückel approximation neglects overlap integrals S and so underestimates the destabilizing influence of antibonding electrons (see Eq. 9.16).

This indicates that the secular equation has the following three roots (listed in order of increasing energy).

$$\text{Roots:} \quad \alpha + \sqrt{2}\beta, \quad \alpha, \quad \alpha - \sqrt{2}\beta \tag{9.26}$$

Exercise 9.3

Use the Hückel approximation to predict the bond formation energy of the linear H_3^+ cation, and compare your result to that for H_2 to determine the bonding energy gained by delocalizing two electrons over three as opposed to two protons.

Solution. The bond formation energy of linear H_3^+ is obtained from the roots of the corresponding secular determinant (Eq. 9.26). This bond energy is equivalent to the difference between the product and reactant energies for the reaction $H + H + H^+ \rightarrow H_3^+$, and thus we predict that $E_B = 2(\alpha + \sqrt{2}\beta) - 2\alpha = 2\sqrt{2}\beta \approx -594$ kJ/mol. This energy is significantly more negative than that of H_2, whose Hückel bond energy is $E_B = -420$ kJ/mol. This implies that the delocalization of two electrons over the linear triatomic chain lowers their energy by ~ -174 kJ/mol with respect to H_2. Note that this stabilization is consistent with the decrease in the ground-state energy of a particle in a box when the size of the box is increased (see problem 1, on page 335).

The following secular determinant of a cyclic triatomic has a slightly different form than that of a linear triatomic, since all the atoms in a cyclic structure are nearest neighbors (because each atom is bonded to the two other atoms).

$$\begin{vmatrix} \alpha - E & \beta & \beta \\ \beta & \alpha - E & \beta \\ \beta & \beta & \alpha - E \end{vmatrix} = 0 \tag{9.27}$$

This secular equation again has three roots, but the two higher-energy roots are now degenerate.

$$\begin{array}{lcc} \text{Roots:} & \alpha + 2\beta & \alpha - \beta \\ \text{Degeneracy:} & 1 & 2 \end{array} \tag{9.28}$$

The above Hückel energy levels may be used to predict the relative energies of various triatomic hydrogen molecules and ions. For example, the triatomic cation H_3^+ contains two electrons, but could have either a linear or a ring structure.

The fact that the cyclic H_3^+ is more stable than the linear H_3^+ can readily be shown using the roots of the corresponding Hückel secular equation (see homework problem 4 on page 336). The cyclic (equilateral triangle) structure of experimental H_3^+ ions has been verified by vibrational and rotational spectroscopic measurements. Cyclic H_3^+ ions have also been detected in the hydrogen-rich upper atmospheres of the planets Jupiter, Uranus, and Saturn. In fact, cyclic H_3^+ is thought to be the *most abundantly produced molecule in the universe*!

Hückel Model for π-Orbital Conjugation and Aromaticity

Exactly the same Hückel approximation method may be used to predict the relative stabilities of various conjugated and aromatic π-bonding structures. In this case the parameter α represents the energy of a lone $2p$ orbital and β represents the resonance energy gained upon formation of the π-bond. Comparison with experimental resonance stabilization energies of various conjugated polyene molecules suggests that $\beta \approx -80$ kJ/mol is a reasonable empirical value of the Hückel resonance integral.

The simplest π-bonded molecule is ethylene, whose two p electrons form a single π bond. Thus, Eq. 9.15 predicts that total π-bonding energy of ethylene is $E_{\pi B} \approx 2\beta \approx -160$ kJ/mol. We may also use the roots of the Hückel secular determinants for linear (Eq. 9.26) and cyclic (Eq. 9.28) triatomics to predict the relative stabilities of the linear allyl cation $CH_2CHCH_2^+$ (in which the π electrons are delocalized over all three carbons) and the cyclopropenium cation c-$C_3H_3^+$ (which is also known as the cyclopropenyl cation). When the latter cation was first synthesized, it was considered to be an exotic curiosity. However, subsequent astronomical observations suggest that it may be *the most abundant organic molecule in the universe*!

The Hückel approximation may be extended to longer linear and cyclic compounds. For example, the π-bonding energies of linear butadiene may be obtained from the four (nondegenerate) roots of the following secular equation.[13]

$$\begin{vmatrix} \alpha-E & \beta & 0 & 0 \\ \beta & \alpha-E & \beta & 0 \\ 0 & \beta & \alpha-E & \beta \\ 0 & 0 & \beta & \alpha-E \end{vmatrix} = 0 \qquad (9.29)$$

$$\text{Roots:} \quad \alpha \pm \left[\frac{1}{2}\left(\sqrt{5}+1\right)\right]\beta \quad \alpha \pm \left[\frac{1}{2}\left(\sqrt{5}-1\right)\right]\beta \qquad (9.30)$$

These roots may be used to predict the stability of the two conjugated double bonds, relative to two uncoupled double bonds (in two separate ethene molecules). The Hückel approximation predicts that the total π-bonding energy of butadiene is $2(\alpha + 1.62\beta) + 2(\alpha \pm 0.62\beta) - 4\alpha = 4.48\beta$, while the total π-bonding energy of two ethene molecules is

[13] See Appendix B regarding the evaluation of 4×4 determinants.

$4(\alpha + \beta) - 4\alpha = 4\beta$ (obtained using Eq. 9.24). Thus, the net *resonance stabilization* energy obtained by conjugating two double bonds is predicted to be about $4.48\beta - 4\beta = 0.48\beta \approx -38$ kJ/mol.

Similarly, the Hückel π-orbital energies of benzene may be obtained from the roots of the following secular equation.

$$\begin{vmatrix} \alpha-E & \beta & 0 & 0 & 0 & \beta \\ \beta & \alpha-E & \beta & 0 & 0 & 0 \\ 0 & \beta & \alpha-E & \beta & 0 & 0 \\ 0 & 0 & \beta & \alpha-E & \beta & 0 \\ 0 & 0 & 0 & \beta & \alpha-E & \beta \\ \beta & 0 & 0 & 0 & \beta & \alpha-E \end{vmatrix} = 0 \qquad (9.31)$$

$$\begin{array}{lcccc} \text{Roots:} & \alpha + 2\beta & \alpha + \beta & \alpha - \beta & \alpha - 2\beta \\ \text{Degeneracy:} & 1 & 2 & 2 & 1 \end{array} \qquad (9.32)$$

These Hückel energies may be used to predict the difference between the π-bonding energy of benzene and that of three uncoupled ethylene molecules. Benzene is predicted to have a π-bonding energy of $2(\alpha + 2\beta) + 4(\alpha + \beta) - 6\alpha = 8\beta$, while three separate double bonds are predicted to have a π-bonding energy of $6(\alpha + \beta) - 6\alpha = 6\beta$. This implies that benzene has a huge aromatic resonance stabilization energy of $8\beta - 6\beta = 2\beta \approx -160$ kJ/mol.

Hückel Polynomials

The Hückel energies of other hydrogen clusters and π-bonded molecules may be obtained by setting up the appropriate secular equation, expanding the determinant, and finding the roots of the resulting polynomial. This procedure can be simplified by first dividing the secular determinant by β and then defining a new variable $x = (\alpha - E)/\beta$. Thus, the secular equation for any linear chain is reduced to the following simpler form.

$$\begin{vmatrix} x & 1 & 0 & \cdots \\ 1 & x & 1 & \cdots \\ 0 & 1 & x & \cdots \\ \vdots & \vdots & \vdots & \ddots \end{vmatrix} = 0 \qquad (9.33)$$

The corresponding determinant for a cyclic chain looks the same, except that the connectivity of the ends of the chain implies that the determinant must have a value of 1 (rather than 0) in the upper-right and lower-left corners. We can also represent various sorts of branched or polycyclic chains by placing values of 1 in each location representing nearest neighbor atoms.

Expanding the resulting determinant yields a polynomial whose roots (x_i) dictate the corresponding Hückel eigenvalues $E_i = \alpha - x_i\beta$. For example, the 2 × 2 Hückel secular determinant (Eq. 9.15) is equivalent to the polynomial $x^2 - 1 = (x + 1)(x - 1)$, whose roots are $x_1 = -1$ and $x_2 = +1$, and so $E_1 = E_+ = \alpha + \beta$ and $E_2 = E_- = \alpha - \beta$, which is consistent with Eq. 9.24. Similarly, the secular determinant for a cyclic trimer (Eq. 9.27) produces a polynomial $x^3 - 3x + 2 = (x + 2)(x - 1)^2$ whose roots are equivalent to those given in Eq. 9.28.

The Hückel eigenvalues of linear and cyclic chains containing n atoms may also be obtained using the following expressions, where k may be any positive integer.

$$\text{Linear: } E_k = \alpha + 2\beta\cos\left[\frac{k\pi}{(n+1)}\right] \qquad \text{Cyclic: } E_k = \alpha + 2\beta\cos\left[\frac{2k\pi}{n}\right]$$

Diagonalizing Hamiltonians

Both the eigenvalues and eigenvectors of the Schrödinger equation for any system may be obtained by diagonalizing the corresponding Hamiltonian matrix. The following is an illustration of how this strategy can be used to determine both the molecular orbital energies E_i and atomic orbital coefficients c_i resulting from the Hückel approximation.

When the Hückel approximation is applied to diatomics (or π bonds) composed of a linear combination of two atomic orbitals, Eq. 9.13 reduces to the following form (since $S_{12} = S_{21} = S = 0$).

$$\begin{pmatrix} H_{aa}-E & H_{ab} \\ H_{ba} & H_{bb}-E \end{pmatrix}\begin{pmatrix} c_a \\ c_b \end{pmatrix} = 0 \qquad (9.34)$$

This expression may be expanded

$$\begin{pmatrix} H_{aa} & H_{ab} \\ H_{ba} & H_{bb} \end{pmatrix}\begin{pmatrix} c_a \\ c_b \end{pmatrix} - \begin{pmatrix} E & 0 \\ 0 & E \end{pmatrix}\begin{pmatrix} c_a \\ c_b \end{pmatrix} = 0 \qquad (9.35)$$

and then rearranged to the following form.[14]

$$\begin{pmatrix} H_{aa} & H_{ab} \\ H_{ba} & H_{bb} \end{pmatrix}\begin{pmatrix} c_a \\ c_b \end{pmatrix} = E\begin{pmatrix} c_a \\ c_b \end{pmatrix} \qquad (9.36)$$

---

[14] The right-hand side of Eq. 9.36 was obtained as follows.

$$\begin{pmatrix} E & 0 \\ 0 & E \end{pmatrix}\begin{pmatrix} c_a \\ c_b \end{pmatrix} = E\begin{pmatrix} 1 & 0 \\ 0 & 1 \end{pmatrix}\begin{pmatrix} c_a \\ c_b \end{pmatrix} = E\begin{pmatrix} c_a \\ c_b \end{pmatrix}$$

Notice that the above expression is equivalent to the Schrödinger equation $\widehat{H}\Psi = E\Psi$ or $\widehat{H}\,|c\rangle = E\,|c\rangle$ where $\Psi = |c\rangle$ is the eigenvector whose atomic orbital coefficients are c_a and c_b, and whose eigenvalue is E. If we multiply $\langle c|$ times $\widehat{H}\,|c\rangle$, we obtain the expectation value of the Hamiltonian $\langle c|\,\widehat{H}\,|c\rangle = \langle c|\,E\,|c\rangle = E\,\langle c|c\rangle = E$. Thus, solving the Schrödinger equation is equivalent to solving the above matrix equation to determine the values of c_1, c_2, and E.

The two eigenvalues E_i and eigenvectors $|c_i\rangle$ of the above 2×2 Hamiltonian give rise to two such equations $\langle c_1|\,\widehat{H}\,|c_1\rangle = E_1$ and $\langle c_2|\,\widehat{H}\,|c_2\rangle = E_2$, which may be combined to form the following matrix equation.

$$\begin{pmatrix} c_{a1} & c_{b1} \\ c_{a2} & c_{b2} \end{pmatrix} \begin{pmatrix} H_{aa} & H_{ab} \\ H_{ba} & H_{bb} \end{pmatrix} \begin{pmatrix} c_{a1} & c_{a2} \\ c_{b1} & c_{b2} \end{pmatrix} = \begin{pmatrix} E_1 & 0 \\ 0 & E_2 \end{pmatrix} \tag{9.37}$$

The three matrices on the left may be expressed as $\widehat{C}^{-1}\widehat{H}\widehat{C}$. Notice that the columns of the $\widehat{C}$ matrix contain the atomic orbital coefficients of the two *ket* eigenvectors $|c_1\rangle$ and $|c_2\rangle$, while the inverse matrix $\widehat{C}^{-1}$ contains rows representing the corresponding *bra* vectors $\langle c_1|$ and $\langle c_2|$.[15] Equation 9.37 indicates that these matrices transform the Hamiltonian to a diagonal matrix containing the eigenvalues of the system. Also, notice that since $\widehat{H}\Psi_1 = E_1\Psi_1$ and $\widehat{H}\Psi_2 = E_2\Psi_2$, the two eigenfunctions take on the following simpler form once the Hamiltonian is diagonalized.

$$\widehat{H}\Psi_1 = \begin{pmatrix} E_1 & 0 \\ 0 & E_2 \end{pmatrix} \begin{pmatrix} 1 \\ 0 \end{pmatrix} = E_1 \begin{pmatrix} 1 \\ 0 \end{pmatrix} = E_1\Psi_1$$

$$\widehat{H}\Psi_2 = \begin{pmatrix} E_1 & 0 \\ 0 & E_2 \end{pmatrix} \begin{pmatrix} 0 \\ 1 \end{pmatrix} = E_2 \begin{pmatrix} 0 \\ 1 \end{pmatrix} = E_2\Psi_2$$

The above procedure may be extended to larger systems by diagonalizing the appropriate Hamiltonian to obtain the corresponding energy eigenvalues and eigenvector coefficients.

9.3 Time-Dependent Processes

Up to this point we have focused primarily on the properties of systems whose Hamiltonians are time-independent. An eigenfunction of such a system is also called a *stationary state* because the system would tend to remain in such a state forever, in the absence of any external perturbation. In other words, stationary states are similar to the normal modes of vibration of a drum, or bell, or guitar string that, in the absence of any energy

---

[15] Since the eigenvectors are normalized (and the atomic orbitals are orthogonal), the product $\widehat{C}^{-1}\widehat{C} = 1$ (where 1 represents a diagonal unit matrix).

dissipation mechanism, would hold the same tone indefinitely. One may also change the state of vibration of such musical instruments by hitting or plucking them. This is similar to what happens when light induces transitions between the eigenstates of a quantum mechanical system. In this section we will consider how interactions between a molecule and its surroundings can induce transitions between molecular eigenstates. As we will see, such changes in the state of quantum mechanical systems may be described using the time-dependent Schrödinger equation (Eq. 6.28 on page 207).

Rabi Oscillations and Fermi's Golden Rule

The Hamiltonian of a system that is subject to an external perturbation can be described as the sum of the unperturbed Hamiltonian $\widehat{H}_0$ plus an additional term $\widehat{H}_1 = V(t)$ that describes the potential energy change induced by the external perturbation.

$$\widehat{H} = \widehat{H}_0 + \widehat{H}_1 = \widehat{H}_0 + V(t) \tag{9.38}$$

The perturbation could have a simple time dependence, such as that of a constant electric or magnetic field that is turned on at some time t_0, or it could have a more complicated time dependence, such as that due to the oscillating electromagnetic field in a beam of light.

The time evolution of a system described by a Hamiltonian such as Eq. 9.38 may in general be described by representing the new time-dependent wavefunction of the system as a superposition of the eigenstates of the unperturbed system Ψ_n, which are now each multiplied by a time-dependent coefficient $c_n(t)$.

$$\Psi(t) = \sum_n c_n(t)\Psi_n \tag{9.39}$$

Under the influence of a time-dependent perturbation the probability of finding the system in an eigenstate Ψ_n is given by the square of the corresponding coefficient.

$$P_n(t) = c_n^*(t)\, c_n(t) = |c_n(t)|^2 \tag{9.40}$$

Recall that, even in the absence of any external perturbation, the stationary eigenstates of any system have a simple oscillatory time dependence with a frequency $\omega_n = E_n/\hbar$ (see Eq. 6.25).

$$\Phi_n(r,t) = \Psi_n(r)e^{-i(E_n/\hbar)t} = \Psi_n e^{-i\omega_n t} \tag{9.41}$$

If we include this oscillatory time dependence in Eq. 9.39, we obtain the following expression for the full time dependence of the wavefunction.

$$\Phi(t) = \sum_n c_n(t)\Psi_i e^{-i\omega_n t} \tag{9.42}$$

If consider only the ground state Ψ_0 and a single excited state Ψ_1 of the system, with an energy difference of $\Delta\varepsilon = \varepsilon_1 - \varepsilon_0$, then the above sum reduces to the following two terms.

$$\Phi(t) = c_0(t)\Psi_0 e^{-i\omega_0 t} + c_1(t)\Psi_1 e^{-i\omega_1 t} \tag{9.43}$$

We may further simplify the problem by assuming that the system is initially in the ground state, so $c_0(0) = 1$ and $c_1(0) = 0$, and that the perturbation V which is turned on when $t = 0$ remains constant after that time. We may then insert $\Phi(t)$ into the time-dependent Schrödinger equation (Eq. 6.28) to obtain (after extensive manipulation) the following *Rabi formula*[16] for the time-dependent probability of finding the system in the excited state.

$$P_1(t) = |c_1(t)|^2 = \left(\frac{4V_{10}^2}{\Delta\varepsilon^2 + 4V_{10}^2}\right)\sin^2\left[\frac{(\Delta\varepsilon^2 + 4V_{10}^2)^{1/2}}{2\hbar}t\right] \tag{9.44}$$

$V_{10} = \langle 1|V|0\rangle$ is the off diagonal matrix element (integral) that couples the ground and excited states. Equation 9.44 predicts that the probability of finding the system in the excited state will undergo time-dependent *Rabi oscillations*, whose magnitude and frequency depend on the values of V_{10} and $\Delta\varepsilon$. Note that when the perturbation is large, so that $V_{10} >> \Delta\varepsilon$, the Rabi formula reduces to the following simpler expression.

$$P_1(t) = \sin^2\left[\left(\frac{V_{10}}{\hbar}\right)t\right] \tag{9.45}$$

This implies that the probability will oscillate back and forth between the ground and excited state with a frequency that is proportional to the coupling strength V_{10}.

Although such Rabi oscillations may seem like another example of strange quantum mechanical behavior, it is completely analogous to what is observed in a classical system composed of two coupled harmonic oscillators, such as two pendulums that hang from the same horizontal string, so that they are coupled to each other. More specifically, if one of the two coupled pendulums is initially swinging, then as time goes on its energy will gradually transfer over to the other pendulum and then back again. So, the interaction (coupling) between two classical pendulums causes energy

[16] The Rabi formula is named after Isidor Isaac Rabi who received the Nobel prize in Physics in 1944 for his accurate measurements of the magnetic properties of atomic nuclei, by refining Otto Stern's atomic beam technique (used in the Stern-Gerlach experiment).

to oscillate back and forth, just as it does in a quantum mechanical two-state system. In both cases, the frequency with which the energy transfers back and forth depends on the strength of the coupling – weakly coupled pendulums will transfer energy slowly and strongly coupled pendulums will transfer energy more rapidly.

The analogy between such quantum mechanical and classical systems does not end there, as it also applies in the weak coupling limit, when $V_{10} << \Delta\varepsilon$. In this case Eq. 9.44 reduces to the following expression.

$$P_1(t) = 4\left(\frac{V_{10}}{\Delta\varepsilon}\right)^2 \sin\left[\frac{1}{2}\left(\frac{\Delta\varepsilon}{\hbar}\right)t\right] \tag{9.46}$$

Since $V_{10} << \Delta\varepsilon$, the probability that the energy will transfer to the excited state will in this case always remain small. This is similar to what happens with a pair of classical coupled pendulums of different frequency (so that $\Delta\varepsilon$ is large in comparison with the coupling energy). In this case the energy placed in one pendulum does not transfer very efficiently to the other one, because the frequency difference between the two pendulums causes them to drift out of phase with each other before much energy is transferred.

Rabi oscillations are only observed in systems with two (or very few) coupled states. In systems with a large number of closely spaced states, the energy will no longer oscillate back and forth, but will instead dissipate into the bath of coupled states. Under these conditions the excited state probability is predicted to become a linear function of time, and so the rate at which energy flows between the coupled states will be constant, and may be predicted using the following *Fermi golden rule* expression.[17]

$$\boxed{\frac{dP_n}{dt} = \frac{d|c_n|^2}{dt} = \frac{2\pi}{\hbar}|\langle n|\, V\, |0\rangle|^2 \rho(\varepsilon)} \tag{9.47}$$

This important result indicates that the transition rate between an initial state $|0\rangle$ and a final state $\langle n|$ is predicted to be proportional to both the coupling strength $|\langle n|\, V\, |0\rangle|^2$ and the density of final states $\rho(\varepsilon)$ (the number of states per unit energy).[18] As we will shortly see, this expression plays a key

[17] This expression was actually first derived by Paul Dirac, but it is named after Enrico Fermi, as he referred to it as a *golden rule*.

[18] Equation 9.47 is appropriate, for example, when describing processes such as ionization in which there is a continuum of final state. However, the same expression also arises much more generally, because light is never perfectly

role in the description of optical absorption processes involving transitions between rotational, vibrational, and electronic quantum states.

9.4 Optical Spectroscopy

Although the relation between quantum state energies and various optical absorption and emission processes have been described in Chapters 1, 7, and 8, we have not yet explicitly considered the time-dependent interaction between molecules and light. Doing so requires applying the time-dependent Schrödinger equation to predict the intensities and selection rules for various spectroscopic processes. Here we will see how this strategy may be used to understand and predict the shapes of molecular rotational, vibrational, and electronic spectra.

Optical Transition Probabilities and Dipole Selection Rules

The interaction between light and a collection of particles of charge q_i located at positions $\vec{r}_i$ may be represented as $V = \mathcal{E} \sum_i q_i \vec{r}_i$, where $\mathcal{E}$ is the electric field strength and $\sum_i q_i \vec{r}_i$ is the net *dipole moment* of the collection of particles.

$$\vec{\mu} \equiv \sum_i q_i \vec{r}_i \tag{9.48}$$

Recall that light of wavelength λ has a frequency of $\nu = c/\lambda$, which corresponds to $\omega = 2\pi c/\lambda$ (radians per second). So, the electric field strength in a beam of light can be expressed as $\mathcal{E}(t) = \mathcal{E}_0 \vec{n} \cos(\omega t)$, where $\mathcal{E}_0$ is the electric field amplitude and $\vec{n}$ is the polarization vector along the optical electric field direction (that is perpendicular to the direction of propagation of the light). Thus, the total energy of interaction between such a beam of light and the collection of charged particles within a molecule may be expressed as follows (where $\vec{n} \cdot \vec{\mu}$ is the vector dot-product that measures the projection of the electric field onto the dipole moment).

$$V(t) = \mathcal{E}_0 |\vec{n} \cdot \vec{\mu}| \cos(\omega t) \tag{9.49}$$

When combined with the Fermi golden rule (Eq. 9.47), the following expression is obtained for the optical transition probability between any initially populated state (i) and any other final state (f), in terms of the corresponding *transition dipole matrix element* $\mu_{fi} = \langle f | \mu | i \rangle$, and the incident light

---

monochromatic and so in any light beam there are always photons present with many slightly different energies, and so $\rho(\varepsilon)$ can also represent this optical density of states.

intensity, expressed either in terms of the electric field intensity $\mathcal{E}_0^2$ or in terms of the corresponding optical energy density ρ_ν.[19]

$$\frac{dP}{dt} = \left(\frac{\mathcal{E}_0^2}{\hbar^2}\right) |\mu_{fi}|^2 \rho(\nu_{fi}) = \left(\frac{1}{6\epsilon_0 \hbar^2}\right) |\mu_{fi}|^2 \rho_\nu \tag{9.50}$$

As a simple illustration of how the above formula may be applied, let's consider an electron confined within a one-dimensional box extending from $x = 0$ to $x = L$, that is irradiated by light whose electric field is polarized along the x-axis. The optical excitation of such an electron from any initial state ($n = i$) to any final state ($n = f$) depends on the following transition dipole matrix element.

$$\mu_{fi} = \langle f|\, ex \,|i\rangle = \int_0^L \Psi_f^* ex \Psi_i dx = e \int_0^L \Psi_f^* x \Psi_i dx \tag{9.51}$$

It is useful to reexpress the position of the electron in terms of its average position $x = \langle x\rangle + (x - \langle x\rangle\,) = \langle x\rangle + \Delta x$. Note that Δx is a variable, while $\langle x\rangle$ is a constant that can be brought out from under the integral.[20]

$$\begin{aligned} \mu_{fi} &= e\,\langle f|\, x \,|i\rangle = e\,\langle f|\,(\,\langle x\rangle + \Delta x)\,|i\rangle \\ &= e\,\langle x\rangle\,\langle f|i\rangle + e\,\langle f|\,\Delta x\,|i\rangle \\ &= e\,\langle f|\,\Delta x\,|i\rangle \end{aligned} \tag{9.52}$$

This result indicates that optical excitation of an electron in a box can only take place if the integral $\langle f|\,\Delta x\,|i\rangle$ is nonzero. Hence, the optical transition will be *forbidden* whenever the integrand is antisymmetric (in the sense that it contains positive and negative lobes of equal area). Notice that the function Δx is itself antisymmetric, and so the transition will be forbidden whenever the product of $\langle f| = \Psi_f^*$ and $|i\rangle = \Psi_i$ is symmetric. More generally, the requirement that $\mu_{fi} \neq 0$ leads to an *optical selection rule* that states that only transitions with $\Delta n = \pm 1$ are *dipole allowed*.

---

[19] The transition dipole matrix element $\mu_{fi} = \langle f|\,\mu\,|i\rangle$ is obtained by integrating $\Psi_f^*(\sum_i q_i \vec{nr_i})\Psi_i$ over the entire system volume. The optical mode density $\rho(\nu_{fi})$ is defined as the number of optical modes per unit frequency near the corresponding quantum state energy difference $\Delta\varepsilon = h\nu_{fi}$. The optical energy density ρ_ν is defined as the optical energy density per unit volume, or the energy of electromagnetic radiation of frequency ν times the number of electromagnetic modes per unit frequency in a unit volume of the sample. When expressed in this way, the resulting transition probability is also equal to the Einstein absorption coefficient times ρ_ν, as further discussed in Section 10.4.

[20] The last equality is obtained by making use of the fact that the eigenfunctions of the particle in a box are orthogonal and so $\langle f|i\rangle = 0$.

Exercise 9.4

Plot the integrands of the appropriate transition dipole integral to graphically prove that an electron in a one-dimensional box may undergo a dipole-allowed transition from $n = 1$ to $n = 2$, while the transition from $n = 1$ to $n = 3$ is dipole forbidden.

Solution.

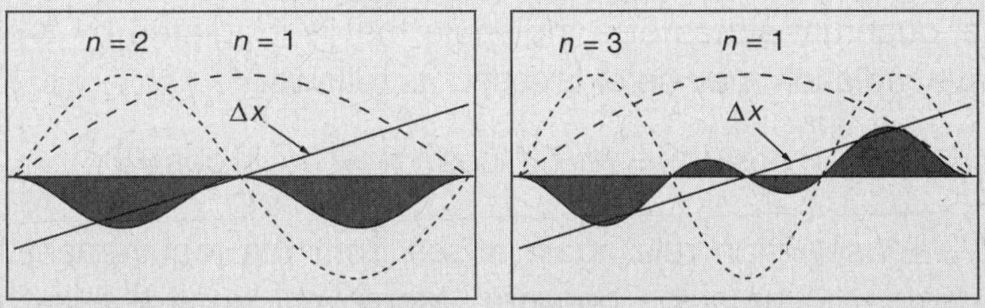

The graph on the left shows that the product of the antisymmetric $n = 2$ wavefunction and the antisymmetric function Δx times the symmetric ground-state $n = 1$ wavefunction produces a symmetric integrand (shaded region), and thus a nonzero transition dipole integral $\langle 2|\,\Delta x\,|1\rangle$. The graph on the right shows that the $n = 1$ to $n = 3$ integrand (shaded region) is antisymmetric, with equal negative and positive regions, and so $\langle 3|\,\Delta x\,|1\rangle = 0$ and thus the transition is dipole forbidden.

As we will see, a similar transition dipole analysis leads to selection rules of $\Delta J = \pm 1$ in molecular rotational microwave spectroscopy and $\Delta v = \pm 1$ in molecular infrared vibrational spectroscopy, as well as other such selection rules.[21] In the following subsections we will consider some of the unique features of each of these spectroscopies, as well as hybrid processes involving simultaneous changes in various molecular quantum mechanical degrees of freedom.

Rotational Microwave Spectroscopy

Molecular rotational quantum state spacings typically correspond to photon energies in the microwave spectral region. However, in addition to requiring that the microwave photon energy must match the corresponding rotational quantum state spacing, we will see that microwave absorption can only occur when a molecule has a permanent dipole moment (and the optical transition satisfies the $\Delta J = \pm 1$ section rules).

---

[21] Recall that it is conventional to use the symbol v instead of n for molecular vibrational quantum numbers, just as J is used instead of ℓ when describing molecular rotations.

The permanent dipole moment of a diatomic molecule such as HCl may be described by approximating its charge distribution by partial positive δq^+ and negative δq^- charges (of equal magnitude) separated by a distance equal to the diatomic bond length R. Such a charge displacement has a dipole moment vector $\vec{\mu}$ whose magnitude is $|\mu| = \delta q R$. As the molecule rotates, the projection of its dipole moment along the z-axis depends on its orientation θ with respect to that axis, $\mu_z = \delta q R \cos\theta = |\mu| \cos\theta$. Thus, the dipole moment matrix element corresponding to a transition between two rotational quantum states of a molecule that is irradiated by z-polarized microwave radiation may be expressed as follows.[22]

$$\mu_{fi} = \langle f|\, \delta R \cos\theta \,|i\rangle = \delta q R \,\langle f|\cos\theta\,|i\rangle = |\mu|\,\langle f|\cos\theta\,|i\rangle \tag{9.53}$$

The $\Delta J = \pm 1$ selection rule again arises from the requirement that the integral $\langle f|\cos\theta\,|i\rangle$ must be nonzero. Moreover, since the last equality in Eq. 9.53 indicates that $\langle f|\cos\theta\,|i\rangle$ is multiplied by $|\mu|$, it is evident that only molecules possessing a nonzero permanent dipole moment can produce a rotational microwave spectrum. Thus, homonuclear diatomic molecules (whose symmetry dictates that $|\mu| = 0$) cannot produce a rotational microwave absorption spectrum, while heteronuclear diatomics and other polar molecules can give rise to rotational microwave spectra.

The probability that a rotational transition will occur depends on the square of the transition dipole matrix element (see Eq. 9.50), and the latter matrix element is composed of the corresponding x-, y-, and z-components of μ_{fi}, such that $|\mu_{fi}|^2 = |\mu_x|^2 + |\mu_y|^2 + |\mu_z|^2$. When the μ_{fi} integral is evaluated (assuming that the molecules have a random orientational distribution), the following expressions are obtained for the $\Delta J = +1$ or $\Delta J = -1$ rotational transition dipole matrix elements.

$$|\mu_{fi}|^2 = |\mu|^2 \left(\frac{J+1}{2J+1}\right) \quad \text{for absorption from } J \text{ to } J+1 \tag{9.54}$$

$$|\mu_{fi}|^2 = |\mu|^2 \left(\frac{J}{2J+1}\right) \quad \text{for emission from } J \text{ to } J-1 \tag{9.55}$$

The corresponding transition energy differences are

$$E_{J+1} - E_J = \left(\frac{\hbar^2}{2I}\right)[(J+1)(J+2) - J(J+1)] = 2B(J+1) \tag{9.56}$$

$$E_J - E_{J-1} = \left(\frac{\hbar^2}{2I}\right)[J(J+1) - (J-1)J] = 2BJ \tag{9.57}$$

---

[22] Note that the the second equality in Eq. 9.53 was obtained by making use of the fact that the rotational eigenvectors are spherical harmonics $Y_J^{m_J}$, which depend on the angles θ and ϕ but are independent of R, which is why R could be taken outside the integral.

where $B = \hbar^2/(2I)$ is the rotational constant (and the moment of inertia of a diatomic molecule of reduced mass μ is $I = \mu R^2$). Since the transition energies are linear functions of J, the molecule's microwave spectrum is predicted to contain an evenly spaced sequence of absorption and emission lines (arising from states with different initial J values).[23] In the next subsection, we will see that the same progression of rotational lines is also predicted to appear in the vibrational spectra of heteronuclear diatomic molecules in the gas phase (as shown in Figure 9.3).

It is interesting to compare the rotational absorption frequency of a diatomic molecule with the corresponding classical molecular rotational frequency. Note that at large J both Eqs. 9.56 and 9.57 indicate that the transition energy approaches $2BJ = (\hbar^2/I)J$. When this energy difference is equated with the energy of a photon $\hbar\omega$, we obtain $\omega = (\hbar/I)J$, which is the same as the rotational frequency of a molecule whose angular momentum is $I\omega = \hbar\sqrt{J(J+1)} \approx \hbar J$. Thus, at large J the frequency of an absorbed microwave photon is the same as the molecule's classical rotational frequency.

Vibrational Infrared Spectroscopy

The vibrational quantum states of diatomic molecules typically give rise to absorption bands in the mid-infrared spectral region. The intensity of the resulting infrared spectrum may again be understood by considering the corresponding transition dipole matrix element. The following analysis indicates that (in addition to the $\Delta v = \pm 1$ selection rule) only vibrational modes that give rise to a change dipole moment are infrared active. Moreover, the frequencies of the corresponding infrared photons are again the same as the corresponding classical vibrational frequencies.

The vibrational motion of a diatomic molecule takes place on a vibrational potential energy surface that looks something like the Morse potential shown in Fig. 7.5 on page 245 (or the H_2^+ potential shown in Fig. 9.1). For small displacements about the potential energy minimum, the potential can be approximated by that of a harmonic oscillator $V(R) = \frac{1}{2} f \Delta R^2$ where f is the harmonic force constant and $\Delta R = R - R_e$ where R_e is the location of the potential energy minimum (or classical equilibrium bond length).

We may again express the dipole moment of a diatomic as $|\mu| = \delta q R$. If we assume that the magnitude of partial charge δq does not change when the bond length R is displaced, then the above expression implies that δq is equivalent to the derivative of $|\mu|$ with respect to R.

$$\left(\frac{\partial|\mu|}{\partial R}\right) = \delta q \tag{9.58}$$

---

[23] The experimentally observed spacing between lines in microwave spectra are often not precisely evenly spaced, due to centrifugal distortion, as further discussed on page 259.

Thus, we may obtain the following expression for the vibrational transition dipole matrix element by equating $\widehat{\mu} = \delta q R = \delta q(R_e + \Delta R)$.[24]

$$\mu_{fi} = \langle f|\,\widehat{\mu}\,|i\rangle = \delta q R_e \langle f|i\rangle + \delta q \langle f|\,\Delta R\,|i\rangle = \left(\frac{\partial|\mu|}{\partial R}\right) \langle f|\,\Delta R\,|i\rangle \tag{9.59}$$

The requirement that $\langle f|\,\Delta R\,|i\rangle \neq 0$ gives rise to the $\Delta v = \pm 1$ selection rule. Moreover, the factor of $(\partial|\mu|/\partial R)$ in front of the integral in Eq. 9.59 indicates that infrared active vibrations must have a nonzero dipole moment derivative. This is why homonuclear diatomics are predicted to be infrared inactive (as their symmetry demands that $\delta q = \partial|\mu|/\partial R = 0$), while heteronuclear diatomics are infrared active.

Since diatomic molecules can both vibrate and rotate, both the $\Delta v = \pm 1$ and $\Delta J = 0, \pm 1$ selection rules apply. Thus, the rotational spectrum of an isolated (gas phase) heteronuclear diatomic molecule is predicted to give rise to a sequence of rotational-vibrational, or *rovibrational*, peaks as shown in Figure 9.3 (and discussed further below).

The $\Delta v = +1$ transitions in the rovibrational spectrum of a heteronuclear diatomic give rise to an *R-branch* arising from $\Delta J = +1$ rotational transitions and a *P-branch* arising from $\Delta J = -1$ rotational transitions. Each of these branches consists of a sequence of approximately evenly spaced lines arising from rotational states with different initial values of J.[25] The intensity of each of the R-branch and P-branch lines depends on the populations of the corresponding initial rotational state (in the ground vibrational state), and so is determined by the following Boltzmann factor.[26]

$$P(J) \propto g_J e^{-\beta E_J} = (2J+1)e^{-BJ(J+1)/k_B T} \tag{9.60}$$

The relative intensities of the R- and P-branch peaks are also influenced by the rotational transition dipole matrix element, which is the same as that associated with the corresponding pure rotational transitions (Eqs. 9.54 and 9.55). Thus, each of the R-branch lines must be multiplied by $(J+1)/(2J+1)$, while each of the P-branch lines must be multiplied by $J/(2J+1)$. Note that the latter $2J+1$ factors exactly cancel those in

---

[24] Note that this result is obtained in a way that is very similar to that used to obtain Eq. 9.52. Again, $\langle f|i\rangle = 0$ because the harmonic oscillator eigenfunctions are orthogonal to each other.

[25] The *Q-branch* associated with $\Delta J = 0$ is usually dipole forbidden (except for some molecules with a nonzero electronic ground-state angular momentum) and so does not appear in vibrational spectrum of diatomics such as HCl.

[26] Note that this Boltzmann factor is the same as that given by Eq. 1.17 on page 25, where g_J and E_J are the degeneracy and energy of the corresponding initial rotational state.

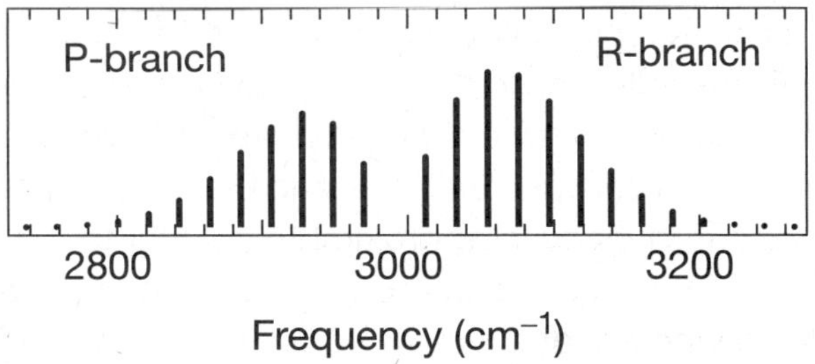

FIGURE 9.3 The predicted $\Delta v = +1$ rovibrational absorption spectrum of HCl at 295 K contains rotational bands arising from J to $J+1$ (R-branch) and J to $J-1$ (P-branch) transitions, whose intensities are obtained using Eq. 9.61.

Eq. 9.60, and so the intensities of the R- and P-branch lines are predicted to have the following dependence on J (and T).

$$I_R(J) \propto (J+1)e^{-BJ(J+1)/k_BT}$$
$$I_P(J) \propto Je^{-BJ(J+1)/k_BT} \quad (9.61)$$

Figure 9.3 shows the predicted rovibrational absoption spectrum of HCl (obtained as further explained in the following exercise).

Exercise 9.5

Use Eq. 9.61 to predict the positions and relative intensities of the $\Delta v = 1$ rovibrational R- and P-branches of HCl at 295 K, given that HCl has a vibrational frequency of $\tilde{\nu} \sim 2991$ cm^{-1} and rotational constant of $\tilde{B} \sim 10.59$ cm^{-1}.

Solution. The R-branch rotational transition energies are $E_{J+1} - E_J = 2B(J+1)$, while the P-branch transition energies are $E_J - E_{J-1} = 2BJ$, and so the predicted frequencies are $2991 + 2(10.59)(J+1)$ for the R-branch and $2991 - 2(10.59)J$ for the P-branch (both expressed in cm^{-1} units). The relative intensities of rotational bands in the R- and P-branches may be predicted using Eq. 9.61. Note that $B = hc\tilde{B}/hc$ so $B/k_BT = hc\tilde{B}/k_BT = \tilde{B}/(k_BT/hc)$ and $k_BT/hc \sim 205$ cm^{-1} at 295 K, so $e^{-BJ(J+1)/k_BT} = e^{-10.59J(J+1)/205}$. Figure 9.3 shows the resulting predicted shape of the $\Delta v = +1$ rovibrational spectrum of HCl.[27]

Rotational subbands, such as those shown in Figure 9.3, often appear in gas phase spectra but only very rarely in liquid or solid spectra. This

[27] This predicted spectrum is quite similar to the corresponding experimental spectrum. However, centrifugal distortion produces small changes in the bond length of HCl, and thus leads to a deviation from the perfectly even spacing shown in Fig. 9.3, as further discussed on page 259. There are also additional contributions to the intensity of the rovibrational lines arising from the anharmonicity of the vibrational potential and the bond length dependence of HCl's dipole moment.

is because molecules typically cannot freely rotate in liquid or solid samples. In other words, collisions in a liquid rapidly reorient molecules so they all have approximately the same thermally averaged angular momentum. This process leads to a collapse of the R- and P-branches to a single vibrational peak whose frequency is approximately equal to the average of all the R- and P-branch frequencies. The same is true in polyatomic liquids and solids, whose vibrational spectra contain single peaks corresponding to different molecular vibrational normal modes (with no rotational subbands). Equation 9.59 further implies that vibrational peaks will only appear in an infrared spectrum if the corresponding vibrational normal mode gives rise to a change in the molecule's dipole moment. Thus, for example, the symmetric stretch vibration of CO_2 is not infrared active, as it does not change the molecule's dipole moment, while the asymmetric stretch and bend vibrations of CO_2 are infrared active, as they do produce a change in dipole moment.

More generally, any molecule with a *center of inversion symmetry* can only give rise to infrared absorption peaks corresponding to asymmetric vibrations.[28] We will next see that another kind of vibrational spectroscopy, called Raman scattering, gives rise to vibrational spectra whose selection rules are complementary to infrared absorption, as the Raman spectra of molecules with a center of inversion symmetry only contain peaks corresponding to symmetric vibrations (which preserve, rather than break, the molecule's symmetry).

Raman Spectroscopy

Raman scattering arises when a photon is inelastically scattered by a molecule.[29] This occurs when a photon exchanges some of its energy with molecular vibrational or rotational degrees of freedom. This is similar to what happens in a classical inelastic scattering process, in which a particle loses (or gains) kinetic energy during a collision. The following analysis indicates that Raman scattering occurs when vibrational or rotational motions give rise to a change in molecular polarizability, and thus both homonuclear and heteronuclear diatomics can produce Raman spectra.

---

[28] A molecule is defined as having a center of inversion symmetry if it would look identical if each of its atoms were reflected through the molecule's center of symmetry. Thus, in addition to CO_2, all homonuclear diatomics, as well as molecules such as ethene and benzene, have centers of inversion symmetry.

[29] A brief introduction to Raman scattering, as well as its nearly simultaneous independent discovery in Calcutta and Moscow in 1928, is described in footnote 11 on page 18, and the associated text.

One of the important features of Raman scattering is that the energy of the incoming photon does not have to match a difference in energy between two vibrational quantum states. Thus, one may obtain a vibrational Raman spectrum by shining visible (rather than infrared) light onto molecules in a gas, liquid, or solid sample. Although Raman scattering was originally observed using sunlight, it is much easier to detect when using a more intense (and monochromatic) laser light source.[30]

Most of the light scattered by a condensed phase sample, such as a transparent liquid, emerges with exactly the same color as the incident light (and is referred to as *Rayleigh scattering*) and only a very small fraction of the incident light (typically $\sim$1 part in 10^6) is inelastically scattered to produce a Raman spectrum.

Energy conservation requires that the vibrational transition energy $\Delta\varepsilon = h\nu$ (where ν is the molecular vibrational frequency) must be exactly equal to the difference between the energies of the incident laser ($h\nu_L$) and Raman scattered ($h\nu_R$) photons.

$$\Delta\varepsilon = h\nu_L - h\nu_R = h(\nu_L - \nu_R) = h\nu \tag{9.62}$$

Thus, the difference between the frequencies of the incident and scattered photons must exactly match the frequency of the corresponding molecular vibrational normal mode, $\nu = \nu_L - \nu_R$.

The intensity of the peaks that appear in a Raman spectrum derive from the interactions between the electric field of the incident laser light $\mathcal{E}$ and the molecule's electronic *polarizability* α. Molecular polarizability arises from the fact that a molecule's electronic probability density ("electron cloud") can be distorted by an applied field and thus give rise to an increase in the molecule's dipole moment. More specifically, the total dipole moment of a molecule is the sum of its permanent dipole moment μ_0 (in the absence of an applied electric field) and the dipole moment that is induced by the applied field $\mu_I = \alpha\mathcal{E}$.

$$\mu(\mathcal{E}) = \mu_0 + \mu_I = \mu_0 + \alpha\mathcal{E} \tag{9.63}$$

When a diatomic molecule vibrates its bond length changes and thus produces a change in its electron cloud and therefore also its polarizability. As

[30] A monochromatic light source is one that has a narrow frequency spectrum. Lasers are also useful because they produce a beam of light that can be very tightly focused, thus improving the efficiency with which Raman spectra can be collected. Nowadays, when using a detector such as a thermoelectrically cooled charge-coupled-device (CCD) camera, one may obtain high-quality Raman spectra of virtually any liquid or solid within a few seconds using a laser intensity of a few milliwatts (similar to that of a battery-powered laser pointer).

a result, in the presence of the applied electric field, the molecule's induced dipole moment will have a nonzero bond-length derivative.

$$\left(\frac{\partial \mu_I}{\partial R}\right) = \left(\frac{\partial \alpha}{\partial R}\right) \mathcal{E} \tag{9.64}$$

Comparison of this expression with Eqs. 9.58 and 9.59 implies that Raman scattering can take place whenever $(\partial\alpha/\partial R) \neq 0$. Moreover, as in Eq. 9.59, the corresponding vibrational transition dipole matrix element $\langle f|\, \Delta R \,|i\rangle$ must be nonzero, and thus Raman scattering by harmonic vibrational modes is again associated with a $\Delta v = \pm 1$ selection rule.

Recall that infrared absorption could only occur for diatomic molecules that had a permanent dipole moment, and so only heteronuclear diatomics are predicted to have allowed infrared spectra. Raman scattering is more generally applicable as the polarizability of both homonuclear and heteronuclear diatomics change upon vibrational excitation, and the same is true for most polyatomic vibrational modes.

Molecules with a center of inversion symmetry are an important exception, as only the symmetric vibrational normal modes of such molecules give rise to a change in α and thus $(\partial\alpha/\partial R) \neq 0$. As a result, the symmetric stretch of CO_2 is Raman active, while its asymmetric stretch and bend are not. Recall that exactly the opposite is the case for the infrared absorption of CO_2, as only its asymmetric modes give rise to a change in dipole moment. The same is true for all molecules with a center of inversion symmetry; the Raman and infrared spectra of such molecules are always perfectly complementary to each other, as the peaks that don't appear in the infrared spectrum may appear in the Raman spectrum (and conversely).

The most intense type of vibrational Raman scattering, referred to as *Stokes* scattering, occurs when a molecule gains energy from a photon, and thus, $\Delta v = +1$ and the energy of the scatter photon *decreases* by $h\nu$. However, it is also possible for a molecule that is initially in a vibrationally excited state to give rise to *anti-Stokes* scattering, which occurs when a vibrationally excited molecule loses energy, so that $\Delta v = -1$ and the energy of the Raman-scattered photon *increases* by $h\nu$. The probability of such an anti-Stokes scattering process becomes very small as the vibrational frequency of the molecule increases, because the vibrationally excited population is proportional to the Boltzmann factor $e^{-h\nu/k_B T}$, and so becomes negligibly small when $h\nu > k_B T$. Recall that at room temperature $k_B T$ is equivalent to a frequency of $\sim$200 cm^{-1}, and so only vibrational frequencies of this magnitude (or smaller) will give rise to substantial anti-Stokes scattering.

Raman scattering can also take place as a result of energy exchanges between light and molecular rotational states (in the gas phase). This occurs because the degree to which a molecule is polarized by an applied electric field depends on the orientation of the molecule relative to the

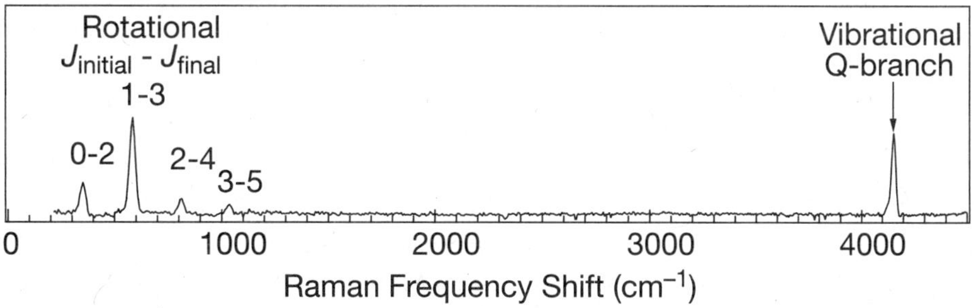

FIGURE 9.4 This experimentally measured Raman spectrum of H_2 gas (at 20°C and 0.1 MPa) reveals peaks arising from both rotational and vibrational Raman scattering transitions.[32] Note that the first rotational peak (corresponding to a transition from $J = 0$ to $J = 2$) is due entirely to *para*-H_2, while the second peak (corresponding to a transition from $J = 1$ to $J = 3$) is due entirely to *ortho*-H_2 (as further explained on pages 280–282).

electric field. In other words, the polarizability of a diatomic is largest along its bond axis, and smallest in the direction that is perpendicular to the bond. So, the polarizability along the applied electric field will vary as a molecule rotates. The selection rule for rotational Raman scattering is $\Delta J = \pm 2$, which derives from the fact that the bond axis of a rotating molecule will align with the applied field twice in every rotational cycle.[31] Thus, the photon energy differences associated with rotational Raman lines are $\Delta E = E_{J+2} - E_J = B[(J+2)(J+3) - J(J+1)] = B(4J+6)$, and so the spacing between rotational Raman lines is predicted to be $4B$ (rather than $2B$ as in rotational microwave absorption spectra).

Molecules in the gas phase may also give rise to rovibrational Raman spectra (similar to rovibrational infrared spectra) that contain rotational structure arising from the Raman selection rules $\Delta v = \pm 1$ and $\Delta J = 0, \pm 2$, thus producing an O-branch ($\Delta J = -2$), Q-branch ($\Delta J = 0$), and S-branch ($\Delta J = +2$). The Q-branch is typically the most prominent as it contains many closely spaced peaks associated with different initial J values, all of which have very similar vibrational frequencies (while the O- and S-branches contain isolated peaks separated by $\sim 4B$).

Figure 9.4 shows the experimentally measured Raman spectrum of H_2 with peaks arising from both pure rotational and vibrational

---

[31] More specifically, the angular periodicity of a molecule's polarizability implies that the Raman scattering integral corresponding to Eq. 9.53 will contain a factor of $\cos^2\theta$ rather than $\cos\theta$, and thus will give rise to a $\Delta J = \pm 2$ rather than a $\Delta J = \pm 1$ selection rule.

[32] The spectrum in Fig. 9.4 was collected by Joel Davis and Blake Rankin in the author's research group at Purdue University, using a home-built Raman system equipped with a 50 mW (514.5 nm wavelength) air-cooled argon-ion excitation laser and thermoelectrically cooled CCD detector.

transitions. Note that since H_2 is a homonuclear diatomic (and so has no dipole moment), its rotational and vibrational spectra could not be observed using microwave or infrared absorption spectroscopy.

Although most molecules, including both homonuclear and heteronuclear diatomics, can give rise to rotational Raman scattering, there are again interesting exceptions. For example, tetrahedral molecules such as CH_4 and CCl_4 have isotropic (spherical) polarizability tensors. This means that the polarizability of such molecules is independent of their orientation, and so they cannot give rise to rotational Raman scattering.

Electronic Absorption and Fluorescence

Optical transitions between molecular electronic states typically occur in the visible (VIS) or ultraviolet (UV) spectral region, and so are referred to as UV-VIS spectra. For example, the particle-in-a-box transition dipole matrix element expression (Eq. 9.52) implies that molecules such as linear polyene chains should give rise to optical absorption, with $\Delta n = +1$, from the HOMO to the LUMO state. Similarly, optical emission (fluorescence), with $\Delta n = -1$, from the LUMO excited state back down to the HOMO state is also optically allowed.

Although one might expect the absorption and fluorescence wavelengths of a given molecule to be identical, in practice molecular fluorescence is typically red-shifted with respect to the corresponding absorption spectrum. One of the reasons for this is that the LUMO excited state has a lower bond order than the HOMO ground state, which gives rise to a slight bond elongation and thus a longer effective box-length in the LUMO than in the HOMO state.

Another factor that contributes to the difference between the absorption and fluorescence maximum wavelengths arises from the influence of vibrational quantum states in the ground and excited electronic states, as illustrated in Fig. 9.5. Molecules in both the ground and excited electronic states have a *manifold* of vibrational states. At thermal equilibrium, most of the vibrational population must be in the ground vibrational state. Thus, electronic absorption occurs primarily from the ground vibrational level of the ground electronic state manifold, but may end in a higher energy vibrational level of the excited electronic state manifold. Fluorescence, on the other hand, occurs primarily from the ground vibrational level of the excited electronic state manifold, but may end in a higher vibrational level of the ground electronic state manifold. This gives rise to a difference between the absorption and fluorescence maximum energies, indicated by the length of the vertical arrows in Figure 9.5.

The intensity of such vibrational-electronic, or *vibronic*, transitions are governed by the *Franck-Condon factor* that is proportional to the overlap

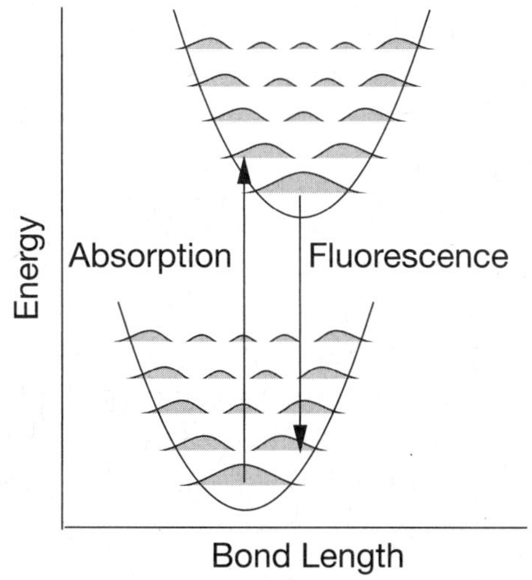

FIGURE 9.5 The relationship between electronic absorption and fluorescence is illustrated schematically for a system with harmonic ground (lower curve) and excited (upper curve) electronic potential energy surfaces. The horizontal displacement (longer bond length) of the excited electronic state potential, as well as the overlap between the ground and excited state vibrational probability densities, dictate the difference between the maximum absorption and fluorescence energies.

between the vibrational probability densities in the ground and excited electronic states. Note that vibrational probability densities are largest near the classical turning point of the corresponding vibrational potential. Thus, the difference between the absorption and fluorescence maximum depends on the degree to which the excited bond lengths are displaced with respect to those in the ground state. The ground and excited states of polyatomic molecules have more complicated vibrational manifolds than those illustrated in Figure 9.5, as they contain many different (anharmonic) vibrational states, pertaining to all of the normal mode vibrations of the molecule.

Electron spin angular momentum also influences the strength of electronic absorption and emission spectra. When an electron is optically excited, the orientation of its spin is typically not changed. If all the electrons in the ground electronic state are paired, then its total spin is $S = 0$, and so it is referred to as being in a *singlet* state (because its spin multiplicity is $2S + 1 = 1$). Optical excitation of such a molecule is thus most likely to give rise to a singlet excited state. However, spin-orbit coupling interactions can give rise to electronic transitions that involve a change in spin multiplicity. For example, if an electron that is paired in the ground state undergoes a spin flip upon optical excitation, then both of the original pair of electrons will end up with the same spin orientation. Thus, the total spin of such a pair of electrons will be $S = \frac{1}{2} + \frac{1}{2} = 1$ and so the excited state will be a *triplet* ($2S + 1 = 3$). Since such *spin forbidden* singlet-triplet transitions may be facilitated by spin-orbit coupling, they are not strictly forbidden, but they are typically much weaker than singlet-singlet (or triplet-triplet) transitions, which do not involve any change in spin multiplicity.

Atomic absorption and emission spectra must also satisfy angular momentum selection rules, which again arise from the symmetry properties of the corresponding transition dipole moment integrand. For example, the transition dipole moment integral of a hydrogen atom is only nonzero when $\Delta\ell = \pm 1$. Moreover, the polarization vector of the light must be correctly oriented with respect to the corresponding atomic orbitals in order to induce a transition. Thus, in order to undergo an optical transition from the ground $1s$ state, to an excited $2p_z$ state, the incident light must be polarized along the z-axis.

It is also interesting to note that the $\Delta\ell = \pm 1$ selection rule is consistent with the fact that photons are Bose particles and so have a spin of $s = 1$. Thus, the $\Delta\ell = \pm 1$ selection rule conserves angular momentum, as the angular momentum of the photon is transferred to the atom when it is absorbed. Such angular momentum conservation also dictates the mechanical properties of classical systems with spherically symmetric potential energy functions, such as the Sun's gravitational potential energy that restricts the motion of planets to elliptical orbits with constant angular momentum.

9.5 Introduction to *Ab Initio* Methods

The electronic structure of atoms and molecules ultimately determines all chemical properties, including the equilibrium constants and rate constants of chemical reactions – this is why the Schrödinger equation plays such a central role in chemistry. Although it is not so hard to write down the Schrödinger equation for any system, solving the equation is often so difficult that it requires making use of various approximation strategies. Paul Dirac famously described this situation in the following words:

> *The underlying physical laws necessary for the mathematical theory of a large part of physics and the whole of chemistry are thus completely known, and the difficulty is only that the exact application of these laws leads to equations much too complicated to be soluble. From P.A.M. Dirac, Quantum Mechanics of Many-Electron Systems, Proceedings of the Royal Society of London. Series A, Vol. 123, p. 714 (1929).*

The Hückel approximation is an example of a particularly simple semiempirical approximation strategy. Although we have seen that the Hückel method can sometimes produce remarkably accurate predictions, it is not very general, as it requires making use of empirical (experimental) information, and its realm of applicability is restricted to systems that are very similar to those for which experimental data is available. A far more general and powerful class of quantum chemical calculation strategies are referred to as *ab initio* ("from the beginning") methods, since they do not require any experimental input information, and so may be applied to any chemical system.

Two widely used *ab initio* techniques are the Hartree-Fock (HF) and density functional theory (DFT) methods. The appeal of both of these methods is that they can provide predictively useful results without requiring as much computational time as many other methods. The HF strategy is only moderately accurate, but serves as a starting point for many other more accurate *ab initio* techniques, including the remarkably accurate and efficient DFT methodology.

Ab initio electronic structure calculations typically rely on the Born-Oppenheimer approximation. In other words, the electronic state of a system is assumed to instantaneously track changes in the locations of its nuclei (which implies that the electronic and nuclear wavefunctions are separable, as in Eq. 9.4). Thus, the key task of quantum chemistry is reduced to determining the electronic structure of the system at various (fixed) nuclear configurations.

Within the Born-Oppenheimer approximation, the exact electronic structure of an N-electron system corresponds to the eigenfunctions of the specific Hamiltonian of interest, which always has the following general form.

$$\widehat{H} = \widehat{K} + \widehat{V} + \widehat{U} = \sum_{i}^{N} -\frac{\hbar^2}{2m_e}\nabla_i^2 + \sum_{i}^{N} V(r_i) + \sum_{i<j}^{N} U(r_i, r_j) \tag{9.65}$$

In addition to the usual electron kinetic energy operator $\widehat{K}$, the electronic potential energy of each electron is separated into two terms $\widehat{V}$ and $\widehat{U}$: where $V(r_i)$ is the negative (attractive) potential energy of interaction between electrons at any location r_i and the nuclei (located at the specified fixed input positions), and $U(r_i, r_j)$ is the positive (repulsive) interaction energy between pairs of electrons located at r_i and r_j. The sums in Eq. 9.65 are carried out over all electrons (and the inequality in the third sum assures that no pair of electron-electron interactions is doubly counted). However, since it is not in general possible to exactly solve the Schrödinger equation (except when $N = 1$), various strategies have been developed in order to transform, and approximately solve, the Schrödinger equation for many-electron systems.

Ab initio methods, including HF and DFT, also typically represent electronic wavefunctions as a linear combination of some set of basis functions (as described on pages 298–301). This assumption need not significantly influence the accuracy of the results, since one can represent any wavefunction arbitrarily accurately by using a sufficiently large number of basis functions. However, adding basis functions increases the time required to carry out quantum calculations. Thus, the accuracy of a given *ab initio* calculation depends both on the level of theory (approximations) used, as well as on the number and type of basis functions that are employed in representing the electronic wavefunctions.

Hartree-Fock Self-Consistent Field Theory

The development of the Hartree-Fock (HF) method began in 1927, soon after Schrödinger first introduced his equation (in 1926), that is, when Douglas Hartree, an English mathematician and physicist, first introduced a procedure he called the *self-consistent field* (SCF) method to (approximately) solve the Schrödinger equation for multielectron atoms. Hartree's idea was motivated by experimental observations that the spectra of many-electron atoms can often be represented using a modified version of Bohr's formula for one-electron atoms and ions (Eq. 8.10) in which the actual charge of the nucleus is partially shielded by the core-electrons.

These observations suggested that multielectron atoms could perhaps be represented by a product of single-electron wavefunctions, each of which is influenced by the *mean field* imposed by the other electrons (as well as the nucleus). In 1930 an American physicist, John Slater, and a Soviet physicist, Vladimir Fock, independently realized that the form in which Hartree originally represented his product of one-electron wavefunctions was not consistent with the requirement that Fermi particle wavefunctions must be antisymmetric with respect to electron exchanges (as described in Section 8.3). This realization led to an important modification of Hartree's method (as described further below) that significantly improved the accuracy of its predictions.

A key feature of the HF (or HF-SCF) method is that the wavefunction for each electron is obtained by solving a one-electron Schrödinger equation in the presence of the average (mean field) potential energy induced by the total electron density of all the other electrons. In other words, the many-electron Schrödinger equation with the Hamiltonian given by Eq. 9.65 is replaced by a set of one-electron Hamiltonians of the following form.

$$\widehat{H_1} = \widehat{K}_1 + \widehat{V}_1 + \widehat{U}_1^{\mathrm{HF}} = -\frac{\hbar^2}{2m}\nabla_1^2 + V_1(r_1) + U_1^{\mathrm{HF}}(r_1) \qquad (9.66)$$

This Hamiltonian represents the total energy of a single electron in the presence of a potential that is the sum of the attractive interaction energy

V_1 of the electron with all the nuclei, and the repulsive interaction energy energy U_1^{HF} between the electron and the mean electron density of all the other electrons in the system. There is a similar Hamiltonian for each electron in the system (although each electron may experience a somewhat different U_1^{HF} potential due to its interactions with all the other electrons).[33]

The resulting one-electron eigenfunctions are obtained by applying the variational principle (see pages 296–298) to minimize the ground-state energy of the system. In practice this requires starting with some assumed electron density for all the electrons (such as a linear combination of atomic orbitals of each of the separate atoms), and then calculating the wave-function of each electron in the presence of an assumed U_1^{HF} potential due to all the other electrons. The resulting one-electron wavefunctions are then used to provide a better estimate of U_1^{HF}. This process is repeated until the results become *self-consistent*, in the sense that the output one-electron wavefunctions are consistent with the input mean field electron densities.

Since electrons are Fermi particles, the Pauli exclusion principle requires that the total wavefunction of any multielectron system must be anti-symmetric with respect to electron exchange (as discussed in Section 8.3). John Slater suggested an elegant way to produce such anti-antisymmetrized wavefunctions, by arranging all the single-electron wavefunctions in a matrix whose determinant (referred to as a *Slater determinant*) is a linear combination of one-electron wavefunctions that changes sign when any pair of electrons is interchanged (and vanishes when any two one-electron wavefunctions are identical to each other).

In order to see how a Slater determinant is constructed, recall that for a two-electron system we obtained $\Psi(r_1, r_2) = \frac{1}{\sqrt{2}}[\psi_a(r_1)\psi_b(r_2) - \psi_a(r_2)\psi_b(r_1)]$ as the antisymmetric linear combination of one-electron wavefunctions, ψ_a and ψ_b (as discussed in Section 8.3). Notice that the latter wavefunction can also be obtained from the following Slater determinant:

$$\Psi(r_1, r_2) = \frac{1}{\sqrt{2}}\begin{vmatrix} \psi_a(r_1) & \psi_b(r_1) \\ \psi_a(r_2) & \psi_b(r_2) \end{vmatrix} = \frac{1}{\sqrt{2}}\{\psi_a(r_1)\psi_b(r_2) - \psi_a(r_2)\psi_b(r_1)\}$$

[33] More specifically, the U_1^{HF} potential includes both diagonal and off-diagonal contributions. The diagonal (coulomb) contributions are proportional to $\langle i|\frac{1}{r}|i\rangle$ and off-diagonal (exchange) contributions are proportional to $\langle i|\frac{1}{r}|j\rangle$, where $|i\rangle$ and $|j\rangle$ represent wavefunctions of all the electrons in the system, and r is the (variable) distance between any two points within the system.

The above procedure may be extended to any number (N) of one-electron wavefunctions, ψ_a, ψ_b, ψ_c,...ψ_N, by defining the corresponding Slater determinant as follows:

$$\Psi(r_1, r_2, r_3, \ldots) = \frac{1}{\sqrt{N!}} \begin{vmatrix} \psi_a(r_1) & \psi_b(r_1) & \psi_c(r_1) & \ldots \\ \psi_a(r_2) & \psi_b(r_2) & \psi_c(r_2) & \ldots \\ \psi_a(r_3) & \psi_b(r_3) & \psi_c(r_3) & \ldots \\ \vdots & \vdots & \vdots & \ddots \end{vmatrix}$$

Such Slater determinants are antisymmetric with respect to electron exchange, because interchanging any two rows of the matrix necessarily changes the sign of the determinant. Moreover, if any two columns of the matrix are identical (or proportional to each other), then the determinant is necessarily equal to zero, thus assuring that no two electrons can occupy identical wavefunctions (all of whose quantum numbers, including spin, are the same).

Although the HF method can yield quite accurate results, representing the many-electron wavefunction as a single Slater determinant implies that correlations between the instantaneous positions of all the electrons are not properly represented. In other words, $|\psi|^2$ for each electron only determines the probability that it will be found in a particular region of space, but does not completely determine how its probability is correlated with that of all the other electrons. Neglecting such correlations can significantly influence the energy of the system, and is thus a key shortcoming of the HF method. Another limitation of the HF method is that it requires solving the Schrödinger equation iteratively until a self-consistent result is obtained. Such calculations scale as the fourth power of the number of basis functions used to describe the system's multielectron wavefunction, and thus become extremely time consuming for systems with large numbers of electrons.

Density Functional Theory

Density functional theory has its roots in the Thomas-Fermi theory developed in 1927 by Llewellyn Thomas and Enrico Fermi. They made use of the known properties of a free electron gas in order to express the kinetic energies of electrons in terms of their probability density $\rho(r)$.[34] The

---

[34] For a single electron $\rho(r) = \Psi^*(r)\Psi(r) = |\Psi(r)|^2$, while for an N-electron system $\rho(r_1, r_2 \ldots, r_N) = |\Psi(r_1, r_2 \ldots, r_N)|^2$. However, one may nevertheless identify the total electron density at any point is space by integrating over the positions of all the electrons except one, $\rho(r) = N \int \rho(r, r_2 \ldots, r_N) dr_2 \ldots dr_N$. Note that this definition of $\rho(r)$ is independent of which electron is selected; this is because the wavefunctions of Fermi particles only changes sign when two electrons are interchanged, and so the probability density will be the same regardless of which electron is selected.

Thomas-Fermi strategy differs from that of many other *ab initio* methods in that only electron probability densities (rather than wavefunctions) are required in order to determine the energy of the system. Although the original Thomas-Fermi theory was not very accurate, it was later improved, and subsequently served as the starting point for modern DFT.

Modern density functional theory originated in 1964, when Pierre Hohenberg and Walter Kohn derived a pair of theorems. These Hohenberg-Kohn (HK) theorems used the variational principle to prove that there is a one-to-one (functional) relationship between the ground state of any system and its total electron density $\rho(r)$, and that the electron density distribution that minimizes the total energy of the ground state is the exact ground-state electron density. This also implies that there is a one-to-one relationship between the exact $\rho(r)$ and the exact Hamiltonian of the system. Thus, all the properties of any system may in principle be derived from its exact ground-state electron density. However, in spite of the elegance and generality of the HK theorems, the nature of the functional relationship between $\rho(r)$ and various observable quantities, including the kinetic and potential energies of the system, are not known exactly, although they can often be approximated with remarkably high accuracy.

A key goal of DFT is to express the energy of a system in terms of its electron density. In other words, the goal is to express the expectation value of each of the terms on the right-hand side of Eq. 9.65 as functionals of $\rho(r)$. It is easy to do that for some of the terms in Eq. 9.65. For example, the attractive interaction of one electron at position r_1 with each of the nuclei of charge Z_k located at r_k depends in the following simple way on the electron's probability density $\rho(r_1)$ and its distance from each nucleus $r_{1k} = |r_k - r_1|$.[35]

$$V[\rho] = -\frac{e^2}{4\pi\epsilon_0}\sum_k \int \frac{Z_k \rho(r_1)}{r_{1k}} d\tau_1 \tag{9.67}$$

Expressing the energy arising from remaining terms in the Hamiltonian as functional of electron density is not quite so simple. Walter Kohn and Lu Sham proposed (in 1965) a clever way to express this functional as the sum of three terms, $K_S[\rho] + U_R[\rho] + U_{XC}[\rho]$, where K_S is the kinetic energy of a fictitious system of non-interacting electrons (the *Kohn-Sham* system) whose density is the same as that of the true interacting-electron system, U_R is the repulsive (coulomb) interaction energy between the electron charge distributions, and U_{XC} arises from electron exchange and correlation. For a two-electron system, U_R depends in the following simple way

---

[35] The locations r_1 and r_k each represent three-dimensional coordinates (vectors), and thus, r_{1k} is the magnitude of the vector extending from r_k to r_1.

on the probability densities of the two electrons $\rho(r_1)$ and $\rho(r_2)$ (where the factor of $\frac{1}{2}$ is included in order to avoid double counting).

$$U_R(r_1, r_2) = \frac{1}{2}\left(\frac{e^2}{4\pi\epsilon_0}\right)\int \frac{\rho(r_1)\rho(r_2)}{r_{12}} d\tau_1 d\tau_2 \tag{9.68}$$

Although the HK theorem implies that U_{XC} should also be expressible in terms of electron density, no exact expression for this functional relationship has yet been discovered. Thus, various approximate expressions for U_{XC} are used in DFT calculations. One of the simplest such relations for the exchange component U_X is the following *local density approximation* (LDA), which is derived from exact results for a uniform gas of electrons.

$$U_X^{LDA}[\rho] = -\frac{3}{4}\left(\frac{3}{\pi}\right)^{1/5} \int \rho(r)^{4/3} d\tau \tag{9.69}$$

The density $\rho(r)$ appearing in the above expression is the total number density of electrons at every location r in the system (and the energy is here expressed in Hartree energy units).[36] The above expression may be improved in various ways. For example, the *generalized gradient approximation* (GGA) makes use of the local electron density and its gradient (derivative with respect r), for both the up and down spin states. Predictions obtained using the GGA functional are far more accurate than those obtained using the HF approximation, and require far less computational time.

The HK theorem implies that one should also be able to express the kinetic energy as a function of the electron density. For example, the following approximate expression for the kinetic energy (used by Thomas and Fermi) is again derived from exact results pertaining to an electron gas of uniform density (and is again expressed in Hartree units).

$$K_S[\rho] = \frac{3}{10}\left(3\pi^2\right)^{2/3} \int \rho(r)^{5/3} d\tau \tag{9.70}$$

Although the above expression (or improvements thereof) can be used in DFT, none of the known kinetic energy functionals are sufficiently accurate for high-level quantum chemical calculations. For this reason molecular DFT calculations often use the exact kinetic energy operator (see Eq. 9.65) to determine the kinetic energy expectation value $\langle K \rangle$, even though doing so requires significant computational time. Nevertheless, the advantages associated with using DFT to calculate all of the contributions to the potential energy make it possible to carry out highly accurate DFT calculations more efficiently than is possible using any other *ab initio* method.

---

[36] Recall that one Hartree of energy $E_h = \hbar^2/(m_e a_0^2)$ is approximately equivalent to 2,626 kJ/mol.

HOMEWORK PROBLEMS

Problems That Illustrate Core Concepts

1. A one-dimensional particle-in-a-box model may be used to illustrate the importance of kinetic energy quantization in covalent bond formation. For example, the electronic energy change associated with the reaction $H + H \rightarrow H_2$ may be modeled by treating each reactant H atom as an electron in a one-dimensional box of length $L_H = 5a_0$ (the 99% electron density diameter of hydrogen), and treating the diatomic H_2 as a one-dimensional box of length $L_{H_2} = R_B + 5a_0$ (where a_0 is the Bohr radius of hydrogen and $R_B \approx 0.74$ Å is the experimental bond length of H_2).

(a) Use the above particle-in-a-box model to predict the bond formation energy of H_2, and compare your result with the experimental value of -436 kJ/mol.

(b) What interactions have been neglected in the above calculation and what does your result imply with regard to the importance of kinetic energy quantization in covalent bond formation?

2. The bond formation reaction $H^+ + H \rightarrow H_2^+$ has an energy of $\Delta U = E_B \approx -210$ kJ/mol and the product H_2^+ cation has a bond length of $R_B \approx 1.06$ Å. Since H^+ has no electron, the entire bond formation reaction energy corresponds to the energy associated with transferring one electron from an isolated H atom to an H_2^+ molecule.

(a) Use the above experimental value of E_B and the energy of a hydrogen atom 1*s* orbital to determine the total *electron-binding* energy of the one electron in the ground state of H_2^+ (in kJ/mol units, this should be a large negative number as it represents the energy gained when a free electron associates with the two protons).

(b) Use the above results to estimate the total *bond formation* energy for the two electrons in H_2 by assuming that the bond formation energies are additive, so the same amount of energy will be gained when another electron is added to H_2^+.

(c) The standard heats of formation of H_2 and H are $\Delta H = 0$ kJ/mol and $\Delta H = +217.2$ kJ/mol, respectively (both in the gas phase at 298K and a pressure of 0.1 MPa).

(i) Use the above heats of formation to predict the bond formation enthalpy of H_2.

(ii) Use the fact that $H = U + PV$, and so $\Delta H = \Delta U + \Delta(PV)$, to calculate the bond formation energy of H_2, and compare your result with the values obtained in 1b. Note: You may assume that the gases are ideal and so $\Delta(PV) = \Delta nRT$ (where Δn is the difference between the number of product and reactant molecules).

3. The Aufbau principle and Hund's rules may be used to predict the bond order and total spin multiplicity ($2S + 1$) of diatomic molecules in their ground states.

(a) Draw an energy level diagram showing the ground-state electronic structures of each of the following diatomics: H_2, N_2, O_2, F_2, and Ne_2.

(b) Use the above diagrams to predict the bond order of each of these diatomics.

(c) Use the above diagrams to predict the total spin multiplicity of these diatomics.

(d) Use the same procedure to predict the bond order of C_2^+ and O_2^+.

(e) The experimental vibrational frequencies of C_2^+ and O_2^+ are 1,350 cm^{-1} and 1,905 cm^{-1}, respectively. Are these consistent with the relative bond orders of these two molecules, as well as with the frequencies of the neutral C_2 and O_2 (which are given on page 304)?

4. Three hydrogen atoms can combine to form various triatomic molecules or ions (with different geometries and/or numbers of electrons).
   (a) Use the experimental bond formation energy of H_2^+, $\Delta U = E_B \approx -210$ kJ/mol, to estimate the Hückel β parameter (resonance integral).
   (b) Expand the cyclic trimer secular determinant given in Eq. 9.27 and verify that the resulting polynomial is equivalent to that obtained using the roots given in Eq. 9.28.
   (c) Use the Hückel expressions for energy levels of linear and cyclic triatomics to predict whether the linear or cyclic forms of H_3^+ should be the most stable, and compare its predicted bond formation energy with the value of -903 kJ/mol obtained from a high-level theoretical calculation.
   (d) The relative energy of H_3^+ and H_2 plays an important role in the capture of protons by H_2. Do you expect the reaction $H_2 + H^+ \rightarrow H_3^+$ (in the gas phase) to be endothermic or exothermic and by how much? In other words, predict the value of ΔH for this reaction.

5. Use the Hückel approximation to answer the following questions, which compare the π-bonding structure of ethene CH_2CH_2 and the allyl cation $CH_2CHCH_2^+$.
   (a) How many π electrons do each of the above molecules have?
   (b) Draw a Hückel π electron energy level diagram for ethene, sketch the ground and excited state π electron wave functions, and label each state with its Hückel π electron energy (expressed in terms of α and β).
   (c) What is the π-bonding energy of ethene (relative to two separated carbon $2p$ electrons)?
   (d) Draw a Hückel π electron energy level diagram for the allyl cation, label each state with its Hückel π electron energy (expressed in terms of α and β), and indicate the location of the π electrons in the ground state.
   (e) What is the π-bonding energy of the allyl cation (relative to two separated $2p$ electrons)?
   (f) What is the π electron delocalization energy of the allyl cation, relative to ethene (expressed as a function of β)?

6. The Hückel approximation may also be applied to the ring compound cyclobutadiene, c-C_4H_4 (which has a square structure).
   (a) Write down the Hückel secular equation for cyclobutadiene.
   (b) Simplify the Hückel secular determinant by dividing each element by β, and then defining $x = (\alpha - E)/\beta$ (as described on pages 309–310).
   (c) Given that the latter determinant is equal to $x^4 - 4x^2$, determine the four roots (x values) of the secular equation, and convert these to the corresponding Hückel energy eigenvalues.
   (d) Use the above results to estimate the total π-bonding energy of cyclobutadiene.
   (e) What is the π delocalization energy of cyclobutadiene, relative to two ethene molecules?
   (f) Sketch the lowest two Hückel π molecular orbitals of cyclobutadiene.

7. The Hückel approximation may also be used to predict the atomic orbital coefficients associated with each molecular orbital.
   (a) Evaluate the matrix product $\widehat{H}\Psi = E\Psi$ for the ground state of a Hückel diatomic (such as H_2 or ethene), to show that the

two atomic orbital coefficients must be equal, $c_a = c_b = c$.

(b) Determine the value of c by using the fact that normalization requires that $c_a^2 + c_b^2 = 1$.

8. Sketch the first three vibrational eigenstates of a *harmonic oscillator* and use them to graphically determine whether the corresponding transition dipole matrix element is or is not equal to zero (in order to confirm the $\Delta v = \pm 1$ selection rule):

(a) Ψ_0 to Ψ_1 (b) Ψ_0 to Ψ_2 (c) Ψ_2 to Ψ_1

9. Consider the following diatomic molecules: FCl, F_2, Cl_2.

(a) Which of the above molecules are infrared active?

(b) Which of the above molecules are microwave active?

(c) Which of the above molecules are vibrational Raman active?

(d) Which of the above molecules are rotational Raman active?

10. Consider the following triatomic molecules: CS_2 and H_2O. Note that CS_2 has a linear structure (like CO_2), while H_2O is bent (with a bond angle of ~104.5°).

(a) Sketch the symmetric and asymmetric stretching vibrational modes of CS_2 and H_2O.

(b) Which of the above modes do you expect to be infrared active (and why)?

(c) Which of the above vibrational modes do you expect to be Raman active (and why)?

11. Draw two horizontal lines, one well above the other, to representing the $v = 0$ and $v = 1$ vibrational energy levels of HCl, and then add additional (more closely spaced) horizontal lines to each vibrational energy, to indicate the first few rotational energy levels in each vibrational state (such that the $J = 0$ level coincides with each vibrational energy and the $J > 0$ levels are higher in energy, with appropriate relative energy spacings).

(a) Draw vertical arrows connecting initial and final energies of various allowed R-branch and P-branch transitions. Note that the lengths of these arrows should scale linearly with J (if your rotational level spacings were drawn realistically).

(b) Sketch the rovibrational spectrum of HCl (as shown in Fig. 9.3) and label each line with the initial and final J values that give rise to that line.

12. Use the experimental position of the first rotational peak in the Raman spectrum of H_2 (see Fig. 9.4) to estimate the rotational constant and bond length of H_2, and compare your results to the tabulated values of $\widetilde{B}$ = 60.9 cm^{-1} and R_o = 0.74 Å.

13. The vertical arrows in Fig. 9.5 indicate the energies of the strongest vibronic bands in the corresponding optical absorption and fluorescence spectra. However, absorption from $v = 0$ in the ground electronic state to other vibrational levels in the excited electronic state (including $v = 0$) often also appear in electronic spectra, with a lower intensity (reflecting the smaller Franck-Condon overlap between the corresponding vibrational probabilities densities). The same is true for fluorescence from $v = 0$ to other ground-state vibrational levels. Moreover, when a molecule is dissolved in a liquid, then the widths of the vibronic subbands in its absorption and fluorescence spectra are often comparable to (or greater than) the vibrational energy spacing.

(a) Sketch the absorption spectrum of such a molecule as a function of wavelength (using a solid curve), and label each vibronic band to indicate the initial and final values of v that produced this band.

(b) Use a dashed curve to draw the corresponding fluorescence spectrum on the same graph, and label each vibronic band to indicate the initial and final values of v that produced this band.

(c) Do you see why electronic absorption and fluorescence spectra often look approximately like mirror images of each other?

Problems That Test Your Understanding

14. Extend the procedure described in problem 1 to estimate the energy change associated with allowing the two electrons in H_2 to delocalize over an H_3^+ linear triatomic (treated as a box of length $5a_0 + 2R_B$), and compare your results with those obtained using the Hückel approximation (see Exercise 9.3 on page 307).

15. The semiempirical Hückel approximation may also be used to describe covalent bond formation.

(a) Use the experimental bond formation energy of H_2 (−436 kJ/mol) to obtain a value for the Hückel β parameter (expressed in kJ/mol units).

(b) Use this value of β to predict the energy of the reaction $3H \rightarrow$ c-H_3, in which the cyclic trimer c-H_3 is formed from three separated hydrogen atoms.

16. Consider the diatomic molecules N_2, N_2^+, and N_2^-.

(a) What is the bond order of each of the above diatomics?

(b) Match the following vibrational frequencies with each of the above diatomics: 1968 cm^{-1}, 2207 cm^{-1}, and 2359 cm^{-1}

(c) Match the following bond lengths with each of the above diatomics:

1.10 Å, 1.12 Å, and 1.19 Å.

(d) Which of the above molecules are infrared active?

(e) Which of the above molecules are microwave active?

(f) Which of the above molecules are vibrational Raman active?

(g) Which of the above molecules are rotational Raman active?

17. The molecular orbital energies of butadiene CH_2=CH–CH=CH_2 can be approximately represented either using the particle-in-a-box model or using the Hückel approximation (although the latter approximation significantly underestimates the energies of antibonding molecular orbitals).

(a) Use the particle-in-a-box model to express the difference between the LUMO and HOMO π-orbital energy levels of butadiene in terms of the box length L (and other fundamental constants).

(b) Use the Hückel approximation to express the difference between the LUMO and HOMO π-orbital energy levels of butadiene in terms of Hückel parameters α and/or β.

(c) If the box length in (a) is assumed to be $L \sim 0.5$ nm, combine your results in (a) and (b) to estimate the value of the Hückel β parameter (expressed in kJ/mol units).

(d) Write down the transition dipole integral pertaining to the HOMO to LUMO transition for the particle-in-a-box model of butadiene, and use the symmetry of the integrand to determine whether the transition is dipole allowed.

18. Use the Hückel approximation to predict the enthalpy change ΔH for the cyclization reaction n-$H_3 \rightarrow$ c-H_3, given that the bond formation energy of H_2 is approximately −440 kJ/mol.

19. The cyclopropenium cation c-$C_3H_3^+$, which has a cyclic (triangular) structure, has the same number of π electrons as ethene

$CH_2{=}CH_2$. A typical C-C single bond has a formation energy of −346 kJ/mol, while a C=C double bond has a formation energy of −602 kJ/mol.

(a) Use the Hückel model to express the π-bonding energy of ethene in terms of the Hückel β parameter.

(b) Use the difference between the above C=C and C-C and bond formation energies to estimate the value of β (in kJ/mol).

(c) Use the Hückel model to predict the π-delocalization energy of c-$C_3H_3^+$ (relative to ethene), and use the above β value to express your answer in kJ/mol.

20. Consider the following molecules: H_2, HBr, HC≡CH, NH_3. Note that HC≡CH is linear and NH_3 is nonplanar.

(a) Which of the above molecules have at least one microwave active rotational transition?

(b) Which of the above molecules have at least one infrared active vibrational transition?

(c) Which of the above molecules have no vibrational modes that are both Raman and IR active?

21. Consider the following atoms, molecules, or ions: H, H_2, H_2^+, c-H_3, n-H_3. Note that n-H_3 is a linear triatomic with two identical bond lengths and c-H_3 is a cyclic triatomic with three identical bond lengths.

(a) Which of the above molecules would you expect to have one or more strongly allowed vibrational infrared absorption transition(s)?

(b) Which of the above molecules would you expect to have one or more strongly allowed vibrational Raman scattering transition(s)?

(c) Which of the above molecules would you expect to have one or more strongly allowed rotational microwave absorption transition(s)?

(d) Which of the above molecules would you expect to have one or more strongly allowed rotational Raman scattering transition(s)?

22. The following figure shows the fluorescence and absorption spectra of a molecule in solution.

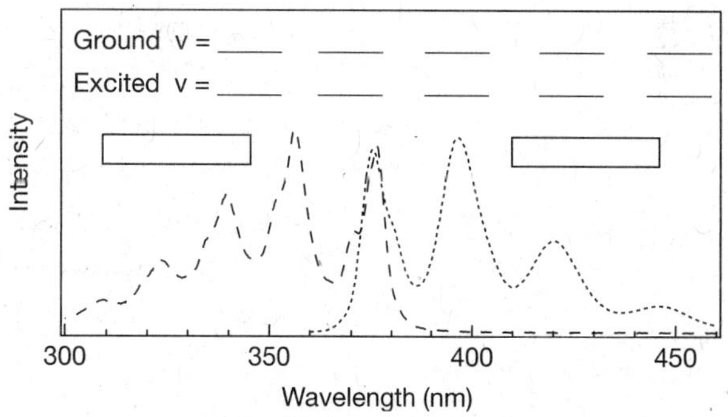

(a) Fill in the boxes in the above figure to indicate which is the fluorescence and which is the absoption spectrum.

(b) Fill in the blanks to assign the vibrational quantum numbers (in the ground and excited electronic states) for the vibrational mode that gives rise to the corresponding prominent peaks in the fluorescence and absorption spectra.

(c) Estimate the frequency of the above vibrational mode in the ground electronic state (in cm^{-1} units).

Chemical and Photon-Molecule Reactions

CHAPTER 10

The equilibrium constants and rates of chemical reactions have much in common with those of other processes, such as the absorption and emission of photons by molecules. Understanding the common features of all such processes requires combining results obtained from thermodynamics, statistical mechanics, and quantum mechanics. For example, statistical thermodynamic identities obtained in Chapters 1 and 5 may be combined with quantum calculations to predict chemical equilibrium constants. A similar strategy may be used to relate the rates of chemical reactions to molecular structures and energy distribution functions. Moreover, the same ideas can be used to understand how elementary processes involving interactions between photons and molecules give rise to linear and nonlinear optics, and to explain how lasers work.

10.1 Gas Phase Reaction Equilibria

Recall that a molecule's partition function is a measure of the number of molecular quantum states that are thermally accessible at a given temperature. The partition function is also required in order to calculate the probability that a given quantum state will be occupied. Since the equilibrium constant of a chemical reaction is a measure of the relative probability of finding the system in the product and reactant states, there is a close connection between partition functions and chemical equilibrium constants. In Chapter 1 this connection was illustrated for reactions of the form $A \rightleftharpoons B$, approximated as two-level systems (see Section 1.4); now we will see how those results may be generalized to other types of reactions.

All chemical processes are driven in the direction that minimizes the difference between the stoichiometric sum of the product and reactant chemical potentials. Thus, at equilibrium the product and reactant chemical potentials become equal to each other (see Section 4.4 and particularly Eqs. 4.52 and 4.53 on page 143).

$$\sum_i^{\text{Reactants}} a_i \mu_i = \sum_i^{\text{Products}} b_i \mu_i$$

The above identity may be expressed more compactly using the generalized stoichiometric coefficients ν_i that are positive for each product species and negative for each reactant species (see Eq. 4.52 on page 143).

$$\sum_i \nu_i \mu_i = 0$$

If we could solve the Schrödinger equation for an entire system, then we could predict μ_i for all of the equilibrium thermodynamic properties of any system (see Section 5.3). However, solving the Schrödinger equation for complex systems such as liquids and biological materials remains a frontier area of current research. Gas phase systems are somewhat more amenable to theoretical analysis as the thermodynamic properties of a gas can be obtained by separately solving the Schrödinger equation for each type of molecule in the system. This is in fact exactly how thermodynamic properties are routinely calculated using various widely available quantum chemistry programs.

Gas Phase Chemical Reactions

Gas phase chemical equilibria may be understood by treating each reactant and product molecule as an independent molecular subsystem. Thus, the chemical potential of each chemical species may be expressed in terms of the number N_i and partition function q_i of such molecules (see Eq. 5.33 on page 172).

$$\mu_i = -k_B T \ln\left(\frac{q_i}{N_i}\right)$$

The following expression for the stoichiometric difference between the product and reactant chemical potentials may be readily obtained using the above equation.

$$\begin{aligned}\sum_i \nu_i \mu_i &= -k_B T \sum_i \nu_i \ln\left(\frac{q_i}{N_i}\right) \\ &= -k_B T \sum_i \nu_i \ln q_i + k_B T \sum_i \nu_i \ln N_i \\ &= -k_B T \left[\ln\left(\prod_i q_i^{\nu_i}\right) - \ln\left(\prod_i N_i^{\nu_i}\right)\right] = 0 \qquad (10.1)\end{aligned}$$

This immediately leads to the following relationship between molecular partition functions q_i and the chemical equilibrium constant K_N, expressed as a stoichiometric ratio of the numbers of product and reactant molecules that are present in the system *at equilibrium* (as indicated by the subscript "eq").[1]

$$K_N = \prod_i N_i^{\nu_i} = \left[\left(\prod_i^{\text{Products}} N_i^{b_i}\right) \Big/ \left(\prod_i^{\text{Reactants}} N_i^{a_i}\right)\right]_{\text{eq}}$$

$$= \prod_i q_i^{\nu_i} = \left(\prod_i^{\text{Products}} q_i^{b_i}\right) \Big/ \left(\prod_i^{\text{Reactants}} q_i^{a_i}\right) \tag{10.2}$$

In other words, the above ratio of the numbers of product and reactant molecules at equilibrium is identical to the corresponding molecular partition function ratio (which in turn represents the ratio of the phase space volume available to the product and reactant species). Equation 10.2, combined with what we now know about molecular quantization, provides a means of predicting chemical equilibrium constants by solving the Schrödinger equation!

Predicting Molecular Partition Functions

In order to facilitate practical calculations of chemical equilibrium constants, it is often appropriate to treat molecular electronic (*e*), vibrational (*v*), rotational (*r*), and translational (*t*) degrees of freedom as independent (uncoupled) coordinates. Thus, we may express the total partition function for any molecule as a product of the partition functions associated with each of these degrees of freedom.

$$\boxed{q = q^e q^v q^r q^t} \tag{10.3}$$

Each of the partition functions on the right represents a sum of the Boltzmann factors $e^{-\beta\varepsilon_i}$ associated with the corresponding quantum state energies ε_i. The quantum state spacings associated with each molecular degree of freedom occur in different energy ranges (see Fig. 1.2). When the spacings are comparable to or greater than k_BT (or RT), as is the case for electronic and vibrational degrees of freedom, then we must explicitly evaluate the partition function sum. However, when the quantum state spacings are much smaller than k_BT, as is typically the case for rotational and translational degrees of freedom, then we may treat energy as a continuous (classical) variable and thus replace the partition function sum by an integral (as described in Section 1.5, and further illustrated below).

---

[1] Recall that Π_i represents a product, just as Σ_i represents the sum, of each of the *i* terms.

Electronic Partition Functions

Electronic partition functions are often quite simple, as excited electronic states typically have energies that differ from the ground state by far more than $k_B T$, and thus have a negligible thermal population. Thus, if the ground electronic state has an energy of ε_0 and a degeneracy of g_0, then the corresponding electronic partition function sum may be equated with its first term (as the Boltzmann factors for the excited state are negligible in comparison with the ground state $e^{-\beta\varepsilon_i} << e^{-\beta\varepsilon_0}$).

$$\boxed{q^e \cong g_0 e^{-\beta\varepsilon_0}} \tag{10.4}$$

Most molecules have nondegenerate ground states ($g_0 = 1$); however, there are some important exceptions, such as the ground state of O_2, which has two unpaired electrons and so has a triplet ground state $g_0 = 2S + 1 =$ (where $S = 1$ is the total electronic spin quantum number) 3. The ground-state degeneracy of an isolated (unbound) atom may be obtained from its ground-state term symbol (see Eq. 8.16), since $g_0 = 2J + 1$ (where J is the total electronic angular momentum quantum number). Additional excited state contributions must be added to Eq. 10.4 when the temperature is high enough that $k_B T$ becomes comparable to the energy gap between the ground and first excited electronic state.[2]

Vibrational Partition Functions

Molecular vibrations can often be well approximated by harmonic (normal mode) vibrations. We have seen that harmonic oscillators have an evenly spaced ladder of vibrational quantum states with $\Delta\varepsilon = h\nu$ (see Eq. 7.20 on page 243) whose partition function sum reduces to $q = 1/(1 - e^{-\beta\Delta\varepsilon})$ (see Eq. 1.21 and the subsequent discussion on page 28). The latter expression for q assumes that the vibrational energies are all referenced to the ground vibrational state ($\varepsilon_0^v = 0$). If we reference the energies to the electronic potential energy minimum, then the vibrational zero-point energy becomes $\varepsilon_0^v = \Delta\varepsilon/2$, which introduces an additional Boltzmann factor $e^{-\beta\varepsilon_0^v}$ to the numerator of the vibrational partition function. Thus, each vibrational normal mode of a molecule is predicted to have a partition

---

[2] Some atoms have low-lying excited electronic states that are significantly populated at room temperature, associated with term symbols corresponding to different electronic m_s and m_ℓ configurations. For example, although the ground electronic state of a carbon atom has a term symbol of $^3P_0$ and so is nondegenerate ($g_0 = 1$), there are two other low-lying electronic configurations with term symbols of $^3P_1$ and $^3P_2$ (and so degeneracies of 3 and 5), whose energies are $\approx$0.20 kJ/mol and $\approx$0.52 kJ/mol above the ground state, respectively. Since $RT \approx 2.5$ kJ/mol at 300 K, these states all contribute to q^e, and so the effective ground-state degeneracy of a carbon atom is approximately $g_0 + g_1 + g_2 = 9$ at room temperature.

function of the following form (when all energies are measured with respect to the electronic potential energy minimum).

$$\boxed{q^v \cong \frac{e^{-\beta\varepsilon_0^v}}{1-e^{-\beta\Delta\varepsilon_0^v}} = \frac{e^{-\frac{1}{2}\Theta_v/T}}{1-e^{-\Theta_v/T}}} \tag{10.5}$$

The second of the above two expressions introduces the *vibrational characteristic temperature*.

$$\Theta_v \equiv \frac{h\nu}{k_B} \tag{10.6}$$

The physical significance of Θ_v becomes evident when we note that $\beta\varepsilon = h\nu/k_BT = \Theta_v/T$, and thus Θ_v represents the temperature at which the corresponding vibrational mode becomes thermally populated.[3]

The vibrational partition function of a polyatomic molecule may be expressed as a product of the corresponding vibrational partition functions for each of its *vibrational normal modes*, $q^v = q_1^v q_2^v \ldots$, each of which has the same form as Eq. 10.5 (but may have different values of Θ_v, corresponding to the different frequencies of each vibrational mode). The number of such normal modes may be inferred from the fact that a collection of n (unbound) atoms will have a total of $3n$ translational degrees of freedom. When the atoms are bound to each other, then three of these degrees of freedom will describe the molecule's center-of-mass translational motion. The rotational motion of a general (nonlinear) polyatomic may be described by three additional angular degrees of freedom. Thus, the remaining $3n - 6$ degrees of freedom represent the number of molecular vibrational normal modes. For example, H_2O has $3(3) - 6 = 3$ vibrational normal modes (one symmetric stretch, one asymmetric stretch, and one bending vibration). Diatomics and other linear molecules (such as CO_2) have only two rotational degrees of freedom, since rotation about the molecular axis does not displace any of the atoms. Thus, linear molecules have $3n - 5$ vibrational degrees of freedom. This is consistent with the fact that diatomics have $3(2) - 5 = 1$ vibrational mode, while CO_2 has $3(3) - 5 = 4$ vibrational normal modes (one symmetric stretch, one asymmetric stretch, and two perpendicular bending vibrations).

Rotational Partition Functions

The rotational motion of a diatomic molecule gives rise to energy eigenvalues of $\varepsilon_J = J(J+1)(\hbar^2/2I) = J(J+1)k_B\Theta_r$, each of which has a degeneracy of $g_J = 2J+1$ where $J = 0, 1, 2, 3, \ldots$ is now the rotational, rather than

---

[3] Note that we might also have absorbed the vibrational zero-point energy into the electronic partition function so that $q^e = g_0 e^{-\beta\varepsilon_0} e^{-\beta\varepsilon_0^v} = g_0 e^{-\beta(\varepsilon_0+\varepsilon_0^v)}$, in which case the vibrational partition function would become $q^v = 1/(1 - e^{\Theta_v/T})$.

electronic, angular momentum quantum number. Thus, the partition function may in general be expressed as $q^r = \sum g_J e^{-\beta\varepsilon_J} = \sum g_J e^{-J(J+1)\Theta_r/T}$, where the *rotational characteristic temperature* again represents the temperature at which the corresponding rotational quantum states become significantly thermally populated.[4]

$$\Theta_r = \frac{hc\widetilde{B}}{k_B} = \frac{\hbar^2}{2Ik_B} \tag{10.7}$$

Since Θ_r is typically much smaller than 300 K, a very large number of rotational quantum states are expected to contribute to q^r. Under these conditions we may evaluate the partition function sum more easily by approximating it by an integral (which assumes that there is a continuous distribution of rotational states, weighted by the appropriate degeneracy factor).[5]

$$\begin{aligned} q^r &= \sum_{J=0}^{\infty} (2J+1)e^{-J(J+1)(\Theta_r/T)} \\ &\approx \int_0^{\infty} (2J+1)e^{-J(J+1)(\Theta_r/T)} dJ \\ &= \int_0^{\infty} e^{-J(J+1)(\Theta_r/T)} d[J(J+1)] \\ &= \frac{T}{\Theta_r} \end{aligned} \tag{10.8}$$

The above expression for q^r accurately represents the experimental partition function of heteronuclear diatomics (such as CO), but experiments indicate that homonuclear diatomics (such as O_2 and N_2) have half as many thermally populated rotational states as the above formula would suggest, so $q^r = \frac{1}{2}(T/\Theta_r)$. This factor of 1/2 arises from the indistinguishability of the two atoms in a homonuclear diatomic – for much the same reason that the total partition function of a collection of N identical molecules must be divided by a factor of $N!$ (see Eq. 5.31). In other words, the additional factor of $\frac{1}{2}$ is linked to the fact that a homonuclear diatomic appears identical when rotated by $180° = \pi$, while a heteronuclear diatomic must be rotated

---

[4] Recall that $\widetilde{B} = B/hc$ where $B = \hbar^2/2I$ is the rotational constant (see Eq. 7.60 on page 259).

[5] The third equality in Eq. 10.8 is obtained by noting that $d[J(J+1)]/dJ = 2J+1$, and thus $d[J(J+1)] = (2J+1)dJ$. The final result is obtained by defining $x = J(J+1)$, and noting that $\int_0^{\infty} e^{-ax}dx = 1/a$.

by $360° = 2\pi$ in order to return to its original orientation.[6] More fundamentally, the factor of 1/2 is a consequence of nuclear spin statistics, which requires that homonuclear diatomics have half as many rotational quantum states as heteronuclear diatomics (as explained on pages 280–282).

Thus, the rotational partition function of any diatomic may be expressed as follows, where the *symmetry number* σ is equal to 1 for a heteronuclear diatomic and 2 for a homonuclear diatomic.

$$\boxed{q^r = \frac{T}{\sigma\Theta_r} = \frac{k_B T}{\sigma hcB} = \frac{2Ik_BT}{\sigma\hbar^2}} \tag{10.9}$$

Equation 10.9 may also be applied to linear polyatomics, as they too have a single rotational constant B. Nonlinear polyatomics may in general have up to three different rotational constants $A = \hbar/(4\pi cI_1)$, $B = \hbar/(4\pi cI_2)$, and $C = \hbar/(4\pi cI_3)$. Highly symmetric molecules, such as CH_4 and SF_6, have three equivalent moments of inertia $I_1 = I_2 = I_3$, and so are called *spherical rotors*. Somewhat less symmetric molecules such as CH_3Cl and C_6H_6 are called *symmetric rotors* (or symmetric tops) as they have two equivalent moments of inertia $I_1 = I_2 \neq I_3$.[7] The rotational partition functions of such nonlinear molecules may be predicted using the following (high-temperature) expression.

$$\boxed{q^r = \frac{1}{\sigma}\left(\frac{k_B T}{hc}\right)^{3/2}\left(\frac{\pi}{ABC}\right)^{1/2}} \tag{10.10}$$

The symmetry number σ of a polyatomic molecule again depends on the number of indistinguishable ways in which it can be reoriented. For example H_2O, NH_3, and CH_4 have symmetry numbers of 2, 3, and 12, respectively.[8]

---

[6] The factor of two difference between q^r for heteronuclear and homonuclear diatomics may also be understood by recalling that a rotating diatomic is equivalent to a particle confined to the surface of a sphere whose radius is equal to the bond length of the diatomic. The partition function (or thermally accessible phase space volume) of such a particle is proportional to the surface area of that sphere. Since a homonuclear diatomic reproduces its original orientation when it has rotated half way around the sphere, it effectively has half as much accessible surface area as a heteronuclear diatomic, and thus q^r for a homonuclear diatomic is half that of a heteronuclear diatomic.

[7] Prolate symmetric rotors, such as CH_3Cl, have $I_3 < I_1$, while oblate rotors, such as C_6H_6, have $I_3 > I_1$.

[8] More generally, a molecule's symmetry number is equivalent to its total number of rotational symmetry elements C_n, including the identity (C_1). Each C_n represents

Translational Partition Functions

The translational partition function q^t may be obtained by considering a particle confined within three-dimensional box. Recall that when a molecule is trapped in a container of volume V, its translational kinetic energy is quantized as a result of its confinement. The translational eigenstates of such a molecule are solutions of the particle-in-a-box Schrödinger equation (see pages 229–231). Since the three spatial directions represent independent degrees of freedom, we may derive the full translational partition function from the following one-dimensional partition function of a molecule confined in a box of length L, whose energy eigenvalues are $\varepsilon_n = n^2h^2/(8mL^2)$ (see Eq. 7.6).

$$q_1^t = \sum_{n=1}^{\infty} e^{-\beta\varepsilon_n^t} = \sum_{n=1}^{\infty} e^{-\beta n^2h^2/(8mL^2)} = \sum_{n=1}^{\infty} e^{-(n/n^*)^2} \tag{10.11}$$

The parameter n^* is a measure of the number of translational quantum states that are thermally occupied at a given temperature.

$$n^* \equiv \frac{L\sqrt{8mk_BT}}{h} \tag{10.12}$$

Note that when $n << n^*$, each term in the sum is approximately equal to 1, while when $n > n^*$, the terms rapidly become insignificantly small. Thus, we expect q_1^t to be approximately equal to n^*.

Exercise 10.1

Estimate the number of translational quantum states in the x-direction that are thermally accessible to N_2 when it is confined within a box of $L = 1-$m length at a temperature of 300 K?

Solution. The molar mass of N_2 (M $\sim$ 28 g/mol) corresponds to a molecular mass (in kg units) of $m \sim$ [28(g/mol)/$\mathcal{N}_A$(1/mol)]1000(kg/g) $\sim 5 \times 10^{-20}$ kg. Thus, Eq. 10.12 predicts that $n^* \approx 6 \times 10^{13}$ translational quantum states in the x-direction will be thermally accessible to N_2.

---

a distinct n-fold rotational symmetry. Thus, NH_3 has two distinct C_3 symmetry elements (one involving rotation by $2\pi/3 = 120°$ and one by $4\pi/3 = 240°$). Thus, in addition to C_1, NH_3 has three rotational symmetry elements and so $\sigma = 3$. Similarly, CH_4 has eight C_3 symmetry elements (two about each C-H bond axis) and three C_2 symmetry elements (corresponding to $\pi = 180°$ rotations about the three axes that bisect the tetrahedral H-C-H bond angles). These eleven rotational symmetry elements, plus the identity element, produce a symmetry number of $\sigma = 12$.

Because the number of thermally accessible translational states is so enormous, we may again replace the sum in Eq. 10.11 by an integral to obtain the following expression for q_1^t.[9]

$$\begin{aligned} q_1^t &= \int_1^\infty e^{-(n/n^*)^2} dn \\ &= n^* \int_{1/n^*}^\infty e^{-x^2} dx \cong n^* \int_0^\infty e^{-x^2} dx = n^* \frac{\sqrt{\pi}}{2} \\ &= L\sqrt{\frac{2\pi m k_B T}{h^2}} = \frac{L}{\Lambda} \end{aligned} \tag{10.13}$$

The parameter Λ is called the *thermal wavelength*, as it is closely related to the de Broglie wavelength of a gas molecule.[10]

$$\Lambda \equiv \sqrt{\frac{h^2}{2\pi m k_B T}} \tag{10.14}$$

The partition function of a gas contained in a three-dimensional box of volume $V = L_x L_y L_z$ is simply the product of three one-dimensional partition functions $q^t = (q_1^t)^3 = L_x L_y L_z/\Lambda^3$.

$$\boxed{q^t = \frac{V}{\Lambda^3} = V\left(\frac{2\pi m k_B T}{h^2}\right)^{3/2}} \tag{10.15}$$

Predicting Chemical Equilibrium Constants

Equations 10.3–10.15 may be used to calculate the partition function for any molecule. For example, the partition function of a diatomic molecule may be expressed in terms of ε_0, g_0, Θ_v, Θ_r, and Λ. An isolated atom has an even simpler partition function, because it only has electronic and translational degrees of freedom (which is equivalent to setting $q^v = q^r = 1$) and thus, its partition function only depends on ε_0, g_0, and Λ.

$$\text{Diatomic:}\quad q = g_0 e^{-\beta\varepsilon_0}\left(\frac{e^{-\frac{1}{2}\Theta_v/T}}{1-e^{-\Theta_v/T}}\right)\left(\frac{T}{\sigma\Theta_r}\right)\left(\frac{V}{\Lambda^3}\right) \tag{10.16}$$

$$\text{Atomic:}\quad q = g_0 e^{-\beta\varepsilon_0}\left(\frac{V}{\Lambda^3}\right) \tag{10.17}$$

---

[9] In the second line of Eq. 10.13, we have changed variables by defining $x = n/n^*$, and made use of the fact that $dx/dn = 1/n^*$, so $dn = n^* dx$ and the lower limit of the integral becomes $x_{min} = n_{min}/n^* = 1/n^*$. Since $n^* > 10^{10}$, we may set the lower bound of the integral to zero without sacrificing any accuracy.

[10] Recall that the most probable translational velocity of a molecule is $v_{max} = \sqrt{2k_B T/m}$ (see Eq. 1.33 on page 37 and Fig. 1.4). Thus, the de Broglie wavelength of such a molecule is $\lambda = h/p = h/(mv_{max}) = \sqrt{h^2/2mk_B T} = \sqrt{\pi}\Lambda$.

These expressions may be used to obtain the equilibrium constant for a diatomic dissociation reaction such as $A_2 \rightleftharpoons A + A$

$$K_N = \left(\frac{N_A^2}{N_{A_2}}\right)_{\text{eq}} = \frac{q_A^2}{q_{A_2}} \tag{10.18}$$

or for a binding reaction of the form $A + B \rightleftharpoons AB$

$$K_N = \left(\frac{N_{AB}}{N_A N_B}\right)_{\text{eq}} = \frac{q_{AB}}{q_A q_B} \tag{10.19}$$

or for an atom exchange reaction such as $A_2 + B_2 \rightleftharpoons 2\,AB$.

$$K_N = \left(\frac{N_{AB}^2}{N_{A_2} N_{B_2}}\right)_{\text{eq}} = \frac{q_{AB}^2}{q_A q_B} \tag{10.20}$$

The electronic ground-state energy difference between the products and reactants $\Delta\varepsilon_0 = 2\varepsilon_0^A - \varepsilon_0^{A_2}$ in a bond dissociation reaction of the form $A_2 \rightleftharpoons A + A$ produces a Boltzmann factor that strongly disfavors dissociation, $e^{-\beta\Delta\varepsilon_0} = e^{-\beta D_e} << 1$. The remaining contributions to K_N include the electronic degeneracies g_0 of the reactant and product species as well as the vibrational, rotational, and translation contributions to the partition function $q^v q^r q^t$. These contributions combine to dictate the total number of thermally accessible quantum states or *effective degeneracy* $g' \equiv g_0 q^v q^r q^t$ of the product and reactant species.

More generally, we may express the equilibrium constant K_N of any reaction as an electronic Boltzmann factor $e^{-\beta\Delta\varepsilon_0}$ multiplied by a pre-factor representing the ratio of the effective degeneracies of the products and reactants.[11]

$$\boxed{K_N = \left(\frac{\prod_j N_j^{\nu_j}}{\prod_k N_k^{\nu_k}}\right)_{\text{eq}} = \left(\frac{\prod_j g_j'}{\prod_k g_k'}\right) e^{-\beta\Delta\varepsilon_0}} \tag{10.21}$$

Notice that the equilibrium constant K_N is in general a function of both the temperature T and volume V of the system. The volume dependence arises because $q^t = V/\Lambda^3$. Thus, if the number of product and reactant species differ by Δn, then the equilibrium constant will contain a factor of $V^{\Delta n}$.[12] This also implies that K_N will not depend on volume for any reaction that involves the same number of reactant and product species (as $V^{\Delta n} = 1$

[11] The energy difference appearing in the Boltzmann factor $\Delta\varepsilon_0 = \sum \nu_i \varepsilon_{0i}$ represents the difference between the ground electronic state energies of the product and reactant molecules.

[12] The factor of $V^{\Delta n}$ is responsible for driving the dissociation of molecules as V increases. This is why, for example, molecules tend to dissociate in outer space.

when $\Delta n = 0$). Thus, K_N is predicted to depend only on temperature whenever $\Delta n = 0$.[13]

If we wish to express the equilibrium constants for any reaction in a form that depends only on temperature, we may simply divide the number of molecules of each species by the system volume. This amounts to expressing the product and reactant populations in number density units $\rho_i = N_i/V$, which is equivalent to multiplying K_N by $V^{-\Delta n}$ (in order to exactly cancel the factor of $V^{\Delta n}$ in K_N). Thus, the resulting equilibrium constant K_ρ is predicted to be a function of temperature only for any gas phase reaction.[14]

$$\boxed{K_\rho = \left(\frac{\prod_j \rho_j^{\nu_j}}{\prod_k \rho_k^{\nu_k}}\right)_{\text{eq}} = \left(\frac{\prod_j (q_j/V)^{\nu_j}}{\prod_k (q_k/V)^{\nu_k}}\right) = \frac{K_N}{V^{\Delta n}}} \tag{10.22}$$

Equilibrium constants are more commonly expressed in molar rather than molecular concentration units $[c] = n/V = \rho/\mathcal{N}_A$ (where $\mathcal{N}_A$ is Avogadro's number and $n = N/\mathcal{N}_A$).[15]

$$\boxed{K_c = \left(\frac{\prod_j [c_j]^{\nu_j}}{\prod_k [c_k]^{\nu_k}}\right)_{\text{oq}} = \frac{K_N}{(\mathcal{N}_A V)^{\Delta n}}} \tag{10.23}$$

Similarly, we may express the equilibrium constant of such a gas phase reaction in terms of the ratios of the partial pressures of the products and reactants. Since $P = nRT/V = Nk_BT/V$, we may obtain K_P by multiplying K_ρ by $(k_BT)^{\Delta n}$, which is equivalent to multiplying K_c by $(RT)^{\Delta n}$.[16]

$$\boxed{K_P = \left(\frac{\prod_j P_j^{\nu_j}}{\prod_k P_k^{\nu_k}}\right)_{\text{eq}} = K_N\left(\frac{k_BT}{V}\right)^{\Delta n} = K_c(RT)^{\Delta n}} \tag{10.24}$$

---

[13] Reactions for which $\Delta n = 0$ include isomerization reactions such as $A \rightleftharpoons B$ as well as reactions such as $A + B \rightleftharpoons C + D$ or $A_2 + B_2 \rightleftharpoons 2AB$ or $A + 2B \rightleftharpoons 2C + D$.

[14] In higher-density systems such as nonideal gases, liquids, and solids, K_ρ is no longer expected to be a function of temperature only, as the partition function of such systems depends on intermolecular interactions, which in turn depend on the volume of the system.

[15] If we wish to express concentrations in mol/L units, then we must include the appropriate conversion factor of 10^3 L/m^3 in Eq. 10.23. More specifically, if molecular SI units are used in calculating K_N and we wish to predict K_c in molar units, then $K_c = K_\rho(\mathcal{N}_A 10^3)^{-\Delta n} = K_N(\mathcal{N}_A V 10^3)^{-\Delta n}$.

[16] Again, care must be taken to use the appropriate units for RT. For example, if K_c is expressed in molar units and we wish to express P in atm units, then we should express R in L atm/(K mol) units.

Finally, it is sometimes convenient to express equilibrium constants in mole fraction units $\chi_i = N_i/N_T$, where $N_T = \sum N_i$ represents the total number of molecules in the system at equilibrium.[17]

$$\boxed{K_\chi = \left(\frac{\prod_j \chi_j^{\nu_j}}{\prod_k \chi_k^{\nu_k}}\right)_{\text{eq}} = \frac{K_N}{N_T{}^{\Delta n}}} \tag{10.25}$$

10.2 Principles of Reaction Dynamics

Early studies of the rates of chemical reactions by the Swedish chemist Peter Waage and Norwegian mathematician Cato Guldberg (between 1864 and 1879) led to the formulation of the *law of mass action*, which predicts that the rates of chemical reactions are proportional to the concentrations of the reacting species. More specifically, the rate of a reaction such as $A + B \rightarrow C$ is predicted to be proportional to $[A][B]$.

$$\frac{d[C]}{dt} = -\frac{d[A]}{dt} = -\frac{d[B]}{dt} = k[A][B] \tag{10.26}$$

The constant of proportionality k is called the reaction *rate constant*. More generally, a reaction such as $\sum a_i A_i \rightarrow \sum b_i B_i$ is predicted to have a reaction rate of the following form (where k_1 designates the rate constant for the forward reaction).

$$\frac{1}{b_i}\left(\frac{d[B_i]}{dt}\right) = -\frac{1}{a_i}\left(\frac{d[A_i]}{dt}\right) = k_1 \prod_i [A_i]^{a_i}$$

The corresponding back reaction $\sum a_i A_i \leftarrow \sum b_i B_i$ is predicted to have the following reaction rate (with a back-reaction rate constant of k_{-1}).

$$\frac{1}{a_i}\left(\frac{d[A_i]}{dt}\right) = -\frac{1}{b_i}\left(\frac{d[B_i]}{dt}\right) = k_{-1} \prod_i [B_i]^{b_i}$$

At equilibrium the rates of the forward and reverse reactions must be equal to each other, so $k_1 \prod_i [A_i]^{a_i} = k_{-1} \prod_i [B_i]^{b_i}$. This leads immediately to the following important relation between the rates and equilibrium constants of chemical reactions.

$$\boxed{\frac{k_1}{k_{-1}} = \left(\frac{\prod_i [B_i]^{b_i}}{\prod_i [A_i]^{b_i}}\right)_{\text{eq}} = K_c} \tag{10.27}$$

[17] Note that the value of N_T does not depend on the volume of the system and so K_χ has the same volume dependence as K_N.

Multiple-Step Reaction Kinetics

Chemical reactions often take place in a sequence of elementary steps. The experimentally observed time-dependence of a given reaction depends on the relative rates of these steps. The slowest elementary step in a sequence of reactions often dictates its overall rate. For example, the gas phase reaction $2H_2 + 2\,NO \rightarrow N_2 + 2\,H_2O$ has an experimental rate that is proportional to $[H_2][NO]^2$, which does not appear to be consistent with the stoichiometry of the overall reaction. This is because the reaction takes place in the following three steps:

$$\begin{array}{cc} 2\,NO \rightleftharpoons N_2O_2 & \text{(fast equilibrium)} \\ N_2O_2 + H_2 \rightarrow N_2O + H_2O & \text{(slow)} \\ N_2O + H_2 \rightarrow N_2 + H_2O & \text{(fast)} \end{array}$$

Since the second reaction is the slowest, it is the rate-limiting step (bottleneck) that dictates the observed reaction rate. The rate of this elementary reaction is $d[N_2O]/dt = k[N_2O_2][H_2]$. The initial fast equilibrium further implies that $[N_2O_2] = K_c\,[NO]^2$, which is why the experimentally observed rate is proportional to $[H_2][NO]^2$.

Some multistep reactions occur in such a way that the apparent order of the reaction depends in a less obvious way on the relative rates of the corresponding elementary steps. For example, the collision-induced dissociation of a dimer $A_2 \rightarrow A + A$ may have an apparent reaction rate that is second-order at low pressure and first-order at high pressure. In order to understand how this could occur, it is instructive to consider what would happen if the reaction involved the following three elementary steps.

$$\begin{aligned} A_2 + A_2 &\xrightarrow{k_1} A_2^* + A_2 \\ A_2^* + A_2 &\xrightarrow{k_{-1}} A_2 + A_2 \\ A_2^* &\xrightarrow{k_2} A + A \end{aligned} \tag{10.28}$$

The first step represents the collision-induced excitation of the dimer A_2 to an activated state A_2^* whose energy is above the dissociation threshold. The collisionally activated state A_2^* may then either lose its energy as indicated in the second step (which is the inverse of the first elementary reaction), or dissociate as indicated in the third step.

The third elementary step implies that the decomposition rate is proportional to $[A_2^*]$. However, the concentration of A_2^* depends on the relative rates of all three of the above elementary steps. The influence of these competing rates can be elucidated by invoking the *steady-state approximation*, which amounts to assuming that the concentration of A_2^* is approximately constant.

Since each of the three elementary steps may either increase or decrease the concentration of A_2^*, the overall rate of change of $[A_2^*]$ is determined by the following sum of three terms pertaining to each of the three elementary steps in Eq. 10.28.

$$\frac{d[A_2^*]}{dt} = k_1[A_2]^2 - k_{-1}[A_2][A_2^*] - k_2[A_2^*] \tag{10.29}$$

Under steady-state conditions $d[A_2^*]/dt \approx 0$, in which case we may rearrange the right-hand side of Eq. 10.29 to solve for $[A_2^*]$.

$$[A_2^*] = \frac{k_1[A_2]^2}{k_{-1}[A_2] + k_2} \tag{10.30}$$

This result, combined with the third elementary step in Eq. 10.28, leads to the following predicted rate of dissociation.

$$-\frac{d[A_2^*]}{dt} = k_2[A_2^*] = \frac{k_2k_1[A_2]^2}{k_{-1}[A_2] + k_2} \tag{10.31}$$

Note that when $[A_2]$ is small (so $k_{-1}[A_2] << k_2$), then $[A_2^*] \approx (k_1/k_2)[A_2]^2$, while when $[A_2]$ is large (so $k_{-1}[A_2] >> k_2$), then $[A_2^*] \approx (k_1/k_{-1})[A_2]$. Thus, the reaction rate is predicted to change from $k_1[A_2]^2$ at low concentration to $(k_2k_1/k_{-1})[A_2]$ at high concentration. In other words, the reaction is predicted to become second-order in A_2 at low concentration (when the first step is rate limiting) but to become first-order in A_2 at high concentration (when the third step becomes rate limiting).

The existence of such collision-induced unimolecular reaction mechanisms was first suggested in 1921 by Frederick Lindemann and later elaborated by Cyril Hinshelwood (and so such reactions are referred to as having *Lindemann* or *Lindemann-Hinshelwood* reaction mechanisms).[18] In Section 10.4 we will see how Einstein had previously, in 1916, carried

[18] Frederick Alexander Lindemann (1886–1957) was a German-born English physicist (professor of "experimental philosophy") at Oxford University. Early in his career he performed experiments that confirm Einstein's 1907 quantum theory of the heat capacities of solids, and was the youngest invited attendee of the famous 1911 Solvay conference devoted to *Radiation and the Quanta* (which included Planck, Einstein, and other leading physicists). During World War II, he became an influential scientific advisor to Winston Churchill. Sir Cyril Norman Hinshelwood (1897–1967) was a younger physical chemist at Trinity College in England when, in 1927, he significantly elaborated Lindemann's general suggestion regarding the consequences of a delay between the activation and dissociation steps in chemical reactions. In 1956, Hinshelwood was awarded a Nobel prize in Chemistry along with Nikolay Semenov "for their researches into the mechanism of chemical reactions." Subsequent theoretical developments of Lindemann and Hinshelwood's

out a very similar analysis of the rates of reactions between photons and molecules in order to both obtain an alternative derivation of Planck's blackbody radiation formula and discover stimulated emission, which led to the subsequent invention of lasers.

Temperature Dependence of Reaction Rate Constants

In the late nineteenth century, the Dutch organic chemist Jacobus van't Hoff and Swedish physicist Svante Arrhenius discovered that the rates of many chemical reactions have an approximately exponential temperature dependence.[19] More specifically, they found that reaction rate constants may be expressed in the following temperature-dependent form.

$$\boxed{k = Ae^{-E_a/RT}} \tag{10.32}$$

The constant E_a is called the *activation energy* of the reaction, and A is sometimes referred to simply as the *pre-exponential factor*, or more descriptively as a *collision frequency factor*.

Although experiments performed over a limited temperature range may often be well described by Eq. 10.32, we should not in general expect the parameters A and E_a to be strictly temperature-independent constants. In the next section, we will see why that is the case, when we consider how a molecular detailed picture of chemical reactions may be used to theoretically predict chemical reaction rate constants, as well as to explain the form of the Arrhenius expression (Eq. 10.32) and the law of mass action (Eq. 10.26).

10.3 Prediction of Reaction Rate Constants

The prediction of chemical reaction rates and the dynamics of complex chemical process – such as those occurring in our upper atmosphere and during automotive combustion – remain challenging subjects of current research. This section describes two physically appealing starting points for understanding and predicting chemical reaction rates. One approach considers the rate of collisions between molecules in a gas, and various factors that may influence the probability that a pair of colliding molecules will undergo a reaction. The other approach builds on the close connection

---

ideas led to more sophisticated and accurate treatments of unimolecular reactions, such as the Rice-Ramsperger-Kassel-Marcus (RRKM) theory.

[19] Jacobus van't Hoff was subsequently awarded the first Nobel prize in Chemistry (in 1901) for his "discovery of the laws of chemical dynamics and osmotic pressure in solutions." Svante Arrhenius also received a Nobel prize in Chemistry two years later (in 1903) for his "electrolytic theory of dissociation."

between reaction rate constants and equilibrium constants to obtain a statistical thermodynamic description of chemical reaction rates. Although the former description may be intuitively easier to grasp, the latter description can provide a more practical, and widely applicable, means of accurately predicting chemical reaction rates from quantum mechanical calculations of molecular energies and partition functions.

Collision Theory of Chemical Reaction Rates

In order for a chemical reaction to take place, the reacting molecules must first collide with each other. Thus, the simplest possible theoretical description of chemical reaction dynamics is one that equates the rate of an elementary reaction with the corresponding collision rate.

The rate at which molecules collide should depend on their velocity as well as on their size. The role of molecular size can be understood by approximating molecules as hard-sphere particles. The rate at which such a hard-sphere particle will collide may be expressed in terms of its *mean free path* λ, which represents the average distance traveled by a molecule before it undergoes a collision. The mean free path of a hard sphere of diameter σ is determined by its *collision cross section* $\pi\sigma^2$, which represents the area of a circular target associated with a collision between two spheres of diameter σ (as illustrated by the large excluded circle in Fig. 5.1 on page 155).

In a gas composed of randomly arranged hard spheres of diameter σ and number density ρ, each sphere is predicted to have the following mean free path.

$$\lambda = \frac{1}{\sqrt{2}(\pi\sigma^2)\rho} \tag{10.33}$$

The frequency at which each molecule in such a gas undergoes a collision may be obtained by dividing the average velocity of each molecule (see Eq. 1.34 on page 37) by its mean free path.

$$\frac{\langle v\rangle}{\lambda} = \frac{\sqrt{\frac{8k_BT}{\pi m}}}{\frac{1}{\sqrt{2}(\pi\sigma^2)\rho}} = \sigma^2\sqrt{\frac{16\pi k_BT}{m}}\,\rho \tag{10.34}$$

We are interested in predicting the overall rate at which molecules will collide (per second per unit volume). The total number of collisions in a unit volume is equal to the product of the number of collisions per molecule $\langle v\rangle/\lambda$ times the number of molecules per unit volume $\rho = N/V$. However, since each collision involves two molecules, this would overcount the number of collisions by a factor of 2. Thus, the total collision frequency in a single component gas is Z_{AA} (collisions per second per unit volume).

$$Z_{AA} = \frac{1}{2}\left(\frac{\langle v \rangle}{\lambda}\right)\rho = \frac{1}{2}\sigma^2\sqrt{\frac{8\pi k_B T}{\mu}}\rho^2 \tag{10.35a}$$

In the second equality, the mass of the colliding pair of molecules has been expressed in terms of their reduced mass $\mu = mm/(m+m) = m/2$.

The above result may be generalized to obtain the following expression for the total number of collisions in a gas composed of a mixture of molecules A and B.

$$\sqrt{8\pi k_B T}\left[\frac{\sigma_{AB}^2}{\sqrt{\mu_{AB}}}\rho_A\rho_B + \frac{1}{2}\frac{\sigma_{AA}^2}{\sqrt{\mu_{AA}}}\rho_A^2 + \frac{1}{2}\frac{\sigma_{BB}^2}{\sqrt{\mu_{BB}}}\rho_B^2\right]$$

Note that the factor of $\frac{1}{2}$ only appears in terms involving collisions between molecules of the same type (in order to avoid overcounting). The average diameter of a colliding pair of molecules is $\sigma_{ij} = (\sigma_i + \sigma_j)/2$ and their reduced mass is $\mu_{ij} = m_i m_j/(m_i + m_j)$.

For reactions between molecules A and B, we only need to consider the first term in the above expression, as this represents the frequency of collisions between the two reactant molecules.

$$Z_{AB} = \sigma_{AB}^2\sqrt{\frac{8\pi k_B T}{\mu_{AB}}}\rho_A\rho_B \tag{10.35b}$$

If every collision led to a reaction, then the rate of a reaction such as $A + B \rightarrow C$ would be determined exactly by the above collision frequency.[20]

$$\frac{d[C]}{dt} = \frac{Z_{AB}}{\mathcal{N}_A} = \left(\mathcal{N}_A\sigma_{AB}^2\sqrt{\frac{8\pi k_B T}{\mu_{AB}}}\right)[A][B] = k_0[A][B] \tag{10.36}$$

For a reaction between two molecules of the same type $A + A \rightarrow B$, Eq. 10.35a leads to the following slightly different expression (because of the extra factor of $\frac{1}{2}$).

$$\frac{d[B]}{dt} = \frac{Z_{AA}}{\mathcal{N}_A} = \left(\mathcal{N}_A\sigma_{AA}^2\sqrt{\frac{2\pi k_B T}{\mu_{AA}}}\right)[A][A] = k_0[A]^2 \tag{10.37}$$

---

[20] We must divide Z_{AB} by Avogadro's number $\mathcal{N}_A$ in order to express the collision rate in molar units. Also, note that $\rho_A\rho_B$ is equal to $[A][B]\mathcal{N}_A^2$, which accounts for the appearance of Avogadro's number on the right-hand side of Eq. 10.36.

Equations 10.36 and 10.37 imply that k_0 would be the reaction rate constant if every collision produced a reaction.

More realistically, we would not expect every collision to lead to a reaction, but only those collisions that have sufficient energy. In other words, if a given reaction has an *activation energy* of ε_a, then we would only expect collisions with an average energy greater than ε_a to be capable of producing a reaction.

The number of collisions between pairs of molecules that have a relative kinetic energy greater than ε_a may be obtained from the average kinetic energy of a single particle in *two dimensions*. The reason for this may be understood by noting that the trajectories of two molecules that are going to undergo a collision may be represented by two intersecting lines. Such a pair of lines defines a plane in which the two molecules are moving. Thus, the relative velocity of the two colliding molecules is given by the following two-dimensional kinetic energy probability distribution function (obtained from Eq. 1.40 on page 40, when $\mathrm{D} = 2$).

$$P_2(\varepsilon)d\varepsilon = \frac{1}{k_B T} e^{-\varepsilon/k_B T} d\varepsilon \qquad (10.38)$$

The probability that a collision will have an energy greater than ε_a is equivalent to the area under the probability distribution function, integrated from ε_a to ∞.

$$P(\varepsilon > \varepsilon_a) = \int_{\varepsilon_a}^{\infty} \frac{1}{k_B T} e^{-\varepsilon/k_B T} d\varepsilon = e^{-\varepsilon_a/k_B T} \qquad (10.39)$$

We may now obtain the rate of an activated reaction simply by multiplying Eq. 10.36 by the above probability (and equating $\varepsilon_a/k_B = E_a/R$).

$$\boxed{-\frac{d[A]}{dt} = -\frac{d[B]}{dt} = \frac{d[C]}{dt} = k_0 e^{-E_a/RT}[A][B] = k[A][B]} \qquad (10.40)$$

Thus, we have obtained an expression for the rate constant k that has the same form as the empirical expression given in Eq. 10.32. Moreover, comparison of Eqs. 10.32 and 10.40 reveals that we have also obtained a theoretical expression for the pre-exponential (or collision frequency) factor $A = k_0$.[21]

The predictions of Eq. 10.40 are often in reasonably good agreement with experimental results, which is impressive considering the fact that this theoretical expression for k has been obtained by treating colliding molecules

[21] Notice that the empirical pre-exponential factor A is often treated as a temperature-independent constant, while the theoretical expression for k_0 suggests that it should be proportional to $\sqrt{T}$. However, under typical experimental conditions the range of absolute temperatures is sufficiently small that the temperature dependence of k_0 may be neglected in comparison with other sources of experimental uncertainty.

as hard spheres. In other words, Eq. 10.40 has been obtained without considering the dependence of chemical reaction rates on the shapes or relative orientations of the colliding molecules. One might expect that some reactions would require that molecules not only have sufficient energy, but also collide with a favorable orientation. We could take such an orientational restriction into account by introducing an additional *steric* or *transmission* factor κ to the expression for the reaction rate.

$$\boxed{k = \kappa k_0 e^{-E_a/RT}} \tag{10.41}$$

Since an orientational restriction on the probability of reaction would tend to reduce the experimentally observed reaction rate, we expect $\kappa \leq 1$, as is often found to be the case. However, some reactions are observed to have steric factors that are *larger* than 1, which would seem to be physically impossible, as that would imply a reaction rate that is greater than the predicted rate of collisions. A famous example of this apparently paradoxical situation is the reaction of halide diatomics with alkali atoms, such as $Br_2 + K \rightarrow KBr + Br$, which has a steric factor of $\kappa \approx 4$. A beautifully simple explanation for this interesting observation is provided by the *harpoon mechanism*, proposed by the Canadian physical chemist John Polanyi.[22] The harpoon mechanism derives its name from the fact that when a K atom approaches a Br_2 molecule, the valence electron of K may jump over to Br_2, like a harpoon. After such an electron exchange, the resulting oppositely charged K^+ and Br_2^- ions experience a strong coulombic attraction that reels them in towards each other. In other words, the reason the rate of such reactions is higher than k_0 is because the coulombic harpoon pulls in the reactants that would otherwise not have collided. After K^+ and Br_2^- are brought together, they rapidly react to form KBr and Br.

Activated Complex Theory of Chemical Reaction Rates

A quite different approach to the theoretical prediction of reaction rates focuses on the activated complex (or transition state) that separates the reactants and products. Thus, this approach is referred to as *activated complex theory* (or *transition state theory*) of chemical reaction rates. For example, the activated complex of a reaction that may be described using a simple one-dimensional potential energy surface, such as that shown in Figure 10.1, is defined as the structure that exists at the top of the barrier. Activated complex theory may also be applied to reactions with multidimensional potentials, by identifying the activated complex with the

---

[22] Polanyi went on to win a Nobel prize in Chemistry in 1986, along with Dudley Herschbach and Yuan Lee, "for their contributions concerning the dynamics of chemical elementary processes."

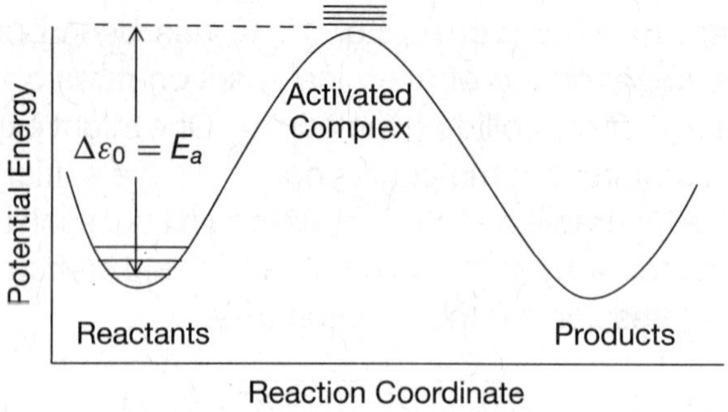

FIGURE 10.1 This is a schematic representation of the potential energy surface along a chemical reaction coordinate that passes through an activated complex (transition state) on its way from reactants to products. The population of the activated complex, and thus also the reaction rate, is determined by the quantum structures of the reactants and activate complex, and the difference between their zero-point energies $\Delta\varepsilon_0 = E_a$.

high-energy saddle point (or bottleneck) that must be traversed in order for the reaction to take place.

The key idea underlying activated complex theory is that the rate of a chemical reaction depends on the instantaneous concentration of the activated complex. The theory further *assumes* that the latter concentration may be predicted using equilibrium statistical thermodynamics (which is often a remarkably good assumption).

Recall that Eq. 10.21 implies that the equilibrium constant of reaction of the form $A + B \rightleftharpoons C$ may be expressed as follows (where the effective degeneracies of each species are again defined as $g' \equiv g_0 q^v q^r q^t$).

$$K_N = \left(\frac{g'_C}{g'_A g'_B}\right) e^{-\beta\Delta\varepsilon_0}$$

If we replace the product C by the activated complex C^*, then the above expression may used to predict the equilibrium constant pertaining to the formation of the activated complex $A + B \rightleftharpoons C^*$ (where $\Delta\varepsilon_0$ is now the energy difference between the activated complex and the reactants, as illustrated in Figure 10.1).

$$K_N^* = \frac{N_C^*}{N_A N_B} = \left(\frac{g'_{C^*}}{g'_A g'_B}\right) e^{-\beta\Delta\varepsilon_0} \tag{10.42}$$

We may rearrange the above expression to obtain an expression for the number of activated complex species.

$$N_C^* = K_N^* N_A N_B = \left(\frac{g'_{C^*}}{g'_A g'_B}\right) e^{-\beta\Delta\varepsilon_0} N_A N_B$$

A key result of transition state theory is the following relation between N_{C^*} and the reaction rate.

$$-\frac{dN_A}{dt} = \left(\frac{k_B T}{h}\right) N_C^* = \left(\frac{k_B T}{h}\right)\left(\frac{g'_{C^*}}{g'_A g'_B}\right) e^{-\beta\Delta\varepsilon_0} N_A N_B \qquad (10.43)$$

Note that the factor $k_B T/h$ (where h is Planck's constant) has units of frequency, since $k_B T(\mathrm{J})/h(\mathrm{J\ s}) = k_B T/h\ (\mathrm{s}^{-1})$. This simple frequency factor arises as a consequence of separating the motion along the reaction coordinate from the other degrees of freedom of the activated complex.[23]

If we express the reactants and products in concentration units (and equate $\beta\Delta\varepsilon_0 = E_a/RT$), then Eq. 10.43 reduces to the following form, which looks somewhat similar to Eq. 10.40.

$$-\frac{d[A]}{dt} = \frac{d[C]}{dt} = \mathcal{N}_A \left(\frac{k_B T}{h}\right)\left(\frac{g'_{C^*}}{g'_A g'_B}\right) e^{-E_a/RT}[A][B] = k[A][B] \qquad (10.44)$$

A more direct connection between Eqs. 10.40 and 10.44 may be made by assuming that the reactants are atoms with hard-sphere diameters σ_A and σ_B (and non-degenerate electronic ground states). We may also reasonably identify the activated complex as the diatomic structure formed at the instant when the two reactant atoms have collided (but before their bond length has decreased to form a covalent bond). Thus, in the activated complex the two atoms are separated by their hard-sphere contact distance $\sigma_{AB} = (\sigma_A + \sigma_B)/2$. Under these conditions Eq. 10.44 becomes identical to Eq. 10.40 (as shown in Exercise 10.2).

---

[23] Equation 10.43 may be derived in various ways. One of the simplest derivations is obtained by treating the motion of the activated complex along the reaction coordinate as a harmonic vibration of frequency ν. If that frequency is sufficiently low that $h\nu << k_B T$ (which is realistic, as long as the activated complex is weakly bound), then the partition function corresponding to vibration along the reaction coordinate is simply $k_B T/h\nu$ (see Eq. 1.21 and footnote 19 on page 28). Note that ν also repesents a realistic estimate of the frequency at which the activated complex crosses over to form products, and so the reaction rate is predicted to become $k = \nu N_{C^*} = \nu(k_B T/h\nu)K_N^* N_A N_B$, which is equivalent to Eq. 10.43. Moreover, since $k_B T/h\nu$ is the vibrational partition function associated with motion along the reaction coordinate, the vibrational partition function of the activated complex q^v (which contributes to its effective degeneracy $g'_{C^*} \equiv g_0 q^v q^r q^t$, in Eq. 10.42) should include all the other vibrational motions of the activated complex, *excluding* the motion along the reaction coordinate.

Exercise 10.2

Demonstrate that Eq. 10.44 is equivalent to Eq. 10.40. (under the conditions described in the previous paragraph).

Solution. Recall that each of the effective degeneracies appearing in Eq. 10.44 is defined as $g' = g_0 q^v q^r q^t$. For the two reactant atoms $g_0 = 1$, $q^v = 1$, and $q^r = 1$ (since atoms have no rotational or vibrational degrees of freedom and their electronic ground states are assumed to be nondegenerate). Thus, we may use Eq. 10.15 to obtain

$$g'_A = q^t_A = \frac{V}{\Lambda_A^3} \quad \text{and} \quad g'_B = q^t_B = \frac{V}{\Lambda_B^3}$$

Since the activated complex has a diatomic structure it may both rotate and translate so $g'_{C*} = q^r q^t$, and thus the effective degeneracy of the activated complex may be obtained using Eqs. 10.9 and 10.15.[24]

$$g'_{C*} \equiv q^t_{C*} q^r_{C*} = \left(\frac{V}{\Lambda_{C*}^3}\right)\left(\frac{k_B T}{\Theta_r^{C*}}\right)$$

We may now multiply $k_B T/h$ times the effective degeneracy ratio, as in Eq. 10.44

$$\left(\frac{k_B T}{h}\right)\left(\frac{g'_{C*}}{g'_A g'_B}\right) = \left(\frac{k_B T}{h}\right)\left(\frac{\Lambda_{C*}}{\Lambda_A \Lambda_B}\right)^3 \left(\frac{k_B T}{\Theta_r^{C*}}\right)$$

and then expand Λ and Θ_r using Eqs. 10.9 and 10.15

$$= \left(\frac{k_B T}{h}\right)\left[\left(\frac{h^2}{2\pi k_B T}\right)\left(\frac{m_{C*}}{m_A m_B}\right)\right]^{3/2}\left(\frac{8\pi^2 I_{C*} k_B T}{h^2}\right)$$

and equate $m_{C*}/m_A m_B = m_A + m_B/m_A m_B = 1/\mu_{AB}$ and $I_{C*} = \mu_{AB}\sigma_{AB}^2$ (since σ_{AB} is the bond length of the activated complex contact diatomic structure), to obtain the following result.

$$= \left(\frac{k_B T}{h}\right)\left[\left(\frac{h^2}{2\pi k_B T}\right)\left(\frac{1}{\mu_{AB}}\right)\right]^{3/2}\left(\frac{8\pi^2 \mu_{AB}\sigma_{AB}^2 k_B T}{h^2}\right) = \sigma_{AB}^2\sqrt{\frac{8\pi k_B T}{\mu_{AB}}}$$

When the result on the far right is plugged into Eq. 10.44, we obtain the following expression, which is identical to Eq. 10.40, thus demonstrating that collision theory and activated complex theory predict the same reaction rate (under the above conditions).

$$-\frac{dN_A}{dt} = \mathcal{N}_A \sigma_{AB}^2 \sqrt{\frac{8\pi\sigma_{AB}^2 k_B T}{\mu_{AB}}} e^{-E_a/RT}[A][B]$$

---

[24] Note that there is no vibrational contribution to g'_{C*}, since the vibrational motion of the diatomic activated complex is identical to the reaction coordinate, which is already represented by the $k_B T/h$ term in Eq. 10.44.

Note that the final expression obtained in Exercise 10.2 is also equivalent to $k_0 e^{-E_a/RT}[A][B]$, where k_0 is the collision rate constant defined by Eq. 10.36. If we had assumed that the two reactant atoms were identical, then we would have obtained a value of k_0 that is equivalent to that in Eq. 10.37 (see homework problem 8).

The fact that activated complex theory is derived from statistical thermodynamics also leads in a natural way to the identification of energetic (or enthalpic) and entropic contributions to chemical reaction rates. More specifically, note that Eq. 10.42 implies that the reaction rate is proportional to the activated complex equilibrium constant, and so Eq. 10.44 leads to the following expression for the binary reaction rate constant (since $K_c^* = [C^*]/[A][B] = e^{-\Delta G^*/RT}$).

$$k = \left(\frac{k_B T}{h}\right) K_c^* = \left(\frac{k_B T}{h}\right) e^{-\Delta G^*/RT} \tag{10.45}$$

We may now use the thermodynamic expression $\Delta G^* = \Delta H^* - T\Delta S^*$ to infer the enthalpic and entropic contributions to the reaction rate constant.

$$\boxed{k = \left(\frac{k_B T}{h}\right) e^{-\Delta G^*/RT} = \left[\left(\frac{k_B T}{h}\right) e^{\Delta S^*/R}\right] e^{-\Delta H^*/RT}} \tag{10.46}$$

The latter result makes it clear that there is a close connection between ΔH^* and the activation energy E_a as well as between ΔS^* and the effective degeneracy ratio in Eq. 10.44.[25]

The above thermodynamic interpretation of reaction rate constants is also self-consistent with the fact that the equilibrium constant for any chemical

---

[25] We may make this connection even more explicit using $\Delta H^* = \Delta U^* + P\Delta V^*$ to obtain $\Delta H^* = E_a + RT$ by identifying $\Delta U^* = E_a$ and using $P\Delta V^* = -RT$, since $P\Delta V^* = \Delta n RT$ for any low-density gas phase reaction and $\Delta n = -1$ for the $A + B \leftrightharpoons C^*$ process. Thus,

$$k = \left(\frac{k_B T}{h}\right) e^{-\Delta G^*/RT} = \left[\left(\frac{k_B T}{h}\right) e^{1+\Delta S^*/R}\right] e^{-E_a/RT} = \left[\left(\frac{k_B T}{h}\right)\left(\frac{\mathcal{N}_A g'_{C^*}}{g'_A g'_B}\right)\right] e^{-E_a/RT}$$

and so $\mathcal{N}_A g'_{C^*}/g'_A g'_B = e^{1+\Delta S^*/R}$ or $\Delta S^* = \ln(\mathcal{N}_A g'_{C^*}/g'_A g'_B) - 1$.

reaction must be equivalent to the ratio of the corresponding rate constants for the forward k_1 and reverse k_{-1} reactions (see Eq. 10.27).

$$\boxed{K_c = \frac{k_1}{k_{-1}} = \frac{K_1^*}{K_{-1}^*} = \frac{([C^*]/[A][B])}{([C^*]/[C])} = \frac{[C]}{[A][B]} = \frac{e^{-\Delta G_1^*/RT}}{e^{-\Delta G_{-1}^*/RT}} = e^{-\Delta G/RT}} \tag{10.47}$$

The last equality was obtained by noting that the Gibbs energy change for the overall reaction is necessarily equal to the difference between the activation-free energies of the forward and reverse reactions $\Delta G = \Delta G_1^* - \Delta G_{-1}^*$.

10.4 Photon-Molecule Reactions

The interaction between light and chemical substances invariably reveals the particle-like properties of photons. This is clearly evident, for example, in the photoelectric effect and Compton scattering (as described on pages 15–18), as well as in the absorption and emission of light by atoms and molecules. All such processes indicate that photons behave like particles of energy $h\nu$ and momentum h/λ.[26]

Since both molecules and photons behave like particles, one might expect to be able to treat optical absorption and emission processes in a way that is similar to that used to describe chemical reaction equilibria and dynamics. The utility of this approach was first demonstrated by Albert Einstein, in his 1916 paper entitled *On the Quantum Theory of Radiation*.[27] This paper is most famous for introducing the concept of *stimulated emission*, which ultimately led to the development of lasers.[28] Here we will see how Einstein's method may be used to describe a wide variety of linear and

[26] The appearance of ν and λ in the expressions for the energy and momentum of photons clearly reveals the intimate link between the particle and wave properties of light. However, the fundamental mechanism that connects photons and electromagnetic waves is not known. In other words, the fundamental problem that Einstein raised in the quotation at the bottom of page 15 has not yet been satisfactorily resolved.

[27] We have previously encountered this paper in Chapter 1 (see the quotation at the bottom of page 15).

[28] The word *laser* is an acronym for *light amplification by stimulated emission of radiation*. The existence of stimulated emission was first confirmed experimentally in 1928 by Rudolf Ladenburg, a German atomic physicist who emigrated to the United States in 1932 and became a professor at Princeton University. In the late 1950s and early 1960s the race was on to produce the first optical laser. The first documented design of a working laser (as well as the first use of the term *laser*) was described in the notebook of a physics graduate student named

nonlinear optical processes, including optical emission that is stimulated by *vacuum fluctuations*.

Linear Optical Processes

The intensity I_ν (J/s m^2) of light of frequency ν is equal to its energy density ρ_ν (J/m^3) multiplied by the speed of light c (m/s). Since each photon has an energy of $h\nu$ (J), the molar concentration of photons $[p_\nu]$ in a beam of light of frequency ν is related as follows to its energy density and intensity.

$$[p_\nu] = \frac{\rho_\nu}{\mathcal{N}_A h\nu} = \frac{I_\nu}{\mathcal{N}_A hc\nu} \tag{10.48}$$

The absorption of a photon p_ν by a molecule M to produce an optically excited molecule M^* is equivalent to the following elementary reaction.

$$M + p_\nu \xrightarrow{k} M^* \tag{10.49}$$

The rate of this reaction may be expressed exactly in the same way as the rate of any other such binary reaction.

$$\frac{d[M^*]}{dt} = k[M][p_\nu] \tag{10.50}$$

The stoichiometry of the above reaction also requires that the rate of appearance of the products is equal to the rate of photon absorption.

$$\begin{aligned} -\frac{d[p_\nu]}{dt} &= k[M][p_\nu] \\ -\frac{1}{\mathcal{N}_A hc\nu}\left(\frac{dI_\nu}{dt}\right) &= k[M]\frac{I_\nu}{\mathcal{N}_A hc\nu} \\ -\frac{dI_\nu}{dt} &= k[M]I_\nu \end{aligned} \tag{10.51}$$

Thus, we have obtained an expression for the rate at which the intensity of a beam of light will decrease as it passes through a sample containing optically absorbing molecules of concentration $[M]$. If the light is moving in the x-direction, then we may express the speed of light as $c = dx/dt$, and so $dI_\nu/dt = (dI_\nu/dx)(dx/dt) = c(dI_\nu/dx)$. If we further assume that the light intensity is sufficiently low that it does not significantly change the concentration $[M]$, then we may rearrange and integrate the above differential

---

Gordon Gould in 1960, after he had met Charles Townes from Bell Labs and had a brief discussion about stimulated emission. Gould's subsequent application for a patent was initially denied and instead Charles Townes and Arthur Schawlow at Bell Labs were awarded a patent for the idea. After years of court battles Gould's rights to the original invention, and substantial royalties, were recognized in 1987.

equation to obtain the following explicit expression for the intensity of light transmitted through a sample of thickness l.

$$-\frac{dI_\nu}{I_\nu} = \frac{k[M]}{c}dx$$

$$-\int_{I_0}^{I}\left(\frac{1}{I_\nu}\right)dI_\nu = \int_0^{l}\frac{k[M]}{c}dx$$

$$-\ln\left(\frac{I}{I_0}\right) = \left(\frac{k}{c}\right)[M]l \tag{10.52}$$

The above result is equivalent to *Beer's law* (also known as the Beer-Lambert-Bouguer law) that may be expressed as $-\ln(I/I_0) = \alpha[M]l$ or equivalently $I = I_0 e^{-\alpha[M]l}$. Thus, the above analysis has revealed the simple connection between the optical absorption coefficient α appearing in Beer's law and the rate constant for the corresponding elementary absorption process $\alpha = k/c$.[29] Beer's law is more commonly expressed in terms of the *absorbance* $A \equiv -\log_{10}(I/I_0) = -\ln(I/I_0)/\ln(10)$, *transmittance* $T = I/I_0$, and the *molar absorptivity* $\epsilon \equiv \alpha/\ln(10)$ of the sample.

$$\boxed{A = -\log_{10} T = \epsilon[M]l = \left(\frac{\alpha}{2.303}\right)[M]l = \left(\frac{k}{2.303\,c}\right)[M]l} \tag{10.53}$$

Nonlinear Optical Processes

Nonlinear optical processes are analogous to higher order chemical reactions involving a simultaneous collision of more than two molecules. Although the probability of such collisions is typically very small, it can become significant if the associated concentrations are sufficiently high. Thus, we expect that the following two-photon absorption processes could occur at high light intensities (and when the combined energy of the two photons is resonant with an excitation energy of the molecule).[30]

$$M + 2p_\nu \xrightarrow{k} M^* \tag{10.54}$$

The stoichiometry of this elementary nonlinear optical process implies that the absorption rate must have the following form.

$$\frac{d[M^*]}{dt} = k[M][p_\nu]^2 \tag{10.55}$$

---

[29] Note that if we express the concentration in units of mol/m^3 and the distance in m, then the units of α become m^2/mol (since $\alpha[M]l$ is dimensionless). Thus, α represents the effective photon absorption cross-sectional (target) area of the absorbing molecule, expressed in molar units.

[30] More specifically, in order for such a two-photon absorption process to occur the molecular transition must not only conserve energy, but must also be symmetry allowed, in the sense discussed in Section 9.4.

Since the probability that such a process will occur is typically quite low, we may safely assume that the concentrations of both the absorbing molecules and the incident photons are not significantly changed when the sample is irradiated with a pulse of light of duration τ. Thus, we may rearrange and integrate the above differential equation to obtain the following expression for the concentration of optically excited molecules produced by such a light pulse.

$$\int_0^{[M^*]} d[M^*] = \int_0^{\tau} k[M][p_\nu]^2 dt$$

$$[M^*] = k[M][p_\nu]^2 \tau$$

$$[M^*] = \frac{k\tau}{(\mathcal{N}_A hc\nu)^2} [M] \, I_\nu^2 \qquad (10.56)$$

This predicts that the number of excited molecules produced in an elementary two-photon absorption process should be proportional to the square of the incident light intensity I_ν^2. The above analysis may readily be extended to higher-order n-photon absorption processes to obtain $[M^*] = \beta_n[M]I_\nu^n$, where β_n is the corresponding nonlinear optical absorption coefficient. This result suggests that one may determine the number of photons that are simultaneously absorbed by plotting $\ln[M^*]$ as a function of $\ln(I_\nu)$, since the slope of such a plot is predicted to be exactly equal to n. This is precisely the procedure that is used to experimentally distinguish linear from nonlinear absorption processes, and to determine the corresponding number of simultaneously absorbed photons.

The Dawn of Stimulated Emission

The year before Einstein published his paper *On the Quantum Theory of Radiation* (in which he first proposed the existence of stimulated emission), he wrote a letter to his friend Michele Besso in which he stated that "a splendid light has dawned on me about the absorption and emission of radiation." His subsequent paper reveals that what he had in mind was a kinetic analysis of photon-molecule reactions. As we will see, he used this analysis to demonstrate that stimulated emission is required in order to obtain agreement between the experimentally observed properties of *blackbody* radiators[31] and formulate a fundamental description of reactions between photons and molecules.

---

[31] A blackbody is defined as a system in which light and matter are in thermal equilibrium with each other. Thus, it is similar to a hot iron bar, whose color indicates its temperature. The spectrum of such a glowing iron bar is essentially identical to that of an ideal blackbody of the same temperature.

Planck's Formula

Recall that Planck's discovery of his blackbody radiation formula played a key role in the development of quantum mechanics (see Section 1.3). The following is a brief summary of the classical theory of blackbody radiation and how Planck improved upon it by introducing energy quantization.

Classical electromagnetic waves confined within a volume V have $8\pi\nu^2/c^3 d\nu$ optical modes (per unit volume per unit frequency).[32] If these electromagnetic vibrations behaved like classical harmonic oscillators, then we would expect each mode to have an average energy of $\langle\varepsilon\rangle = k_B T$ at thermal equilibrium (see Eq. 1.43 and its application to a classical harmonic oscillator on page 41). This leads immediately to the following classical expression for the energy density (spectrum) of a blackbody.

$$\rho_\nu^{\text{classical}} = \frac{8\pi\nu^2}{c^3}\langle\varepsilon\rangle = \frac{8\pi\nu^2}{c^3}k_B T \tag{10.57}$$

This classical formula does not agree with experimental blackbody spectra, except at very low frequencies, as illustrated in Fig. 10.2. Planck discovered that perfect agreement with the experimental spectra may be achieved if one assumes that light of frequency ν is composed of quanta of energy $h\nu$.[33] In other words, an electromagnetic (light) wave of frequency ν behaves like a quantized harmonic oscillator, with an evenly spaced ladder of energy levels separated by $\Delta\nu = h\nu$. The average energy of such a quantized harmonic oscillator may be obtained from the corresponding Boltzmann probability density $P(\varepsilon_n)$, which yields the following expression for $\langle\varepsilon\rangle = \sum_n \varepsilon_n P(\varepsilon_n)$ (see Eq. 1.24 on page 29).

$$\langle\varepsilon\rangle = \frac{\Delta\varepsilon}{e^{\beta\Delta\varepsilon} - 1} = \frac{h\nu}{e^{h\nu/k_B T} - 1} \tag{10.58}$$

Note that at low frequencies (when $h\nu << k_B T$ and so $e^{h\nu/k_B T} \approx 1 + h\nu/k_B T$) the above expression reduces to the classical result $\langle\varepsilon\rangle = k_B T$. The famous formula that Planck first announced in 1900 may be obtained

---

[32] Optical modes are like the vibrations of a guitar string, except that they are electromagnetic standing waves (of light) within an optical resonator of volume V (and at each frequency light has two possible polarization states).

[33] Actually, Planck interpreted the quanta of energy $h\nu$ as arising from the material oscillators in the walls of the blackbody resonator. Einstein later demonstrated that it is the light itself that is quantized.

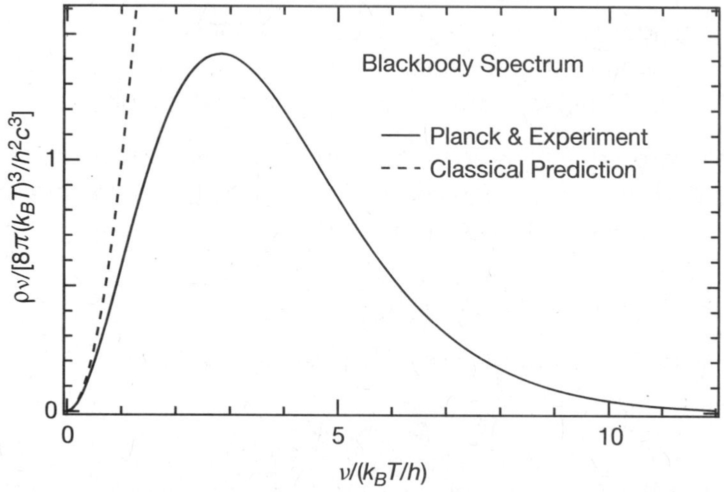

FIGURE 10.2 Comparison of classical (dashed) and quantum (solid) blackbody spectra. The frequency ν and energy density ρ_ν are plotted in reduced units, so this graph represents results at any temperature.

directly from Eq. 10.57, simply by using the correct quantum mechanical expression for $\langle\varepsilon\rangle$ (Eq. 10.58).

$$\boxed{\rho_\nu = \left(\frac{8\pi\nu^2}{c^3}\right)\frac{h\nu}{e^{h\nu/k_BT}-1}} \tag{10.59}$$

Einstein's Kinetic Analysis of Optical Equilibria

The new idea that dawned on Einstein in 1916 is that Planck's formula may be used to uncover additional fundamental information about optical absorption and emission processes. More specifically, he discovered that one could determine both the forms and relative rates of elementary optical processes by requiring that the resulting kinetic predictions agree with Planck's formula.

Recall that the equilibrium concentrations of molecules in the ground and excited state are determined by the corresponding Boltzmann probabilities. More specifically, if we assume that the ground and excited state molecules have the same structures and degeneracies, then the equilibrium constant K_c for the process $M \rightleftharpoons M^*$ would simply be equal to the following Boltzmann factor.[34]

$$K_c = \frac{[M^*]}{[M]} = e^{-\beta\Delta\varepsilon} = e^{-h\nu/k_BT} \tag{10.60}$$

[34] Although Eq. 10.60 does not consider differences in the degeneracies (or the internal vibrational and rotational partition functions) of the ground and excited molecules, this could readily be included, without altering the general conclusions.

The second identity is obtained by noting that the energy difference between the ground and excited states $\Delta\varepsilon$ must be equal to the energy of a photon $h\nu$ associated with such an optical excitation.

In order to more explicitly incorporate the rates of various photon-molecule reactions into the above equilibrium, Einstein suggested that one must consider the following three elementary optical processes.

$$M + p_\nu \xrightarrow{k_1} M^*$$
$$M^* \xrightarrow{k_{-1}} M + p_\nu$$
$$M^* + p_\nu \xrightarrow{k_2} M + 2p_\nu \quad (10.61)$$

The first two reactions are the expected photon absorption and spontaneous emission processes. The third reaction is rather unexpected, as it represents a new kind of process in which an excited molecule reacts with a photon to produce (stimulate) the emission of another photon. The reason that Einstein found it necessary to postulate the existence of such a stimulated emission process will shortly become clear.[35]

The total rate of change of the excited state concentration $[M^*]$ is determined by the sum of the rates arising from each of the above three elementary reactions.

$$\frac{d[M^*]}{dt} = k_1[M][p_\nu] - k_{-1}[M^*] - k_2[M^*][p_\nu] \quad (10.62)$$

At equilibrium all concentrations must achieve a steady state, and so the time derivative on the left-hand side must be equal to zero.

$$0 = [p_\nu](k_1[M] - k_2[M^*]) - k_{-1}[M^*] \quad (10.63)$$

Einstein next pointed out that at high temperatures the concentration of photons should become sufficiently large that $k_1[M][p_\nu]$ and $k_2[M^*][p_\nu]$ will far exceed $k_{-1}[M^*]$. Moreover, Eq. 10.60 implies that $[M^*] \approx [M]$, when $k_B T >> h\nu$. Thus, at high temperature one expects that $k_1[M] - k_2[M] = 0$,

[35] Einstein argued that the existence of such a process is to be expected even in a classical mechanical system, because the direction of energy flow between two coupled oscillators (such as light and a molecule) can go either way, depending on the relative phases of the oscillators. More specifically, we may consider one of the oscillators to be an external driving force (such as light) and the other to be the driven system (such as a molecule). If the driving oscillator is in phase with the driven oscillator, then the driven oscillator will gain energy – which is analogous to optical absorption. However, if the two oscillators are out of phase, then the driven oscillator will transfer energy back to the driving oscillator – which is analogous to stimulated emission.

and so k_1 and k_2 must be equal.

$$\boxed{k_1 = k_2} \tag{10.64}$$

Although this identity was obtained using a high temperature argument, Einstein implicitly recognized that the rate constants associated with optical absorption and emission processes should be temperature-independent, and so the above identity should apply at all temperatures. Thus, at thermal equilibrium (at any temperature) one may combine Eqs. 10.63 and 10.64 to obtain the following expression for the equilibrium (steady-state) photon concentration.

$$[p_\nu] = \frac{k_{-1}[M^*]}{k_1\left([M] - [M^*]\right)} = \left(\frac{k_{-1}}{k_1}\right)\frac{1}{\frac{[M]}{[M^*]} - 1} \tag{10.65}$$

We may now make use of Eqs. 10.48 and 10.60 to obtain the following expression for the optical energy density.

$$\boxed{\rho_\nu = \left(\frac{\mathcal{N}_A k_{-1}}{k_1}\right)\frac{h\nu}{e^{h\nu/k_B T} - 1}} \tag{10.66}$$

This expression is clearly very similar to Planck's formula (Eq. 10.59). Notice that if stimulated emission had not been included as one of the three elementary reactions, then Einstein would not have obtained a result consistent with Planck's formula. Thus, Einstein was compelled to postulate the existance of stimulated emission in order to reconcile fundamental statistical thermodynamic and kinetic results with the experimental spectra of blackbody radiators.

Comparison of Eqs. 10.59 and 10.66 further indicates that the spontaneous emission rate constant k_{-1} must be related to the absorption rate constant as follows.

$$\boxed{k_{-1} = k_1 \frac{8\pi\nu^2}{\mathcal{N}_A c^3}} \tag{10.67}$$

The rate constants k_1, k_{-1}, and k_2 are equivalent to the *Einstein coefficients* for optical absorption, spontaneous emission, and stimulated emission, respectively.

Vacuum Fluctuations

The quantity $8\pi\nu^2/(\mathcal{N}_A c^3)$ appearing in Eq. 10.67 may be equated with the effective concentration of photons $[p_\nu]_0$ arising from *vacuum fluctuations* (which is another term for the zero-point energy of electromagnetic harmonic oscillators). Thus, Eq. 10.67 may be expressed as $k_{-1} = k_1[p_\nu]_0$. This also implies that the rate of an elementary spontaneous emission process (Eq. 10.61) may be expressed as $d[M]/dt = k_{-1}[M^*] = k_1[M^*][p_\nu]_0 = k_2[M^*][p_\nu]_0$. Note that $k_2[M^*][p_\nu]_0$ is equivalent to the stimulated emission rate induced by a photon concentration of $[p_\nu]_0$. This suggests that spontaneous emission may be viewed as emission that is stimulated by vacuum fluctuations.

If $[p_\nu]_0$ represents an actual photon concentration, that would imply that we are currently bathed in an infinite energy of such photons, extending up to arbitrarily high frequencies. Although this bath of light is not directly visible around us, the following remarkable experiments provide indirect confirmation of its existence. These experiments rely on the suppression of vacuum fluctuations within a cavity that is bounded by two mirrors. The separation between the mirrors determines the allowed optical modes within such a cavity (which look just like the eigenstates of a particle in a box). Thus, vacuum fluctuations should be suppressed at frequencies that do not match the cavity resonances.

In 1948 the Dutch physicist Hendrik Casimir predicted that the suppression of vacuum fluctuations inside (but not outside) of such a cavity should give rise to a measurable force pushing the cavity mirrors towards each other. The existence of such a *Casimir* force was subsequently confirmed experimentally. Moreover, if molecular spontaneous emission is indeed stimulated by vacuum fluctuations, then one might expect the spontaneous emission rates of atoms to decrease inside a cavity (if the atomic emission frequency does not match one of the cavity modes). This too has been experimentally confirmed, and has led to an active field of research referred to as *cavity quantum electrodynamics*.

HOMEWORK PROBLEMS

Problems That Illustrate Core Concepts

1. N_2 and CO have approximately the same rotation characteristic temperatures of Θ_r= 2.9 K and 2.8 K, respectively. Predict the number of thermally populated rotational states in these two molecules at 300 K.

2. Cl_2 and Br_2 have vibrational characteristic temperatures of Θ_v = 560 K and 323 K, respectively. At what temperature would Cl_2 have the same vibrational partition coefficient as Br_2 does at 300 K?

3. F_2 has a bond length of 1.42 Å, a vibrational force constant of 445 N/m, a dissociation energy of D_e = 154 kJ/mol, and a nondegenerate ground state.

(a) What is the rotational characteristic temperature Θ_r of F_2 (in K units)?

(b) What is the vibrational characteristic temperature Θ_v of F_2 (in K units)?

(c) What is the thermal wavelength Λ of F_2 at 300 K (in both Å and m units)?

(d) What is the thermal wavelength Λ of an F atom at 300 K (in both Å and m units)?

(e) What is the ground-state term symbol and degeneracy of an F atom?

(f) Predict the equilibrium constant K_c for the reaction $F_2 \rightleftharpoons F + F$.

4. Consider a gas containing a mixture of He and Xe each at a concentration of 1 M and a temperature of 300 K. The atomic weights of He and Xe are ~4 g/mol and ~131 g/mol, and their diameters are ~2.1Å and ~3.9 Å, respectively.

(a) Predict the ratio of the average velocities of Xe and He.

(b) Predict the ratio of the average kinetic energies of Xe and He.

(c) Predict the rate of collisions between He and Xe, in units of L/(mol s).

5. Consider the chemical reaction equilibrium $H_2 + D_2 \rightleftharpoons 2\ HD$ established at a temperature $\Theta_r << T << \Theta_v$ (given that all three diatomics have essentially identical bond lengths of r = 0.74 Å and vibrational force constants f = 540 N/m).

(a) Predict the equilibrium constant of the above reaction *assuming* that all three diatomics have the same bond dissociation energy (so the equilibrium constant is entirely determined by translation and rotational quantum states of the reactant and product molecules).

(b) Since the electronic Hamiltonians of the three molecules are identical, only vibrational zero-point energy is expected to contribute to the reaction energy. Predict the resulting energy difference $\Delta\varepsilon$ between the reactants and products, and compare your result with the experimental heat of formation +0.32 kJ/mol of HD (from $\frac{1}{2}H_2 + \frac{1}{2}D_2$).

(c) Combine the results you obtained in (a) and (b) to predict the equilibrium constant of the above reaction at a temperature of 500 K.

6. Consider two reactant molecules A and B, where molecule A has a diameter of 5 Å and a mass of 100 g/mol, and molecule B has a diameter of 7 Å and a mass of 300 g/mol. Predict the rate constant for $A + B \rightarrow C$ at 300 K under the following conditions.

(a) If there is no activation energy for the reaction.

(b) If there is an activation energy of 10 kJ/mol.

(c) If there is an additional steric factor that implies that only 5% of the molecules that collide with sufficient energy also have the right orientation.

7. Consider a collision-induced unimolecular decomposition reaction of the form $A \rightarrow B + C$.

(a) Write down the three elementary steps in a Lindemann-Hinshelwood mechanism pertaining to such a decomposition reaction.

(b) Use the steady-state approximation to obtain an expression for the rate of production of the products B and C, when the concentration of A is sufficiently high that $k_{-1}[A] >> k_2$.

(c) Express the rate you obtained in (b) in terms of the equilibrium constant pertaining to the activation process $A \rightleftharpoons A^*$.

(d) Obtain an expression for the latter equilibrium constant in terms of the energies ε_A and ε_{A^*}, and effective degeneracies g'_A and g'_{A^*}, of A and A^*, respectively.

(e) What do your results imply regarding the relation between the constants k_2, g'_A g'_{A^*}, ε_A, and ε_{A^*}, and the experimental pre-exponential factor and activation energy of the reaction (in the high concentration limit)? What additional assumption have you made in reaching your conclusion?

8. Return to Exercise 10.2 on page 362 and consider how the final result would change if the two reactant atoms were identical, so that the activated complex has a homonuclear rather than a heteronuclear diatomic structure. How does the value of k_0 you obtained compare with that in Eq. 10.37?

9. Consider a process in which one or more photons p_ν interact with a molecule A to produce an optically excited molecule A^*.

(a) Use the standard chemical kinetic expression for the rate of a two-body reaction to show that the rate of optical absorption, $\partial[A^*]/\partial t$, for a linear optical process in which a single photon is absorbed by a molecule should be proportional to the incident photon concentration, $[p_\nu]$ (or light intensity, I_ν).

(b) Use the standard chemical kinetic expression for the rate of a three-body reaction to show that the rate of optical absorption for a two-photon (nonlinear) absorption process should be proportional to the square of the photon concentration $[p_\nu]^2$ (or I_ν^2).

(c) How could you use a plot of $\ln[\partial[A^*]/\partial t]$ vs. $\ln[p_\nu]$ (or $\ln[A^*]$ vs. $\ln I_\nu$) to determine whether A^* was produced by a linear or nonlinear optical absorption process? More specifically, how is the slope of such a plot related to the number of photons simultaneously absorbed?

(d) How would Einstein's steady-state formula have changed if he had not included the stimulated emission process? In other words, demonstrate that stimulated emission is required in order to obtain agreement with Planck's blackbody radiation formula.

Problems That Test Your Understanding

10. Consider the gas phase molecules H_2 and Cl_2 (contained in a box of the same volume and temperature). In answering the following question, you may assume that both molecules have the same bond length and force constant (and the masses of the H and Cl atoms are 1 g/mol and 35 g/mol, respectively).

(a) Which molecule has the largest rotational characteristic temperature?

(b) Which molecule has the largest vibrational characteristic temperature?

(c) Which molecule has the largest translational partition coefficient?

11. The diatomic molecule HCl has a rotational characteristic temperature of 15.2 K, a vibrational characteristic temperature of 4,304 K, and a molecular weight of 36 g/mol. The following questions pertain to HCl gas contained in a 1 m^3 volume at 300 K.

(a) How many vibrational quantum states of HCl are thermally populated?

(b) How many rotational quantum states of HCl are thermally populated?

(c) How many translational quantum states of HCl are thermally populated?

12. Consider the isotopic variants of a hydrogen diatomic: H_2, D_2, and HD (where $M_H = 1$ g/mol and $M_D = 2$ g/mol). In answering the following questions you may assume that all three diatomics have the same bond length, force constant, and electronic structure. You may also assume that they are each contained in a box of the same volume V and temperature $T >> \Theta_r$.

(a) Which of the above molecules has the smallest thermal wavelength at 300 K?

(b) Which of the above molecules has the smallest symmetry number?

(c) Which of the above molecules has the largest number of thermally accessible rotational states?

(d) Which of the above molecules has the smallest vibrational zero-point energy?

13. The atmospheric reaction $O + CH_3 \rightarrow CH_2O + H$ has an experimental rate constant that is nearly temperature independent and has a pre-exponential factor of 8.4×10^{10} $(M\ s)^{-1}$.

(a) Use the above results to estimate the steric factor for the reaction, given that the calculated collision rate constant is $\sim 1.9 \times 10^{11}$ (M s)$^{-1}$.

(b) What does the fact that the experimental rate constant is nearly temperature-independent imply regarding the reaction's activation energy?

14. Consider an optical 2-photon absorption process H + p_ν+ p_ν → H*, where H is a ground-state hydrogen atom, p_ν is a photon, and H* is an excited hydrogen atom.

(a) How much would the rate of production of H* change if the concentration of H atoms were doubled?

(b) How much would the rate of production of H* change if the intensity of the light were doubled?

15. Consider the following two-photon absorption and stimulated emission processes:

$$\mathrm{M} + 2p_\nu \xrightarrow{k_1} \mathrm{M}^*$$

$$\mathrm{M}^* + 2p_\nu \xrightarrow{k_2} \mathrm{M} + 3p_\nu$$

Derive an expression that relates k_1 and k_2, by making use of the fact that at equilibrium, in the high temperature limit, the total rate of change of [M*] resulting from the above two processes must be equal to zero.

Appendices

Answers to Problems That Test Your Understanding

APPENDIX A

Chapter 1

$\boxed{16}$ (a) ~0.1, (b) ~10.5; $\boxed{17}$ ~29 J/(K mol); $\boxed{18}$ (a) ~4, (b) ~1.47; $\boxed{19}$ (a) ~−2626 kJ/mol, (b) ~328 kJ/mol; $\boxed{20}$ (a) 1 J/m, (b) 1 J, (c) $\frac{1}{4}$ J; $\boxed{21}$ ~400 m/s; $\boxed{22}$ Use $\langle\varepsilon\rangle = \frac{1}{2}m\langle v^2\rangle = \frac{3}{2}k_BT$, or use $\int_0^\infty x^4 e^{-ax^2}dx = \frac{3}{8}\sqrt{\frac{\pi}{a^5}}$; $\boxed{23}$ $\langle\varepsilon\rangle = (\varepsilon_1 e^{-\beta\varepsilon_1} + \varepsilon_2 e^{-\beta\varepsilon_2})/(1 + e^{-\beta\varepsilon_1} + e^{-\beta\varepsilon_2})$; $\boxed{24}$ (a) $10t$ m/s, (b) 2500 J, (c) −2500 J, (d) −2490 J, (e) $H = \frac{1}{2}mv^2 + V(x)$ ~10 J; $\boxed{25}$ (a) ~3.7 kJ/mol, (b) ~443 m/s; $\boxed{26}$ (a) ~1.3×10^{-20} J/photon, (b) ~1 (using $\beta\Delta\varepsilon \sim 10.4$), (c) ~$1.4\times10^4$ K; $\boxed{27}$ (a) ~3.5 kJ/mol, (b) ~600 K, (c) ~0, (d) ~2.

Chapter 2

$\boxed{10}$ (a) $\Delta S = k_B \ln(V_2/V_1) \sim 19.1$ J/K, (b) $\Delta S = k_B \ln\Omega_2 - k_B \ln\Omega_1 = k_B \ln(\Omega_2/\Omega_1) = k_B \ln(V_2/V_1)$, which implies that $\Omega = cV$; $\boxed{11}$ (a) $S = k_B \ln 1 = 0$, (b) ~144 K, (c) $q \sim \Omega \sim k_BT/\Delta\varepsilon$, (d) $S = k_B \ln(k_BT/\Delta\varepsilon)$; $\boxed{12}$ (a) ~0 J, (b) ~288 J, (c) ~0.575 J/K, (d) 0 J/K (because the process is reversible); $\boxed{13}$ (a) 0 J (because no work is exchanged), (b) 0 J (because no heat is exchanged), (c) ~0.575 J/K, (d) ~0.575 J/K; $\boxed{14}$ (a) ~315 K, (b) ~−230 J, (c) ~−230 J, (d) $\Delta S = nR\ln[(V_2/V_1)(T_2/T_1)^{3/2}] = nR\ln 1 = 0$ J/K (which is consistent with the fact that any reversible adiabatic process is isentropic); $\boxed{15}$ (a) ~5.8 J/K mol, (b) The entropy change is path independent, because entropy is a state function, (c) ~-5.8 J/K mol, (d) ~1.7 kJ/mol, (e) ~5.8 J/K mol; $\boxed{16}$ (a) ~16.6 kPa, (b) ~2900 J, (c) ~0.05 m^3, (d) ~-830 J, (e) ~3700 J, (f) ~15 J/K; $\boxed{17}$ (a) ~2494 Pa (=J/m^3), (b) 0 J, (c) ~-2245 J, (d) ~+2245 J, (e) ~+19.14 J/K, (f) ~+11.7 J/K; $\boxed{18}$ (a) ~−11200 J, (b) ~−11200 J, (c) 0 J/K, (d) 0 J/K; $\boxed{19}$ (a) $C_V = \frac{7}{2}R \sim 29.1$ is equivalent to the amount of heat (in J) required to increase the temperature of the system by 1 K, (b) ~ 28.4 J; $\boxed{20}$ $-W = \frac{1}{2}Q_{in} = \frac{1}{2}nRT\ln(V_2/V_2) \sim 144J$ (where $Q_{in} = Q_2$).

Chapter 3

[17] (a) nR/P, (b) $-nRT/V^2$, (c) $\frac{3}{2}nR$; [18] (a) nR/V, (b) $\frac{5}{2}nR$, (c) $\frac{5nR}{2T}$, (d) $\frac{5V}{2T}$; [19] $C_P = \left(\frac{D}{2}+1\right)R \sim 10$; [20] $A = G - (\partial G/\partial P)_T P = G - PV$; [21] (a) $(\partial H/\partial S)_P = +T$, (b) $(\partial G/\partial P)_T = V$, (c) $(\partial A/\partial V)_T = -P$, (d) $(\partial U/\partial V)_S = -P$, (e) $(\partial S/\partial U)_V = \frac{1}{T}$; [22] (a) $\sim$−3.14 J/K, (b) 0 K, (c) 0 J, (d) 0 J, (e) $\sim$1111 J, (f) $\sim$1111 J; [23] $H = U - (\partial U/\partial V)_{S,N_i}V = U + PV$; [24] (a) $(\partial U/\partial S)_V = +T$, (b) $(\partial G/\partial P)_T = +V$, (c) $(\partial A/\partial V)_T = -P$, (d) $(\partial H/\partial P)_S = +V$, (e) $(\partial S/\partial V)_U = +\frac{P}{T}$; [25] (a) $dU = (\partial U/\partial T)_P dT + (\partial U/\partial P)_T dP$, (b) $dU = \frac{3}{2}nRdT + 0dP = \frac{3}{2}nRdT$.

Chapter 4

[10] (a) $-V\alpha_P$, (b) S/V; [11] (a) α_P/κ_T, (b) $-P + (T\alpha_P/\kappa_T)$; [12] (a) $TV\alpha_P/C_P$, (b) $-1/\kappa_T$; [13] (a) $\sim$1000 Pa/K, (b) $\sim$0.0083 m^3; [14] (a) $\sim$−5.3 kJ/mol, (b) $\sim$+5.3 kJ/mol; [15] (a) α_P/κ_T (Pa/K), (b) $-T\alpha_P/(C_V\kappa_T)$ (K/m^3), (c) $PV\kappa_T$ (m^3), (d) $T(\alpha_P/\kappa_T) - P$ (Pa); [16] $(C_V/T) - S\alpha_P/(P\kappa_T)$; [17] $(\partial P/\partial n_1)_{T,V,n_0} = -(\partial V/\partial n_1)_{T,P,n_0}/(\partial V/\partial P)_{T,n_0,n_1} = \overline{V}_1/(V\kappa_T)$; [18] [Ag]$\sim 10^{-10}$ when [Ab]= [AgAb]; [19] (a) $\Delta G^0 = -RT\ln(0.5) \sim 1.7$ kJ/mol, (b) $\Delta H^0 = \Delta G^0 + T\Delta S^0 \sim 8.9$ kJ/mol, (c) $(\partial \Delta G^0/\partial T)_{P,N_i} = -\Delta S^0 = -24$ J/(K mol) $= -0.024$ kJ/(K mol); [20] (a) Graph A, (b) Curve c, (c) $\sim$0.007 M–0.008 M, (d) $\sim\frac{1}{2} = 0.5$ M; [21] (a) B ($n^* \sim 100$), (b) A (largest σ), (c) The distribution will broaden as temperature increases, (d) The distributions will all disappear (only free monomers will be present) below [cmc]; [22] (a) $C^*_{\text{cmc}} \approx 0.02$ M, (b) $b = n^* \approx 120$, (c) $\sigma \approx 10$, (d) $\sigma \approx \sqrt{k_BT/(2\kappa^* n^*)} \approx \sqrt{k_BT/(2ab)} \approx 10$, so $a \approx k_BT/(2 \times 120 \times 10^2) \approx 1.7 \times 10^{-25}$ J or $\sim$0.1 J/mol.

Chapter 5

[14] (a) false, (b) false, (c) true, (d) false, (e) true; [15] (a) $(\partial P/\partial T)_V = R/(\overline{V} - \mathrm{b})$, (b) $(\partial U/\partial V)_T = T(\partial P/\partial T)_V - P = RT/(\overline{V} - \mathrm{b}) - [RT/(\overline{V} - \mathrm{b}) - \mathrm{a}/\overline{V}^2] = \mathrm{a}/\overline{V}^2$; [16] (a) $\Delta G^0_s = RT(\beta\mu^\times)$ so, $\beta\mu^\times = \Delta G^0_s/RT \sim 3.2$, (b) $K_C = e^{-\Delta G^0_s/RT} = e^{-\beta\mu^\times} \sim 0.04$; [17] (a) $d = 0.36/0.54 \sim 0.67$, $\tau\sqrt{550 \times 2030}$ $\sim$1057 K, (b) $\rho \sim 6.1$ (molecules/nm^3), (c) $\eta = (\pi/6)6.1\ (0.54)^3 \sim 0.5$, (d) when $\eta \sim 0.5$, $\beta\mu^\times \sim -0.5$; [18] (a) $K_C = 9.2/0.0052 \sim 1770$, (b) $\beta\mu^\times = -\ln(K_C) \sim -7.5$, (c) $\langle e^{-\beta\Psi}\rangle_0 = e^{-\beta\mu^\times} = K_C \sim 1770$ (d) Use Eq. 5.14 with $\eta = 0.5$, and solve to obtain $\tau \sim 1825$ K; [19] (a) $\sim$11.5 J/(K mol), (b) 0 J/(K mol); [20] (a) $\Delta G^0_s = -RT\ln 0.1 \sim 5.7$ kJ/mol, (b) $\beta\mu^x = -\ln 0.1 \sim 2.3$, (c) $\Delta G^0_s = RT(\beta\mu^x) \sim 5.0$ kJ/mol.

Chapter 6

[9] (a) yes, $-a$, (b) yes, a^2; [10] (a) 6.6×10^{-25} kg m/s, (b) 6.3×10^9 rad/m, (c) 2.4×10^{-19} J; [11] (a) 4673 nm, (b) 1140 N/m; [12] (a)

$\int_{-1}^{1}\Psi^*\Psi dx = \int_{-1}^{1}\frac{15}{16}(1-x^2)^2 dx = \frac{15}{16}(\frac{1}{5}x^5 - \frac{2}{3}x^3 + x|_{-1}^{1} = 1$, (b) 0 by symmetry, or $\int_{-1}^{1}\Psi^* x\Psi dx = \int_{-1}^{1}\frac{15}{16}x(1-x^2)^2 dx = \frac{15}{16}(\frac{1}{6}x^6 - \frac{2}{4}x^4 + \frac{1}{2}x^2|_{-1}^{1} = 0$; (c) $\int_{-1}^{1}\Psi^* x^2\Psi dx = \int_{-1}^{1}\frac{15}{16}x^2(1-x^2)^2 dx = \frac{15}{16}(\frac{1}{7}x^7 - \frac{2}{5}x^5 + \frac{1}{3}x^3|_{-1}^{1} = \frac{1}{7}$, so $\sigma_x = \sqrt{\langle x^2\rangle - \langle x\rangle^2} = \sqrt{1/7} = 0.38$; [13] $[\widehat{A},\widehat{B}]\Psi = (\widehat{A}\widehat{B} - \widehat{B}\widehat{A})\Psi = -\Psi$, so $[\widehat{A},\widehat{B}] = -1$ and $\sigma_A\sigma_B \geq \frac{1}{2}|\langle[\widehat{A},\widehat{B}]\rangle| = \frac{1}{2}$; [14] (a) $p = h/\lambda \sim 1.3 \times 10^{-27}$ J s/m, (b) $\varepsilon = h\nu = hc/\lambda = pc \sim 4 \times 10^{-19}$ J or kg(m/s)^2, (c) $\nu = \varepsilon/h \sim 6 \times 10^{14}$ s^{-1} or Hz, (d) $\widetilde{\nu} = 1/\lambda = 1/500 \times 10^{-9}$ (m^{-1}) $\times 10^{-2}$ (m/cm) $\sim 2 \times 10^4$ cm^{-1}; [15] $\sigma_x\sigma_p \geq \hbar/2$ and $\sigma_x = a$, so $\langle p^2\rangle = \sigma_p^2 \geq \hbar^2/(4a^2)$; [16] (a) $\int_{-a/2}^{a/2}\psi^2 dx = \frac{a}{2}$ so $\Psi = \sqrt{\frac{2}{a}}\sin[(2\pi/a)x]$, (b) Two lobes, equal to zero at $x = -a/2$, 0, and $a/2$ (see the $n = 2$ shaded curve in Fig. 7.1), (c) 0.5 by symmetry; [17] (a) $\langle K_x\rangle = \langle p_x^2\rangle/2m = [\hbar^2(\pi/a)^2]/2m$, (b) $\langle p_x^2\rangle = 2m\langle K_x\rangle = \hbar^2(\pi/a)^2$, (c) $\langle x\rangle = a/2$ by symmetry; [18] $\langle x\rangle = 0$ by symmetry and $\langle x\rangle = a^2$ so $\sigma_x = a = 1$ Å, and $\sigma_x\sigma_p = \hbar/2$ (exactly equal for a Gaussian wavefunction) so $\sigma_p = \hbar/2a = \hbar/2\text{Å} \sim 5.5 \times 10^{-25}$ kg m/s; [19] (a) $b = 2ac$, (b) ac^2.

Chapter 7

[22] (a) The wavefunction has no nodes (see Fig. 7.1), (b) 2 nm, (c) 4×10^{-20} J, (d) 4×10^{-20} J; [23] (a) There are 8 π electrons so n_{HOMO}=4, n_{LUMO}=5, (b) $\Delta E = 5.3 \times 10^{-19}$ J so $\lambda = hc/\Delta E = 370$ nm; [24] (a) $h^2/(3m_eL^2)[(1^2 + 1^2 + 2^2) - (1^2 + 1^2 + 1^2)] \sim 2 \times 10^{-18}$ J, (b) $E_{211} = E_{121} = E_{112} = 3$; [25] (a) $\langle 2|\widehat{H}|2\rangle = \frac{5}{2}h\nu$ (see Eqs. 7.35 and 7.37), (b) $\hat{a}|3\rangle = \sqrt{3}|2\rangle$ (see Eq. 7.38); [26] (a) 6.2×10^{-48} kg m^2, (b) $J = 2$ and degeneracy$= 2J + 1 = 5$, (c) 2.7×10^{-34} J s, (d) 0, $\hbar$, $-\hbar$, $2\hbar$, $-2\hbar$, (e) $\Delta E \sim 1.9 \times 10^{-21}$ J so $\Delta E/hc = 98 \times 10^2$ m$^{-1} \sim 98$ cm^{-1}; [27] From the virial theorem $\langle V\rangle = \langle K\rangle = \frac{1}{2}\langle H\rangle = \frac{1}{2}h\nu(3 + \frac{1}{2}) \sim 7.4 \times 10^{-21}$ J; [28] $E_{222} = E_{123} = 7h^2/(8m)$, degeneracy$= 2$; [29] Finite walls produce tunneling tails and a lower energy; [30] (a) $|2\rangle$ is a column vector, $\langle 2|2\rangle = 0 + 0 + 1 + 0 \cdots = 1$, (b) $\widehat{H}|2\rangle = \frac{5}{2}h\nu|2\rangle$; [31] $\widetilde{\nu} = 1/(2\pi(c)\sqrt{f/\mu}$, thus $\widetilde{\nu}_{HD}/\widetilde{\nu}_{H_2} = \sqrt{\mu_{HD}/\mu_{H_2}} \sim 0.87$, and so $\widetilde{\nu}_{HD} \sim 0.87 \times 4400$ cm$^{-1} \sim 3810$ cm^{-1}; [32] (a) $\mu \sim 6.86$ g/mol $\sim 1.14 \times 10^{-26}$ kg, (b) $r_0 = \sqrt{h/(8\pi^2\mu c\widetilde{B})} \sim 1.13 \times 10^{-10}$ m $= 1.13$ Å, (c) $\sqrt{\langle L^2\rangle} = \hbar\sqrt{2}$ and $\langle L_z\rangle = \hbar m_\ell = -\hbar$, 0, $+\hbar$; [33] (a) (1,1,1) and (2,1,1) (b) $\Delta E = E_{211} - E_{111} = 2 \times 10^{-18}$ J; [34] (a) 4, (b) degeneracy $= 4$ ($E_{2111} = E_{1211} = E_{1121} = E_{1112}$); [35] $\Psi_{10} = N\cos\theta$ so $\frac{\partial^2}{\partial\phi^2}\Psi_{10} = 0$, and $\sin\theta\frac{\partial}{\partial\theta}\left(\sin\theta\frac{\partial}{\partial\theta}\right) = \sin\theta[\cos\theta\frac{\partial}{\partial\theta} + \sin\theta\frac{\partial^2}{\partial\theta^2})$, so $\widehat{K}\Psi_{10} = -(\hbar^2/2I)(-\Psi_{10} - \Psi_{10}) = (\hbar^2/I)\Psi_{10}$.

Chapter 8

[14] (a) $n = 2$, $\ell = 1$, $m_\ell = -1,0,1$, $s = 1/2$, $m_s = -1/2$, $+1/2$, (b) $S = s = 1/2$, $L = \ell = 1$, $J = 1/2$ or $3/2$, so there are two possible term

symbols $^2P_{1/2}$, $^2P_{3/2}$, (c) $^2P_{1/2}$ will produce $2J+1=2$ spots; 15 (a) $\Psi=\alpha_1\beta_2-\alpha_2\beta 1$, (b) $\Psi=\alpha_1\alpha_2-\alpha_2\alpha_1=0$; 16 (a) $\langle H\rangle=E_1=hcR_z/1^2\sim-(6^2)2.18\times10^{-18}$ J $\sim-7.8\times10^{-17}$ J $\sim$−47,200 kJ/mol, (b) Use $\langle H\rangle=\langle K\rangle+\langle V\rangle$ and $\langle V\rangle/\langle K\rangle=2/n=-2$ to obtain $\langle K\rangle=-E_1$ and $\langle V\rangle=2E_1$, (c) $\Delta E=E_2-E_1=h\nu=hc/\lambda$ and so $\Delta E\sim5.9\times10^{-17}$ J, and $\lambda=hc/\Delta E\sim3.4\times10^{-9}$ m $\sim$ 3.4 nm; 17 (Ne)$3s^23p^2$ so $S=1$, $L=1$, and $J=0$, and $3P_0$; 18 $\int_0^{a_0}P_{1s}(r)dr=(4/a_0^3)\int_0^{a_0}r^2e^{-2r/a_0}dr\sim0.32$; 19 (a) $n=3$, $\ell=2$, $m_\ell=0,\pm1,\pm2$ and $s=\frac{1}{2}$, (b) $\langle H\rangle=E_n=-hcR_H/2^3\sim-2.42\times10^{-19}$ J, (c) use the virial theorem to obtain $\langle V\rangle=2\langle H\rangle\sim-4.84\times10^{-19}$ J and $\langle K\rangle=-\langle H\rangle\sim+2.42\times10^{-19}$ J, (d) use Eq. 8.11 to obtain $\langle r\rangle=a_0\times10.5\sim5.56$ Å; 20 (a) $1s^22s^22p^5$, (b) $^2P_{3/2}$; 21 (a) (Kr)$5s^24d^9$, (b) $^2D_{5/2}$, (c) $2J+1=6$ spots; 22 (a) $P(r)=Ar^2e^{-2Zr/a_0}$, $dP/dr=A\,2re^{-2Zr/a_0}(1-Zr/a_0)=0$ when $r=r_{max}=a_0/4$, (b) $\int_{a_0/4}^{\infty}P(r)dr=1-\int_0^{a_0/4}P(r)dr=5e^{-2}\sim0.68$; 23 $\widehat{S}_x^2=\widehat{S}_y^2=\widehat{S}_z^2=\frac{\hbar^2}{4}\begin{pmatrix}1&0\\0&1\end{pmatrix}$, $\widehat{S}^2=\frac{3}{4}\hbar^2\begin{pmatrix}1&0\\0&1\end{pmatrix}$, $|+\rangle=\begin{pmatrix}1\\0\end{pmatrix}$, $|-\rangle=\begin{pmatrix}0\\1\end{pmatrix}$ so $\widehat{S}^2|\pm\rangle=\frac{3}{4}\hbar^2|\pm\rangle=s(s+1)\hbar^2|\pm\rangle$.

Chapter 9

14 The particle-in-a-box stabilization energy 2(213–318) $\sim$−207 kJ/mol is similar to Hückel predictions (where ~213 kJ/mol and ~318 kJ/mol are the ground-state energies of one electron in a box the size of H_3^+ and H_2, respectively); 15 (a) $E_B=2(\alpha+\beta)-2\alpha=2\beta$ so $\beta\sim-220$ kJ/-mol, (b) $2(\alpha+2\beta)-\alpha-\beta=3\beta\sim-660$ kJ/mol; 16 The diatomics N_2, N_2^+, and N_2^- have (a) bond orders of 3, 2.5, and 2.5, (b) vibrational frequencies of 2359 cm^{-1}, 2207 cm^{-1}, and 1968 cm^{-1}, (c) bond lengths of 1.10 Å, 1.12 Å, and 1.19 Å, respectively (note that N_2^- has more antibonding electrons than N_2^+), (d) None, (e) None, (f) All, (g) All; 17 (a) $\Delta E=E_{LUMO}-E_{HOMO}=\frac{h^2}{8mL^2}(3^2-2^2)=\frac{5h^2}{8mL^2}$, (b) $\Delta E=E_{LUMO}-E_{HOMO}=\alpha-0.62\beta-(\alpha+0.62\beta)=-1.24\beta$, (c) $\beta=-\frac{1}{124}\frac{5h^2}{8mL^2}\sim9.65\times10^{-19}$ J $\sim$ 581 kJ/mol, (d) $\mu_{32}=\int\Psi_3^*\Delta x\Psi_2dx$, and since Ψ_3 is symmetric, Δx is antisymmetric, and Ψ_3 is antisymmetric, their product is symmetric, so the integral is nonzero (and the transition is dipole allowed); 18 $\beta\sim-220$ kJ/mol (see problem 15 a), so $\Delta H=\Delta U+\Delta nRT=\Delta U=2(\alpha+2\beta)+\alpha-\beta-[2(\alpha+\sqrt{2}\beta+\alpha]=(4-2\sqrt{2})\beta\sim1.17\beta\sim-258$ kJ/mol; 19 (a) $E_B=2(\alpha+\beta)-2\alpha=2\beta$, (b) $2\beta=-602-(-346)\sim-256$ kJ/mol, so $\beta\sim-128$ kJ/mol, (c) E_B(c-$C_3H_3^+$) $=4\beta$ and E_B(CH_2=CH_2) $=2\beta$, so the π-delocalization energy is $4\beta-2\beta=2\beta\sim-256$ kJ/mol; 20 (a) HBr, NH_3, (b) HBr, HC≡CH, NH_3, (c) HC≡CH; 21 (a) c-H_3, n-H_3, (b) H_2, H_2^+, c-H_3, n-H_3, (c) none, (d) H_2, H_2^+, c-H_3, n-H_3; 22 (a) Absorption (left box), fluorescence (right box), (b) ground v = 0, 0, 0, 1, 2, excited v = 2, 1, 0, 0, 0, (c) $\tilde{\nu}=\Delta\varepsilon/hc=1/\lambda_1-1/\lambda_2\approx(1/375\,\text{nm}-1/396\,\text{nm})\,10^7\,(\text{nm/cm})\approx$

$1500\,\text{cm}^{-1}$ (note that the ground-state vibrational frequency gives rise to the separation between the peaks in the fluorescence spectrum).

Chapter 10

10 (a) H_2, because $\Theta_R \propto 1/I \propto 1/\mu$, (b) H_2, because $\Theta_V \propto \nu \propto 1/\mu$, (c) Cl_2, because $q^T \propto m^{3/2}$; **11** (a) $q^V \sim 1$, (b) $q^R \sim 20$, (c) $q^T \sim 2 \times 10^{32}$; **12** (a) D_2 ($\Lambda \propto 1/\sqrt{m}$), (b) HD ($\sigma = 1$), (c) HD ($q^r \propto \mu/\sigma$), (d) D_2 ($\frac{1}{2}h\nu \propto 1/\sqrt{\mu}$); **13** (a) $P_s = A/k_0 \sim 0.44$, (b) $Ae^{-E_a/RT_1} \approx Ae^{-E_a/RT_2}$ so $T_1 \approx T_2$; **14** (a) $d[H^*]/dt = k[H][p_\nu]^2 \propto [H]$, rate change $\times 2$, (b) $d[H^*]/dt = k[H][p_\nu]^2 \propto [p_\nu]^2 \propto I^2$, rate change $\times 4$; **15** At high temperature $[M] \sim [M^*]$ and $d[M^*]/dt = k_1[M][p_\nu]^2 - k_2[M^*][p_\nu]^2 = [p_\nu]^2[M](k_1 - k_2) = 0$, so $k_1 = k_2$.

Fundamental Constants and Mathematical Identities

APPENDIX B

Fundamental Constants (to three significant figures)

Name	Symbol	Value (in SI units)
Avogadro's Number	$\mathcal{N}_A$	6.02×10^{23} molecules/mol
Boltzmann's Constant	k_B	1.38×10^{-23} J/K
Gas Constant	$R = \mathcal{N}_A k_B$	8.31 J/(K mol)
Standard Gravity	g	9.81 m/s^2
Speed of Light	c	3.00×10^{8} m/s
Electron Mass	m_e	9.11×10^{-31} kg
Proton Mass	m_p	1.67×10^{-27} kg
Electron Charge	e	1.60×10^{-19} C
Faraday's Constant	$\mathcal{F} = \mathcal{N}_A e$	9.65×10^{4} C/mol
Planck's Constant	h	6.63×10^{-34} J s
	$\hbar = h/2\pi$	1.05×10^{-34} J s
Vacuum Permittivity	ϵ_0	8.85×10^{-12} C^2/(J m)
Rydberg Constant	$R_H = \frac{m_e e^4}{8h^3 c \epsilon_0^2}$	1.10×10^{7} m^{-1} (109,737 cm^{-1})
	hcR_Z	Z^2 1,313 kJ/mol
Bohr Radius	$a_0 = \frac{4\pi\epsilon_0 \hbar^2}{m_e e^2}$	0.0529 nm (0.529 Å)
Hartree	$E_h = \frac{\hbar^2}{m_e a_0^2}$	4.36×10^{-18} J 2,626 kJ/mol

$\pi = 3.14\ldots$ $e = 2.72\ldots$ $i = \sqrt{-1}$ $e^{i\pi} = -1$

Units and Conversion Factors

SI units: $1\ \text{J} = 1\ \text{N m} = 1\ \text{kg (m/s)}^2$ $1\ \text{N} = 1\ \text{J/m} = 1\ \text{kg m/s}^2$
$1\ \text{Pa} = 1\ \text{J m}^3 = 1\ \text{N/m}^2$ $1\ \text{W} = 1\ \text{J/s}$

Conversion factors: $1\ \text{MPa} = 10^6\ \text{Pa} = 1\ \text{kJ/L} = 1\ \text{J/cm}^3$

$1\ \text{eV} = 1.60 \times 10^{-19}\ \text{J} = 98.5\ \text{kJ/mol} = 23.6\ \text{kcal/mol} = 8{,}230\ \text{cm}^{-1}$

$\beta = 1/k_B T$ or $1/RT$, $RT = \mathcal{N}_A k_B T$ ($RT \sim 2.5$ kJ/mol at 300 K)

$k_B/hc = 0.695\ \text{cm}^{-1}/\text{K}$ ($k_B T/hc \sim 210\ \text{cm}^{-1}$ at 300 K)

$1\ \text{Å} = 100\ \text{pm} = 0.1\ \text{nm} = 10^{-10}\ \text{m}$ $0°\text{C} = 273.15\ \text{K}$

Mathematical Identities

$$\ln(ab) = \ln a + \ln b \quad \ln(a/b) = \ln a - \ln b \quad \ln(a^b) = b\ln a \quad \ln(1/a) = -\ln a$$

$$e^{a+b} = e^a e^b \qquad e^{a-b} = \frac{e^a}{e^b} \qquad e^{ab} = (e^a)^b \qquad e^{a/b} = (e^a)^{1/b}$$

$$e^{\pm ix} = \cos x \pm i \sin x \qquad \cos x = \frac{1}{2}(e^{ix} + e^{-ix}) \qquad \sin x = \frac{1}{2i}(e^{ix} - e^{-ix})$$

$$e^{\pm x} = \cosh x \pm \sinh x \quad \cosh x = \frac{1}{2}(e^x + e^{-x}) \quad \sinh x = \frac{1}{2}(e^x - e^{-x})$$

$$\lim_{x\to 0} e^{ax} = 1 + ax \qquad \lim_{x\to 0} \ln(1+ax) = ax$$

$$\lim_{x\to 0} (1+x)^a = 1 + ax \quad \text{(for any real number } a\text{)}$$

Derivatives

$$\frac{d}{dx}[ax^n] = nax^{n-1} \qquad \frac{d}{dx}\left[\frac{a}{x^n}\right] = -n\frac{a}{x^{n+1}}$$

$$\frac{d}{dx}\sin(ax) = a\cos(ax) \qquad \frac{d}{dx}\cos(ax) = -a\sin(ax)$$

$$\frac{d}{dx}[e^{ax}] = ae^{ax} \qquad \frac{d}{dx}[e^f] = \frac{d}{dx}[f]\,e^f$$

$$\frac{d}{dx}[\ln(x)] = \frac{1}{x} \qquad \frac{d}{dx}[\ln(f)] = \frac{1}{f}\frac{d}{dx}[f]$$

$$\frac{d}{dx}\left[\frac{1}{f}\right] = -\frac{1}{f^2}\frac{d}{dx}[f]$$

$$\frac{d}{dx}[fg] = f\frac{d}{dx}[g] + g\frac{d}{dx}[f]$$

$$\frac{d}{dx}\left[\frac{f}{g}\right] = \frac{g\frac{d}{dx}[f] - f\frac{d}{dx}[g]}{g^2}$$

Indefinite Integrals

$$\int ax^n dx = a\left(\frac{x^{n+1}}{n+1}\right) + C \qquad \int \frac{a}{x}dx = a\ln x + C \qquad \int e^{ax}dx = \frac{e^{ax}}{a} + C$$

$$\int x^2 e^{ax} dx = e^{ax}\left(\frac{x^2}{a} - \frac{2x}{a^2} + \frac{2}{a^3}\right) + C \qquad \int x^n e^{ax} dx = \left(\frac{\partial}{\partial a}\right)^n \frac{e^{ax}}{a} + C$$

$$\int \sin^2 ax\, dx = \frac{x}{2} - \frac{1}{4a}\sin 2ax + C \qquad \int \cos^2 ax\, dx = \frac{x}{2} + \frac{1}{4a}\sin 2ax + C$$

$$\int x\sin^2 ax\, dx = \frac{x^2}{4} - \frac{x}{4a}\sin 2ax - \frac{1}{8a^2}\cos 2ax + C$$

$$\int x\cos^2 ax\, dx = \frac{x^2}{4} + \frac{x}{4a}\sin 2ax + \frac{1}{8a^2}\cos 2ax + C$$

$$\int x^2 \sin^2 ax\, dx = \frac{x^3}{6} - \left(\frac{x^2}{4a} - \frac{1}{8a^3}\right)\sin 2ax - \frac{x}{4a^2}\cos 2ax + C$$

$$\int x^2 \cos^2 ax\, dx = \frac{x^3}{6} + \left(\frac{x^2}{4a} - \frac{1}{8a^3}\right)\sin 2ax + \frac{x}{4a^2}\cos 2ax + C$$

Definite Integrals

$$\int_0^\infty e^{-ax^2} dx = \frac{1}{2}\sqrt{\frac{\pi}{a}} \qquad \int_0^\infty x e^{-ax^2} dx = \frac{1}{2a}$$

$$\int_0^\infty x^2 e^{-ax^2} dx = \frac{1}{4}\sqrt{\frac{\pi}{a^3}} \qquad \int_0^\infty x^3 e^{-ax^2} dx = \frac{1}{2a^2}$$

Note that integrals of the form $\int_{-\infty}^{\infty} x^n e^{-ax^2} dx$ are either equal to zero or to $2\int_0^\infty x^n e^{-ax^2} dx$ (depending on the symmetry of the integrand). Other integrals of the form $\int_0^\infty x^n e^{-ax^2} dx$ or $\int_0^\infty x^n e^{-ax} dx$ can be evaluated using the following more general expressions.

$$\int_0^\infty x^b e^{-cx^2} dx = \frac{\Gamma\left(\frac{b+1}{2}\right)}{2\, c^{\left(\frac{b+1}{2}\right)}} \qquad \int_0^\infty x^b e^{-cx} dx = \frac{\Gamma(b+1)}{c^{(b+1)}}$$

$$(\text{real } x,\ b > -1, c > 0)$$

$$\Gamma(n+1) = n! \quad (\text{for any integer } n \geq 1)$$

$$\Gamma(s+1) = s\Gamma(s) \quad (\text{for any real number } s)$$

$$\Gamma(1/2) = \sqrt{\pi}, \quad \Gamma(3/2) = \sqrt{\pi}/2, \quad \Gamma(5/2) = 3\sqrt{\pi}/4, \ldots$$

Matrix Multiplication

Evaluating the (inner or dot) product of two matrices **AB** requires that the number of columns in matrix **A** must be equal to the number of rows in matrix **B**. More specifically, the product of an $n \times k$ matrix and a $k \times m$ matrix is a $n \times m$ matrix whose nm'th element is $(AB)_{nm} = \sum_{i=1}^{k} A_{ni} \times B_{im}$.

For example, the product of a 1×2 matrix (row vector) by a 2×1 matrix (column vector) is a 1×1 matrix (scalar).

$$(a\ b)\begin{pmatrix}\alpha\\ \beta\end{pmatrix} = (a\alpha + b\beta) = a\alpha + b\beta$$

The following is an illustration of the multiplication of a 2×3 matrix by a 3×1 column vector to produce a 2×1 column vector.

$$\begin{pmatrix}a & b & c\\ d & e & f\end{pmatrix}\begin{pmatrix}\alpha\\ \beta\\ \gamma\end{pmatrix} = \begin{pmatrix}a\alpha + b\beta + c\gamma\\ d\alpha + e\beta + f\gamma\end{pmatrix}$$

The following is an illustration of the multiplication of a 3×2 matrix by a 2×2 matrix to produce a 3×2 matrix.

$$\begin{pmatrix} a & b \\ c & d \\ e & f \end{pmatrix} \begin{pmatrix} \alpha & \beta \\ \gamma & \delta \end{pmatrix} = \begin{pmatrix} a\alpha + b\gamma & a\beta + b\delta \\ c\alpha + d\gamma & c\beta + d\delta \\ e\alpha + f\gamma & e\beta + f\delta \end{pmatrix}$$

Determinants

The determinants of 2×2 and 3×3 matrices may be evaluated as follows:

$$\begin{vmatrix} a_1 & b_1 \\ a_2 & b_2 \end{vmatrix} = a_1 b_2 - b_1 a_2$$

$$\begin{vmatrix} a_1 & b_1 & c_1 \\ a_2 & b_2 & c_2 \\ a_3 & b_3 & c_3 \end{vmatrix} = a_1 b_2 c_3 + b_1 c_2 a_3 + c_1 a_2 b_3 - c_1 b_2 a_3 - a_1 c_2 b_3 - b_1 a_2 c_3$$

A 3×3 determinant can also be expressed in terms of the following sum of 2×2 determinants (called cofactors).

$$\begin{vmatrix} a_1 & b_1 & c_1 \\ a_2 & b_2 & c_2 \\ a_3 & b_3 & c_3 \end{vmatrix} = a_1 \begin{vmatrix} b_2 & c_2 \\ b_3 & c_3 \end{vmatrix} - a_2 \begin{vmatrix} b_1 & c_1 \\ b_3 & c_3 \end{vmatrix} + a_3 \begin{vmatrix} b_1 & c_1 \\ b_2 & c_2 \end{vmatrix}$$

More generally, any $n \times n$ determinant can be expressed as a sum of lower-order cofactors obtained by eliminating all the numbers in row i and column j of the original determinant. The parent determinant may then be evaluated by summing all the cofactors derived from *any one column* of the parent matrix, each multiplied by $(-1)^{i+j}$ times the value of the a_{ij} element in the parent matrix.

For example, here is how the determinant of a 4×4 matrix can be obtained from four 3×3 cofactors.

$$\begin{vmatrix} a_1 & b_1 & c_1 & d_1 \\ a_2 & b_2 & c_2 & d_2 \\ a_3 & b_3 & c_3 & d_3 \\ a_4 & b_4 & c_4 & d_4 \end{vmatrix} =$$

$$a_1 \begin{vmatrix} b_2 & c_2 & d_2 \\ b_3 & c_3 & d_3 \\ b_4 & c_4 & d_4 \end{vmatrix} - a_2 \begin{vmatrix} b_1 & c_1 & d_1 \\ b_3 & c_3 & d_3 \\ b_4 & c_4 & d_4 \end{vmatrix} + a_3 \begin{vmatrix} b_1 & c_1 & d_1 \\ b_2 & c_2 & d_2 \\ b_4 & c_4 & d_4 \end{vmatrix} - a_4 \begin{vmatrix} b_1 & c_1 & d_1 \\ b_2 & c_2 & d_2 \\ b_3 & c_3 & d_3 \end{vmatrix}$$

Notice that one may use the cofactors of *any* column of the parent matrix in order to evaluate the determinant. The most convenient column to choose is that which has the largest number of elements that are equal to zero (as the associated cofactors will not appear in the sum).

Periodic Table

APPENDIX C

PERIODIC TABLE OF THE ELEMENTS

Key: Atomic number → 6; Symbol → C; Name (IUPAC) → Carbon; Atomic mass → 12.011

IUPAC recommendations → (upper group number)
Chemical Abstracts Service group notation → (lower group label)

1 IA	2 IIA	3 IIIB	4 IVB	5 VB	6 VIB	7 VIIB	8 VIIIB	9 VIIIB	10 VIIIB	11 IB	12 IIB	13 IIIA	14 IVA	15 VA	16 VIA	17 VIIA	18 VIIIA
1 H Hydrogen 1.0079																	2 He Helium 4.0026
3 Li Lithium 6.941	4 Be Beryllium 9.0122											5 B Boron 10.81	6 C Carbon 12.011	7 N Nitrogen 14.007	8 O Oxygen 15.999	9 F Fluorine 18.998	10 Ne Neon 20.180
11 Na Sodium 22.990	12 Mg Magnesium 24.305											13 Al Aluminum 26.982	14 Si Silicon 28.086	15 P Phosphorus 30.974	16 S Sulfur 32.065	17 Cl Chlorine 35.453	18 Ar Argon 39.948
19 K Potassium 39.098	20 Ca Calcium 40.078	21 Sc Scandium 44.956	22 Ti Titanium 47.867	23 V Vanadium 50.942	24 Cr Chromium 51.996	25 Mn Manganese 54.938	26 Fe Iron 55.845	27 Co Cobalt 58.933	28 Ni Nickel 58.693	29 Cu Copper 63.546	30 Zn Zinc 65.409	31 Ga Gallium 69.723	32 Ge Germanium 72.64	33 As Arsenic 74.922	34 Se Selenium 78.96	35 Br Bromine 79.904	36 Kr Krypton 83.798
37 Rb Rubidium 85.468	38 Sr Strontium 87.62	39 Y Yttrium 88.906	40 Zr Zirconium 91.224	41 Nb Niobium 92.906	42 Mo Molybdenum 95.94	43 Tc Technetium (98)	44 Ru Ruthenium 101.07	45 Rh Rhodium 102.91	46 Pd Palladium 106.42	47 Ag Silver 107.87	48 Cd Cadmium 112.41	49 In Indium 114.82	50 Sn Tin 118.71	51 Sb Antimony 121.76	52 Te Tellurium 127.60	53 I Iodine 126.90	54 Xe Xenon 131.29
55 Cs Cesium 132.91	56 Ba Barium 137.33	57 *La Lanthanum 138.91	72 Hf Hafnium 178.49	73 Ta Tantalum 180.95	74 W Tungsten 183.84	75 Re Rhenium 186.21	76 Os Osmium 190.23	77 Ir Iridium 192.22	78 Pt Platinum 195.08	79 Au Gold 196.97	80 Hg Mercury 200.59	81 Tl Thallium 204.38	82 Pb Lead 207.2	83 Bi Bismuth 208.98	84 Po Polonium (209)	85 At Astatine (210)	86 Rn Radon (222)
87 Fr Francium (223)	88 Ra Radium (226)	89 #Ac Actinium (227)	104 Rf Rutherfordium (261)	105 Db Dubnium (262)	106 Sg Seaborgium (266)	107 Bh Bohrium (264)	108 Hs Hassium (277)	109 Mt Meitnerium (268)	110 Ds Darmstadtium (281)	111 Rg Roentgenium (272)	112 Cn Copernicium (285)	113 Uut (284)	114 Fl Flerovium (289)	115 Uup (288)	116 Lv Livermorium (293)	117 Uus (294)	118 Uuo (294)

Series														
*Lanthanide Series	58 Ce Cerium 140.12	59 Pr Praseodymium 140.91	60 Nd Neodymium 144.24	61 Pm Promethium (145)	62 Sm Samarium 150.36	63 Eu Europium 151.96	64 Gd Gadolinium 157.25	65 Tb Terbium 158.93	66 Dy Dysprosium 162.50	67 Ho Holmium 164.93	68 Er Erbium 167.26	69 Tm Thulium 168.93	70 Yb Ytterbium 173.04	71 Lu Lutetium 174.97
# Actinide Series	90 Th Thorium 232.04	91 Pa Protactinium 231.04	92 U Uranium 238.03	93 Np Neptunium (237)	94 Pu Plutonium (244)	95 Am Americium (243)	96 Cm Curium (247)	97 Bk Berkelium (247)	98 Cf Californium (251)	99 Es Einsteinium (252)	100 Fm Fermium (257)	101 Md Mendelevium (258)	102 No Nobelium (259)	103 Lr Lawrencium (262)

Index